C0-CFG-088

Aeolian sand and sand dunes

TITLES OF RELATED INTEREST

Aeolian geomorphology
W. G. Nickling (ed.)

Catastrophic flooding
L. Mayer & D. Nash (eds)

Discovering landscape in England and Wales
A. S. Goudie & R. Gardner

Environmental change and tropical geomorphology
I. Douglas & T. Spencer (eds)

Experiments in physical sedimentology
J. R. L. Allen

The face of the Earth
G. H. Dury

Geomorphological techniques
A. S. Goudie (ed.)

Geomorphology: pure and applied
M. G. Hart

Geomorphology in arid regions
D. O. Doehring (ed.)

Geomorphology and engineering
D. R. Coates (ed.)

Geomorphological hazards in Los Angeles
R. U. Cooke

Glacial geomorphology
D. R. Coates (ed.)

Hillslope processes
A. D. Abrahams (ed.)

History of geomorphology
K. J. Tinkler (ed.)

Image interpretation in geology
S. Drury

Karst geomorphology and hydrology
D. C. Ford & P. W. Williams

Models in geomorphology
M. J. Woldenberg (ed.)

Planetary landscapes
R. Greeley

Principles of physical sedimentology
J. R. L. Allen

Rock glaciers
J. Giardino *et al.* (eds)

Rocks and landforms
J. Gerrard

Sedimentology: process and product
M. R. Leeder

Tectonic geomorphology
M. Morisawa & J. T. Hack (eds)

Aeolian sand and sand dunes

Kenneth Pye
Postgraduate Research Institute for Sedimentology, University of Reading.

Haim Tsoar
Department of Geography, Ben-Gurion University of the Negev.

London
UNWIN HYMAN
Boston Sydney Wellington

Published by the Academic Division of
Unwin Hyman Ltd
15/17 Broadwick Street, London W1V 1FP, UK

Unwin Hyman Inc.
955 Massachusetts Avenue, Cambridge, MA 02139, USA

Allen & Unwin (Australia) Ltd
8 Napier Street, North Sydney, NSW 2060, Australia

Allen & Unwin (New Zealand) Ltd
in association with the Port Nicholson Press Ltd
Compusales Building, 75 Ghuznee Street, Wellington 1, New Zealand

First published in 1990

British Library Cataloguing in Publication Data
Pye, Kenneth
Aeolian sand and sand dunes.
1. Sand
I. Title II. Tsoar, Haim III. Series
553.622

ISBN 0–04–551125–X

Library of Congress Cataloging-in-Publication Data
Pye, Kenneth.
Aeolian sand and sand dunes/Kenneth Pye, Haim Tsoar.
p. cm.
Includes bibliographical references.
ISBN 0–04–551125–X
1. Sand dunes. 2. Sand. I. Tsoar, Haim. II. Title.
GB631.P9 1990
551.3′75 – dc20 90–33246
CIP

Typeset in 9 on 11 point Times by Computape (Pickering) Ltd,
North Yorkshire
and printed in Great Britain by Cambridge University Press

For Diane and Sarah

Preface

It is almost exactly half a century since the publication of R. A. Bagnold's classic book *The physics of blown sand and desert dunes*, and it is a tribute to the quality of Bagnold's work that many of the fundamental principles which he developed remain valid today. His book continues to be essential reading for any serious student of aeolian processes. However, the past two decades have seen an explosion in the scale of research dealing with aeolian transport processes, sediments, and landforms. Some of this work has been summarized in review papers and edited conference proceedings, but this book provides the first attempt to review the whole field of aeolian sand research. Inevitably, it has not been possible to cover all aspects in equal depth, and the balance of included material naturally reflects the authors' own interests to a significant degree. However, our aim has been to provide as broad a perspective as possible, and to provide an entry point to an extensive multi-disciplinary scientific literature, some of which has not been given the attention it deserves in earlier textbooks and review papers. Many examples are drawn from existing published work, but the book also makes extensive use of our own research in the Middle East, Australia, Europe, and North America.

The book has been written principally for use by advanced undergraduates, postgraduates, and more senior research workers in geomorphology and sedimentology. The emphasis is therefore on physical processes and sediment properties rather than ecology, human usage, and management. However, we believe that the book will also prove useful to many botanists, agriculturalists, engineers, and planners who have an interest in sand dunes.

Following a short introductory chapter which outlines the nature and importance of aeolian sand research, the physical background to airflow is discussed in Chapter 2. The basic properties and formation of sand grains, together with the textural and mineralogical properties of aeolian sediments, are considered in Chapter 3. Chapters 4, 5, and 6, which lie at the heart of the book, deal with the mechanisms of aeolian sand transport, the formation of sand seas, and the dynamics of aeolian bedforms, respectively. Chapter 7 provides a summary of the internal structures found in aeolian sand deposits, and weathering and early post-depositional modification of dune sands are considered in Chapter 8. Chapter 9 examines the interactions between the physical properties of sand and dune vegetation, the problems arising from the human use of dune areas, and techniques used to stabilize windblown sand. The final chapter (Chapter 10) provides further information about some of the main techniques currently used in aeolian research.

Where possible, SI units of measurement have been used in the text. However, since many older papers present data in c.g.s. units, a conversion table is presented in Appendix 1.

Acknowledgements

Many people and organizations have provided practical and financial assistance during the preparation of this book. Useful comments on earlier drafts of some chapters were made by J. D. Iversen, D. Skibin, A. Danin, V. Goldsmith, N. Lancaster, A. S. Issar, B. B. Willetts, P. Nalpanis and D. Hartmann. Any remaining errors or inaccuracies are, of course, the sole responsibility of the authors. A. Cross and J. Watkins provided invaluable assistance with the drafting and photography. Special thanks are due to Roger Jones of Unwin Hyman for his patience and fortitude during the book's long gestation period, and to Andy Oppenheimer for assistance during the final stages of preparation. Most of all, we wish to thank our long-suffering families for their unfailing support in trying circumstances.

The following organizations, individuals and publishers kindly provided illustrative material or gave permission to reproduce copyrighted figures and tables (some in modified form):
J. R. Riley (Fig. 2.9); D. H. Krinsley (3.24); J. Shelton (6.17, 6.19, 6.22); K. W. Glennie (6.23); D. Ball (6.35); A. Warren (6.27); V. P. Wright (8.18, 8.25); V. Goldsmith (6.21); D. Blumberg (10.8); W. G. Nickling (10.2); Beach Protection Authority of Queensland (9.19); C.S.I.R.O. (Australia) (10.10); Endecotts Ltd. (10.15); American Society of Civil Engineers (4.19, 4.20, 9.11, 9.28, 9.31), U.S. Army Coastal Engineering Research Center (9.29); O. E. Barndorff-Nielsen (3.7, 6.30, 6.32, 10.9); Department of National Mapping (Australia) (6.44, 6.45, 6.48); U.S. Geological Survey (3.13, 3.14, 5.4, 7.20); Surveys and Mapping Department of South Africa (6.53); D.S.I.R. (New Zealand) (5.8); Royal Meteorological Society (2.24); Geological Society of America (5.11, 5.12, 5.14, 6.6); Soil Science Society of America (4.12); International Association of Sedimentologists (4.10, 6.26, 6.32, 6.36, 7.2, 7.3, 7.4, 7.6, 7.7, 7.12, 7.15, 7.29, 7.30); Geografisker Annaler (6.17, 6.46, 6.47, 7.9, 7.10, 7.11, 7.13, 7.14); Society of Economic Paleontologists and Mineralogists (3.10, 3.20, 6.5, 6.41, 8.26); The Royal Society (3.7); Episodes Secretariat (5.19); American Association of Petroleum Geologists (8.19); Gebruder Borntraeger (5.10); Longman Ltd. (6.58, 6.59); Macmillan Journals Ltd. (4.7, 4.22, 6.20, 8.12); Blackwells Ltd. (8.16); Oxford University Press (3.16, 3.17, 3.18); John Wiley Ltd. (7.31); University of Chicago Press (3.8, 3.30, 6.50); Elsevier Scientific Publishers (3.21, 3.22, 5.2, 5.5, 5.21, 6.32, 10.11); Unwin Hyman Ltd. (5.7, 5.16, 6.56, 7.19, 8.1, 8.11, 9.4, 9.25); Van Nostrand Reinhold (2.5); McGraw-Hill Book Company (2.19, 2.21); Martinus Nijhoff Publishers (2.26); Chapman and Hall Ltd. (4.15, 7.5); Edward Arnold Ltd. (4.21); D. Reidel (2.25, 9.20); SPB Academic Publishers (9.27); W. H. Freeman and Company (6.17, 6.19); Pergamon Press (5.18).

Contents

List of tables

1
The nature and importance of aeolian sand research

1.1 Definitions

Processes described as aeolian (derived from *Aeolus*, the Greek god of the winds), may be loosely defined as those which involve wind action, that is, erosion, transport, or deposition arising from movement of air over the Earth's surface. Air is one of two important fluids, the other being water, which are mainly responsible for transporting sediment over the Earth's surface (a fluid is defined as a substance that cannot sustain shear stress and which is deformed by it with limitless continuity). Water and air have completely different physical properties, and the nature of sediment transport differs significantly in the two fluids. In water, as a liquid, cohesive forces hold the individual molecules together, thereby imparting volume but not shape to the water body. Air, as a gas, is composed of non-cohesive molecules which experience constant random movement and tend to disperse unless confined. Air can be compressed much more readily than water, and the density of air at 18 °C and at sea level ($1.3\,\mathrm{kg\,m^{-3}}$) is about 800 times smaller than that of water ($1000\,\mathrm{kg\,m^{-3}}$). The viscosity of air ($1.8 \times 10^{5}\,\mathrm{N\,s\,m^{-2}}$) is also about two orders of magnitude lower than that of water ($1.06 \times 10^{-3}\,\mathrm{N\,s\,m^{-2}}$). As a result, a current of water can entrain, and keep suspended, much larger sediment particles than a current of air flowing with the same velocity.

With relatively few exceptions, aeolian dunes and ripples are composed of grains in the sand-size range (defined as 2 mm–0.063 mm according to the Udden–Wentworth grain size scale; see Chapter 3). In air, grains of this size are transported mainly by saltation (bouncing) or surface creep (rolling). Smaller individual particles of silt and clay are transported in suspension and may be dispersed over a wide area. Such fine particles generally do not form aeolian ripples or dunes unless the grains are aggregated into pellets of sand size.

There are three main groups of aeolian processes which are responsible for erosion, transport, and sedimentation (Fig. 1.1). Erosional processes are of several types and include (a) deflation of loose sediment due to direct wind drag, (b) entrainment of loose sediment by impacting grains in the windstream, and (c) abrasion of hard surfaces by particles entrained in the flow. Aeolian transport processes include movement of individual grains, by creep, saltation, or suspension, and the migration of bedforms. Sedimentation processes can also be divided into those which involve individual grains and those which involve stabilization of bedforms. A clear distinction between transport and depositional processes cannot be made, however, since sedimentation may occur simultaneously with bedform migration as, for example, during the formation of climbing ripple lamination (Hunter 1977a).

Although the entrainment of sand particles is discussed in Chapter 4, the formation of wind erosion and abrasion forms is outside the scope of this book. For a recent review of these aspects, see Breed *et al.* (1989).

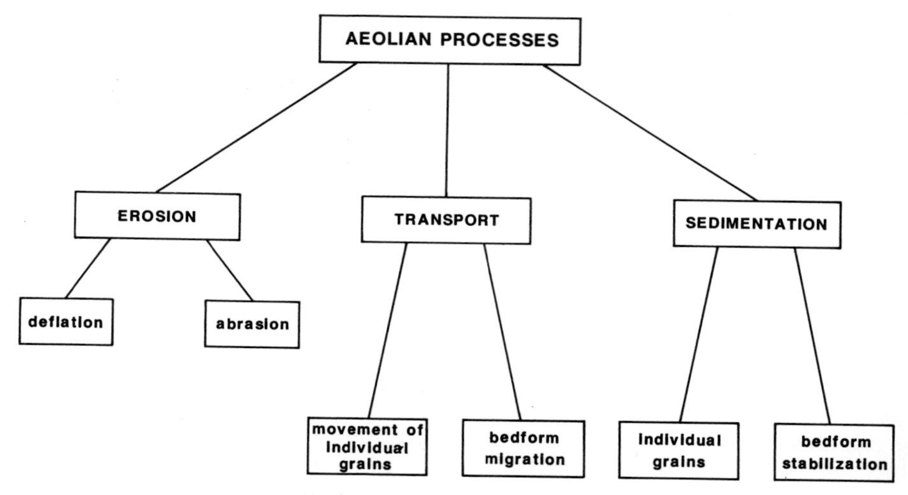

Figure 1.1 The nature of aeolian processes.

Aeolian deposits fall into three main categories: *sand dunes*, *sand sheets*, and *loess blankets*. An aeolian sand dune can be defined simply as a mound or ridge formed by wind deposition of loose sand. Dunes range in size from less than 1 m to several kilometres. They can occur either as isolated ridges or be grouped together to form dunefields. Dunes are found in many different settings and can be classified, according to their geographical occurrence, as inland or continental dunes, coastal or sea-shore dunes, riverbank dunes, and lake-shore dunes.

Sand sheets are accumulations of windblown sand which have a level or gently undulating surface without significant development of dune topography. The term *coversand*, which is used in parts of Western Europe, has both a morphological and a stratigraphic meaning. It refers to sand sheet deposits of late Pleistocene (mainly Weichselian) age which blanket large areas with a more or less uniform thickness, forming relief which does not vary by more than 5 m and with slope angles predominantly less than 6° (Koster 1982). The term *drift sand* is used in Europe to describe later (Holocene) sand sheet or dune deposits which have formed by partial reworking of Pleistocene coversands (Koster 1982, Castel *et al.* 1989).

Loess blankets are deposits of windblown dust, consisting principally of silt-size particles, which mantle a pre-existing land surface. Depending on the nature of the underlying topography, the surface of a loess deposit may be almost flat, gently undulating, or deeply dissected [see Pye (1987) for a review of loess characteristics].

Fluvio-aeolian deposits are interbedded or reworked mixtures of fluvial and aeolian sediments. They can form either by partial aeolian reworking of the upper surface of an exposed fluvial deposit, or by fluvial reworking and re-deposition of aeolian sediments during floods (Glennie 1970, Mader 1982, Good & Bryant 1985, Langford 1989).

Niveo-aeolian deposits are mixtures of wind-transported sediment (usually sand) and snow which are commonly found in polar regions and in some temperate regions, especially at higher altitudes (Cailleux 1978, Ballantyne & Whittington 1987, Koster & Dijkmans 1988).

1.2 Previous work

Aeolian sand research is carried out in many different branches of the physical sciences, earth sciences, life sciences, and development studies including agriculture (Fig. 1.2). Although there is considerable overlap, engineers have tended to concentrate on the mechanics of sand transport and practical measures aimed at stabilizing blowing sand, while geomorphologists and geologists have focused mainly on the classification and morphometric analysis of dune forms, on measurements of aeolian processes, and on the interpretation of sediment characteristics and internal structures.

Until the end of the nineteenth century, most geologists considered wind transport of sediment to be much less important than sediment transport by water or glaciers. Early recognition of the effects of aeolian processes included work by Ehrenberg (1847), who described airborne dust transported from Africa to Europe, Blake (1855), who was one of the first to recognize the extensive development of wind erosion forms in deserts, and von Richthofen (1882), who recognized the primary aeolian origin of the vast loess deposits which blanket much of northern China. However, most nineteenth century geologists regarded wind as relatively unimportant in comparison with water as an agent of sediment transport. For example, Udden (1894, p. 320) expressed the view that 'wind erosion becomes geologically important only in certain localities, and the conditions favouring it are a dry climate and a topography of abrupt and broken reliefs'. However, he also felt (Udden 1894, p. 318) that 'the work performed by the winds in the atmosphere appears hardly to have received its fair share of attention', and subsequently undertook some of the first detailed sedimentological studies of windblown sand and dust (Udden 1896, 1898, 1914).

The early twentieth century saw slightly greater interest in aeolian processes and sediments. During this period, wind erosion of soils emerged as a matter of concern in the Midwestern United States (Free 1911), and several books and papers dealing with the formation of inland and coastal sand dunes were published (Sokolow 1894,

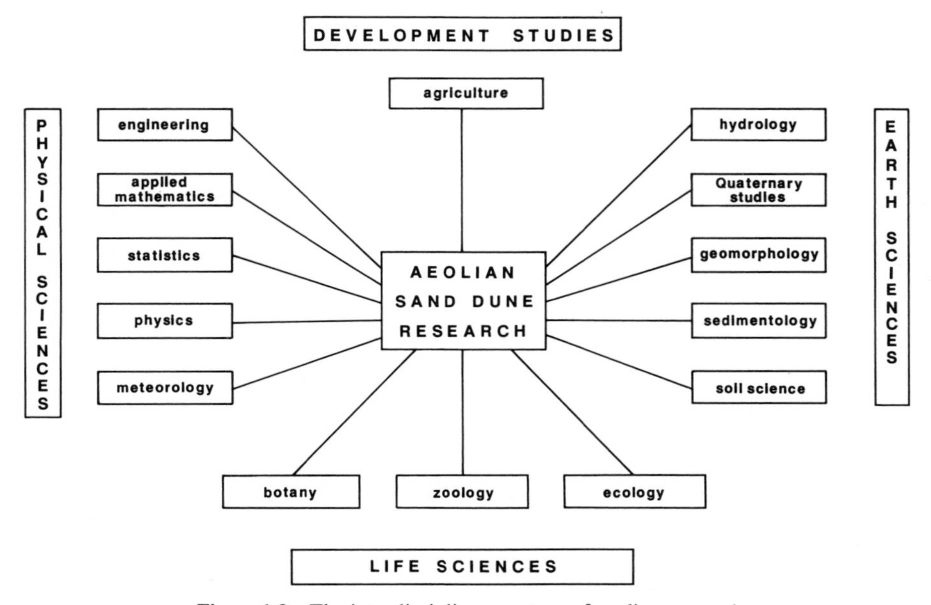

Figure 1.2 The interdisciplinary nature of aeolian research.

Figure 1.3 Brigadier Ralph Alger Bagnold FRS: pioneer in aeolian transport research (photograph by C. R. Thorne).

Cornish 1897, 1900, 1914, Beadnell 1910, Case 1914, Hogbom 1923, Townsend 1925, Cressey 1928, Aufrère 1931, Enquist 1932, van Dieren 1934). However, much of this early work was descriptive, and it was not until the mid-1930s that major advances were made in understanding the mechanics of aeolian transport and dune formation. By far the most important single contribution in this area was made by R. A. Bagnold (Fig. 1.3), an engineer and soldier who made several sorties into the Libyan desert during the early 1930s (Bagnold 1931, 1933, 1935a). He subsequently carried out a number of fundamental experimental studies of sand movement by wind (Bagnold 1935b, 1936, 1937a,b). By virtue of his training, Bagnold was able to apply and extend many of the fundamental principles of fluid mechanics established by von Kármán (1934, 1935), Prandtl (1935) and Shields (1936). Bagnold's work, summarized in *The physics of blown sand and desert dunes* (Bagnold 1941), provided an important theoretical basis which has influenced all subsequent studies of aeolian sand transport and dune formation.

Significant contributions to the understanding of the mechanics of soil erosion by wind were also made independently in the USA, following the 'dust bowl' years of the 1930s, by Chepil and his associates (Chepil 1941, 1945a,b, Chepil & Milne 1939, 1941, Zingg & Chepil 1950). Much of this work was usefully summarized by Chepil & Woodruff (1963).

The 1970s witnessed a significant growth of interest in aeolian studies, particularly amongst geomorphologists and sedimentologists. Several factors contributed to this situation. First, the exploitation of several new oil and gas provinces, including the Middle East and southern North Sea, generated interest in present-day sand seas as analogues for ancient aeolian reservoir sandstones (Glennie 1970, 1983a, 1987, Fryberger *et al.* 1983, 1984). Variations in the texture, internal structure, and degree of early diagenesis of aeolian sands have exerted a significant influence on the productive capacity of hydrocarbon reservoirs (Weber 1987, Lindquist 1988, Richardson *et al.* 1988, Chandler *et al.* 1989). Second, the Mariner 9 and Viking 1 and 2 spacecraft missions undertaken by NASA, which revealed that aeolian processes are important on Mars, stimulated research into possible terrestrial analogues and the fundamental mechanisms of aeolian sediment transport which might be relevant to other planets (Breed & Grow 1979, Greeley *et al.* 1974a,b, 1981, Iversen & White 1982). Much of this work was summarized by McKee (1979a) and Greeley & Iversen (1985). Third, the serious droughts which affected sub-Saharan Africa and parts of Asia in the early 1970s, coupled with a growing recognition that sand deserts were much more extensive during some earlier periods of the Pleistocene (Grove & Warren 1968, Grove 1969, Sarnthein 1978), led to increased concern about the processes and consequences of desertification in arid regions (Rapp 1974, Hagedorn *et al.* 1977, El-Baz & Hassan 1986). Even more recently, concern about the possible effects of greenhouse warming and sea level rises on coastal erosion and flooding risks has led to increased interest in coastal dune dynamics. Beach–dune interaction has emerged as an issue of great practical importance to coastal engineers and planners [see, e.g., papers in Psuty (1988) and van der Meulen *et al.* (1989)].

Some of the research carried out in the last two decades has been compiled in edited volumes and conference proceedings (Brookfield & Ahlbrandt 1983, Nickling 1986, El-Baz & Hassan 1986, Kocurek 1988a, Hesp & Fryberger 1988, Nordstrom *et al.* 1990). More general collections of papers on arid zone processes were edited by Frostick & Reid (1987) and Thomas (1989a), and coastal environments have been reviewed by Carter (1988). Bibliographies of desert dunes have been compiled by Warren (1969), Lancaster & Hallward (1984), and Lancaster (1988d); an annotated bibliography of sand stabilization literature was provided by Busche *et al.* (1984).

1.3 Future research requirements

We now have a relatively good understanding of the mechanics of aeolian grain transport, grain–bed interactions and the formation of small bedforms such as ripples. Satellite imagery and other remote sensing techniques have also provided much information about the geographical distribution and morphological variety of dunes at the regional scale. However, major uncertainties still surround the mechanisms by which dunes are initiated, grow, and migrate in equilibrium with the airflow and pattern of sediment transport over them. Very few detailed field studies have been carried out to measure wind velocity and direction, surface shear stress, and rates of sand transport on different parts of major dunes. The micrometeorological studies which have been undertaken to date (e.g. Knott 1979, Tsoar 1978, Lettau & Lettau 1978, Livingstone 1986, Lee 1987, Mulligan 1987, Lancaster 1989b) refer to relatively small dunes or have been hampered by a lack of adequate instrumentation at an appropriate scale on large dunes. Consequently there is a requirement for more detailed studies of spatial variations in shear stress and sand transport rates, both over individual dunes and at the dunefield scale. The relationship between instantaneous turbulent flow velocities and sediment entrainment also requires clarification, as does the effect of high-magnitude, low-frequency winds in controlling dune morphology.

The past five years have seen a rapid increase in the use of numerical modelling techniques in aeolian studies (Walmsley & Howard 1985, Hunt & Nalpanis 1985, Anderson 1987a,b, Anderson & Hallet 1986, Anderson & Haff 1988, Fisher & Galdies 1988). These studies have already contributed significantly to the understanding of grain transport and depositional processes over plane beds and individual simple dunes. For the future, the challenge lies in expanding the terrain complexity and temporal scales which can be modelled, and in verifying models at all scales by field data.

Knowledge about the thickness, mineral composition, age structure and environmental history of sand deposits in many of the world's major sand seas, particularly in Africa and Central Asia, is still relatively limited. There is therefore a pressing requirement for further broad-scale field studies, involving geophysical surveys, drilling, and supporting programmes of sediment analysis and dating. Only in this way can the relationships between sand sea formation and regional geology, tectonics, and climatic changes be fully documented and understood. The relative importance of factors which control the onset of dune activation at the regional scale, including wind energy, rainfall, and evaporation regime, needs to be clarified if we are to make adequate predictions about the possible effects of greenhouse warming and other future climatic changes.

In coastal environments, further work is required to elucidate the relationships between phases of dune construction and changes in sea level, sediment supply, and wind and wave climate. This can probably best be accomplished through a combination of morpho-stratigraphic and dating studies, laboratory and field experiments, and numerical modelling. It is our hope that this book will contribute significantly to the ultimate attainment of these goals by acting as a stimulus for further research.

2
The nature of airflow

2.1 Physical properties of air and the Earth's atmosphere

2.1.1 The nature of air as a gas

For the purposes of discussion, the atmosphere can be regarded as an envelope of air in which pressure (p), density (ρ) and temperature (T) depend on height (z) above the surface. The atmospheric pressure at any point is equal to the weight of a column of air above that point. Consider a horizontal slice of air with a thickness of δz within an air column of unit cross section. If δz is very small, the overall density in the slice may be considered as constant and its mass is $\rho\delta z$. Its weight is therefore $g\rho\delta z$, where the g is the acceleration due to gravity, and is equal to the pressure difference (δp) between the lower boundary (p_1) and the upper boundary (p_2) of the slice (Fig. 2.1):

$$-\delta p = p_2 - p_1 = \delta z \rho g \tag{2.1}$$

The negative sign indicates that pressure decreases proportionately with increasing height. As δz tends to be infinitesimal, the following differential equation applies:

$$\partial p/\partial z = -\rho g \tag{2.2}$$

Equation 2.1, known as the *hydrostatic equation*, is fundamental for quantifying the vertical pressure distribution in a static atmosphere (Panofsky 1982). It implies that the

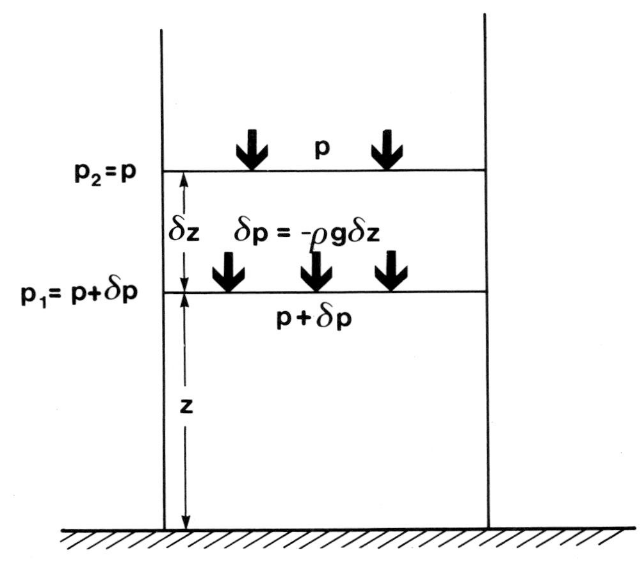

Figure 2.1 Relationship between pressure and height in the atmosphere. See text for explanation.

forces of gravity and vertical pressure gradient seek a mutually compensatory equilibrium. As indicated above, air is compressible and changes its density in proportion to pressure and inversely in proportion to temperature. Air density is usually not measured directly but is derived from the gas law equation for dry air:

$$p = R\rho T \tag{2.3}$$

where R is the specific gas content for dry air and has a value of 287.26 J $kg^{-1}K^{-1}$. By eliminating density from Equation 2.3 using the hydrostatic equation (Eqn 2.2), the *hypsometric equation* is obtained (Panofsky 1982):

$$dp/p = -g dz/RT \tag{2.4}$$

Equation 2.4 states that the pressure falls more slowly with height in warmer air than in colder air.

2.1.2 *Composition of the lower atmosphere*

The composition of the Earth's atmosphere has changed over geological time. At present, the main gaseous constituents are (by volume) 78% nitrogen (N_2), 21% oxygen (O_2), < 1% argon (Ar), and traces of other gases including ozone (O_3) and carbon dioxide (CO_2). Carbon dioxide is the principal constituent of the atmospheres of Venus and Mars, and may once also have been the dominant constituent of the Earth's atmosphere. Much of it is now dissolved in the oceans or 'locked up' in carbonate rocks, shales, and coal deposits. The present carbon dioxide concentration in the atmosphere is about 350 ppm (Kasting *et al.* 1988).

Water vapour is concentrated in the lower levels of the atmosphere. The amount of water vapour carried by air depends on temperature and therefore varies according to geographical location. In the humid tropics the moisture content cannot exceed 4% by volume, while in cold high latitudes it is less than 0.04% (Shaw 1936, p. 39). The degree of moisture saturation has only a small effect on air density in comparison with temperature (Table 2.1).

2.1.3 *Vertical gradient of temperature and stability of the atmosphere*

The atmosphere is heated mainly by solar radiation absorbed on the Earth's surface, the result being a general fall in temperature with increasing height within the lower atmosphere. The rate of temperature decrease with height is known as the *atmospheric lapse-rate*. It is highly variable from place to place and from time to time, but averages about 6.6 K km^{-1}. It never exceeds 10 K per 100 m, with the exception of air layers immediately above surfaces such as bare soil or sand. The lower part of the atmosphere, in which temperature decreases with height, is known as the *troposphere*. The troposphere, which has an average thickness of 12 km, is characterized by strong vertical mixing which brings about the formation of clouds, precipitation, and winds.

When a volume of air undergoes a change in pressure, temperature, or volume (Eqn 2.4) without any heat being either added to it or withdrawn from it, the process involved is defined as *adiabatic*. A common adiabatic atmospheric phenomenon occurs when an air parcel is forced to rise. Since the ambient pressure decreases with height, the air parcel expands while gaining height and its temperature decreases. The rate at which atmospheric temperature changes take place according to this process is known as the *dry adiabatic lapse-rate* and averages about 9.8 K km^{-1} (Panofsky 1982).

The vertical temperature gradient is a vital factor which controls atmospheric wind structure and weather conditions. The relationship between the prevailing atmospheric

Table 2.1 Variations in density of dry air, saturated air, and water as a function of temperature. (Data from Oke 1978).

Temperature		Density		
		Dry air	Saturated air	Water
T (°C)	T (K)	ρ_a (kg m^{-3})	$\rho_{a^*(T)}$ (kg m^{-3})	ρ_w (kg m^{-3} × 10^3)
−5	268.2	1.316	1.314	0.9992
0	273.2	1.292	1.289	0.9999
5	278.2	1.269	1.265	0.9999
10	283.2	1.246	1.240	0.9997
15	288.2	1.225	1.217	0.9991
20	293.2	1.204	1.194	0.9982
25	298.2	1.183	1.169	0.9971
30	303.2	1.164	1.145	0.9957
35	308.2	1.146	1.121	0.9941
40	313.2	1.128	1.096	0.9923

lapse-rate and dry adiabatic lapse-rate determines the stability conditions of the atmosphere. The temperature of a raised parcel of dry air will change according to the dry adiabatic lapse-rate. If the atmospheric lapse-rate is lower than the dry adiabatic lapse-rate, then the raised parcel of air will be colder and denser than the ambient atmosphere and will tend to sink back to its original level. This atmospheric condition is known as a *stable atmosphere*. If the atmospheric lapse-rate equals the dry adiabatic lapse-rate, then the displaced air parcel will be equal in density and temperature to the surrounding atmosphere at the same level; it will neither sink nor rise but will remain in its displaced position. This is the case of a *neutral atmosphere*. If the atmospheric lapse-rate exceeds the dry adiabatic lapse-rate, the displaced air parcel will consequently be warmer and less dense than its surroundings and will continue its convection as long as these differential conditions prevail. This condition is described as an *unstable atmosphere*.

The degree of atmospheric stability can be described by relating the temperature that a parcel of dry air, at any level, would have if brought adiabatically from that level to the 1000 mb pressure level. This temperature is referred to as the *potential temperature* (θ). If $d\theta/dz = 0$ the atmosphere is neutral, when $d\theta/dz < 0$ the atmosphere is unstable and when $d\theta/dz > 0$ it is stable.

Just above the surface of bare desert soil during hot days the atmosphere is unstable, causing it to overturn and create a new, more stable, lapse-rate. This phenomenon contributes greatly to the small-scale turbulence which is characteristic of air flow during day time in hot deserts. With stable conditions, small-scale turbulence is generally weak. A particular case of an absolutely stable atmosphere occurs in deserts at night time when the surface starts to cool and an *inversion* develops at the point where the atmospheric temperature begins to increase with height. Below the inversion the stable atmosphere produces calm conditions with almost no wind at the surface.

An important example of rising air not being cooled according to the dry adiabatic lapse-rate occurs during cloud formation, when the release of latent heat through condensation may slow the cooling of the air parcel. For every gram of water vapour condensed, about 143 J of heat are added to the rising air parcel, making it less dense (Shaw 1936, p. 133). This low rate of cooling, known as the *wet adiabatic lapse-rate*, normally ranges between 5 and 6 K km^{-1}.

If a parcel of air in a typical stable atmosphere (lapse-rate $6.6\,\mathrm{K\,km^{-1}}$) that is not fully saturated is forced to rise, it first cools adiabatically without condensation until it reaches the dew-point temperature. If this air parcel continues to rise, a cloud is formed and the air continues to cool more slowly than in the dry adiabatic process owing to the addition of latent heat. The prevailing atmospheric lapse-rate ($6.6\,\mathrm{K\,km^{-1}}$) is now greater than the wet adiabatic lapse-rate (e.g. $5.5\,\mathrm{K\,km^{-1}}$) and the situation will have changed from a stable to an unstable condition. This *conditional instability*, which depends on the presence of water vapour in the atmosphere, accounts for most atmospheric disturbances, such as the creation of cumulus clouds, thunderstorms, and showers. However, this is a way by which energy is added to the atmosphere; the more latent heat which is suddenly released, the more violent will be the resulting storm.

2.2 Nature and types of air motion

2.2.1 Horizontal air motion

Vertical convection leading to the release of latent heat is often compensated by large-scale horizontal air motion. There is a wide range of horizontal motions ranging from global air circulation down to local, small-scale motions associated with individual storms. The main driving force behind every wind is a pressure gradient arising from temperature differences. However, other forces which influence airflow arise owing to the Earth's rotation, the curvature of its surface, and frictional drag.

The pressure gradient force (F_p) acts in the direction of higher or lower pressure, and varies according to the change in pressure with distance perpendicular to the isobars ($\mathrm{d}p/\mathrm{d}n$) (Holton 1979, p. 6):

$$F_p = -(1/\rho)(\mathrm{d}p/\mathrm{d}n) \tag{2.5}$$

The minus sign indicates that the force orientates itself from high to low pressure.

Owing to the Earth's rotation and surface curvature, there is an apparent force which deflects moving air from a straight-line path (Holton 1979, p. 13). This deviation, known as the *Coriolis effect*, causes all free-moving objects, including winds, to be deflected to the right in the northern hemisphere and to the left in the southern hemisphere. The magnitude of this deviation force (D) can be calculated from (Brunt 1939, p. 166):

$$D = 2U\Omega\sin\phi \tag{2.6}$$

where U is the wind veolocity and Ω is the angular velocity of rotation of the Earth ($= 2\pi\,\mathrm{day}^{-1}$) and ϕ is the latitude.

When a horizontal pressure difference exists in the atmosphere, air moves down the gradient according to Equation 2.5. As it acquires velocity, it comes increasingly under the influence of the Coriolis force. The resulting deflection does not last infinitely and within a relatively short time the pressure gradient force and the Coriolis force come into balance. For frictionless air flow with straight isobars, this balance creates a wind known as the *geostrophic wind*, which blows parallel to the isobars (Brunt 1939, p. 189). The velocity of the geostrophic wind (U_g) can be calculated by equalizing the two forces of Equations 2.5 and 2.6:

$$U_g = -(\mathrm{d}p/\mathrm{d}n)/2\Omega\rho\sin\phi \tag{2.7}$$

The stronger the pressure gradient, the greater is the Coriolis force exerted to acquire balance with the pressure gradient force and, hence, the stronger is the geostrophic wind. It follows from Equation 2.7 that, for a given horizontal pressure gradient, the geostrophic wind is stronger at low than at high latitudes.

The effect of the Coriolis force is reduced by surface friction as air moves over the Earth's surface. Consequently, the pressure gradient force becomes stronger than the Coriolis force and the wind therefore tends to blow across the isobars from higher to lower pressure. The angle at which winds cross the isobars depends on the magnitude of friction; it is about 30–40° over land and 20° over the sea (Byers 1959, p. 217). The layer of the atmosphere where friction and resulting wind shifts occur is known as the *Ekman layer*. The effect of friction decreases with height, and it is usually found that, above a height of 500–1000 m, the flow closely approximates that of the geostrophic wind.

If the isobars are not straight but curved, as in most cyclones and anticyclones, the airflow will be subjected to a centrifugal force (U^2/r, where r is the radius of curvature of the flow path) since it is being constrained to move in a circular path. In order to allow for the centrifugal force, a correction must be introduced into Equation 2.7. Thus, for steady frictionless flow in a curved path, the equation of motion for a geostrophic wind is

$$(1/\rho)(\mathrm{d}p/\mathrm{d}n) = 2\Omega U \sin\phi \pm (U^2/r) \tag{2.8}$$

Such a wind, associated with a three-way balance between pressure gradient, Coriolis force, and centrifugal force, is referred to as the *gradient wind* (Brunt 1939, p. 189). The centrifugal force term in Equation 2.8 is positive in the case of cyclones and negative in the case of anticyclones since in the former both the Coriolis and the centrifugal forces act outwardly, whereas in anticyclones they are oppositely directed (Fig. 2.2).

Since the Coriolis force is dependent on latitude (Eqn 2.6), the deflection at high latitudes is much more significant than the centrifugal force. Near the equator the Coriolis force is negligible while the centrifugal force is of great importance. This is especially the case in cyclones of small radius such as hurricanes. According to Equation 2.8, when the radius of curvature (r) is small, as it is near the anticyclone centre, the value

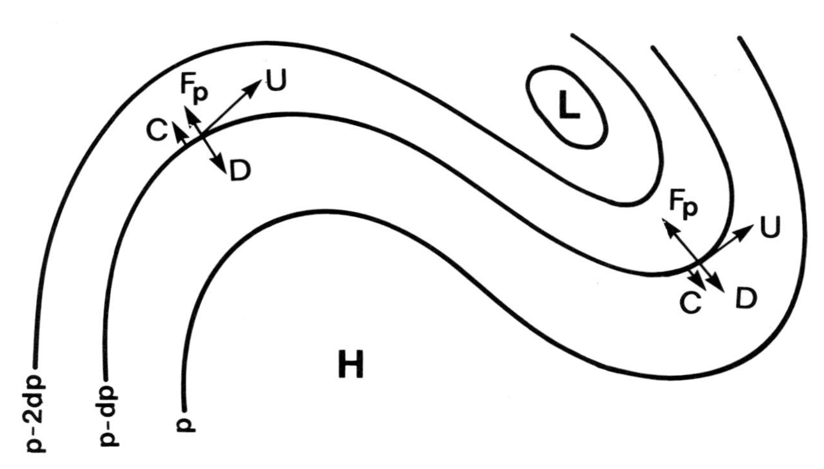

Figure 2.2 Balance of three forces in an anticyclone (H) and a cyclone (L) in the northern hemisphere. D = the Coriolis force; F_p = the horizontal pressure gradient force; C = the centrifugal force; U = the direction of the gradient wind ensuing from these three forces.

of U^2/r, which has a negative sign, becomes very large. Therefore, the pressure gradient there must be very small. For this reason, anticyclones, which are the dominant pressure systems found over subtropical deserts, have a very small pressure gradient near their centres, bringing about lighter winds as the centre is approached.

With cyclones the situation is reversed. Near the centre, the centrifugal force is very large, resulting in high values on the right-hand side of Equation 2.8 and a corresponding steep pressure gradient, causing the winds to become stronger near the centre of the system. The reinforcement, in cyclones, of the Coriolis force by the centrifugal force allows the balance of forces (Eqn 2.8) to be achieved with a wind velocity smaller than that which would be required if the Coriolis force were to act by itself. Hence, in this case, it is possible to maintain, at lower levels of the atmosphere, a wind flow parallel to the isobars despite the effects of friction. In anticyclones the centrifugal force opposes the Coriolis force so that a wind much stronger than the prevailing one is required to maintain a flow parallel to the isobars. A balance between the three forces will be attained with a wind flow deviating much more toward the low-pressure zone than in the case of a geostrophic wind.

2.2.2 The global atmospheric circulation

The general global circulation arises as the atmosphere seeks to achieve a balance between low-latitude energy accumulation and a deficit of energy at high latitudes. A non-rotating Earth with a homogeneous surface would have in each hemisphere a single simple meridional circulation cell, known as a *Hadley cell*, with rising warm air at the equator, poleward flow in the upper levels of the atmosphere, sinking cool air at the poles, and equatorward flow at the surface (Rossby 1941, Lorenz 1970). Such a simple single cell circulation pattern is not possible on a rotating planet, such as the Earth, because meridional movement is subject to the law of conservation of angular momentum. The angular momentum is proportional to the angular velocity and the square of the distance from the rotational axis of the Earth (Byers 1959, p. 198). At the equator, where the atmosphere is at a great distance from the axis of rotation of the Earth, the angular movement has a high value. At high latitudes it is smaller, becoming zero at the poles.

Air flowing poleward should increase its velocity with latitude as the distance from the axis of rotation becomes less, and the angular velocity must be greater to maintain the same momentum. For air moving equatorward from the poles, the opposite should occur, namely a rapid decrease in velocity. The consequence is that poleward flow at high levels and equatorward flow at near-surface levels will be deflected to the right in the northern hemisphere and to the left in the southern hemisphere, creating a system of upper westerlies and lower easterlies (Fig. 2.3) (Rossby 1941).

Another result of the meridional poleward flow is a piling-up of air at a latitude of about 30°N and S, where the angular momentum is already 15% less than at the equator. This accumulation of air results in a high-pressure belt at this latitude, best exemplified by the subtropical anticyclones over the oceans. The piling-up of air is compensated by subsidence of a portion of it, the remainder continuing as a westerly flow aloft. Another explanation for the subtropical anticyclones relates their existence to radiation cooling following the density increase of the air aloft and the latter's subsequent sinking (Rossby 1941). Likewise, the pressure–wind relationship applied to the surface circulation requires a belt of high pressure centred at about 30°N and S. Near the poles, subsiding air produces a surface flow that moves equatorward and is deflected by the Coriolis force into the polar easterlies. Lower pressure is required between the westerlies and the polar easterlies and a zone of relatively sharp change in density, known as the polar front, is developed where the cold easterly flow meets the warmer poleward drifting air (Fig. 2.3).

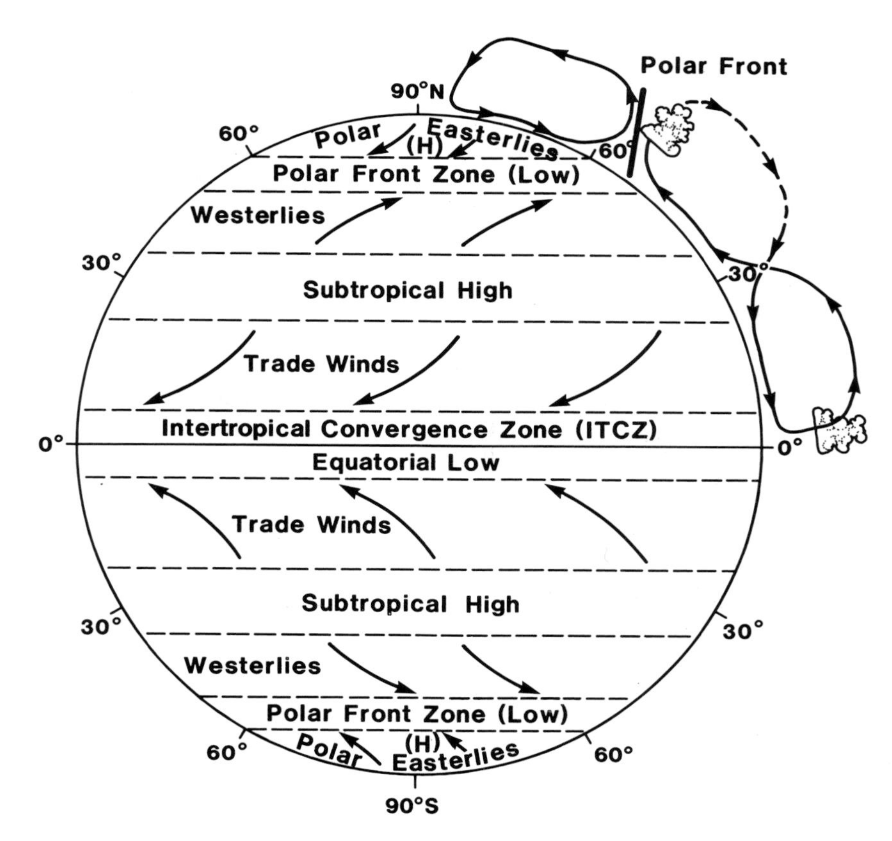

Figure 2.3 Schematic pattern of surface pressure fields and related surface winds over the Earth. The meridional circulation is shown in the vertical cross-section around the upper-right profile of the Earth. (Adapted from Rossby 1941).

In this way, the Hadley cell breaks down into three smaller meridional circulation cells: tropical, mid-latitude, and polar (Fig. 2.3). They give rise to three wind belts at the surface in each hemisphere: the trade winds of low latitudes, the middle latitude westerlies, and the polar easterlies (Rossby 1941). Rising warm air near the equator releases latent heat and provides the energy to drive the tropical cell. The subsiding cold air near the poles drives the polar cell. The mid-latitude cell is not so clearly defined as the other two. It is thermally indirect and would need to be driven by the other two cells. According to the angular momentum consideration, this cell should give rise to easterly winds aloft, yet observations demonstrate the existence of strong upper westerlies. In these middle latitudes, transfer of heat and momentum both poleward and equatorward is accomplished predominantly by the movement of near-surface highs and lows acting in conjunction with their interdependent wave pattern aloft.

This simple global circulation pattern is, in reality, modified by the irregular distribution of continents, oceans and mountain ranges which have different thermal and dynamical properties with strong seasonal variations. Except for the easterlies of the trade-wind belt, the surface wind systems are quite variable. The continents develop thermally induced wind and pressure systems with relatively high pressure over land areas in winter and low pressure in summer. The wind systems shown in the schematic representation of the general circulation (Fig. 2.3) are best developed over the oceans, particularly in the southern hemisphere. The low-pressure areas and fronts around latitude 60° are the most variable. It is an average representing the effect of all moving cyclones in middle and high latitudes. The polar highs are also variable in time, space, and intensity, especially in the northern hemisphere.

At the intertropical convergence zone (ITCZ), known sometimes as the meteorological equator, the Coriolis force is weak or absent, the air flows directly across the isobars, and no cyclones can develop. The three cells shown in Figure 2.3 move north and south following the sun in its seasonal movement. At certain times the ITCZ is displaced 10–15° away from the equator. Under these circumstances cyclones can develop in the zone of displacement.

The high-pressure cells over the oceans in subtropical latitudes are the most permanent features of the general circulation. The Pacific and Azores anticyclones are large areas of subsiding air. The circulation is such that the subsidence effects are most noticeable in the eastern parts of these highs. In the western parts of the oceanic highs, convergence and ascent appear to be more prevalent.

The west coasts of the continents near latitudes 20–30°N and S are markedly deficient in rainfall. In these areas there are surface cold oceanic currents parallel to the coastline which are drawn away from the shore by the Coriolis effect. Surface water has a greater celerity than that at greater depths, and it is therefore deflected at a rate greater than the latter, allowing the upwelling of colder water to the surface (Hartline 1980). As a result, the surface water temperature is abnormally low, which increases the thermal stability of the atmosphere above. Some of the most arid regions of the world are found in such places, including the Western Sahara Desert, Namib Desert, and the coastal deserts of Northern Chile, Peru, and Southern California. The extreme warmth and dryness of the subtropical inland desert regions are the result of large-scale air subsidence to a point where convection is almost entirely suppressed.

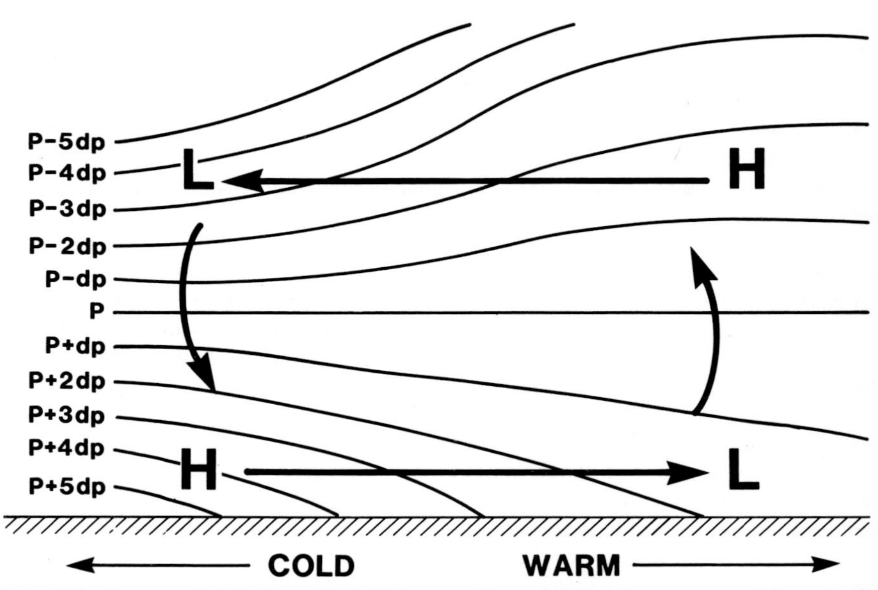

Figure 2.4 Pressure distribution with height over adjacent cold and warm areas. The potential energy is transformed into kinetic energy resulting in air flow as depicted by the arrows. H = high pressure (at that level) and L = low pressure at the same level.

2.3 Storm types that generate sand-transporting winds

2.3.1 The energy of violent storms

The energy behind atmospheric motions originates indirectly from the sun in two ways: firstly, by the sun's direct heating of the surface which, in turn, transfers heat to the atmosphere through conduction, convection, and radiation, and secondly, by the release of heat when atmospheric water condenses into fog or clouds.

When a surface and the air immediately overlying it are heated by solar radiation, this does not increase the air's kinetic energy of motion but does increase its temperature. On being heated the air expands vertically, adding to its potential energy. The amount of solar energy received at the surface varies globally. If it were uniform everywhere the potential energy could not be discharged as kinetic energy. However, the equator receives more than twice the annual solar energy at the poles. Differences in the albedo and thermal properties of particular surfaces also create large horizontal temperature differences, thus creating thermodynamic imbalances. Air motions are the unavoidable result as the atmosphere attempts to recover its thermodynamic equilibrium.

Figure 2.4 illustrates schematically the conversion of the potential energy into kinetic energy. A juxtaposition of contrasting thermal environments results in the development of horizontal pressure gradient forces. Cold areas include the higher latitudes, oceans during the summer daytime, and the continents during winter time. Warm areas include, amongst others, the oceans surrounding the polar regions, the land during summer daytime, and the oceans during winter. According to Equations 2.2 and 2.4 the pressure falls more slowly with height in a warm than in a cold air column. This generates a pressure gradient aloft from the warm to the cold areas. The consequent mass transfer of air aloft creates a surface high over the cold area which is followed by a flow on that surface from the cold to the warm area. Circulation patterns induced by the rise of warm air, the reactive relation of cold air, and the resulting horizontal flow across the isobars toward lower pressure zones are referred to as *thermally induced circulations* (Fig. 2.4). Isobar P in Figure 2.4 differentiates the surface pressure regime and the one aloft. The total amount of kinetic energy that can be transferred to the atmosphere in this way is restricted by the maximum thermal difference that has been created.

Figure 2.5 shows the global distribution of wind energy. As a general rule, wind energy is highest on coasts and at the poleward extremes of the continents (Ash & Wasson 1983). High-velocity winds over the oceans result from their lower surface roughness compared with land surface and from a relatively greater thermal instability above the oceans, which is evident mainly in winter time and at night (Hsu 1971a, Moore 1979). Thermally induced winds are also characteristic of coasts in summer when differences in temperature between land and sea during daytime produce a pressure gradient which results in a sea-breeze (Fig. 2.4). The opposite temperature gradient causes the nocturnal land-breeze which reaches its peak during winter. Active sand dunes form on many coasts owing to the presence of effective onshore winds reinforced by the sea-breeze effect.

The highest average wind velocities on Earth occur near the Antarctic Circle where, at Mawson Coast, the average annual velocity is $22\,\mathrm{m\,s^{-1}}$ (about five times the average annual velocity in Europe); during July it maintains an average of $48\,\mathrm{m\,s^{-1}}$ (Mawson 1930, p. 326). This long-term high average velocity stems from intense radiational cooling of surface air that flows at great velocities down the steep ice-slope as a katabatic wind to the coastline (Mather 1969). Also, a very sharp thermal gradient between the Antarctic and the encircling Southern Oceans favours cyclogenesis and the formation of strong gradient winds in latitudes 40–60°S (Phillpot 1985).

The high wind velocities which prevail around the poles mean that, in such areas,

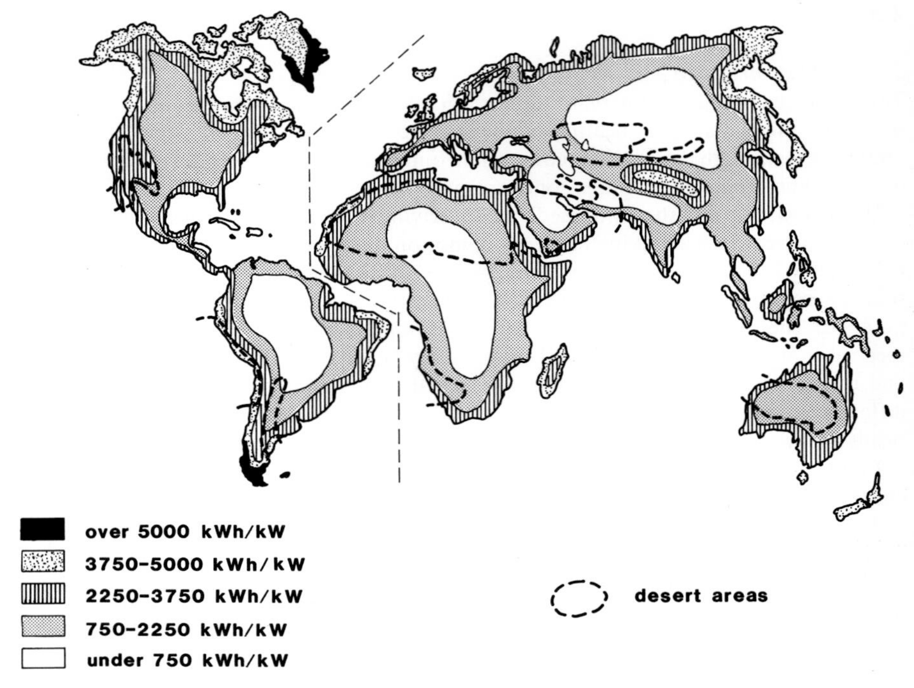

Figure 2.5 Global distribution of wind energy expressed in terms of estimated wind energy in $kW\,h\,yr^{-1}$ per output power (in kW) of a wind machine operating at a constant velocity of $11\,m\,s^{-1}$. (After Eldridge, *Wind machines*, 2nd edn. Copyright by Van Nostrand Reinhold, 1980).

aeolian processes are very important (Péwé 1960, 1974, Lindsay 1973, Calkin & Rutford 1974). The biggest dunefield on the planet Mars, found close to the north pole, is due to a high-speed wind regime of similar origin (Tsoar *et al.* 1979, Haberle 1986). During glacial periods of the Pleistocene, when ice caps extended equatorward, large areas in Europe and North America experienced a periglacial climate and a strong wind regime. Active sand dunes were widespread at such times (Sundborg 1955, Cailleux 1969, Seppala 1972, Lindroos 1972).

Most desert areas, where sand seas are extensive and aeolian processes important, have low windiness compared with humid areas (Fig. 2.5) (Ash & Wasson 1983). In hyperarid areas, receiving up to 100 mm of rainfall per annum, the frequency of dust storms appears to be much lower than that in arid areas with a rainfall between 100 and 200 mm (Goudie 1983a). A possible explanation is that atmospheric conditions such as fronts and convective storms are a rare phenomenon in hyperarid areas.

Rates of sand deposition in coastal dunefields can be ten times higher than those in mid-desert sand seas (Illenberger & Rust 1988). This higher rate of deposition may result from higher sand supply and the much higher energy of coastal winds.

The most violent wind storms, such as tornadoes and hurricanes, acquire their kinetic energy mostly from the release of latent heat. Some of the highest wind velocities recorded (ca $100\,m\,s^{-1}$) have been associated with tornadoes. However, their effect on the long-term average wind velocities in a given area is negligible. When the air is stable and its relative humidity is very low, as is the case in deserts, no latent heat can be released during most storms and the kinetic energy of the wind is relatively small. Some striking examples are demonstrated in Table 2.2. Dust devils and tornadoes produce

Table 2.2 Average estimated kinetic energy of various atmospheric motion systems. Data for the tornado and thunderstorms refer to a total lifetime of kinetic energy. Data for other phenomena refer to kinetic energy at any given moment during maturity, which may be considerably less than the lifetime expenditure. (After Battan 1961, p. 21, and Weather and Climate Modification 1966, pp.35–36).

Wind system	Approximate kinetic energy (J)
gust	10^6
dust devil	10^7
tornado funnel	10^{14}
small thunderstorm	10^{15}
large thunderstorm	10^{16}
hurricane	10^{18}
extratropical cyclone	10^{19}

similar wind systems but on different scales. The former are provoked by fierce direct heating of the surface whereas the latter gather their energy from condensation and release of latent heat leading to strong convection. The total energy of a tornado, acquired mainly by condensation, is several orders of magnitude greater than the energy released by direct solar heating of the surface.

2.3.2 Atmospheric stability and instability in subtropical deserts

More than 99% of the world's active sand dunes are located in deserts which are characterized by a stable, dry atmosphere. Some of the great deserts lie under the subtropical high-pressure belts which give rise to subsidence and high temperatures. The heated air which rises in the equatorial region releases latent heat as it creates masses of clouds. When this air subsides in the subtropics it is heated adiabatically; hence, at an equivalent altitude it is warmer than in the tropics. According to Equations 2.2 and 2.4, a warm anticyclone intensifies with height and becomes a stable permanent system.

Several factors, such as lack of clouds, the low thermal diffusivity of sand (see Section 9.1.1), and sparse vegetation cover, account for a very strong solar input and high summer-daytime temperatures at the desert surface. This is accompanied by near-surface atmospheric lapse-rates that greatly exceed the dry adiabatic rate (Schempf 1943). This temperature gradient, known as the *superadiabatic lapse-rate*, causes instability during daytime in the form of an uprush and overturn of hot and very light surface air, with cooler air aloft seeking to fill the vacuum thus produced (Bagnold 1953b). The air layer in which this compensatory process occurs is called the *mixing layer*, and it is capped by an inversion layer (Fig. 2.6a). Mixing allows the higher momentum of fast-moving upper air layers, which are subject to little frictional retardation, to be brought down to surface level, thereby causing the highly turbulent surface winds that characterize desert dunefields during hot summer afternoons (Schempf 1943). In most cases, this increase in momentum in the mixing layer also raises the velocity of the wind above the threshold, thus producing sand movement.

In the Sahara, this strong thermic mixing reaches altitudes of more than 2000 m during daytime (Peters 1932, Durst 1935) and, in extreme cases, over 4000 m (Dubief 1979). In the arid southwestern United States, the mixing layer during clear spring afternoons can extend up to 5000 m and sometimes 7000 m above mean sea level (Jackson *et al.* 1973). It can be discerned by the altitude of the haze, denoting the presence of dust in the atmosphere, which attains its maximum during the afternoon.

The same factors responsible for high daytime temperatures also cause a drastic cooling of the ground at night. As the temperature of the air exceeds that of the ground a downward heat flux is created, resulting in a stable layer near the surface. The existing upper inversion is lowered and a ground-based inversion develops, thus increasing the stability of the surface air and reducing the wind velocity to almost nil (Townsend 1967). For this reason, desert dunefields have very low wind velocities during summer nights (Fig. 2.6b).

The great amount of heat energy concentrated at the surface during the daytime is the driving force behind most of the summer sand storms in subtropical deserts. The hot surface air disperses the inversion and forms an unstable low-altitude mixing layer in which aeolian processes are operative (Fig. 2.6a) (Warren & Knott 1983).

2.3.3 *Dust devils*

The daytime near-surface air instability characteristic of bare ground in deserts leads to the development of miniature whirlwinds known as *dust devils* (Fig. 2.7). This type of whirlwind generally raises a column of dust 1–300 m in diameter up to heights of 3–300 m, but some cases of gigantic dust devils attaining heights of up to 4000 m have been reported (Jutson 1920, Sinclair 1964).

Dust devils develop over reasonably level ground when there is strong thermal convection through the superadiabatic layer near the ground. Horizontal radial inflow towards the base of the column is generated by the displacement of air in a warm buoyant upcurrent. Horizontal velocities of 18–40 m s^{-1}, well over the threshold values needed to transport sand particles, can be associated with dust devils (Flower 1936, Ives 1947, Brooks 1960) owing to the tendency of the air to conserve its angular momentum as it moves towards the dust devil's axis. The continuous influx of warm surface air into the visible vortex as it moves along helps to sustain the dust devil's motion (Sinclair 1969). The effect of the Coriolis force induced by the Earth's rotation is negligible for these small systems, so that the rotation of winds about the axis may be in either direction (Jutson 1920, Flower 1936, Brooks 1960, Webb 1964, Sinclair 1964).

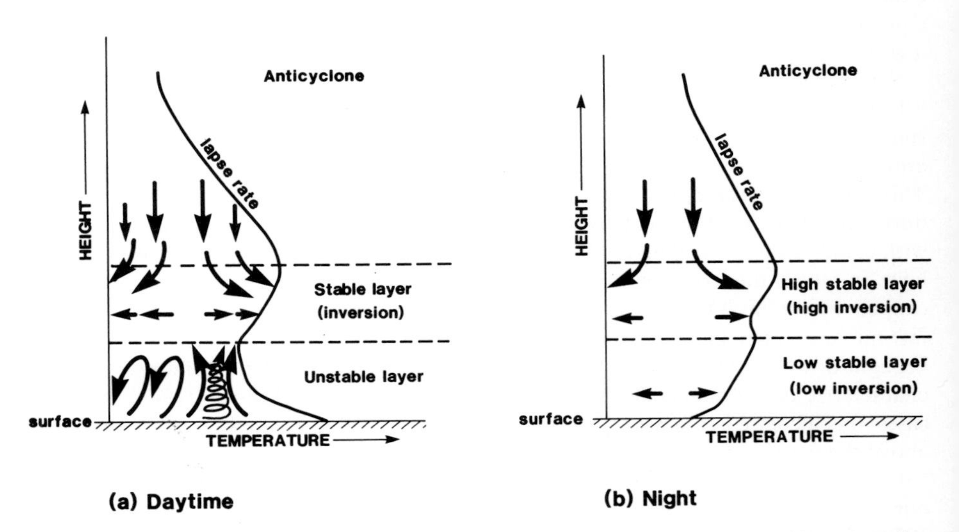

Figure 2.6 Schematic diagram of the lapse rate over a subtropical desert during (a) summer day time and (b) summer night time. Based partly on variations of upper air temperature over Ismailia, Egypt (Peters 1932).

Figure 2.7 Photograph of a dust devil in the Negev desert, Israel.

The life cycle of a dust devil lasts from less than 1 min up to 20 min. The duration typically increases with the size and height of the dust devil (Flower 1936). Dust devils develop mostly in areas without cloud cover during the early afternoon, which is the time of maximum soil surface temperature and convective heat flux. They are most frequent in the hottest months of the dry summer. Dust devil activity is considerably suppressed by windspeeds above $5\,\mathrm{m\,s^{-1}}$, which increase the vertical mixing of the hot surface air layer and thus reduce the temperature gradient near the surface (Sinclair 1969).

In spite of their high frequency, dust devils are, as pointed out above, of low duration and therefore they play a minor role as geomorphic forces in deserts. Dust devils are common on loess plains, valleys, and dry river beds where their spouts act as a channel through which dust is funnelled into the atmosphere. They have not been seen to perform an important role in sand dune development. On the other hand, dust devils launching dust up to 6 km above the surface are considered to play an important role in the initiation of the large Martian dust storms (Greeley *et al.* 1981, Thomas & Gierasch 1985).

2.3.4 Squalls

Thunderstorms have high kinetic energy (Table 2.2) arising from an unstable atmosphere in which masses of warm air rising vertically are associated with the release of large amounts of latent heat during cloud formation. These storms are accompanied by sudden violent surface winds, known as squalls, which in arid areas are known to cause vigorous sand and dust blowing (Idso 1974). They are frequently observed during summer (May–October) over some deserts in the southern Sahara and the southwestern United States (Sutton 1925, 1931, Farquharson 1937, Schempf 1943, Lawson 1971, Idso *et al.* 1972, Idso 1973, Brazel & Hsu 1981). They have also been reported in other areas such as Australia (Lindsay 1933) and southern Ukraine (Shikula 1981).

In front of the storm, buoyant warm air is lifted beyond its condensation level to the tropopause where cumulonimbus clouds form, usually giving rise to heavy rain. Dry air reaching the storm at high levels from the rear is cooled by the evaporation of the precipitation droplets and descends (Fig. 2.8). As this gravity current hits the ground, it is deflected forward and moves out of the cloud in large lobes, forming a high wall of dust that rises up to 1000–2500 m (Fig. 2.9). The lobe expands vertically and horizontally until its forward movement relative to the front decreases and a new lobe arises out of the dying one (Lawson 1971). In Arizona the average maximum velocity of the wind in these lobes was found to be 21 m s^{-1} (Idso *et al.* 1972). In Sudan the winds reach 49 m s^{-1} in severe cases and have an average velocity of 14.5 m s^{-1}. The velocity at the front of the lobe is 9–12 m s^{-1}, which is half of the maximum wind velocity behind it (Sutton 1925, 1931, Lawson 1971).

Violent dust storms of this type are known in Sudan as *haboobs* and have been reported in a 1200-km wide belt around Khartoum. They are rarely observed in Aswan to the north and occur there once in two years. About 22 of them occur each summer in the Khartoum area when the ITCZ lies over the area (Sutton 1931). Vigorous surface

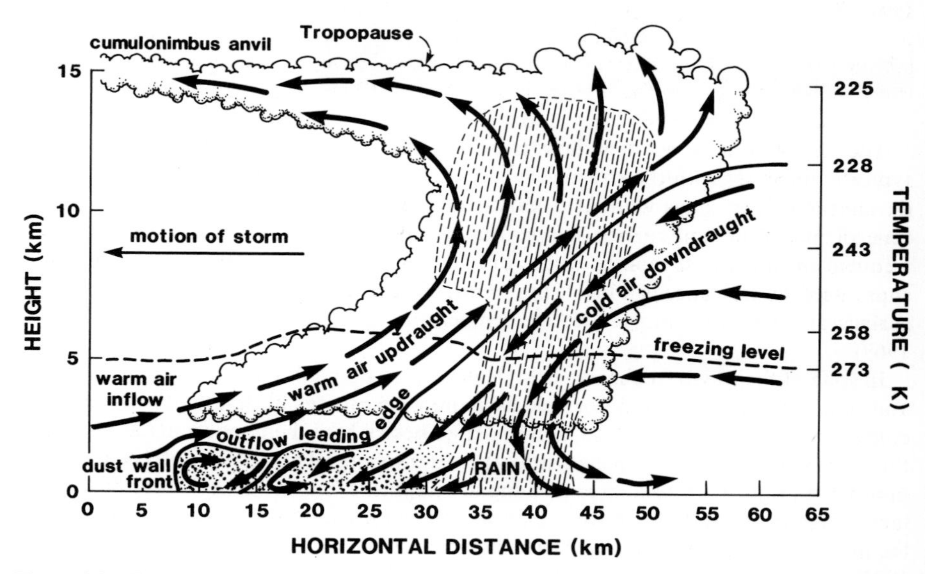

Figure 2.8 Formation of a dust storm (haboob type) depicted schematically in a thunderstorm structure. The arrows indicate the direction of the air currents; the hachured region represents falling or suspended precipitation; the stippled region depicts a dust storm caused by the violent downdraught of cold air. (After Charba 1974, Goff 1976, Idso 1976).

Figure 2.9 A high wall of dust formed by a gust front at Daoga, northern Mali. Photograph taken in October 1978. (Courtesy of J. R. Riley).

convection is possible with light, relatively moist surface westerlies overlain aloft by moist easterlies, allowing the formation of cumulonimbus (Schempf 1943, Lawson 1971). Similar synoptic conditions prevail in the arid southwestern United States, where on average there are about 12 haboob storms each year at Phoenix, Arizona. They are initiated by moist tropical air from the Pacific overlain by an upper flow of moist air from the Gulf of Mexico (Idso 1976, Brazel & Hsu 1981).

As several thunderstorms are often arranged in long lines, the haboobs can give rise to a dust wall front several kilometres long. Dust storms are usually short, lasting 1–3 h, and are sometimes followed by heavy rain which settles the dust and stops sand movement. Often, however, the trailing thunderstorm does not arrive or the rain evaporates before reaching the ground (Sutton 1931). In such circumstances the sand and dust storm can last for several hours (Idso 1976). Despite the violent winds associated with it, the haboob can be considered as a short-lived desert storm of local character.

2.3.5 Wind regimes in the world's deserts

The wind regimes of the world's major deserts are reflected by the pattern and alignment of the sand dunes found in them (Holm 1968, Brookfield 1970, Wilson 1971, Fryberger & Ahlbrandt 1979, Mainguet & Cossus 1980, Fryberger *et al.* 1984). The sand deserts can be divided into two main types, hot subtropical and cold middle latitude deserts. The former include the Sahara, Arabia, Namib and Kalahari, Thar and Rajasthan, Australia, Iran, Peru and southwestern North America. The climate of these deserts is mild to warm in winter and hot to very hot in summer. The middle latitude deserts are found mainly in Central Asia and are cold in winter but warm to hot in summer.

The wind climate of the Sahara desert, which provides a typical example of a hot subtropical desert, is dominated by the existence of a high-pressure cell across the region

(Fig. 2.3). The northern margins of the Sahara are influenced by westerlies during winter and the southern margins by southwesterly monsoon winds during summer. As noted previously (Section 2.2.1), the centre of the anticyclonic belt is characterized by relatively stable air. The central regions of the Sahara, therefore, have fewer sandstorms than the marginal zones and have winds of lower average velocity (Mainguet 1986).

The Mediterranean Sea to the north of the Sahara is warmer during winter than the surrounding continents and supplies enormous amounts of water vapour to the air. As such, it is a source of energy for wind storms which evolve from cyclones. The southern flank of the resulting low-pressure systems affects the northern Sahara during winter by

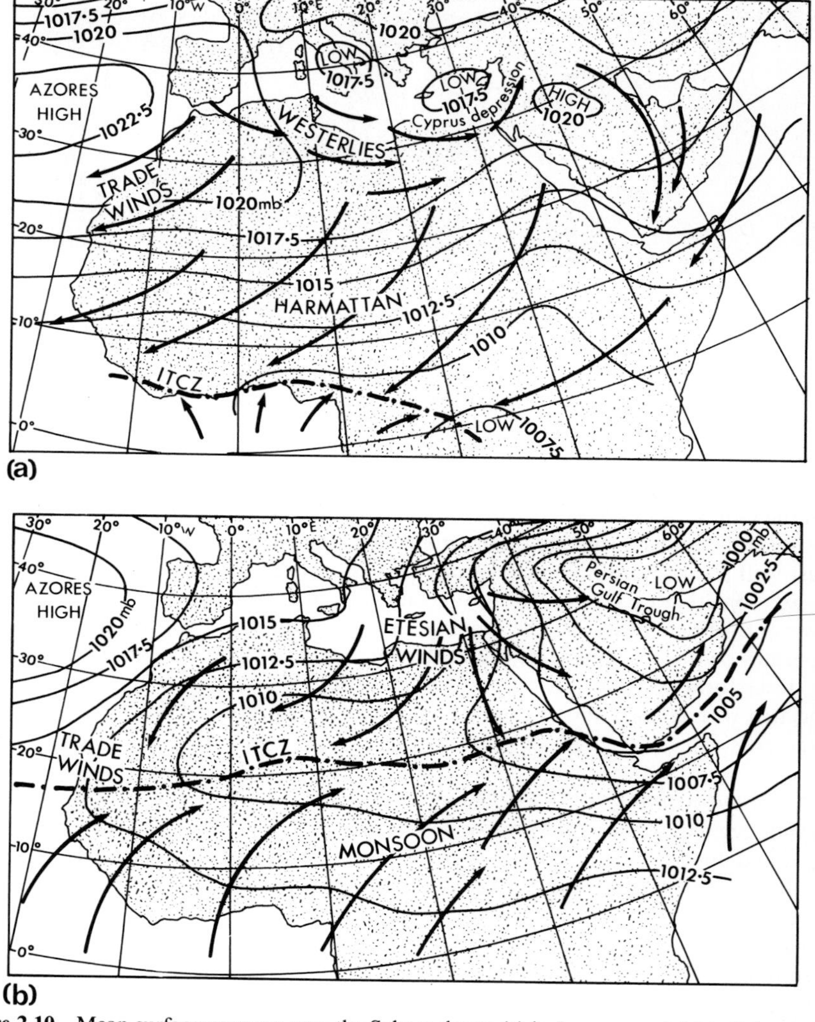

Figure 2.10 Mean surface pressure over the Sahara desert (a) in January and (b) in July; arrows indicate the wind directions. (After Brooks & Mirrlees 1932, Meteorological Office 1962, Griffiths & Soliman 1972, Dubief 1979).

generating strong winds which veer from SW to W and NW (Fig. 2.10a). In some cases the westerly and north-westerly winds are accompanied by rain (Tsoar 1974).

Infrequent tropical depressions known as *Saharo – Sudanese depressions* (Dubief & Queney 1935) develop when the upper trough of westerlies extends equatorward above the tropical trade winds. They lead to the development of cyclonic vortices which travel ahead of the upper air trough northeastward across the Sahara and reach the Mediterranean coast where they continue on a trajectory which is more southerly than that of the polar front depressions (Durward 1936, Dubief 1979, Nicholson & Flohn 1980).

Along the North African coast such depressions are frequent during spring when the contrasts in temperature and stability of the air masses over the Sahara and Europe are most prominent. The resulting unstable conditions lead to the formation of small, hot depressions which move eastward along the coast from Algiers to Israel and Syria. Preceding these depressions, hot storm winds play a major role in the transport of sand. The latter are characterized by southeasterly to southerly winds going in front of the depression and are known as *khamsin* (El-Fandy 1940, Lunson 1950). They are usually devoid of rain since, according to the hypsometric equation (Eqn 2.4), low pressure weakens above hot air.

South of the area affected by the westerlies and the polar front, the winter period (October to April) is characterized by thermal stability and persistent northerly to north-easterly winds known in the southern Sahara as the *harmattan*. Surface wind direction frequencies in Khartoum and in Kano, Nigeria, show persistent northerly winds between November and March (Sutton 1931, Samways 1976). Thermal turbulence provoked by high near-ground temperatures during this dry period in the west-southern Sahara and the Sudanese Sahel boosts the velocity of these winds above the threshold for dust erosion, forming the *harmattan haze* (Samways 1976, Adetunji *et al.* 1979, Dubief 1979, Kalu 1979).

In Northern Nigeria the harmattan wind carries dust in periodic plumes which reduces visibility (down to 150 m) and air temperature (Samways 1976, McTainsh & Walker 1982, McTainsh 1984). The source of the harmattan air includes both the high-pressure belt of the relatively cool Sahara and polar front depressions crossing the Mediterranean which bring a burst of cold air from the north (Hamilton & Archbold 1945). The Azores subtropical high moves southward over the eastern Atlantic during winter, giving rise to cool, moist trade winds blowing in a direction similar to the harmattan (Fig. 2.10a).

In summer the situation is different. Aloft the area is still under the influence of the subtropical high-pressure zone, but at the surface a trough exists over the southern Sahara, Arabia, and the Persian Gulf. This long trough, which is of monsoonal type, gives rise to northerly winds over North Africa and northwesterly to westerly winds in the Middle East, known locally as *etesian* winds (Fig. 2.10b). Along the Mediterranean coast the etesian winds are augmented by sea-breezes during the daytime. During summer the ITCZ moves northward over the southern Sahara and represents the boundary between the hot, dry Saharan air of anticyclonic origin arriving from the north, and the cooler, moister maritime air coming from the south. This area of instability is characterized by depressions and intense convective activity accompanied by haboob dust storms (Sutton 1931).

In Central Australia, as in the Sahara, a continental subtropical high-pressure cell strengthens during winter. It is weakened by low-level thermal activity in the summer. The strongest winds in Australia are not experienced in the centre of the desert but in the southern winter rainfall zone where they are related to the winter westerlies (Ash & Wasson 1983). The way linear dunes are aligned in the Australian deserts (Fig. 2.11) suggests the existence of large-scale wind circulation around a central high-pressure cell

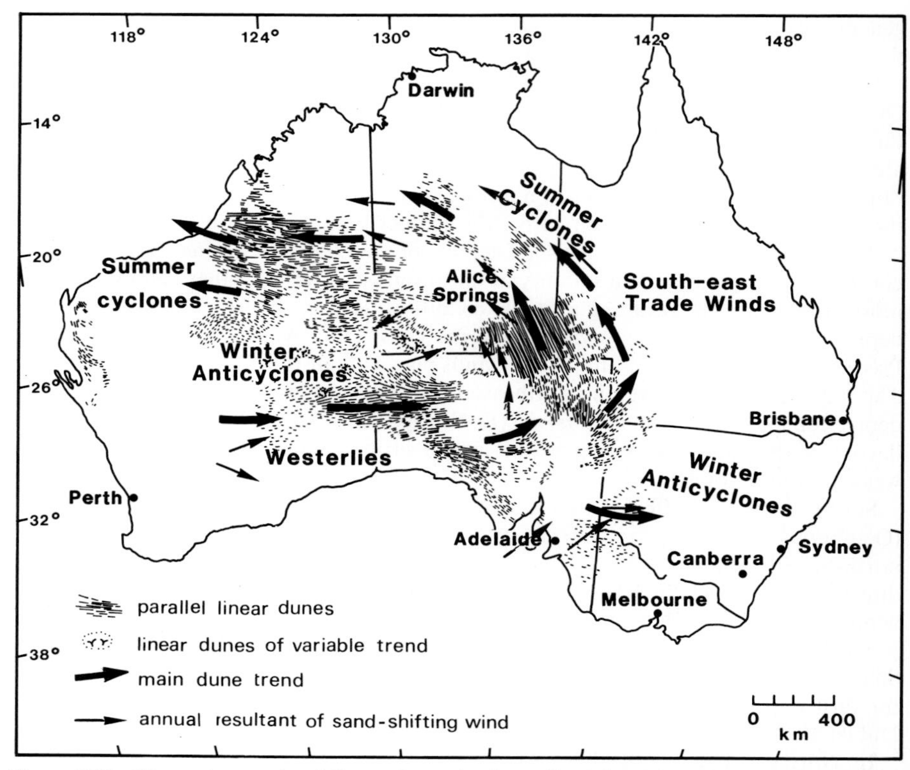

Figure 2.11 The main trends of linear dunes in Australia (after King 1960, Jennings 1968), showing also the annual resultant of sand-moving winds (after Wasson 1986) and the principal synoptic systems effecting sand movement (after Brookfield 1970). The main dune trends in the centre of the map diverge slightly from the resultant drift direction.

(Madigan 1936). However, Brookfield (1970) found that the high-pressure cell which predominates there during winter is much larger than the radius of dune curvature. Further, the anticyclone does not produce strong sand-moving winds like the summer cyclones. Hence, the trend of the sand dunes seems to reflect the resultant of winds from several directions rather than of a single anticyclonic wind system. Most of the linear dunes show good accordance with sand-shifting wind resultants (Wasson 1986), except for the core area of the main dune system (Fig. 2.11).

The mid-latitude sandy deserts of Asia experience a different climatological regime. They stretch eastward for about 5600 km, from the Caspian Sea to western Mongolia. Their dryness results from their great distance from the sea together with the shadow effect created by neighbouring mountain ranges. During winter, Central Asia experiences a pronounced heat loss from the surface leading to the development of a persistent, large, and intensified thermal anticyclone centred over Mongolia and eastern Siberia. Calm, dry air dominates most of the desert areas, with clockwise circulation radiating from the central area of high pressure. This circulating flow moves northerly winds to most of northern China (Walker 1982). These winds are blocked by the Tibetan Plateau which, in turn, causes a divergence of the wind system approximately along longitude 97°E. To the west of this line, NE winds prevail whereas to the east, NW winds are dominant (Petrov 1976, Sung-Chiao 1984). This divergent flow raises clouds of dust and causes sandstorms in the north China deserts (Pye & Zhou, 1989).

In summer the surface of the Asian continent gains heat. The dominant atmospheric systems are thermal depressions which start their development in the southern part of Central Asia. The Persian Gulf trough (Fig. 2.10b) is part of this system. The northwest part of Central Asia is characterized by the predominance of anticyclonic circulation caused by a high-pressure cell over the Ustyurt Plateau, east of the northern Caspian Sea. Northwesterly and westerly winds occur on the northeastern margin of this system, as do easterly winds in the south. The highest average wind velocities (up to 40–50 m s^{-1}) in the Central Asian deserts are recorded in spring when the winter anticyclone dissipates and frontal activity becomes more intense (Zhirkov 1964, Shikula 1981, Nalivkin 1982). A secondary maximum occurs in summer whereas the lowest mean monthly velocities occur in autumn and winter under the influence of the Mongol–Siberian anticyclone (Petrov 1976).

2.3.6 Coastal wind regimes

Coastal sand dunes are found in most climates. Every climatic belt has its own wind regime, and it is therefore impossible to generalize about the wind regimes of the world's coasts. However, they all share some important properties. It was mentioned earlier (Fig. 2.5) that the wind energy of coasts is relatively higher than that of inland areas. Surface roughness is, in general, much more prominent over coastal land surfaces than over the sea. Consequently, there is an abrupt increase in surface shear stress as winds cross the shoreline (Hsu 1971a, Sacré 1981, Greeley & Iversen 1985, p. 46, Illenberger & Rust 1988).

The coastline is also the boundary between two bodies with different thermal properties. Deep water has a low albedo, and exhibits very little thermal response to solar radiation changes. Radiation penetrates to considerable depths and the spread of heat is also enhanced by convective currents. The thermal capacity of the oceans is considerably larger than that of the land. Warming of the surface water of the oceans takes longer than warming of the land surface, but cooling of the latter is also more rapid. The resulting contrast in air temperature and moisture content between sea and land plays an essential part in the creation of local wind and weather systems.

During daytime in summer, water bodies have a lower surface temperature than the adjacent land surface, creating a pressure gradient with high pressure over the water and low pressure over the land (Fig. 2.4). The induced air flow is known as a *sea-breeze*. This local pressure gradient causes a distortion of the general regional pressure field, tending to retard offshore winds and to enhance onshore winds. Sea-breezes are most effective in generating aeolian sand transport on tropical and subtropical coasts (Inman *et al.* 1966, Flohn 1969, Tsoar 1978, Hunter & Richmond 1988).

In the eastern Mediterranean, the sea-breeze coincides with the etesian winds (Fig. 2.10b) and becomes an important summer sand-transporting element in coastal areas. Figure 2.12 shows the winter and summer sand-transporting wind roses for a dune area 35 km south of the Mediterranean shoreline of Sinai. In summer, effective sand-transporting winds are associated almost exclusively with sea-breezes blowing from the NNW, perpendicular to the shoreline. During winter, however, sea-breezes are rare. About 7% of the sand is transported by winds coming from the NNW while most sand-transporting winds are associated with easterly moving depressions (Fig. 2.10a).

The two wind roses in Figure 2.12 reflect the different nature of sea-breeze winds and cyclonic winds. As stated earlier, sea-breezes result from a pressure gradient fixed in space. They show a degree of steadiness as high as 94% [the degree of steadiness is given by $100\,V_r/V_s$, where V_r is the hourly vector mean wind velocity and V_s is the hourly

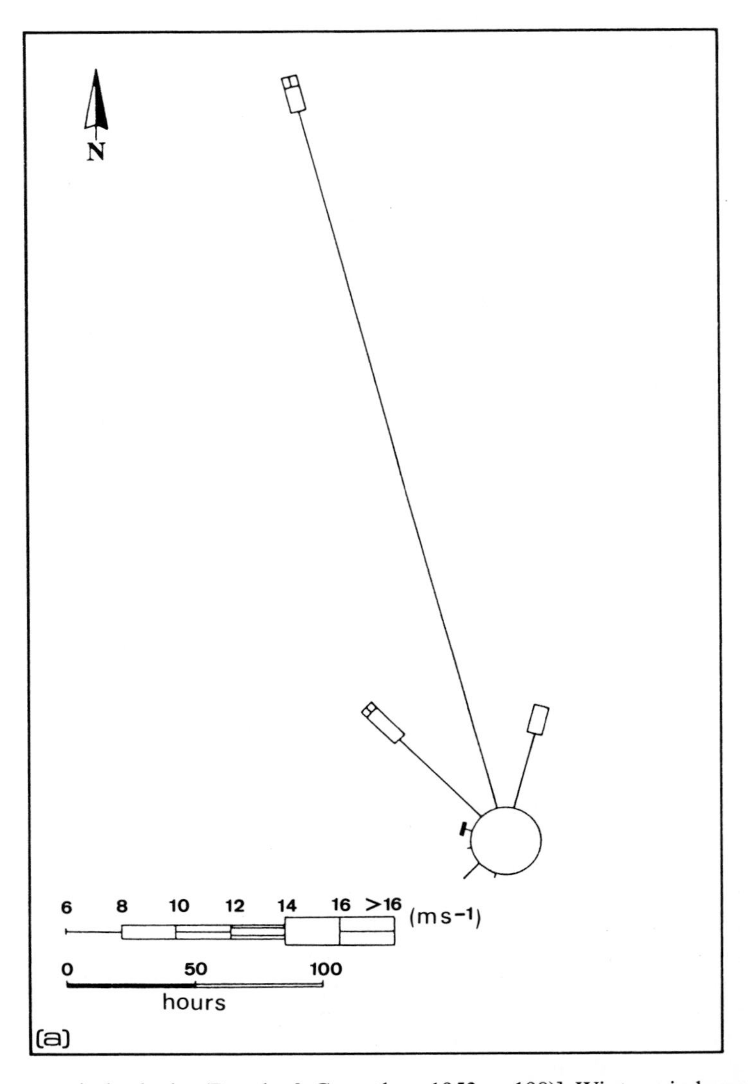

scalar mean wind velocity (Brooks & Carruthers 1953, p. 198)]. Winter winds result from dynamic lows with widely varying trajectories, bringing the degree of steadiness down to 56%.

The strength of the sea-breeze is determined by the temperature difference between land and sea. Therefore, the resultant wind velocities are relatively low. On the coast of Sinai, mean hourly wind velocities never exceed $12\,m\,s^{-1}$ at a height of 3.5 m above the sand dunes. The velocity of winds associated with cyclones is much higher (Fig. 2.12). The coastal sand dunes in Sinai are shaped mainly by the strongest winds but the effect of the sea-breeze is also imprinted on them (Tsoar 1978). In Baja California, the sea-breeze is effective the whole year round; the maximum mean velocity, also of $12\,m\,s^{-1}$, occurs in mid-afternoon during summer; during winter, the pressure gradient is less developed and the afternoon breezes are lighter (Inman *et al.* 1966).

At night, the land surface temperature drops more rapidly than the sea surface

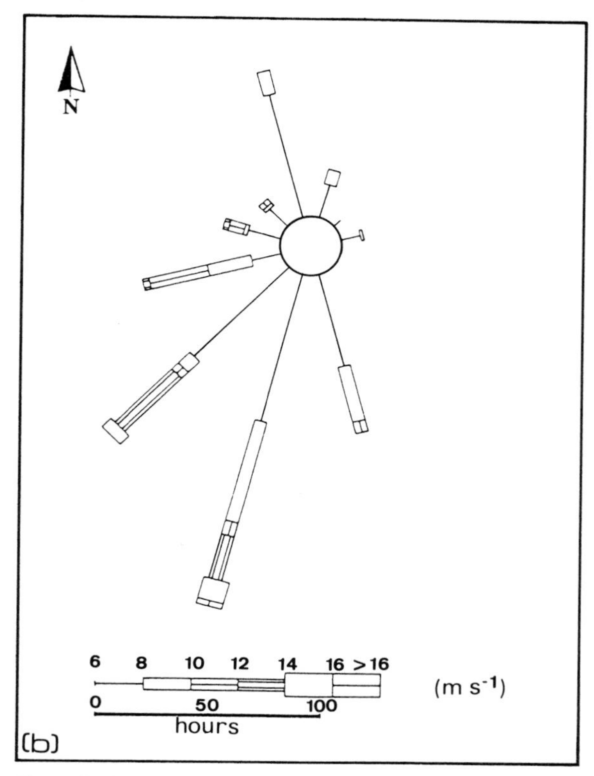

Figure 2.12 Wind roses recorded in a sand dune area 35 km south of the Mediterranean shoreline of Sinai. (a) Summer; (b) Winter. (From Tsoar 1985.)

---

temperature, resulting in the creation of a seaward pressure gradient which induces an offshore land-breeze (Flohn 1969). Since the temperature differences at night are rarely as great as those during daytime, nocturnal land-breezes normally never exceed the threshold velocity needed to transport sand.

2.4 Flow in the atmospheric boundary layer

2.4.1 Viscosity, Reynolds number and their effect on the airflow

A fluid is defined as a substance that cannot sustain shear stress. The viscosity of a fluid is an internal property which is indicative of its ability to resist shear stress. Viscosity arises from the interaction between adjacent layers and the molecular cohesion of the fluid. Air molecules are further apart so that their cohesive force is correspondingly lower than that of water molecules. Therefore, water, with higher viscosity, flows sluggishly in comparison with air.

Consider air flowing over a smooth surface so that any air parcel moves parallel to the surface. Such air motion is assumed to take place in a series of thin, parallel layers, having a thickness, dz, referred to as laminae (Fig. 2.13). A very thin lamina of air at rest

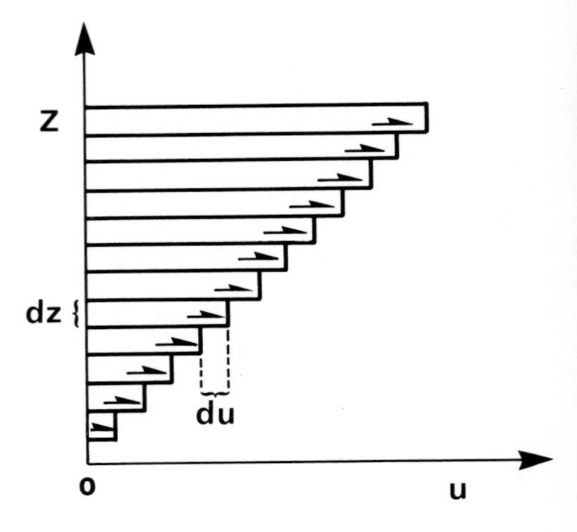

Figure 2.13 Deformation of laminar flow above the ground surface due to viscosity.

adheres to the ground surface. Above this lamina, the velocity of the laminae varies linearly with distance from the ground. The force per unit area required to overcome the viscosity, and to maintain the flow by sliding of the air laminae over each other, is known as the *laminar shear stress* (τ_l).

Now consider only two adjacent laminae of air. The one furthest from the ground surface has a velocity which is higher by dU relative to the lower one due to molecular friction (Fig. 2.13). This velocity gradient, dU/dz, known as the *rate of shearing strain*, causes a shear stress, τ_l, and is inherent to the two laminae involved. The total shear stress, i.e. the friction per laminae unit area, is equal to the product of the rate of shearing strain and the *dynamic viscosity*, μ, of the air:

$$\tau_l = \mu(dU/dz) \tag{2.9}$$

Dynamic viscosity is a characteristic physical parameter of air which increases gradually with temperature and is practically independent of velocity (von Kármán 1934). The dynamic viscosity of air at 15°C at sea level is $1.78 \times 10^{-5}\,\mathrm{kg\,m^{-1}\,s^{-1}}$, whereas its density, ρ, is $1.23\,\mathrm{kg\,m^{-3}}$ (Table 2.1).

The dynamic viscosity divided by density gives the *kinematic viscosity*, $\nu = \mu\rho$, which is $1.45 \times 10^{-5}\,\mathrm{m^2\,s^{-1}}$. In spite of the low viscosity of air, this property has an important effect on the flow near the surface where most of the aeolian processes occur.

Fluids with a constant viscosity [that is, the relation between their shear stress and velocity gradient (Eqn 2.9) can be expected to be linear] are referred to as *Newtonian fluids*. Air and pure water are Newtonian fluids. Other substances that flow in nature such as glacier ice and mud are non-Newtonian and for this reason their movement is more complicated.

According to Newton's first and second laws, flowing air has to sustain a state of motion that continues in the same direction, a condition known as *inertia*. Inertial force has to overcome the viscous forces in order to allow flow over a surface. The ratio of the inertial force (proportional to $\rho U^2/L$; where U is the velocity and L is a length parameter) to the viscous force (proportional to $\mu U/L^2$) provides an important index of the type of flow. This dimensionless parameter is known as the Reynolds number, *Re*:

$$Re = (\rho U^2/L)/(\mu U/L^2) = \rho LU/\mu = LU/\nu \tag{2.10}$$

The numerical value of length parameter (L) depends on the type and scale of flow under investigation. For airflow over sand dunes it is usual to take the dune height as a value for L. Obviously, the characteristic length selected for Equation 2.10 will ultimately determine the numerical magnitude of *Re* for the given set of conditions.

When *Re* is small, viscous effects are dominant. When it is large, the inertial effects predominate. In the first case, we have laminar flow in the atmospheric boundary layer (Fig. 2.13). In the second case, the stratified laminar structure is deformed and destroyed, with random irregular motions in all directions being superimposed on the principal average air flow. This type of flow, known as *turbulent flow*, is characterized by fluctuating pressures and velocities. There is random formation and decay of a multitude of small eddies throughout the turbulent flow stream.

The components of the turbulent wind velocity, at any instant, are u, v, and w in the x, y, and z directions, respectively, of a Cartesian coordinate system in which x is in the direction of the average airflow parallel to the ground and y and z are directions across the flow in the horizontal and vertical plane, respectively. The mean velocities $\bar{u}$, $\bar{v}$, and $\bar{w}$ can be defined for an interval of time. The difference between the mean velocity and the instantaneous velocity is the eddy velocity (u', v', and w', respectively, Sutton 1934):

$$\begin{aligned} u' &= u - \bar{u} \\ v' &= v - \bar{v} \\ w' &= w - \bar{w} \end{aligned} \tag{2.11}$$

One important characteristic of turbulent wind flow is gustiness, that is, bursts of choppy violent wind of very short duration. The components of wind gustiness are defined as $u'/\bar{u}$, $v'/\bar{u}$, and $w'/\bar{u}$ (Sutton 1953, p. 250). The effect of gusts is shown clearly by short-term changes in horizontal wind velocity (U) and direction recorded at meteorological stations (Fig. 2.14).

The critical Reynolds number, at which laminar atmospheric boundary layer flow becomes turbulent, is greater than 6000 (Houghton 1986, p. 126). For air at sea level and 15°C, the Reynolds number is $6.9 \times 10^4 \bar{U}L$ ($\bar{U}$ in m s^{-1} and L in m). It is evident that over small roughness lengths low wind velocities of only 0.1 m s^{-1} are needed to create turbulent airflow. All natural flows involved in aeolian processes are turbulent and only rarely do we find laminar flow in the atmosphere (von Kármán 1937).

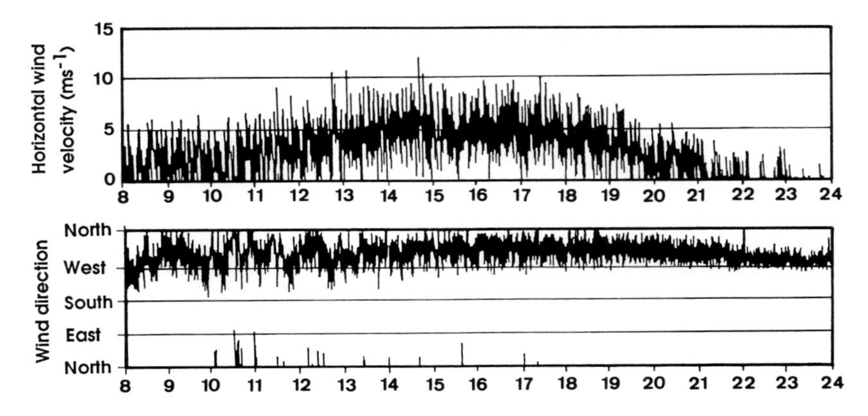

Figure 2.14 A wind chart showing short-term fluctuations in wind velocity and direction at Beer Sheva, northern Negev desert, during a summer day. The turbulent wind flow during the day is characterized by gusts of high velocity and constant change in direction.

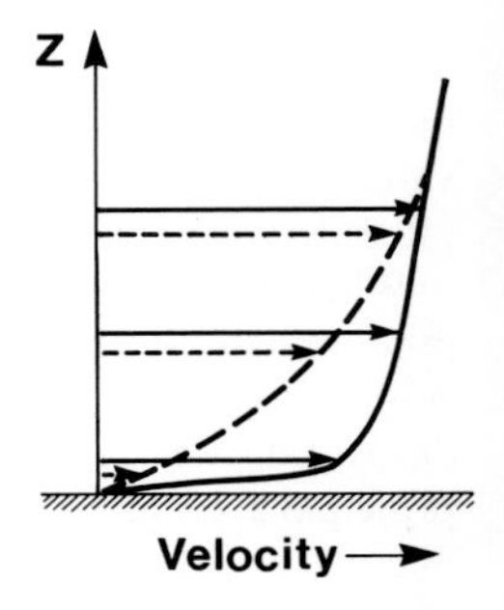

Figure 2.15 Boundary layer velocity profiles for turbulent flow (solid line) and laminar flow (broken line) plotted on the same scale. The arrows are directly proportional to the velocity.

Turbulent wind flow is retarded not only by viscous friction, as in laminar flow (Eqn 2.9), but also by exchanges of momentum from layer to layer due to velocity fluctuations (η), all of which determines the turbulent shear stress ($\bar{\tau}_t$). Thus, the total *mean shear stess* in turbulent flow, $\bar{\tau}$, is (Calder 1949)

$$\bar{\tau} = \tau l + \tau_t = (\mu + \eta)(d\bar{U}/dz) \tag{2.12}$$

η is also known as the *eddy viscosity*, although it is not a property of the fluid but of the rate of turbulence (Stanton 1911). Because of the random movement imparted to the air by the turbulent eddies, one needs to use the average velocity $\bar{U}$ in turbulent flow. When *Re* is large, the laminar (viscous) shear stress (Eqn 2.9) is a negligible part of the total shear stress, with the exception of the layer adjacent to a smooth surface (von Kármán 1934, 1937, Calder 1949, Owen 1960).

2.4.2 *Variation of wind velocity with height*

The viscous shearing action creates a horizontal drag on the moving air in the vicinity of the ground surface (Fig. 2.13). At a certain height above the ground, the wind velocity is almost unaffected by the viscous force, but is affected by the inertial force, the Coriolis force, and the pressure gradient forces. The flow at this level is known as *inviscid flow* and has an undisturbed velocity of U_∞, also known as the free-stream velocity. At the ground surface itself the wind velocity is zero. Near the surface the wind velocity increases sharply with height, and thereafter it increases asymptotically with height. The vertical mass exchange, which characterizes turbulent flow, flattens the velocity profile of the turbulent boundary layer close to the surface as compared with laminar flow (Fig. 2.15). The part of the flow in the lower atmosphere where the wind velocity changes from zero to U_∞ is termed the *atmospheric boundary layer*.

It is conventional to define the upper limit of the boundary layer as the point above the surface where the velocity is 99% of the free stream velocity (the velocity in absence of the surface boundary). The thickness of the atmospheric turbulent boundary layer varies according to the roughness of the ground surface, and can be a few hundred metres up to 1 km or more over large sand dunes during daytime.

The large velocity gradient (Fig. 2.15) in the boundary layer creates a large shear stress adjacent to the ground surface, even in a fluid such as air that has low viscosity. The turbulent shear stress ($\bar{\tau}_t$), equal to the transfer of momentum in unit time per unit area, is given by (von Kármán 1935, Prandtl 1935, p. 127, Sutton 1953, p. 73)

$$\bar{\tau}_t = -\overline{\rho u' w'} = \eta(dU/dz) \tag{2.13}$$

The bar indicates mean values with respect to time, and the negative sign reflects the fact that the product of u' and w' is always negative. The quantity $-\overline{\rho u' w'}$, called the *eddy shearing stress*, is the mathematical expression of the vertical transport of momentum by the velocity fluctuations. This vertical movement along a determined average distance is known as the *mixing length*, l, and exists as long as its momentum is not absorbed (Prandtl 1935). The higher the degree of turbulence, the greater is the mixing length. According to the mixing-length assumption, u' is proportional to $l(\mathrm{d}U/\mathrm{d}z)$ and $|w'|$ is proportional to $|u'|$, so that Equation 2.13 can be expressed in the form (Prandtl 1935, p. 130)

$$\bar{\tau}_t = \rho l^2(\mathrm{d}U/\mathrm{d}z)^2 \tag{2.14}$$

Equation 2.14 is of little practical value as the mixing length is an unknown variable. Prandtl (1935, p. 132) assumed that l is proportional to the distance above the surface, z, since the turbulent exchange increases at greater distance from the surface, whereas at the surface it is zero:

$$l = kz \tag{2.15}$$

where k is known as the *von Kármán universal constant* for turbulent flow. Its value, as determined empirically, varies between 0.33 and 0.41 but it is commonly taken as 0.40 (von Kármán 1935, Tennekes 1973).

Substituting Equation 2.15 in Equation 2.14:

$$\bar{\tau}_t = \rho k^2 z^2(\mathrm{d}\bar{U}/\mathrm{d}z)^2 \tag{2.16}$$

Extraction of the square root gives

$$\mathrm{d}\bar{U}/\mathrm{d}z = (1/kz)(\bar{\tau}_t/\rho)^{1/2} \tag{2.17}$$

The shear stress at the ground surface is denoted by τ_0 (Prandtl 1935, p. 135). The quantity $(\tau_0/\rho)^{1/2}$, which has the dimension of velocity, is known as the *friction velocity*, u_* (von Kármán 1934), although it is actually a measure of the shear stress:

$$u_* = (\tau_0/\rho)^{1/2} \tag{2.18}$$

The friction velocity has an advantage over the actual velocity being independent of the height of the velocity measurement above the surface. $\bar{\tau}_t$ does not vary with distance z close to the ground and is equal to the shear stress, τ_0, at the surface (Sutton 1953, p. 76). Substituting Equation 2.18 in Equation 2.17 and integrating yields:

$$\bar{U}/u_* = (1/k)\ln(z/C) \tag{2.19}$$

where C is the integration constant.

When the surface is rough and the air velocity is sufficiently high, the resistance to the flow is dependent solely on the height, shape, and density of distribution of the surface roughness elements (Deacon 1953). Accordingly, the form of Equation 2.19 has to be

$$\bar{U}/u_* = (1/k)\ln(z/z_0) \tag{2.20}$$

where z_0, the *roughness length* of the surface, was found to be approximately $d/30$, d being the diameter of the sand grains forming the surface, over a flat, homogeneous surface for conditions that prevail under a high Reynolds number in which there is no obvious loose sand movement (Nikuradse 1933, Byrne 1968, Maegley 1976). This value was adopted for airflow over quiescent sand by Bagnold (1936) and Monin & Yaglom (1965, p. 289). Zingg (1953a) found that d/z_0 decreases rapidly with increasing grain size above 0.2 mm.

The roughness length does not depend solely on the diameter of the sand grains but is also influenced by their ground area density (Greeley & Iversen 1985, p. 43). An increase in the roughness length of the surface results in an increase of the friction velocity, whereas a decrease in roughness length brings about a decrease in surface shear stress with an 'overshoot' in the boundary, resulting in a zone of deposition (Elliott 1958, Blom & Wartena 1969, Hsu 1971a, Greeley & Iversen 1987).

Equation 2.20, known as the *von Kármán – Prandtl logarithmic velocity profile law*, is valid in the lowest zone of the boundary layer (Blom & Wartena 1969, Nickling 1978) for a neutrally stratified atmosphere in the absence of sand movement, under strong winds where the velocity profile is independent of viscosity. Under stably stratified (night time) atmospheric conditions, the wind shear is greater than predicted by Equation 2.17, whereas under unstable conditions it is smaller (Thom 1975, Nickling 1983).

The friction velocity is an expression of the velocity gradient. At the surface the velocity is zero, so the gradient starts at the height of the roughness length, z_0. When the height z, in a neutrally stratified atmosphere, is plotted on a logarithmic scale, the velocity gradient becomes linear (Fig. 2.16). A quick way of determining u_* is to measure

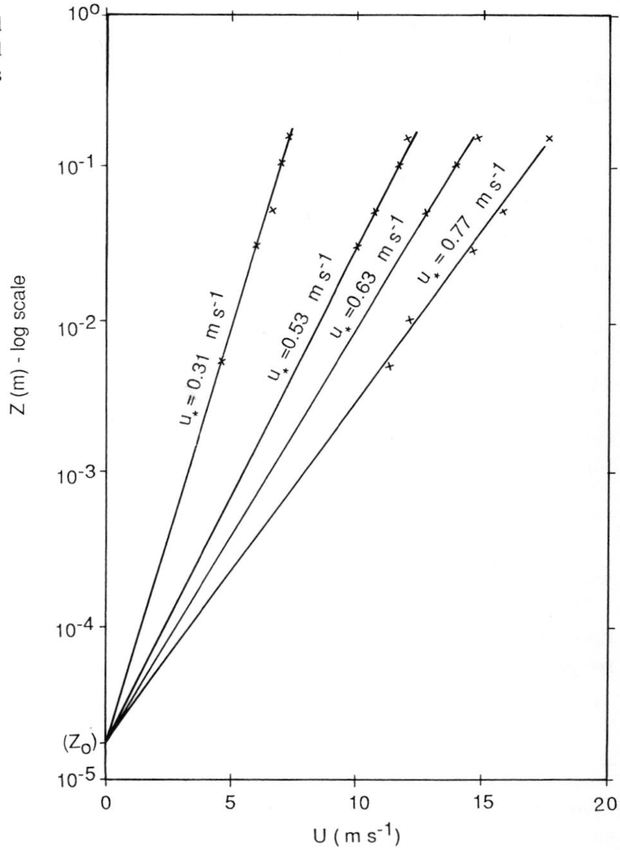

Figure 2.16 Several wind velocity gradients measured over a surface whose roughness length is 1.766×10^{-5}m.

the wind velocity difference (ΔU) at two heights, one at an elevation of one (dimensions can be either centimetres or metres) and the other at an elevation 2.718 times higher. This difference, according to Equation 2.20, is $\Delta Uk = u_*$ (Bagnold 1941, p. 51). The parameter z_0 can be calculated by plotting the measured wind velocity profile (when there is no sand movement) above the ground on semi-logarithmic paper and obtaining the intercept of the profile with the height (z) ordinate at zero wind velocity (Fig. 2.16).

In areas where tall vegetation of height h covers the surface, the datum surface of zero height cannot be taken at ground level but must be displaced upward by a height, d, known as the *zero plane displacement* (Deacon 1953, Monteith 1973, p. 88). The value of d can only be determined from the wind profile. The height above the zero plane, $z - d$, replaces z in Equation 2.20, giving

$$\bar{U}/u_* = (1/k)\ln[(z - d)/z_0] \tag{2.21}$$

The value of d depends on h and the spacing of the canopy; usually it is between $0.6h$ and $0.8h$.

Close to a very smooth ground surface the windspeed decreases (Fig. 2.16) and viscous forces become predominant so that, in a very limited region immediately adjacent to the surface, there is a thin layer in which the flow is laminar, known as the *laminar sub-layer*. As this layer is very thin, it is extremely difficult to carry out any experimental observations on it. According to Rouse (1937),

$$\delta_L = 11.6\nu/u_* \tag{2.22}$$

where δ_L is the thickness of the laminar sub-layer.

Equation 2.10 for the determination of the Reynolds number can be changed slightly by substituting u_* for U and the grain diameter, d, for L:

$$Re_* = u_* \mathrm{d}/\nu \tag{2.23}$$

where Re_* is known as the *particle friction Reynolds number* or the Reynolds number of the grain (Shields 1936). When $d = \delta_L$, $Re_* = 11.6$ (Eqns 2.22 and 2.23). Accordingly, the surface is said to be aerodynamically smooth when $Re_* < 4$ and the average grain size is about one third, or less, of the thickness of the laminar sub-layer. When $Re_* > 60$ and the average grain size is about five times the thickness of the laminar sub-layer, the surface is said to be aerodynamically rough (von Kármán 1935, Webber 1971, p. 92).

Figure 2.17 shows two cases of different relative size between the laminar sub-layer and the grain size. In Figure 2.17a the grains lie well within the flow and prevent any eddies forming on the grains. This surface is considered to be aerodynamically smooth. Figure 2.17b shows that the grains are much larger and only have one fifth of their diameter immersed in the laminar sub-layer. Because of the projections, vortices are formed in the flow and the laminar sub-layer is almost completely disrupted. This surface is considered to be aerodynamically rough (von Kármán 1937).

Figure 2.17a shows that there are no projections into the turbulent flow so the drag exerted by the boundary surface must be transmitted to the air through the laminar sub-layer. The surface roughness (z_0) does not determine the velocity profile which is dependent on viscosity (von Kármán 1935, Deacon 1953):

$$\bar{U}/u_* = (1/k)\ln[(9.05\,u_* z)/\nu] \tag{2.24}$$

Equation 2.20 gives the logarithmic velocity profile over an aerodynamically rough surface, whereas Equation 2.24 is for an aerodynamically smooth surface.

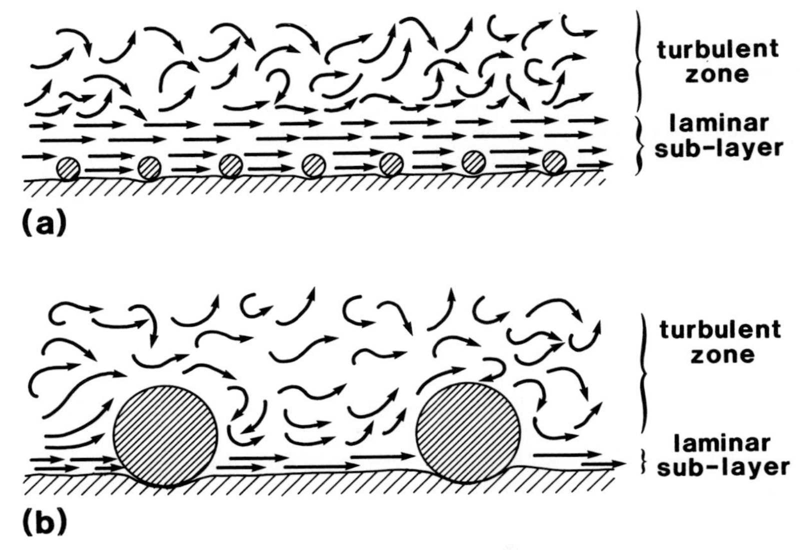

Figure 2.17 The structure of the wind blowing over sand grains at the ground surface. (a) The relatively small grains are immersed deep in the laminar sub-layer, creating an aerodynamically smooth surface. (b) The relatively large grains stand out into the turbulent flow with only one fifth of their diameters immersed in the laminar sub-layer, creating an aerodynamically rough surface. (Modified after Chepil 1958a).

2.4.3 Continuity of air flow: Bernoulli equation and separation of flow

An important property of gases is their compressibility. At wind velocities up to 60 m s^{-1} the compressibility of air is negligible (Ower & Pankhurst 1977, p. 15). Since such high wind velocities are rare in general and uncommon over desert sand dunes, we can accept the postulate that the air of natural winds is incompressible. This allows us to use the principles of the laws of conservation of mass and energy to develop the continuity and Bernoulli equations for air flow. In turbulent flow the parcels of moving air actually have an irregular motion. However, for the purposes of this analysis the turbulent properties can be ignored and only the average direction of the wind in two dimensions is considered. There are actually an infinite number of streamlines in any airflow, but only a selected number of them are used for purposes of demonstration.

Consider flow over a hillock where only the streamlines on the windward slope are shown as they tend to converge over the crest (Fig. 2.18). The volume of air passing area a_1 with velocity U_1 in a given time must be the same as that passing area a_2 with velocity U_2 in the same time. This may be expressed in the general form of the *continuity equation*:

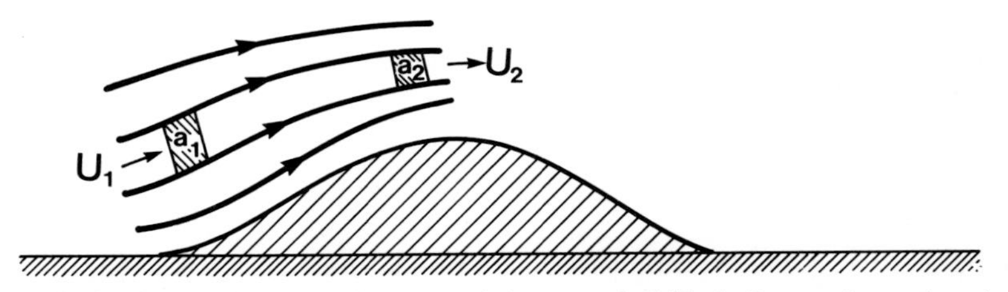

Figure 2.18 Converging streamlines toward the crest of a hillock. See text for explanation.

$$U_1 a_1 = U_2 a_2 = \text{constant} \tag{2.25}$$

It follows from equation 2.25 that converging streamlines are associated with an increase and diverging streamlines with a decrease in velocity.

In a similar way, we can refer to the flow in Figure 2.18 as part of a system in which conservation of energy should be maintained. Velocity, pressure, and weight (potential energy) are the forms of energy in this flow system. The difference in height between the base and the crest is small. Together with the small magnitude of air density, it makes the effect of weight constant for all points along the streamlines. The total energy in the form of velocity and pressure must be the same at the base as at the crest of the hillock:

$$p_t = p_s + \tfrac{1}{2}\rho U^2 = \text{constant} \tag{2.26}$$

where p_t is the *total pressure*, and is also known as the *stagnation pressure* as it also exists when the flowing air is brought to rest. The term p_s represents the *static pressure*; this is the pressure felt on the surface over which the air is flowing. It can be easily measured in a way that does not disturb the flow, as for instance from a hole in the surface. The term $\tfrac{1}{2}\rho U^2$, known as the *dynamic pressure* of the flow, is an expression of the kinetic energy of the flow. Equation 2.26 is a simplified form of the *Bernoulli equation*, which states that the pressure plus the kinetic energy per unit mass of a fluid has a constant value everywhere. This equation is widely used in wind measurements and in understanding aeolian processes.

The presence of obstacles to wind flow, such as sand dunes, causes the energy to be redistributed between the two forms, velocity and pressure. Consider the pattern of streamlines around a two-dimensional semicylinder. The surface of the semicylinder causes retardation of the flow due to the viscous shearing action, thus forming a small boundary layer adjacent to the surface (Fig. 2.19). At the upstream face the streamlines diverge, inducing a pronounced reduction in velocity at point A. According to Equation 2.26, low velocity at point A increases the static pressure on that point so that it approaches the total pressure. Towards the crest of the semicylinder the velocity attains

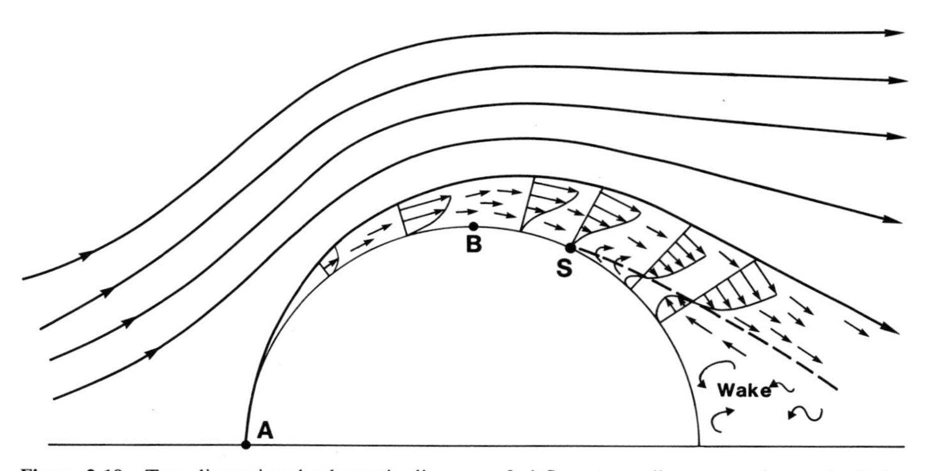

Figure 2.19 Two-dimensional schematic diagram of airflow streamlines around a semicylinder. The streamline adjacent to the semicylinder surface demarcates the upper limit of the boundary layer; the broken line indicates the zero velocity surface beneath which backflow is beginning. The wind profiles and small arrows indicate the flow within the boundary layer. For explanation, see text. (After Mironer 1979, reproduced by permission of McGraw-Hill).

(A)

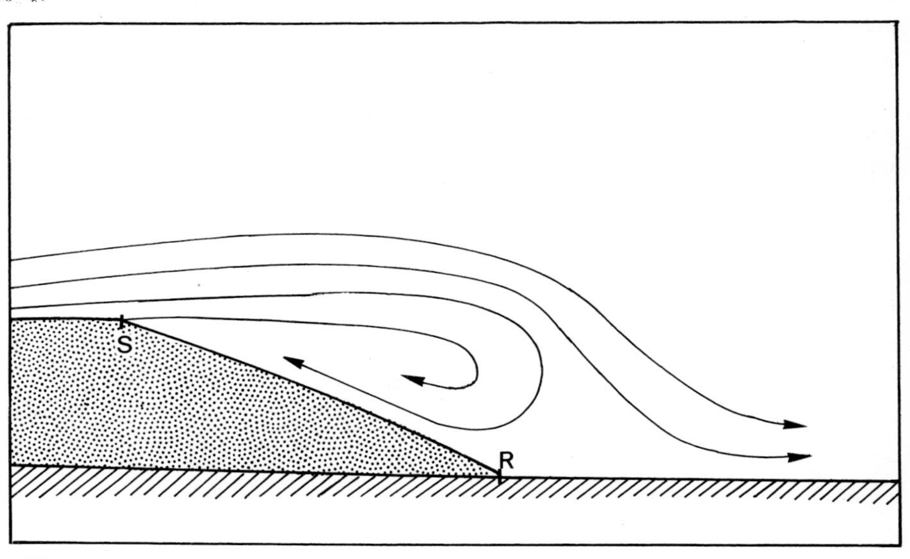

(B)

Figure 2.20 (A) Flow separation above the slip face of a barchan dune demonstrated by smoke. (B) A schematic diagram of the phenomenon shown in A. S = the separation point; R = the reattachment point.

a maximum so the static pressure decreases gradually from A to B, forming a decreasing pressure gradient. This accelerating flow offsets the effect of the viscosity; thus, on the windward side of the semicylinder the boundary layer remains relatively thin (Fig. 2.19).

Downstream of point B the flow is retarded and the static pressure rises. In this case the effect of the viscosity is intensified by the increasing pressure gradient, causing the boundary layer to thicken sharply downstream. Ultimately, at point S the velocity profile near the surface becomes zero. Downstream of point S the surface layer is separated from the ground level and enters the mainstream of the flow as a free shear layer. This phenomenon, known as the *separation of flow*, is very important and governs the processes of sand erosion, transport, and deposition on the lee side of many dunes. The reverse flow downstream of point S is offset by a forward flow, forming on the lee side a region, known as the *wake*, in which mechanical energy is continuously being dispersed into turbulent eddies (Fig. 2.19).

Sharp-edged bodies which have an abrupt change in surface inclination will cause airflow separation because the air is not capable of reaching an infinitely large velocity at the sharp brink (Cooper 1944, Chang 1976, p. 3). Most separation phenomena over sand dunes are of this kind. Figure 2.20 demonstrates the separation of flow over the brink of a barchan dune. In this case the shear layer is separated from the ground at the brink of the slip face where there is a sudden change in the surface inclination. This separated layer returns and re-attaches to the ground at a distance downflow, where it often splits, with part of the flow being deflected upstream into a recirculating backflow region, and the other part continuing downstream. This type of separation, and the lee-side eddies and secondary flows that it produces, represents one of the most important processes governing the formation of sand dunes.

2.4.4 The drag force

The flow around the circumference of the semicylinder (Fig. 2.19) is divided into disturbed viscous flow in a boundary layer formed near the surface and, at some distance from it, undisturbed flow in the free-stream shear layer. According to the Bernoulli equation (Eqn 2.26), the total pressure in the system shown in Figure 2.19 is constant; therefore,

$$p_s + \frac{1}{2}\rho U^2 = p_{s\infty} + \frac{1}{2}\rho U_\infty^2 \qquad (2.27)$$

where p_s = static pressure in the boundary layer flow whose velocity is U and p_{s_∞} = static pressure in the free-stream flow whose velocity is U_∞. Equation 2.27 can be developed into

$$\frac{p_s - p_{s\infty}}{\rho U_\infty^2/2} = 1 - (U/U_\infty)^2 \qquad (2.28)$$

The term on the right is known as the *pressure coefficient*, C_p, and it provides a way to express the dimensionless ratio of pressure force around bodies of different shapes. Figure 2.21 shows the pressure coefficient distribution around the cylinder. Positive C_p is developed where the flow slows down and is then less than that of the free shear layer. At point A the flow is stopped completely and C_p attains the maximum possible value, +1. When the velocity around the cylinder is equal to that of the undisturbed flow, then the two respective static pressures equal each other and hence $C_p = 0$. When the

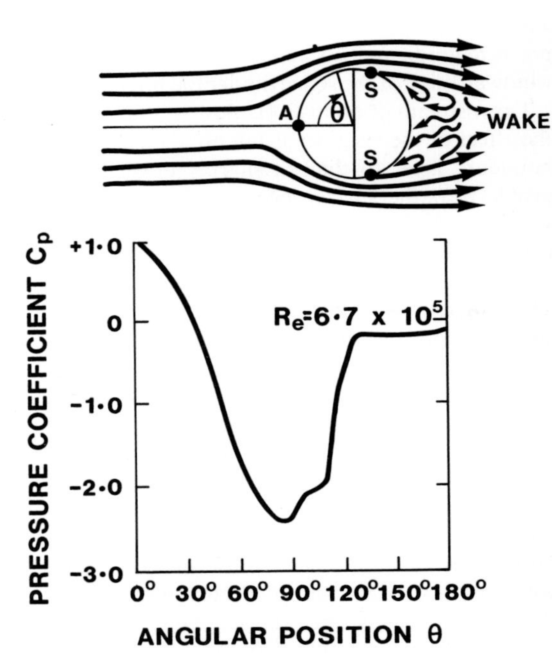

Figure 2.21 Pressure distribution around a cylinder. A indicates the stagnation point and S the separation points. (Adapted from Mironer 1979, reproduced by permission of McGraw-Hill).

disturbed velocity is greater than that of the free shear layer, so that $p_s < p_{s_\infty}$, C_P values will always be negative. Within the wake of the separated flow there is a very low rate of directed flow movement. Therefore, the Bernoulli equation cannot be applied there and the pressure on the surface remains fairly uniform, at about the value of the static pressure at the point of separation (S). Figure 2.21 indicates a variation in the pressure distribution around the cylinder. A vertical line divides the cylinder in Figure 2.21 into two halves, one facing upstream and the other downstream. Integration of the pressure over both sides results in a net pressure force in the direction of the flow. This force, known as *pressure drag*, is effective when a body in flowing air induces separation. This is the main force behind sand grain movement (see Chapter 4).

The force which air exerts tangentially (shear stress) as it flows over the surface (τ_0, Eqn 2.18) is also known as the *skin friction drag*. This force is directly attributable to the viscous shear (Eqn 2.13) and is effective under a turbulent boundary layer. The total drag on a body immersed in air is the combination of both pressure drag and skin friction drag. It is customary to express total drag (F_d) as

$$F_d = C_D \frac{1}{2} \rho U^2 A \tag{2.29}$$

where A is the largest projection area of the body and the product $\frac{1}{2}\rho U^2$ is the dynamic pressure. C_D is a dimensionless drag coefficient that depends on a number of factors, the most important being the shape of the body, but also including the Reynolds number and surface roughness. The shape and projected area of the particles or bedforms exposed to the wind are the factors which principally determine the drag.

The geometry of a body is the factor that determines the formation and size of flow separation. No separation will occur with flow over an ellipsoid shape which tapers downstream to a point. In this case, there would be a gradual increase in the static

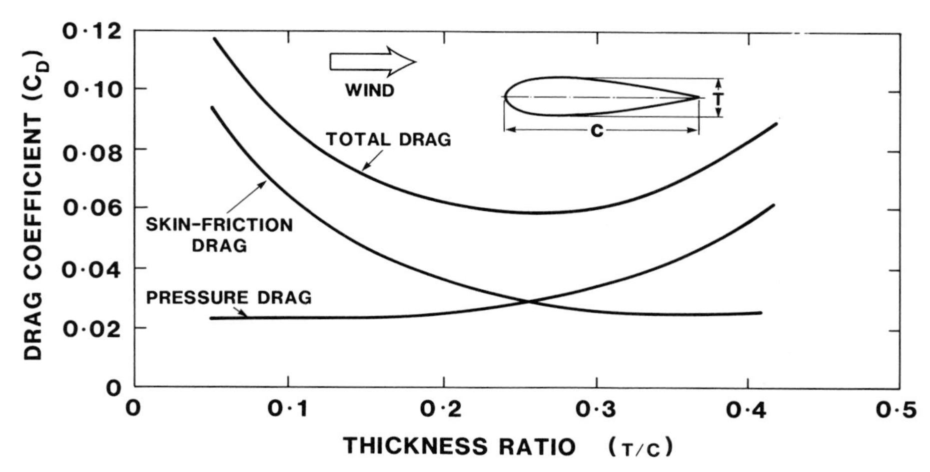

Figure 2.22 Variation in the coefficient of pressure drag, skin friction drag, and total drag as a function of thickness ratio of a streamlined body. (Adapted from Goldstein 1938, p. 403).

pressure downstream which would prevent separation. On the other hand, the long, narrow downstream shape would increase the skin friction drag so that the total drag would still be very high. Figure 2.22 shows that a tapering body, whose width to length ratio is 1:4, has the minimum total drag. This form, known as a *streamlined body*, is almost devoid of separation. All other bodies which induce boundary-layer separation are known as *bluff bodies*. All sand dunes are bluff bodies whereas aeolian erosional forms such as *yardangs* are often shaped by the wind into streamlined shapes (Ward & Greeley 1984).

Erosion of sand dunes and yardangs is the result of shear stress (τ_0) over the surface, that is, skin friction drag. There is, however, an inter-relation between the body shape governing pressure drag and skin friction drag. A bluff shape induces increased velocity where the streamlines converge at the point of separation (Figs 2.19 and 2.21). Here strong skin friction drag erodes sediment which will eventually be deposited in the wake area. The final result will be a tapered body approaching streamlined shape. Wet sand behaves as a cohesive sediment and may be shaped by wind erosion into streamlined bodies, in a manner similar to yardangs (Fig. 2.23) (Hunter *et al.* 1983).

2.4.5 Air flow over isolated hills and complex terrain

Air flow over natural rough terrain, such as dune topography, displays important variations from air flow over a flat surface. Topographic obstacles give rise to perturbations in the flow which in turn generate vertical and spatial variations in shear stress and in turbulence characteristics. Although major advances have been made in understanding and prediction of turbulent airflow over hills in the last 15 years (Taylor & Gent 1974, Jackson & Hunt 1975, Hunt 1980, Smedman & Bergstrom 1984, Jensen & Zeman 1985, Zeman & Jensen 1987, Hunt *et al.* 1988a, b), an entirely adequate predictive model of airflow and surface shear stress variations in complex terrain is still some way off.

In their theoretical analysis of air flow over low hills, Jackson & Hunt (1975) divided the flow in the surface layer (the lowermost part of the atmospheric boundary layer) into two parts, an outer inviscid region and an inner region where perturbation shear stresses affect the perturbed flow. Many of the basic predictions of the Jackson & Hunt theory for neutral conditions have been tested by field observations and numerical simulations

(Sykes 1980, Bradley 1980, 1983, Jensen 1983, Britter *et al.* 1981, Mason & King, 1985), and the theory has been modified for flow over three-dimensional hills (Mason and Sykes 1979, Walmsley *et al.* 1982). Recently, an improved version of the theory has been published (Hunt *et al.* 1988a) in which the outer region is divided into an upper and a middle layer, while the inner region is divided into a shear stress layer and an inner surface layer (Fig. 2.24). The characteristic hill length scale is indicated by L, where $2L$ is the distance between the points corresponding to the half-height of the hill. The upwind reference speed, which determines the pressure perturbations in the flow, and hence the perturbations close to the surface, is taken at the top of the middle layer (h_m). In the middle layer the flow is considered to be inviscid but rotational. The effect of non-logarithmic upwind velocity profiles is also included in the revised theory of Hunt *et al.* (1988a), while Hunt *et al.* (1988b) considered the conditions of stably stratified flow. However, these analyses are still strictly applicable only to convex hills whose length is large relative to their height (i.e. which have gentle slopes). The effects of buoyancy are also assumed to be negligible. The applicability of the theory to flow over steep dune slopes and under conditions of strong solar heating is therefore questionable (for further discussion, see Rasmussen 1990).

Changes in wind velocity over steep hills and escarpments have also been investigated by Jackson (1976), Bowen & Lindley (1977), Pearse *et al.* (1981), and Norstrud (1982). A simple measure of the degree of wind acceleration over the hill is given by the *amplification factor*, Az (Bowen & Lindley 1977), defined as

$$Az = \bar{U}_2/\bar{U}_1 \tag{2.30}$$

Figure 2.23 Erosional streamlined shapes that were formed on a wet dune sand surface, Raabjerg Mile dunefield, Denmark.

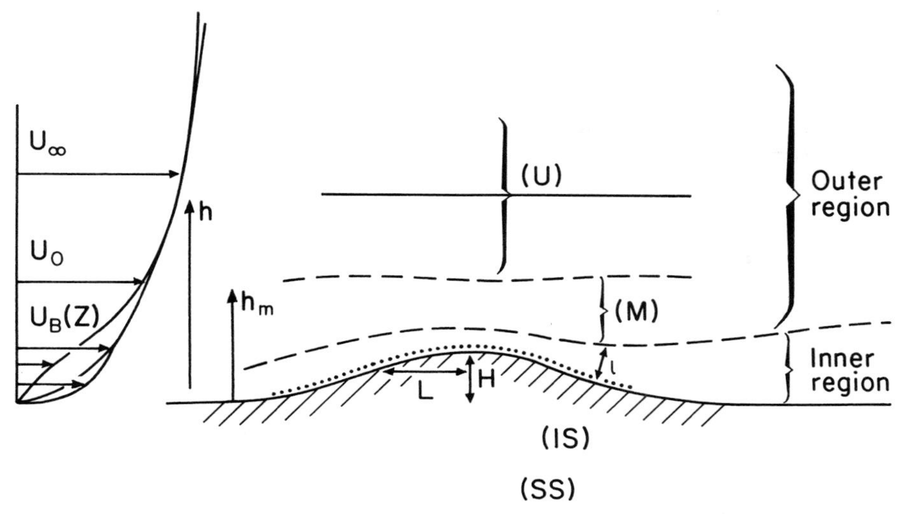

Figure 2.24 Definition sketch of flow over a hill showing the main regions of the flow and their subdivision (after Hunt *et al.* 1988a). The height of the middle layer (*M*) defines the reference velocity (U_0) used in the analysis. Also shown is the range of upwind velocity profiles considered by Hunt *et al.* (1988a). *H* represents the hill height, *L* represents the characteristic hill length, IS is the inner surface layer, and SS is the shear stress layer.

where $\bar{U}_2$ is the mean disturbed flow velocity above the hill, measured at height z, and $\bar{U}_1$ is the mean undisturbed flow velocity measured at the same height over flat ground upwind of the hill.

The hill shape and steepness have an important effect on the magnitude of the amplification factor. Experiments using a wide range of model hill and escarpment shapes have shown that there is a velocity reduction at the base and just upwind of the windward slope (i.e. $Az < 1$), while the velocity increases relative to the undisturbed flow velocity over the escarpment crest ($Az > 1$). An example of flow variation over two different escarpment shapes is shown in Figure 2.25. Flow separation occurs at a certain distance downwind from the top of the escarpment, the distance depending on the slope steepness. The model proposed by Bowen & Lindley (1977) for flow over escarpments appears to provide a close approximation of actual flow conditions in areas where cliff-top dunes are developed (e.g. Marsh & Marsh 1987).

In the case of flow over steeply convex hills, the amplification factor decreases on the upper part of the windward slope near the crest (Norstrud 1982; Walmsley *et al.*, 1982). However, with concave or rectilinear windward slope profiles, the amplification factor increases progressively (but not at a uniform rate) towards the crest (Pearse *et al.* 1981; Norstrud 1982; Tsoar 1986). The equation for the amplification factor over a uniform windward slope given by Jackson (1976) is

$$Az = 1 + \frac{h}{4\pi L} \ln \left\{ \frac{(z/L)^2 + [1 + (x/L)]^2}{(z/L)^2 + [1 - (x/L)]^2} \right\} \tag{2.31}$$

where z is the height above the ground surface at which the velocity measurements are taken, L the horizontal distance from the top of the escarpment to the point of half-maximum escarpment height, h the maximum height of the escarpment, and x the horizontal distance from the point of half-maximum escarpment height. As shown in

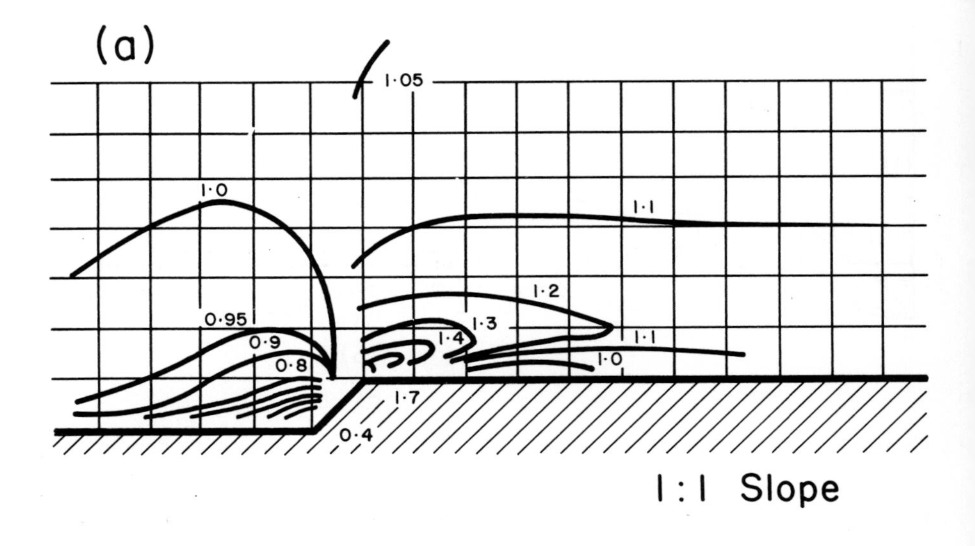

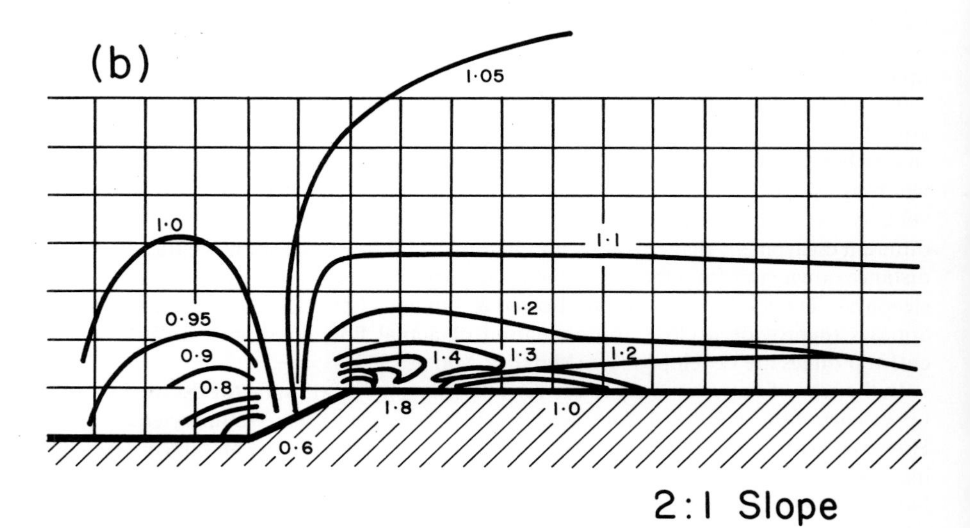

Figure 2.25 Variations in the velocity amplification factor over two slopes of different angles (after Bowen & Lindley 1977). The distance of the zone of flow separation downwind of the escarpment crest increases at lower slope angles.

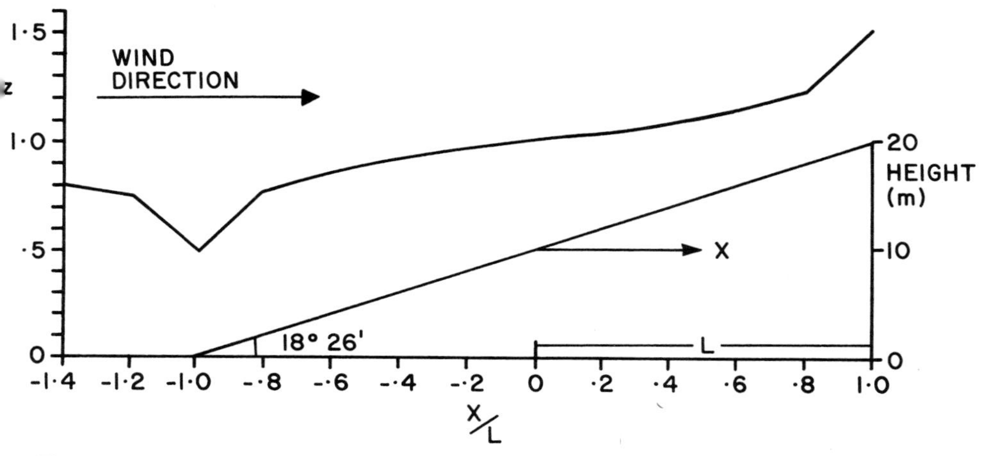

Figure 2.26 Calculated amplification factor (*Az*) over a uniform windward slope having an inclination of 18°26′. After Tsoar (1986), based on Jackson's (1977) equation for flow amplification over a uniform escarpment.

Figure 2.26, this model predicts a non-uniform increase in the amplification factor towards the top of the escarpment.

Since the near-surface wind profile over a hill deviates from the ideal logarithmic profile, the surface shear stress cannot be calculated accurately from wind velocity measurements taken at two points above the surface. This creates difficulties for the accurate prediction of threshold shear velocities and sand transport rates, as discussed further in Chapters 4 and 6.

3
Characteristics of windblown sediments

3.1 General properties of sediment grains

The movement of grains in any fluid is governed partly by the size, shape, and density of the grains and partly by the physical properties of the fluid. Entrainment of grains from the bed is influenced not only by the characteristics of individual grains, but also by bulk sediment properties which include the grain size distribution (sorting), orientation, packing arrangement, porosity, and cohesion. During transport, grains are sorted according to size, shape, and density, and may undergo changes in shape due to inter-particle collisions or contact with the bed. An understanding of the physical characteristics of sand grains and the manner in which the characteristics of grain populations are modified during aeolian transport is therefore essential for correct palaeoenvironmental interpretation of aeolian sediments.

3.1.1 Concepts of grain size

Grain 'size' can be specified and measured in several different ways. Indications of size can be obtained by measuring the caliper dimensions of a particle, by determining its volume or its mass, or by determining its settling velocity. A number of alternative definitions of grain size are given in Table 3.1. All methods of grain size determination have disadvantages and the choice of the most appropriate method is governed by the nature of the sample and the use to which the data are to be put. Four main methods are currently used for size analysis of sands: (a) sieving; (b) settling tube analysis; (c) electro-optical methods, including Coulter Counter analysis and laser granulometry; and (d) computerized image analysis. Some of these methods are discussed further in Chapter 10. The most widely used method is dry sieving, in which a sand sample is shaken through a nest of successively finer mesh sieves (Ingram 1971; McManus 1988). Conventionally the weight of sand retained on each sieve is converted to a percentage of the total sample. Whether or not a grain passes through a particular sieve is governed by its intermediate dimensions relative to the width of the mesh apertures. Several studies have shown that particle shape can have a significant influence on sieve-size data (Komar and Cui 1984; Kennedy *et al.* 1985), and difficulties may be encountered when samples contain a mixture of quartz and platy shell fragments (Carter 1982).

3.1.2 Grain size scales

Sediment particles range in size from several metres to less than 1 μm, and a number of grade scales have been proposed which divide the total distribution into different size classes. Most geologists use the size class divisions and terminology of the Udden–Wentworth scale (Udden 1894, 1914, Wentworth 1922), or a modified version of it (Table 3.2). This is a ratio scale in which the boundaries between adjacent size classes differ by a factor of two. The original Udden–Wentworth scheme placed the boundary

Table 3.1 Some definitions of particle size. (After Allen 1981).

Symbol	Name	Definition	Formula
d_v	volume diameter	diameter of a sphere having the same volume as the particle	$V = \frac{\pi}{6} d_v^3$
d_s	surface diameter	diameter of a sphere having the same surface as the particle	$S = \pi d_s^2$
d_{sv}	surface volume diameter	diameter of a sphere having the same external surface to volume ratio as a sphere	$d_{sv} = \frac{d_v^3}{d_s^2}$
d_d	drag diameter	diameter of a sphere having the same resistance to motion as the particle in a fluid of the same viscosity and at the same velocity (d_d approximates d_s when R_e is small)	$\left\{ F_D = C_D A \rho_f \frac{v^2}{2} \right.$ where $C_D A = f(d_d)$; $\left\{ F_D = 3\pi d_d \eta \nu \right.$ $R_e < 0.2$
d_f	free-falling diameter	diameter of a sphere having the same density and the same free-falling speed as the particle in a fluid of the same density and viscosity	
d_{st}	Stokes' diameter	the free-falling diameter of a particle in the laminar flow region ($R_e < 0.2$)	$d_{st}^2 = \frac{(d_v^3)}{d_d}$
d_a	projected area diameter	diameter of a circle having the same area as the projected area of the particle resting in a stable position	$A = \frac{\pi}{4} d_a^2$
d_p	projected area diameter	diameter of a circle having the same area as the projected area of the particle in random orientation	Mean value for all possible orientations $d_p = d_s$ for convex particles
d_c	perimeter diameter	diameter of a circle having the same perimeter as the projected outline of the particle	$d_F = d_c$
d_A	sieve diameter	the width of the minimum square aperture through which the particle will pass	
d_F	Feret's diameter	the mean value of the distance between pairs of parallel tangents to the projected outline of the particle	
d_M	Martin's diameter	the mean chord length of the projected outline of the particle	

between silt and clay at 4 μm, but soil scientists and most sedimentologists now place the boundary at 2 μm.

In order to simplify the graphical presentation and statistical manipulation of grain size frequency data, Krumbein (1934) proposed that the grade boundaries should be logarithmically transformed into phi (ϕ) values, using the expression

$$\phi = -\log_2 d \qquad (3.1)$$

where d is the diameter in millimetres. Since phi units are dimensionless, it is strictly more correct to state that

$$\phi = -\log_2(d/d_0) \qquad (3.2)$$

Table 3.2 Size scales of Udden (1914) and Wentworth (1922), with class terminology modifications proposed by Friedman & Sanders (1978).

Size: mm	Size: μm	Size: phi	Sediment size class terminology of Wentworth (1922)	Sediment size class terminology of Friedman & Sanders (1978)	
				very large boulders	
2048		−11			
				large boulders	
1024		−10			
				medium boulders	
512		−9	cobbles		
				small boulders	
256		−8			
				large cobbles	
128		−7			
				small cobbles	gravels
64		−6			
				very coarse pebbles	
32		−5			
				coarse pebbles	
16		−4	pebbles		
				medium pebbles	
8		−3			
				fine pebbles	
4		−2			
			granules	very fine pebbles	
2	2000	−1			
			very coarse sand	very coarse sand	
1	1000	0			
			coarse sand	coarse sand	
0.5	500	1			
			medium sand	medium sand	sand
0.25	250	2			
			fine sand	fine sand	
0.125	125	3			
			very fine sand	very fine sand	
0.063	63	4			
				very coarse silt	
0.031	31	5			
				coarse silt	
0.016	16	6			
			silt	medium silt	silt
0.008	8	7			
				fine silt	
0.004	4	8			
				very fine silt	
0.002	2	9			
			clay	clay	clay

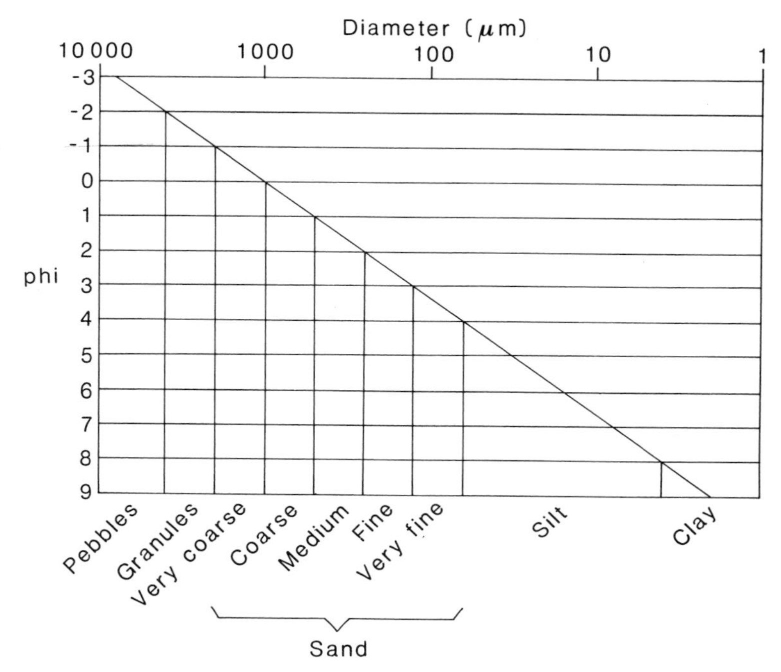

Figure 3.1 Relationship between the phi (φ) scale and the Wentworth grade scale.

where d_0 is the standard grain size of 1 mm (McManus 1963).

The relationship between metric and phi size scales, based on tables published by Page (1955), is shown graphically in Figure 3.1.

3.1.3 Grain mass and density

The behaviour of a grain when acted on by a fluid is often more closely controlled by its mass, rather than by its external dimensions. Mass represents a measure of the inertia of a body, i.e. the resistance that the body offers to having its velocity or position changed by the application of a force. Whereas the mass of a body is constant throughout space, weight varies with gravity. Mass (m) is related to weight (w) by the expression

$$m = w/g \tag{3.3}$$

where g is the acceleration due to gravity. The precise value of g varies over the Earth's surface but for sedimentological purposes the average value of $980\,\text{cm}\,\text{s}^{-2}$ can be regarded as a constant.

For spherical particles mass varies as the cube of the radius. Thus a 10 mm diameter sphere is five times larger than a 2 mm diameter sphere in terms of diameter, but 125 times larger in terms of mass (Leeder 1982). The shear stress required to initiate movement of the 10 mm diameter particle should therefore be 125 times greater than that required to move the 2 mm diameter particle.

The density of a particle is defined as its mass per unit volume expressed in $\text{kg}\,\text{m}^{-3}$. The densities of particles differ considerably depending on their elemental composition and the presence or absence of internal voids, which may be filled with air or another fluid. The densities of some common materials are listed in Table 3.3.

Table 3.3 Density of minerals commonly found in aeolian deposits.

Mineral	Composition	Density ($kg\,m^{-3}$)
Light minerals:		
quartz	SiO_2	2650
albite	$NaAlSi_3O_8$	2620
labradorite	$(Ca,Na)(Al,Si)AlSi_2O_8$	2700
anorthite	$CaAl_2Si_2O_8$	2750
orthoclase	$KAlSi_3O_8$	2560
microcline	$KAlSi_3O_8$	2560
calcite	$CaCO_3$	2710
aragonite	$CaCO_3$	2930
dolomite	$CaMg(CO_3)_2$	2870
gypsum	$CaSO_4.2H_2O$	2320
halite	$NaCl$	2160
anhydrite	$CaSO_4$	2890–2980
Heavy minerals:		
pyroxenes	$(Ca,Mg,Fe)_2(Si,Al)_2O_6$	3200–3550
hornblende	$NaCa_2(Mg,Fe,Al)_5(Si,Al)_8O_{22}(OH)_2$	3000–3470
garnet	$(Fe,Al,Mg,Mn,Ca)_5(SiO_4)_3$	3560–4320
epidote	$Ca_2(Al,Fe)_3O_{12}(OH)$	3250–3500
olivine	$(Mg,Fe)_2SiO_4$	3210–4390
staurolite	$FeAl_4Si_2O_{10}(OH)_2$	3700
kyanite	Al_2SiO_5	3690
andalusite	Al_2SiO_5	3160–3200
sillimanite	Al_2SiO_5	3230–3270
zircon	$ZrSiO_4$	4670
rutile	TiO_2	4250
anatase	TiO_2	3900
apatite	$Ca_5(PO_4)_3(F,Cl,OH)$	3100–3250
tourmaline	$Na(Mg,Fe)_3Al_6(BO_3)_3(Si_6O_{18})(OH)_4$	3030–3100
monazite	$(Ce,La,Y,Th)PO_4$	5270
Clay minerals and micas:		
muscovite	$KAl_2(AlSi_3O_{10})(OH)_2$	2800–2900
biotite	$K(Mg,Fe)_3(AlSi_3O_{10})(OH)_2$	2800–3400
chlorite	$(Mg,Fe,Al)_6(Al,Si)_4O_{10}(OH)_8$	2600–3300
kaolinite	$Al_4Si_2O_5(OH)_4$	2600–2630
illite	$KAl_2(Al,Si_3O_{10})(OH)_2$	2600–2700
palygorskite	$(Mg,Al)_5(Si,Al)_8O_{20}.4H_2O(OH)_2$	2200–2360
montmorillonite	$Na(Al_3Mg)(Si_8O_{20})(OH)_4.H_2O$	2000–2300
ice	H_2O	920

When transported by a fluid, small particles composed of high-density material may display the same behaviour (hydraulic equivalence) as much larger particles composed of low-density material.

3.1.4 Graphical presentation of grain size data

The simplest method of presenting grain size data is in the form of a histogram, with grain diameter (in mm, μm or phi units) plotted on the abscissa and weight or weight per cent plotted on the ordinate (Fig. 3.2). A frequency distribution curve can be drawn by joining the mid-points of the tops of each bar in the histogram. This type of presentation provides a rapid qualitative impression of the nature of a distribution, for example whether it is unimodal or bimodal, well sorted, skewed, or highly peaked. Bagnold (1941, p. 111) pointed out that the shape of the histogram is dependent on the size class interval used, and suggested that each percentage weight should be divided by the size class interval to which it refers. However, most workers have considered that Bagnold's transformation is not necessary when the same set of ratio-scale sieves is used routinely.

Krumbein (1934) pointed out that grain size histogram plots of many natural sediments resemble a 'normal' or Gaussian distribution, and a majority of sedimentologists have subsequently based their interpretations of grain size data on comparisons between actual distributions and the Gaussian curve. Since the size scale is actually logarithmic (to base 2), the distributions approximate log-normal.

Natural grain size distributions have often been compared with the log-normal model by plotting cumulative weight per cent data against grain size on log-arithmetic probability graph paper (Fig. 3.3). The ordinate scaling on this type of paper is derived by dividing the area beneath a Gaussian curve into columnar segments of equal area. An ideal log-normal distribution plots as a straight line, but most natural sediments deviate to a greater or lesser extent. Cumulative frequency curves plotted in this way sometimes display a number of segments whose sedimentary significance has been much debated. Earlier workers (e.g. Visher 1969, Middleton 1976) considered that segmented curves indicate the existence of mixed sediment populations or discrete sediment transport

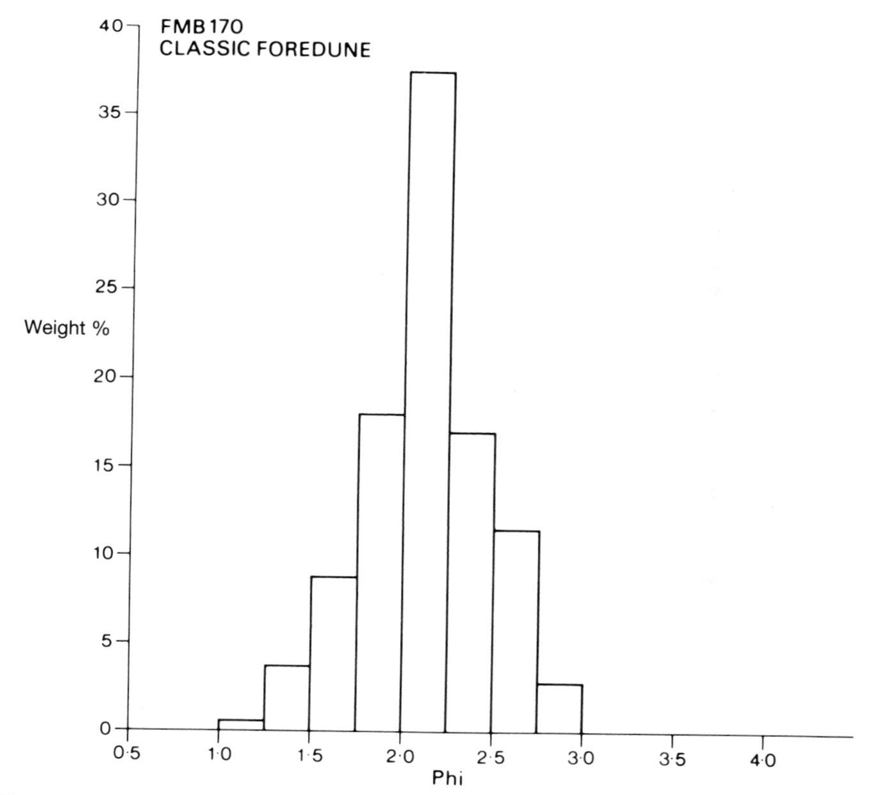

Figure 3.2 Grain size frequency histogram of a sample of foredune sand from North Queensland, sieved at quarter-phi intervals.

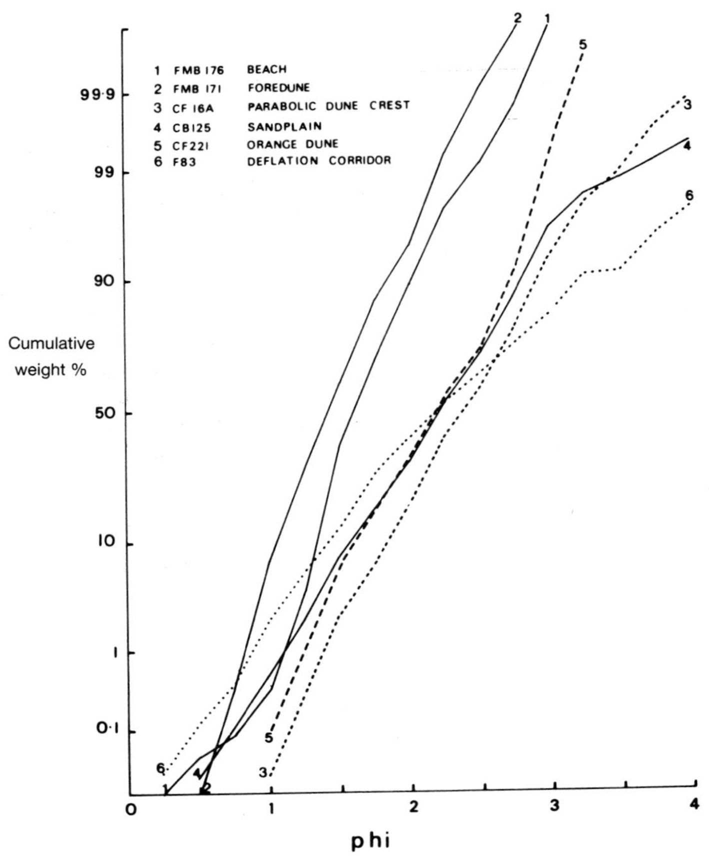

Figure 3.3 Cumulative percentage frequency curves showing differences between samples from different aeolian environments at Cape Flattery, North Queensland. (After Pye 1980a).

modes, but more recently this view has been challenged (Walton *et al.* 1980, Christiansen *et al.* 1984).

3.1.5 Graphical statistical parameters

Quantitative comparisons between natural grain size distributions and the log-normal distribution can be made using a number of statistical parameters computed from the cumulative percentage frequency curve. The parameters proposed by Folk and Ward (1957) have been widely used, particularly in earlier work before the use of computers and microcomputers became widely adopted.

The *graphic mean*, M_z, is given by

$$M_z = \frac{\phi_{16} + \phi_{50} + \rho_{84}}{3} \tag{3.4}$$

where ϕ_{16}, ϕ_{50}, and ϕ_{84} are the phi size values corresponding to the sixteenth, fiftieth, and eighty-fourth percentiles, respectively, read from the cumulative frequency curve.

Table 3.4 Terminology applied to graphical statistical parameter values (modified after Folk & Ward 1957)

Inclusive graphic standard deviation or phi sorting (σ)		Inclusive graphic skewness or phi skewness (Sk_1)		Inclusive graphic kurtosis or phi kurtosis (K_G)	
very well sorted	<0.35	very positively skewed	$+0.3$ to $+1.0$	very platykurtic	<0.67
well sorted	0.35–0.50	positively skewed	$+0.1$ to $+0.3$	platykurtic	0.67–0.90
moderately well sorted	0.50–0.70	symmetrical	$+0.1$ to -0.1	mesokurtic	0.90–1.11
moderately sorted	0.70–1.00	negatively skewed	-0.1 to -0.3	leptokurtic	1.11–1.50
poorly sorted	1.00–2.00	very negatively skewed	-0.3 to -1.0	very leptokurtic	1.50–3.00
very poorly sorted	2.00–4.00				

A measure of the spread about the mean, or sorting, is given by the *inclusive graphic standard deviation*, σ_1, defined as

$$\sigma_1 = \frac{\phi_{84} + \phi_{16}}{4} + \frac{\phi_{95} + \phi_5}{6.6} \tag{3.5}$$

Better sorted sediments have lower values of σ_1 (Table 3.4).

The asymmetry of skewness of the distribution is indicated by the *inclusive graphic skewness*, Sk_1:

$$Sk_1 = \frac{\phi_{16} + \phi_{84} - 2\phi_{50}}{2(\phi_{84} - \phi_{16})} + \frac{\phi_5 + \phi_{95} - 2\phi_{50}}{2(\phi_{95} - \phi_5)} \tag{3.6}$$

Positive values of Sk_1 indicate that the distribution has a more pronounced tail of fine material compared with a log-normal distribution. Conversely, negative values of Sk_1 indicate a tail of coarser particles or a deficiency of fine particles compared with the log-normal distribution (Fig. 3.4A).

The 'peakedness' or kurtosis of a distribution is indicated by the *inclusive graphic kurtosis*, K_G:

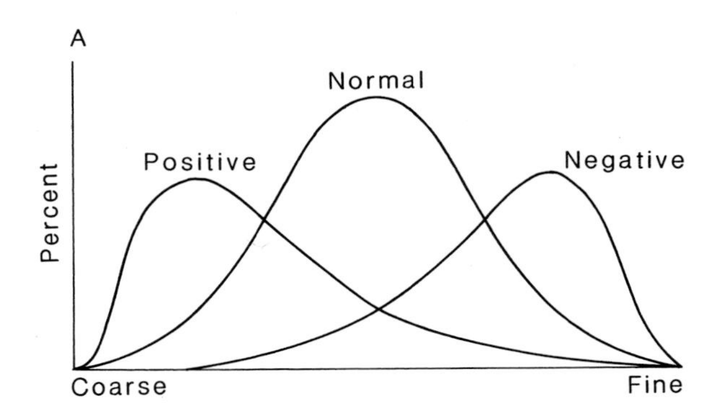

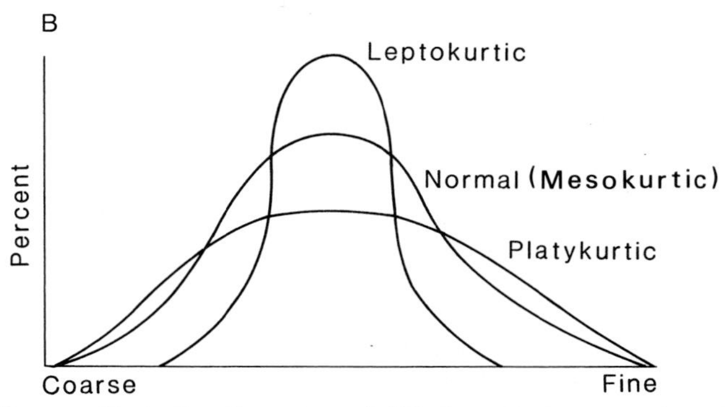

Figure 3.4 Diagrams illustrating the nature of (A) skewness (asymmetry) and (B) kurtosis (peakedness) in grain size distributions.

$$K_G = \frac{\phi_{95} - \phi_5}{2.44(\phi_{75} - \phi_{25})} \tag{3.7}$$

Frequency distributions which are flatter than a normal probability curve are referred to as *platykurtic* and strongly peaked curves are described as *leptokurtic*. Curves which approximate a Gaussian profile are referred to as *mesokurtic* (Table 3.4; Fig. 3.4B).

3.1.6 Moment parameters

A second major method of quantifying the nature of grain size distributions involves the calculation of *moment statistics* (Friedman 1961; McBride 1971). Individual moments are computed from the product of the weight percentage in a given size class and the number of class intervals from the origin of the curve. The moment grain size statistics are defined as follows:

first moment (mean):

$$\bar{x}_\phi = \frac{\Sigma fm}{100} \tag{3.8}$$

second moment (standard deviation):

$$\sigma_\phi = \sqrt{\frac{\Sigma f(m - \bar{x})^2}{100}} \tag{3.9}$$

third moment (skewness):

$$Sk_\phi = \frac{\Sigma f(m - \bar{x})^3}{100\ \sigma^3} \tag{3.10}$$

fourth moment (kurtosis):

$$K_\phi = \frac{\Sigma f(m - \bar{x})^4}{100\ \sigma^4} \tag{3.11}$$

where f is the frequency (weight per cent) in each grain size grade present and m is the mid-point of each class interval (in phi values).

The moment parameters are analogous to the graphical statistical parameters but are widely considered to be more representative because they take into account the entire grain size distribution rather than just that part between the fifth and ninety-fifth percentiles.

3.1.7 Bivariate plots and statistical analysis of grain size parameters

Many attempts have been made to differentiate between sediments from different environments using bivariate scattergram plots of moment or graphical parameters (Friedman 1961, 1967, Schlee *et al.* 1965, Moiola & Weiser 1968, Khalaf 1989). Plots of mean size against sorting and skewness have generally proved most useful, and in some

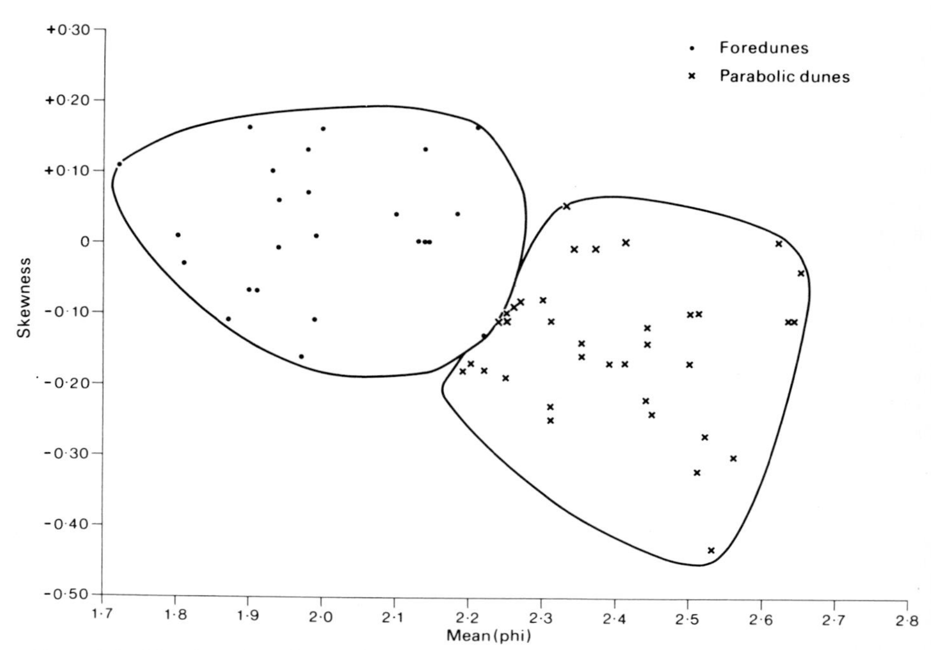

Figure 3.5 Bivariate plot of graphic mean against inclusive graphic skewness of foredune and parabolic dune sand samples from Cape Flattery, North Queensland. (After Pye 1980a).

cases have successfully differentiated betwen different sedimentary environments (Fig. 3.5). Besler (1983) used the term 'response diagram' to describe bivariate plots of mean grain size and sorting which allowed her to discriminate between different desert sediments. On the other hand Vincent (1985) and Thomas (1986, 1987a) concluded that response diagrams are of little value in this context. There is often considerable overlap between grain size parameters from river, beach, dune, and sub-marine sediments (Shepard & Young 1961; Schlee *et al.* 1964). Better discrimination has been reported when the data are analysed by statistical techniques such as multiple discriminant analysis (Moiola *et al.* 1974, Moiola & Spencer 1979) and factor analysis (Klovan 1966). These techniques have the advantage that factors other than grain size, such as mineral composition, can also be incorporated in the analysis. However, multiple discriminant analysis does not always clearly discriminate between sedimentary environments, even within a single region (e.g. Thomas 1987a). This may be due in part to local reworking and mixing of sediments. In other instances it arises because the source sediment characteristics exert a strong influence on the properties of other sediments derived from them.

3.1.8 Log-hyperbolic parameters

As long ago as the 1930s, Bagnold (1937a) realised that more information can be gained about the tails of a grain size distribution if both the grain size and grain frequency scales are transformed logarithmically (Bagnold used logarithms to base 10). When the data are plotted on a log–log diagram, the resulting curve forms a hyperbola. Bagnold proposed four parameters which characterize the log-hyperbolic curves: the *coarse grade coefficient*, *small grade coefficient*, *peak diameter*, and *width* of the distribution (see Bagnold 1941, pp.115–16).

Bagnold's method of plotting sand size distributions was largely ignored until the late

1970s, when interest was revived by statisticians who believed that many natural distributions more closely approximate a log-hyperbolic distribution than a log-normal distribution (Barndorff-Nielsen 1977). The similarities between the log-hyperbolic distribution and the mass size distributions of windblown sediments have subsequently been investigated in a number of studies (Bagnold & Barndorff-Nielsen 1980; Barndorff-Nielsen *et al.* 1982, Barndorff-Nielsen & Christiansen 1988, Vincent 1986, McArthur 1987, Hartmann 1988).

The geometrical interpretations of the main log-hyperbolic parameters are shown in Figure 3.6. The parameters ϕ and γ describe the slopes of the left and right asymptotes of the log-hyperbolic probability function and correspond to Bagnold's 'coarse grade coefficient' and 'small grade coefficient', respectively. The abscissa of the intersection point of the two asymptotes is denoted by μ, which is equivalent to Bagnold's 'peak diameter'. However, a better measure of peak size is provided by the mode, ν which is given by

$$\nu = \mu + \delta \qquad (3.12)$$

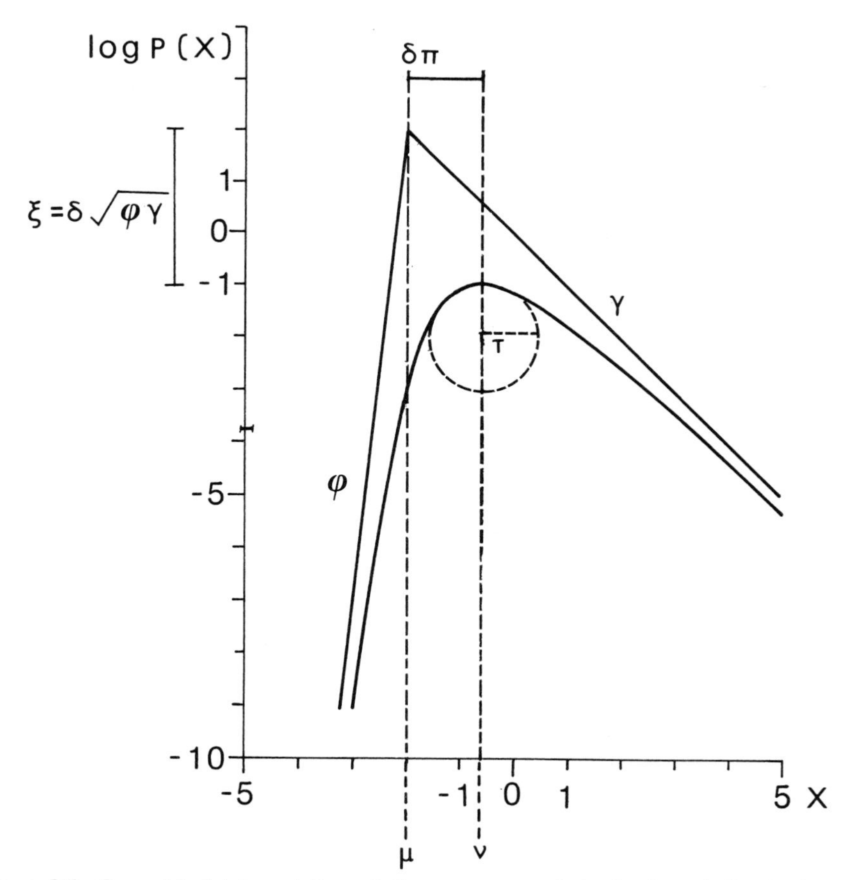

Figure 3.6 Geometrical interpretation of the parameters of the log-hyperbolic distribution. (Modified after Bagnold & Barndorff-Nielsen 1980, Hartmann & Christiansen 1988). See text for explanation of parameters.

where δ has no direct interpretation in Figure 3.6 but

$$\zeta = \delta\sqrt{\phi\gamma} \tag{3.13}$$

where ζ is the difference between the maximum ordinate of the log-hyperbolic curve and the ordinate of the intersection point of the asymptotes.

The spread of the distribution can be described by several different parameters. Near the mode it is described by

$$\tau = \zeta\delta^{-2}(1 + \pi^2)^{-1} \tag{3.14}$$

which represents the curvature of the hyperbola at that point. High values are indicative of good sorting. δ, ζ, and κ are also measures of spread, where

$$\kappa = \sqrt{\phi\gamma} \tag{3.15}$$

The asymmetry or skewness of the distribution is given by the derived parameter π, where

$$\pi = \tfrac{1}{2}(\phi - \gamma)/\sqrt{\phi + \gamma} \tag{3.16}$$

and by the parameter χ (Barndorff-Nielsen *et al.* 1985a, Barndorff-Nielsen & Christiansen 1988), where

$$\chi = (\phi - \gamma)/(\phi + \gamma)\zeta \tag{3.17}$$

The peakedness can be expressed by

$$\xi = (1 + \delta\sqrt{\phi\gamma})^{-\frac{1}{2}} \tag{3.18}$$

The domain of variation of χ and ξ is a triangle (Barndorff-Nielsen *et al.* 1985a), which is referred to as the *log-hyperbolic shape triangle* (Fig. 3.7). It can be seen from Figure 3.7 that the log-normal distribution represents only one limiting case of the log-hyperbolic model, and the advocates of the latter model maintain that it offers much greater flexibility for interpretation of natural grain size distributions (Vincent 1986, McArthur 1987, Hartmann 1988, Hartmann & Christiansen 1988). Comparisons between log-normal and log-hyperbolic parameters computed using the same sieve-size data have shown that the latter yield more environmentally sensitive information (Christiansen 1984). Hyperbolic shape triangle (χ, ξ) plots have also been used to develop a sensitive sediment erosion and deposition model (Barndorff-Nielsen & Christiansen 1988).

In a preliminary investigation it may be adequate to estimate parameters from the log–log graphical plot (Barndorff-Nielsen 1977), but more sophisticated investigation requires the use of FORTRAN sub-routines such as those available in the NAG library (Numerical Algorithms Group 1982). This requirement for a relatively high level of computing sophistication has been regarded as a disadvantage by some (Wyrwoll & Smyth 1985). Recently, PC-based programs have been published in an attempt to simplify the procedure (Jensen 1988, Christiansen & Hartmann 1988a). However, some sedimentologists remain unconvinced that use of the log-hyperbolic distribution and its variants offers significant new insights into the operation of sedimentary processes. For example, Wyrwoll and Smyth (1985) found that both the log-normal and log-hyperbolic distributions gave good approximations of the grain size distributions of dune crest and

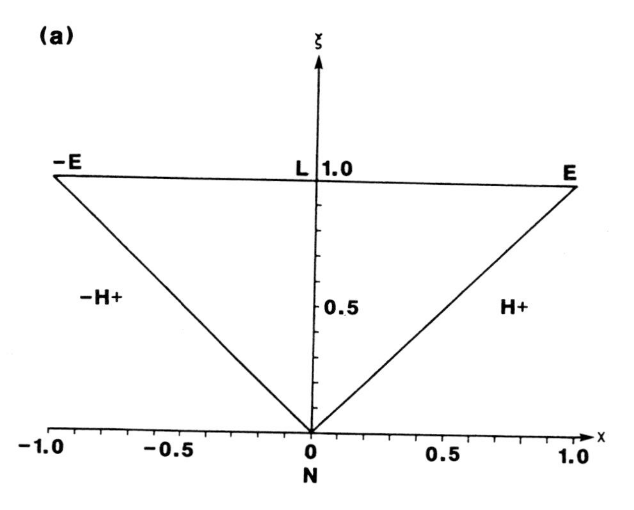

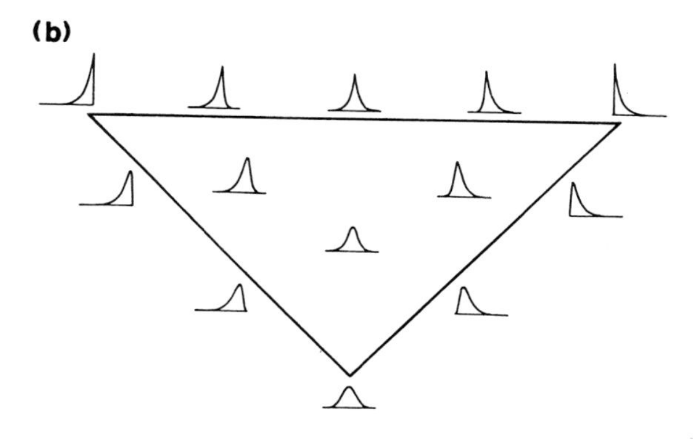

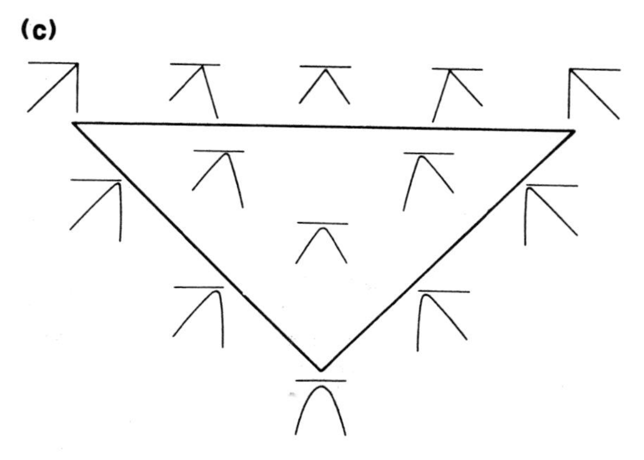

Figure 3.7 (a) The shape triangle, i.e. the domain of variation of the invariant parameters χ and ζ of the hyperbolic distribution. The letters at the boundaries show how the normal distribution (N), the positive and negative hyperbolic distributions (H^+ and $-H^+$), the Laplace distribution (symmetrical or skew) (L), and the exponential distributions (E) are limits of the hyperbolic distribution. (b) Representative probability functions corresponding to selected (χ, ζ) values, including limiting forms of the hyperbolic distribution. The distributions have been selected so as to have variance = 1. (c) Log probability functions corresponding to (b). (After Barndorff-Nielsen & Christiansen 1988).

dune side samples in northwestern Australia. It was also argued by these authors that an inability to model bimodality, which is common in desert sediments, is a shortcoming common to both distributions (for further debate, see Christiansen and Hartmann 1988b, Wyrwoll & Smyth 1988, Vincent 1988).

3.2 Grain shape

The terms 'shape' and 'form' have previously been used in a variety of ways (Pryor 1971, Whalley 1972, Barrett 1980, Orford 1981, Winkelmolen 1982, Willetts & Rice 1983), but as used here 'shape' includes all aspects of the external morphology of a grain, including the gross form (sphericity), the roundness (sharpness of edges and corners), and the surface texture (roughness or smoothness).

Grain shape can sometimes be described qualitatively in terms of resemblance to readily recognizable geometric shapes or organic analogues, using such terms as cubic, hexagonal, conical, globular, vermiform, and reniform. However, such descriptions are subjective and of little value when grains have no clearly defined form. Consequently, a number of numerical parameters have been developed to allow quantitative descriptions of grain shape and statistical comparisons of grain shape data.

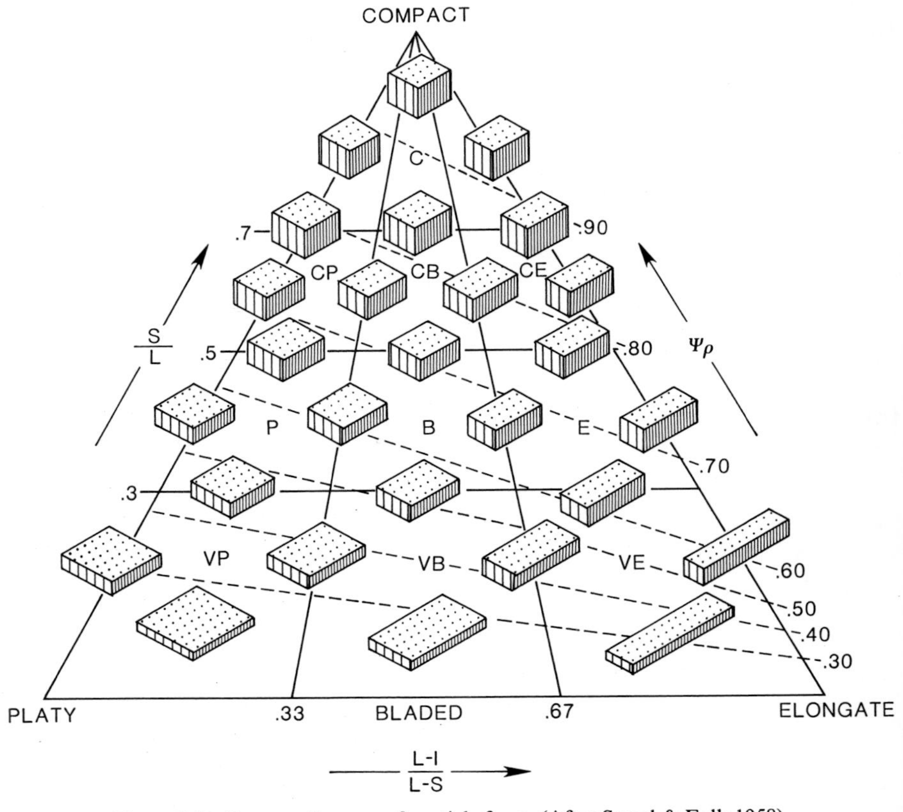

Figure 3.8 Ternary diagram of particle form. (After Sneed & Folk 1958).

3.2.1 Grain form

Form indices provide a measure of equi-dimensionality and are most frequently obtained in one of two ways: (a) by measuring the mutually perpendicular long (L), intermediate (I), and short (S) axes of a grain and by calculating their ratios, or (b) by assessing the degree of deviation from some geometrical standard, such as a sphere. Wadell (1933) defined the first measure of grain sphericity, ψ, as

$$\psi = s/S \tag{3.19}$$

where s is the surface area of a sphere of the same volume of the particle and S is the actual surface area of the particle. However, the surface area of sediment grains is difficult to measure directly, and a more widely used measure of sphericity is the maximum projection sphericity, ψ_ρ, introduced by Sneed and Folk (1958) and defined by

$$\psi_\rho = (S^2/LI)^{1/3} \tag{3.20}$$

where S, L, and I are the short, long, and intermediate caliper diameters, respectively. This expression takes into account the fact that grains settling in still water tend to orientate themselves with the maximum projected area normal to the direction of movement, and Sneed & Folk considered that it provides a measure of sphericity which is more relevant to the settling behaviour of grains. It is not necessary to calculate the values of ψ_ρ, and the ratios of the a, b, and c axes can simply be plotted on a ternary diagram (Fig. 3.8).

Sneed & Folk's (1958) method of calculating sphericity was originally developed for pebbles, but can also be applied to sand grains. The L and I axial dimensions can easily be determined using a binocular microscope and graduated eyepiece. The dimensions of the S axis can be determined using a horizontal Perspex plate attached to a mirror inclined at 45° (Willetts & Rice 1983).

For estimates of grain sphericity in thin sections, where only two-dimensional data are available, the projection sphericity ψ_r proposed by Riley (1941) can be determined:

$$\psi_r = d_i/D_c \tag{3.21}$$

where d_i is the diameter of the largest inscribed circle and D_c is the diameter of the largest circumscribing circle.

Where a more rapid estimate is required, grain projections can be compared with a visual comparator such as that developed by Powers (1953). The Powers classification has two classes of sphericity combined with six classes of grain roundness (Fig. 3.9).

3.2.2 Grain roundness

Wadell (1933) proposed a measure of degree of roundness, P_d:

$$P_d = \frac{\Sigma(r/R)}{N} \tag{3.22}$$

where r is the curvature radius of individual grain corners, N the number of grain corners including corners whose radii are zero and R the radius of the maximum inscribed circle.

Because measurements of this type are time consuming most workers have estimated roundess by reference to the Powers or some other visual comparator. Numerical values are assigned to each Powers roundness class using Wadell's formula (Table 3.5).

Based on the observation that many grain roundness frequency distributions approximate a log-normal distribution, Folk (1955) proposed a logarithmic transformation of

Table 3.5 Degree of roundness class terminology and numerical indices.

Powers roundness class name	Corresponding Wadell (1933) class intervals	Corresponding values of Folk's rho scale (Folk 1955)
very angular	0.12 – 0.17	0 – 1.0
angular	0.17 – 0.25	1.0 – 2.0
subangular	0.25 – 0.35	2.0 – 3.0
subrounded	0.35 – 0.49	3.0 – 4.0
rounded	0.49 – 0.70	4.0 – 5.0
well rounded	0.70 – 1.00	5.0 – 6.0

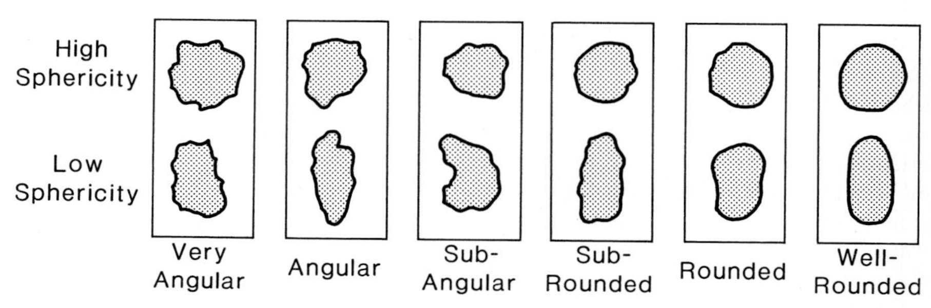

Figure 3.9 Grain roundness and sphericity classes [grain outlines drawn from the visual comparator of Powers (1953)].

roundness values analogous to the phi scale for grain size, for which he suggested the term rho (ρ) scale. Rho value range from 0 for very angular particles to 6 for well rounded particles (Table 3.5.). The results of roundness counts can be plotted on arithmetic probability paper. Mean roundness (M_ρ) and standard deviation of roundness (σ_ρ) can then be calculated in the same manner as for graphical grain size parameters.

3.2.3 Grain surface texture

In simple terms, grain surface texture can be described as the degree of surface smoothness or roughness (i.e. the degree of development of microrelief). In favourable circumstances, specific microfeatures may provide clues to the source, transport, and weathering history of the grains.

Some surface textural features, such as polish or frosting, can be seen with the naked eye or with the aid of a binocular microscope. Polish, or gloss, is related to the quality of light reflection. Grains which have a coating of secondary silica, or which have been very gently abraded, often have a high degree of polish. Frosting, on the other hand, is related to the scattering or diffusion of light due to the presence of closely spaced surface irregularities which may be caused by violent abrasion or by chemical etching (Kuenen & Perdok 1962).

More detailed analysis of grain surface microfeatures is normally undertaken using a scanning electron microscope. Early studies were confined to identifying the presence or absence of particular features on individual grains (e.g. Margolis & Krinsley 1971, Krinsley & Doornkamp 1973), but in recent years more quantitative approaches have been adopted (e.g. Culver *et al.* 1983, Elzenga *et al.* 1987).

3.2.4 Two-dimensional analysis of digitized grain outlines

With the development of automatic image analysers, it is now possible to perform quantitative shape analysis on large numbers of projected two-dimensional grain outlines within a short period of time. Once grain outlines have been digitized and converted into a series of *x–y* coordinates, detailed variations in shape can be analysed using techniques such as Fourier analysis (Ehrlich & Weinberg 1970, Ehrlich *et al.* 1974, 1980, 1987, Clark 1987) and fractal analysis (Serra 1982, Orford & Whalley 1983, 1987). These techniques have made a significant contribution to provenance and sediment transport studies in the past decade. However, to date they have been applied only to a limited extent in aeolian research (e.g. Mazzullo *et al.* 1986, Bui *et al.* 1989).

In Fourier grain shape analysis, the maximum projected grain profile is compartmentalized into a series of standard shape components (harmonics) which converge to reproduce the natural grain shape (Ehrlich & Weinberg 1970). The grain perimeter, $R(\theta)$, is expressed as a Fourier series expansion of the grain radius as a function of the polar angle about the centre of gravity of the grain:

$$R(\theta) = R_0 + R_n \cos(n\theta - \theta_n) \qquad (3.23)$$

where R_n is the harmonic amplitude, θ the polar angle, R_0 the grain radius, n the harmonic number, and θ_n the phase angle.

Close reproductions of the natural grain shape is normally achieved by the first 20 harmonics, although 24 may be used (Mazzullo *et al.* 1986). The gross characteristics of the observed shape are measured by the low-order harmonics and increasingly smaller scale features are measured by the higher order harmonics (Fig. 3.10).

For routine analyses 200–300 grains are normally taken from a single size fraction (e.g. 125–250 μm) to eliminate the possibility of shape variation due to grain size. For accurate results, analysis should be restricted to a single mineral (usually quartz).

The data can be presented graphically by plotting frequency of occurrences as a function of each harmonic amplitude. Such histograms are known as shape–frequency histograms (Ehrlich *et al.* 1980) (Fig. 3.11). The interval boundaries in these distributions are defined by maximum entropy concept described by Full *et al.* (1984). Each sample can be represented by a series of up to 23 shape–frequency distributions, one for each harmonic.

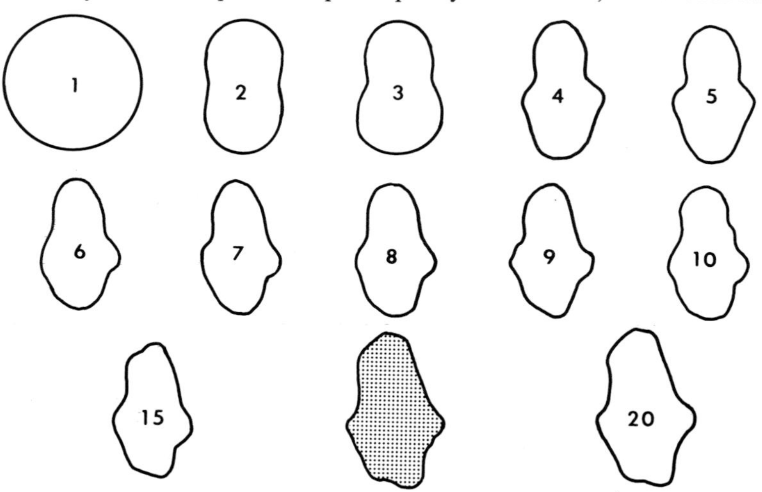

Figure 3.10 Regeneration of a grain shape (stippled) by addition of successive harmonics computed by the Fourier series. Note that the lower harmonics reflect the gross shape of the particle and the higher harmonics add increasingly finer detail. (After Ehrlich *et al.* 1980).

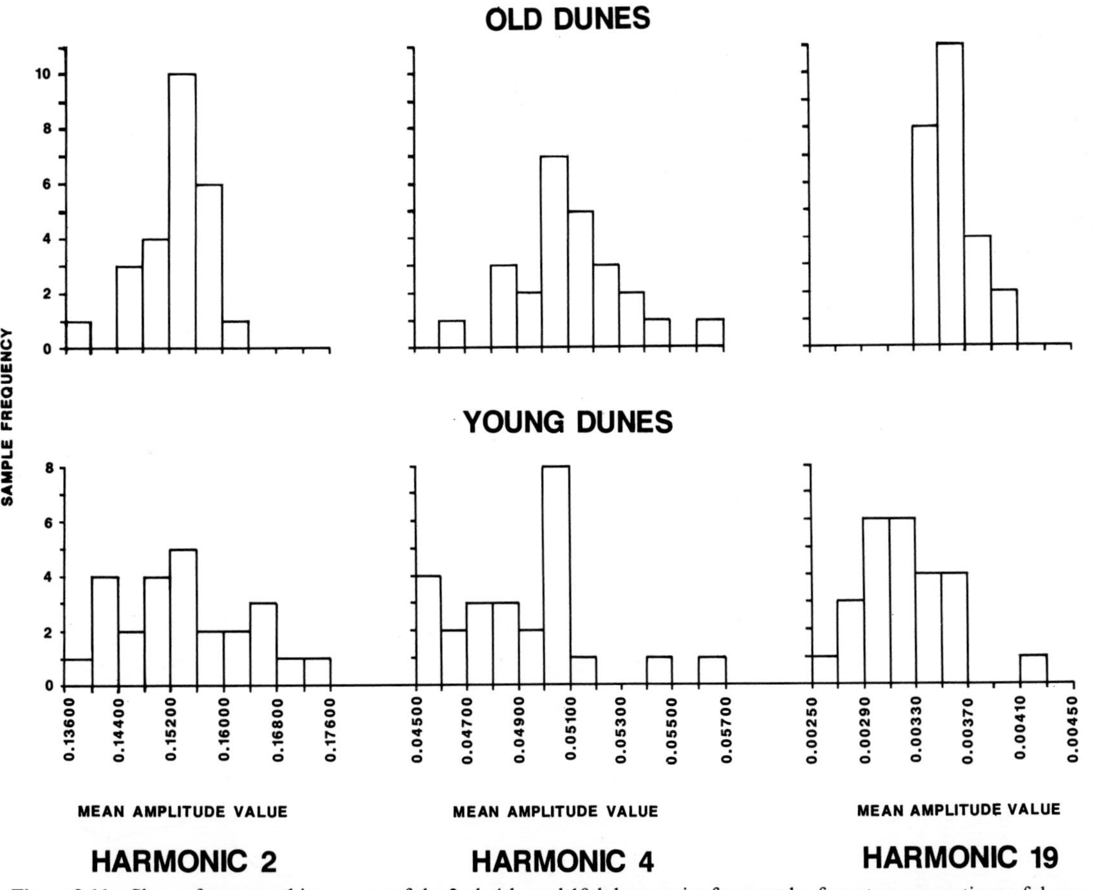

Figure 3.11 Shape–frequency histograms of the 2nd, 4th, and 19th harmonics for samples from two generations of dunes at Ramsay Bay, Hinchinbrook Island, North Queensland. Greater grain angularity in the older, more weathered dunes is indicated by the higher order harmonics (analysis by J. Mazzullo).

Since the quantity of data generated is large (23 harmonics × 300 grains × up to 100 samples), further statistical analysis is restricted to a number of selected harmonics which display the greatest inter-sample variability. The most informative harmonics are identified by relative entropy analysis of the amplitude frequency distributions of each harmonic (Full *et al.* 1984).

The amplitude–frequency distributions of the selected harmonics are often polymodal. The number of shape sub-populations present and their relative contributions are usually determined by application of a *Q*-mode algorithm (Full *et al.* 1981).

3.2.5 Behavioural indicators of grain shape

Laboratory experiments have shown that grain shape has an important influence on the hydraulic behaviour of grains (Carrigy 1970, Willetts 1983, Willetts & Rice 1983, Li & Komar 1986). Generally, the greater the departure of a grain from a spherical shape, the greater is the reduction in its settling velocity due to the increased drag it exerts on the fluid (Komar & Reimers 1978, Cui *et al.* 1983). Roundness and surface texture have a lesser effect on settling velocity and entrainment than does the overall particle form (Williams 1966, Baba & Komar 1981). However, roundness and surface roughness do have a significant effect on some properties such as the angle of repose. For example, Allen (1985) demonstrated that the angle of repose of finely polished glass beads was 5–10° lower than that of similar beads which had been etched with hydrofluoric acid.

A number of authors have devised behavioural indices of grain shape. These include the concept of 'rollability' developed by Winkelmolen (1971) and the 'dynamic shape factor' of Briggs *et al.* (1962), which reflects the effect of shape on grain settling velocity. Rollability is a functional shape property measured by the time taken for grains to travel down the length of a rotating inclined cylinder. In a comparative study, Willetts & Rice (1983) found that both the Sneed & Folk (1958) form parameter and rollability could discriminate clearly between the shape characteristics of three sands over the size range 150–500 μm, but the dynamic shape factor did not discriminate reliably between the shapes of grains smaller than 300 μm, because the differences between drag on spheres and other shapes become very small at Reynolds numbers corresponding to grains smaller than 300 μm.

3.2.6 Controls on the shape of sand grains

The shape of sand grains is determined by their composition, origin, and transport history. Coralline and algal sand are often highly porous, the voids being due to the former presence of living organisms. Shelly sand is typically highly variable in shape, and includes large numbers of thin, curved, platy grains. Quartz grains which have recently been released by weathering processes or formed by crushing are typically angular or sub-angular, but grains which have experienced several cycles of erosion and deposition are often sub-rounded to well rounded. Feldspars and non-biogenic carbonate grains may initially have less irregular, more blocky shape than quartz on account of their better developed cleavage, which influences the nature of grain fracturing. However, during transport these grains tend to become rounded more rapidly than quartz on account of their lesser hardness.

3.3 Porosity, permeability, and packing of sands

The entrainment of sediment grains from the bed when acted on by a fluid is determined by the bulk properties of the sediment as well as by the physical properties of the individual grains. The bulk sediment characteristics also have an important influence on the properties of sands as aquifers, hydrocarbon reservoirs, and agricultural soils. The most important bulk sediment properties are porosity, packing, and permeability.

Porosity is defined as the percentage of the total volume of the bulk sediment which is occupied by voids. The porosity (n) of an undisturbed sand sample can be estimated from the density of the individual grains (ρ_s) and the bulk density (γ):

$$n = \frac{1 - (\gamma/\rho_s)}{100} \tag{3.24}$$

For example, Tsoar (1974) found that the bulk density of a sample of quartz dune sand from Sinai was $1.65 \times 10^3\,\mathrm{kg\,m^{-3}}$. Given that the density of quartz is $2.65 \times 10^3\,\mathrm{kg\,m^{-3}}$, it follows that the porosity of this sand is approximately 38%.

Another parameter commonly used to characterize sediments, particularly in the engineering literature, is the *voids ratio* (e), which is defined as the ratio of the volume of voids (V_v) to the volume of solids (V_s):

$$e = V_v/V_s \tag{3.25}$$

The *packing arrangement* of the grains has a strong influence on the porosity, voids ratio, permeability, and ease with which sand is entrained by the wind. Spherical grains can have several different packing arrangements (Graton & Fraser 1935). A cubic packing arrangement gives rise to a maximum porosity of 48%, while a rhombohedral packing arrangement gives a minimum porosity of 26% (Fig. 3.12). Beard & Weyl (1973) showed that the porosity of sands with similar packing is almost independent of grain size, but is strongly dependent on sorting (Figs 3.13 and 3.14).

Permeability is the property of a sediment or rock which allows a fluid to pass through it without damage to the sediment structure. The *coefficient of permeability*, K, which is alternatively known as the hydraulic conductivity (Section 9.1.2), is measured in darcies. A sediment is said to have a permeability of 1 darcy when it yields $1\,\mathrm{cm^3}$ of fluid (viscosity $10^{-3}\,\mathrm{N\,s\,m^2}$) per second through a cross section of $1\,\mathrm{cm^2}$ under a pressure gradient of 1 atm ($1.01325 \times 10^5\,\mathrm{N\,m^{-2}}$) per cm of length.

Permeability is determined principally by the size, connectivity, and roughness of the pores (including fractures and macropores in addition to intergranular pores). Intergranular porosity is influenced by several sediment properties, including the packing arrangement, grain size distribution, grain shape, and orientation (Krumbein & Monk 1942). A sediment or rock with a high intergranular porosity is not necessarily highly permeable, as shown by many mudrocks and chalks. However, most recent sands have high permeabilities ranging from 10 to 100 darcies. Permeability is often higher parallel than transverse to the bedding. This may be due partly to a grain orientation effect and partly to size grading between laminae (finer sand laminae are generally less permeable than coarser laminae, although they may have a higher porosity). Bagnold (1938b) noted that when dune sand is wetted in the field, and then a shallow excavation made, the finer sand layers form vertical faces while the coarser sand layers fall away. Based on this observation, Bagnold suggested that water moves faster through the fine sand layers. However, the greater cohesion of the fine laminae is more likely to be due to more water retention and greater surface tension caused by the greater grain surface area and number of intergrain contacts.

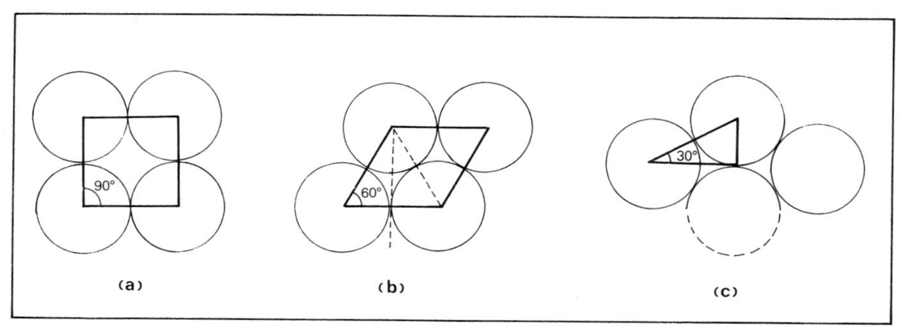

Figure 3.12 Three alternative packing arrangements for uniformly sized spheres: (a) cubic packing arrangement; (b) orthorhombic packing arrangement, and (c) tilted rhombic packing arrangement which creates a surface slope of 30°

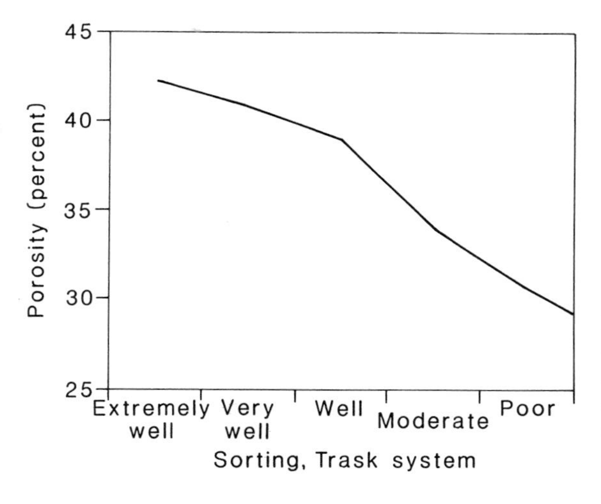

Figure 3.13 Relationship between porosity and sorting for wet packed sand (data of Beard & Weyl 1973, after Ahlbrandt 1979). The measure of sorting is that proposed by Trask (1930).

3.4 Grain size characteristics of aeolian sediments

3.4.1 The nature of aeolian sediments

Aeolian sediments can be divided into four main groups: (a) aeolian lag deposits, sand sheets, and interdune deposits, which often consist of poorly sorted, coarse sediments from which the medium and fine sand fractions have been partly removed by winnowing, (b) moderately to well sorted dune sands which consist mainly of grains in the size range 50–70 μm, (c) aeolian silt (loess), which is composed mainly of 10–70 μm particles, and (d) far-travelled, fine-grained aeolian dust composed mostly of material finer than 10 μm (Fig. 3.15). As discussed in Chapter 4, each of these sediment types is formed by a different dominant transport process. Although examples of these distinct sediment types are widespread in nature, sediments of intermediate composition do exist, reflecting the variable effectiveness of aeolian sorting processes and, in some cases, mixing of sediments deposited at different times by different processes. For example, a gradation from aeolian sands to loess is seen at the margins of late Pleistocene coversands in northwest Europe (Fig. 3.16). A transition zone of sandy silt, termed *sandloess*, occurs between the coversand and true loess deposits. Some sandloess samples are bimodal and clearly represent mixtures of coversand and loess (Fig. 3.17). Aeolian sandy silts have also been described from some high mountain environments where the extent of aeolian action has not completely separated the sand and silt fractions present in weathering debris (e.g. Pye & Paine 1984).

Particles larger than sand size can be moved by very strong winds. For example, Newell & Boyd (1955) reported ripples composed of very coarse sand and granules with a modal diameter of about 3 mm in coastal Peru. Extremely strong winds are required to transport solid grains of this size. However, if the sediment grains are porous and have a low density, as in the case of pumice fragments, they may be blown considerable distances, although rarely in sufficient numbers to form distinct aeolian bedforms. Very large rocks can also slide across flat terrain under the influence of the wind if they are supported by a film of ice, water, or mud (Clements 1952, Sharp & Carey 1976, Wehmeier 1986). The movement of rocks across playa surfaces often leaves distinct trails, and they are consequently known as 'playa scrapers'. However, not all playa scrapers move as a result of aeolian action; alternative causes can be tilting of the supporting sediment due to dehydration, salt crystal growth or neotectonics.

Active dune sands normally do not contain large amounts of silt and clay because

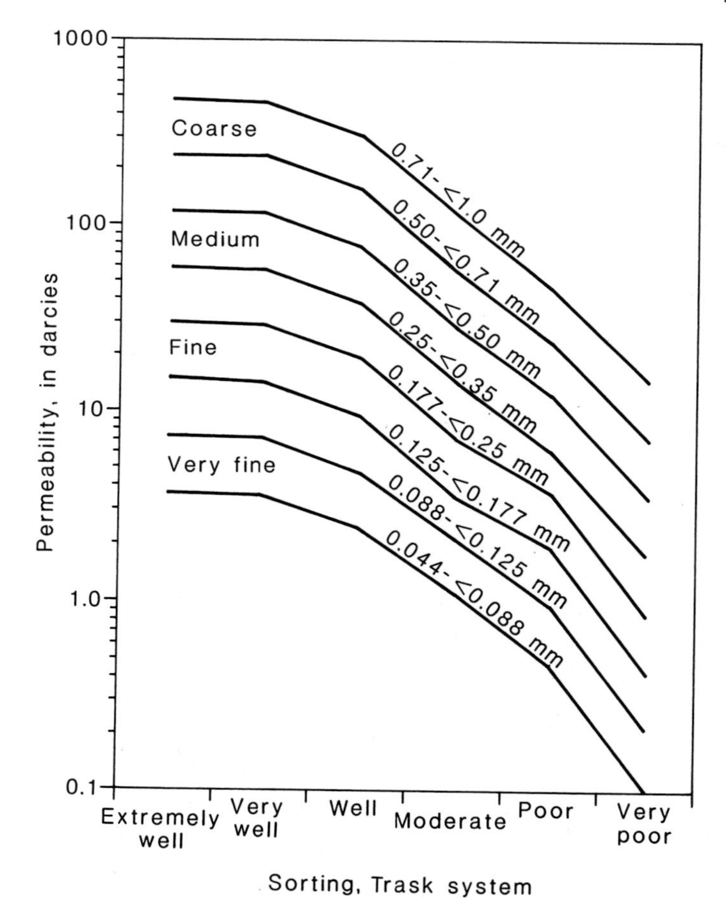

Figure 3.14 Relationship between sorting and permeability for different sizes of wet, packed sands (data of Beard & Weyl 1973, after Ahlbrandt 1979).

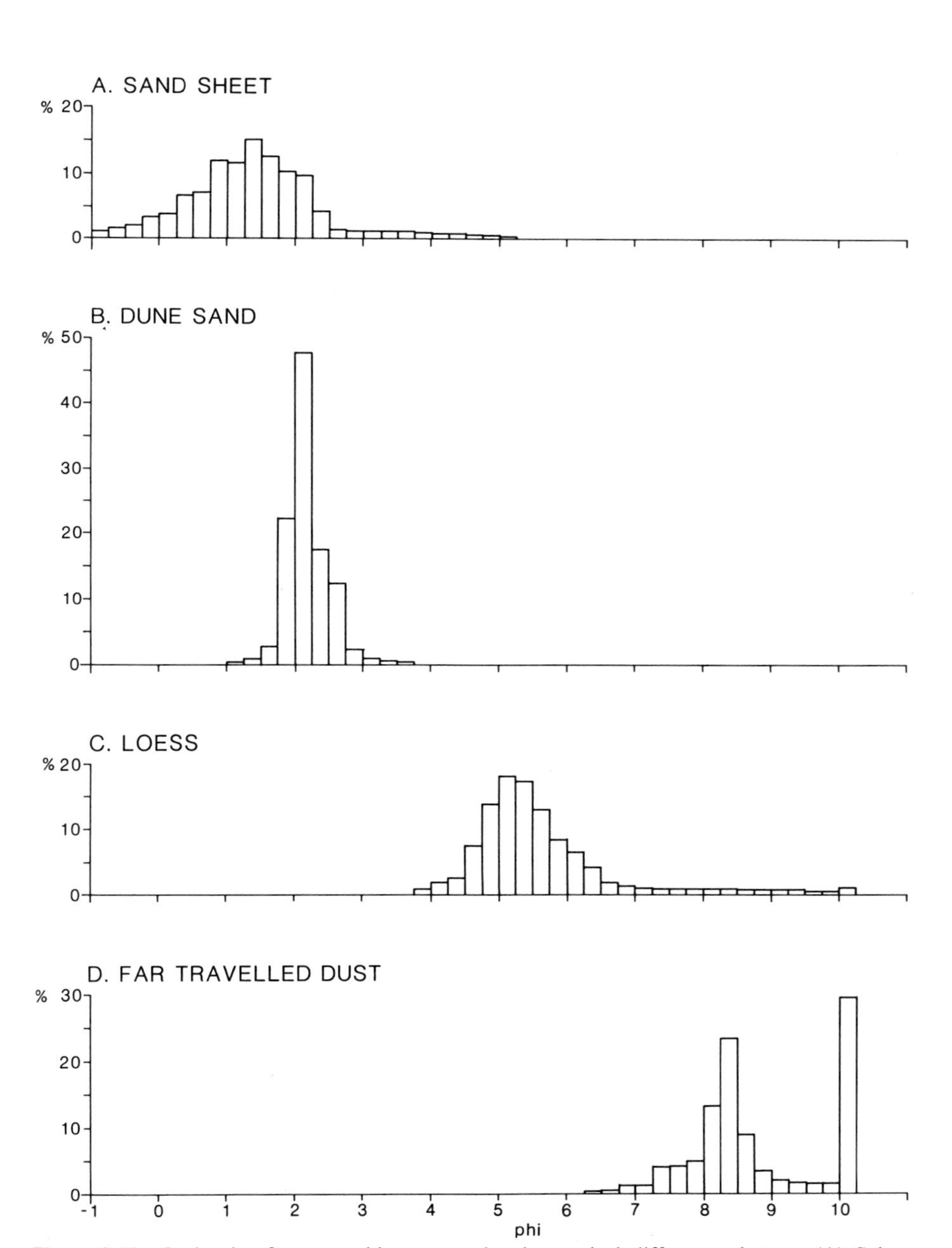

Figure 3.15 Grain size–frequency histograms showing typical differences betwen (A) Sahara Desert aeolian sand sheet, (B) aeolian dune sand from North Queensland, (C) Mississippi Valley loess, and (D) Saharan dust collected at Barbados. Material finer than 10ϕ in (D) is aggregated into a single size class ($< 10\phi$).

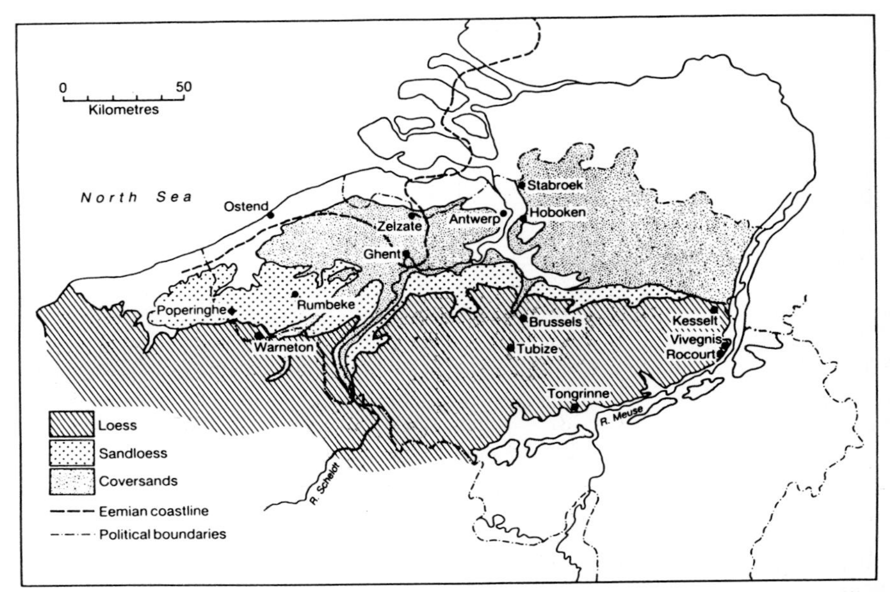

Figure 3.16 Distribution of aeolian deposits in Belgium, showing the presence of the sandloess belt between coversands in the north and loess in the south. (After Catt 1986, Paepe & Vanhoorne 1967).

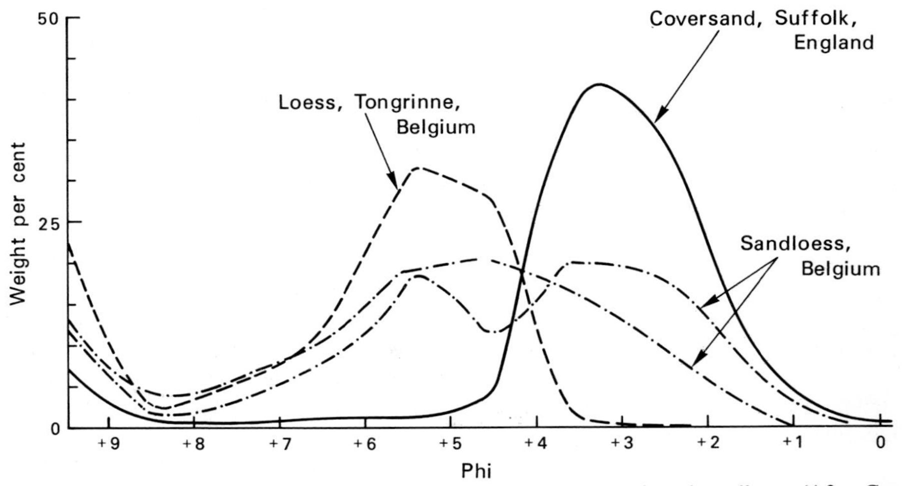

Figure 3.17 Grain size curves of typical samples of loess, coversand, and sandloess. (After Catt 1986).

during saltation fine particles are raised into the air and carried away in suspension. Once stabilized, however, the silt and clay content of dune sands may increase rapidly by accumulation of allochthonous dust and fines formed by *in situ* weathering of the sand grains (see Section 8.5.2).

3.4.2 *Differentiation between aeolian dune and other environments*

There has been much debate about whether aeolian sands can be distinguished from those deposited in other environments. Folk & Ward (1957), Mason & Folk (1958), Friedman (1961), Visher (1969) and Vincent (1986) concluded that aeolian sediments can be distinguished, at least locally, from beach and fluvial sands on the basis of mean size, sorting, and skewness, but other authors (Udden 1914, Shepard & Young 1961, Moiola & Weiser 1968, Bigarella 1972) judged textural parameters to be less diagnostic. There is widespread agreement that kurtosis is a weak discriminant (Friedman 1961, Tsoar 1976, Chaudhri & Khan 1981).

Ahlbrandt (1979) compiled mean size and sorting data for 464 inland and coastal dune samples and showed that the mean size lies predominantly in the fine sand range (125–250 μm). The sands are moderately well sorted (ϕ sorting values of 0.5–0.71). It is not possible to give single representative size or sorting values for individual sand seas or dune types because in most cases there is wide intra-dunefield, and sometimes intra-dune, variability (Lancaster, 1981b, Watson, 1986, Goudie *et al.* 1987, Buckley 1989). There is no simple relationship between dune type and grain size, although a number of authors have suggested a possible relationship between dune height, spacing, and grain size (Ch. 6).

The coastal dune samples considered in Ahlbrandt's (1979) analysis (many from Brazil) were composed of very well sorted fine sand which is predominantly positively skewed and platykurtic to mesokurtic (Fig. 3.18a). The inland dune sand samples, by contrast, showed a much greater range in size, skewness and sorting, while the kurtosis values tended to be mainly mesokurtic or platykurtic (Fig. 3.18b). The size distributions of 40 interdune and serir sands were found to be even more variable (Fig. 3.18c), with poor sorting values in many cases being associated with bimodality (Folk 1968, Scoček & Saadallah 1972, Warren 1972, Binda & Hildred 1973, Lancaster & Teller 1988).

The poorer sorting of the inland desert dune samples compared with the coastal dune samples can be explained by the fact that the former are derived mainly from poorly sorted fluvial sediments or weathered bedrock and, in the case of the basin and range deserts of the United States, are relatively young in age and have not been transported great distances from the source. For these reasons it has proved difficult to differentiate inland dune sand from its source sediments (Moiola & Weiser 1968). The variability of the textural parameters reflects differences in source sediment composition, varying transport and depositional processes, and post-depositional changes which in some cases have modified the primary textures. Skewness and kurtosis values show particularly wide scatter. Skewness appears to be partly dependent on mean grain size. For example, Ahlbrandt (1975) showed that negative skewness increases with decreasing mean grain size in the Killpecker dunefield of Wyoming (Fig. 3.19). A similar relationship was observed in the Namib linear dunes by Lancaster (1981b) and in North Queensland coastal dunes by Pye (1982a).

The grain size distribution of the source sediments also exerts a strong influence on the grain size distribution of coastal dune sands. Bigarella *et al.* (1969a) pointed out that, where beach sands are fine grained, aeolian transport is not selective and there are no significant differences in the mean size, sorting, and skewness between the beach and adjacent dune sands. Where beach sands are coarse and poorly sorted, adjacent dune sands are usually finer and better sorted.

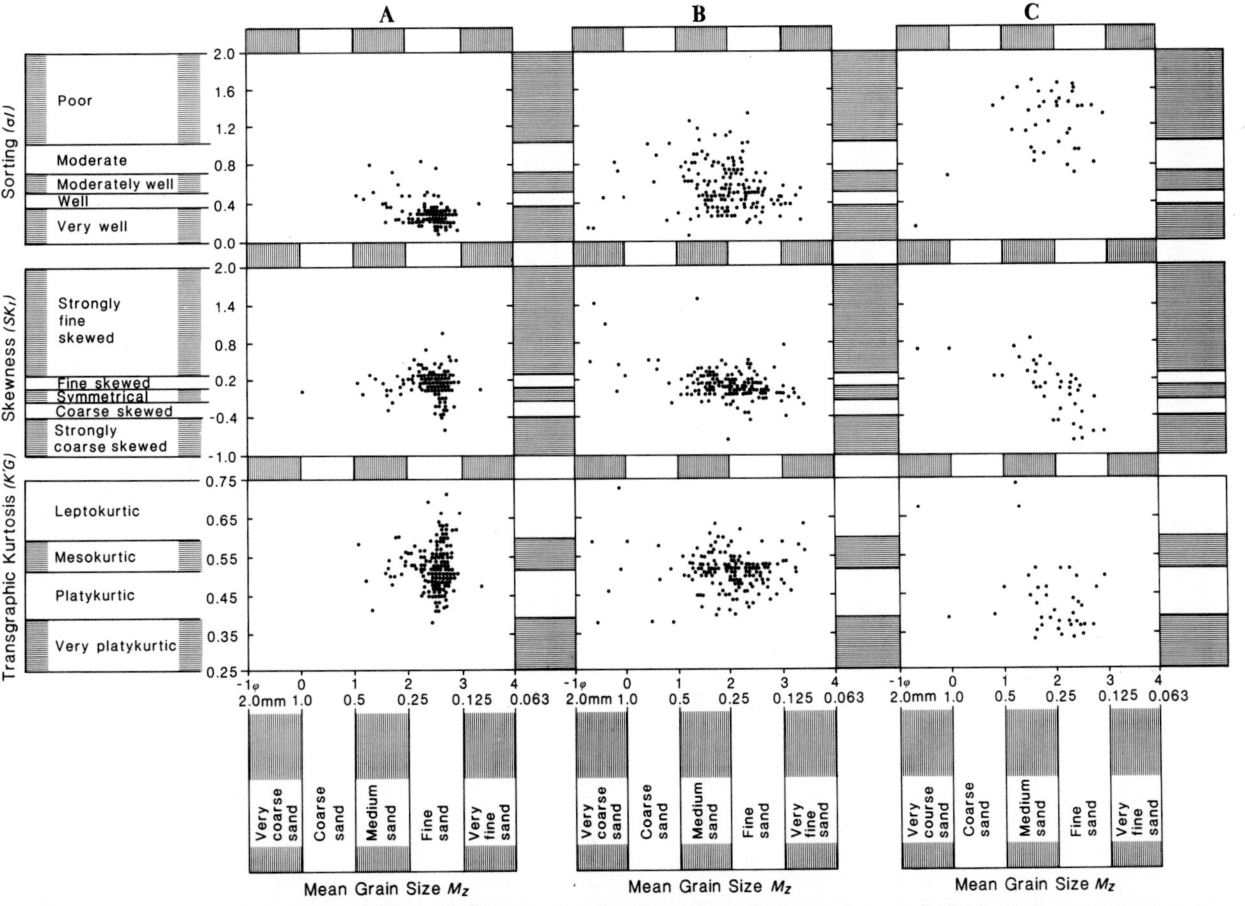

Figure 3.18 Bivariate plot of graphical statistical parameters for (A) coastal dune, (B) inland dune, and (C) interdune and serir sands from different parts of the world. (Modified after Ahlbrandt 1979).

3.4.3 Grain size variations within dunefields and on individual dunes

Variations in grain size and sorting characteristics exist at three scales: (a) on individual dunes; (b) between different dune types and interdune areas in the same part of a dunefield; and (c) regionally across the dunefield. On individual dunes there may be notable differences between crest and base, flanks, and slip face, between rippled surfaces and non-rippled surfaces, and between individual laminae. Often there are seasonal or shorter-term changes related to fluctuations in wind regime (Livingstone 1987, 1989b). These variations can present difficulties in choosing the most appropriate scale, timing and method of sampling in order to obtain 'representative' results.

Bagnold (1941) identified three different sand populations on Libyan seif dunes: a basal zone (not always present) consisting of very poorly sorted sand, an accretion zone (plinth) in which the median size is relatively constant but sorting improves with height, and the crest, where slip face formation and avalanching carry coarser grains downwards towards the plinth, resulting in fining and better sorting towards the crest. Tsoar (1978) and Lancaster (1981b) also found a clear distinction between plinth and crest sands on seif dunes in Sinai and the Namib, respectively. Lancaster (1981a) reported that the finest sands occur on the mid-slip face. However, Watson (1986) reported a much more gradual transition of grain size across another Namib linear dune. On poorly vegetated linear dunes elsewhere, including parts of the Simpson Desert (Folk 1971a), western Kansas (Simonett 1960), the Kalahari (Lancaster 1986), and the Thar Desert (Chaudhri & Khan 1981), sands have been reported to become coarser (but still better sorted) towards the crest. Crestal coarsening occurs principally on vegetated linear dunes which have limited slip-face development (and therefore limited opportunity for grain size sorting by avalanching). The coarser character of the crest sands on such dunes may also be enhanced by selective winnowing of the finer grains (Chaudhri & Khan 1981).

Many authors have observed that lee slope deposits are generally finer than windward slope deposits (Cornish 1897, Folk 1971a, Barndorff-Nielsen *et al.* 1982). Both the windward slope and the horns of barchans often tend to be coarser than the crest and slip face sands (Finkel 1959, Hastenrath 1967, Lindsay 1973, Warren 1976, Lancaster 1982a). This may be partly explained by the fact that the coarser sands are moved up the

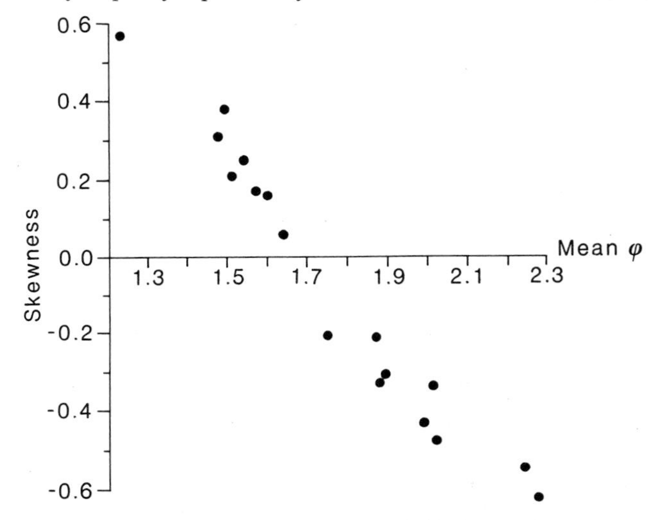

Figure 3.19 Relationship between graphic mean and inclusive graphic skewness for dome dune sand samples from the Killpecker dunefield, Wyoming. (Data of Ahlbrandt 1975).

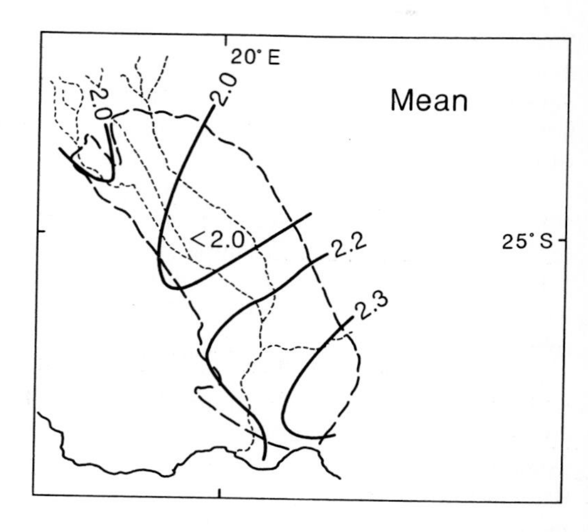

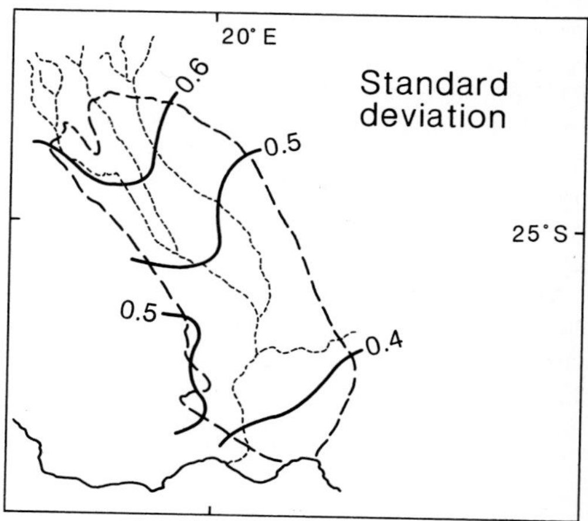

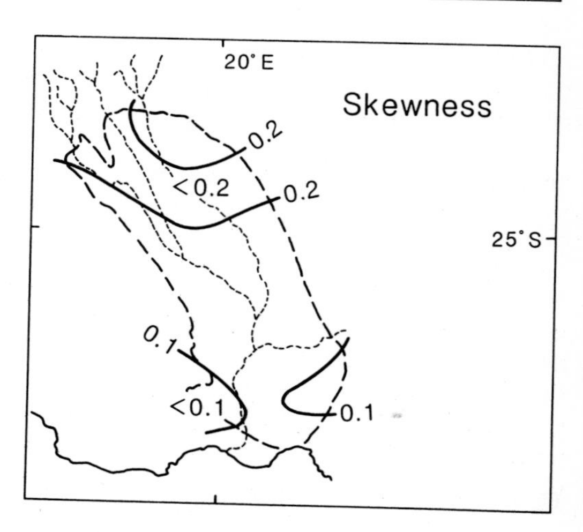

Figure 3.20 Regional trends in grain size and sorting parameters (phi units) in the southwest Kalahari. (After Lancaster 1986).

windward slope with more difficulty, and tend to be deflected round the dune base along the line of least resistance. An alternative, or complementary, explanation is that the coarser sands around the base of the dune represent basal foreset sands, in which coarser grains that have become exhumed during forward movement of the dune, are concentrated by rolling and avalanching on the slip face, (Bagnold, 1941, pp.226–9, Sharp 1966, Lindsay 1973).

Sneh & Weissbrod (1983) reported that the flanks of seif and other dunes that are influenced by ripple migration possess grain size characteristics that can clearly be differentiated from slip-face sands, which are dominated by avalanching processes. On the Sinai dunes which they studied, flank sands were found to be better sorted upslope whereas the slip-face sands were better sorted downslope.

Lancaster (1982a) found that barchan dunes in the Skeleton Coast dunefield of Namibia are slightly finer and better sorted than neighbouring transverse ridges and barchanoid ridges, while flat or gently undulating sand sheets are composed of poorly sorted coarse sands. At the regional scale, the sands were found to become finer and better sorted southwest to northeast across the dunefield, in the direction of downwind transport. Fining and improved sorting in the direction of transport is also seen in the Kalahari (Lancaster 1986) (Fig. 3.20). Lancaster suggested that the finer sand fraction is preferentially transported downwind and becomes concentrated in dunes, whereas the coarser grains remain as a lag closer to the source and are particularly concentrated in interdune areas [but see Thomas & Martin (1987) for a critical discussion]. By contrast, Buckley (1989) found that regional differences within the central Australian dunefields are relatively slight. This is consistent with the view that these sands have not experienced long-distance transport and grain-size sorting.

3.5 Shape characteristics of aeolian dune sands

Many early papers (Sorby 1877, 1880, Phillips 1882) gave the impression that desert dune sands are typically near-spherical and well rounded, and the term 'millet-seed' has been widely used to describe such grains. To some extent this view was supported by the results of laboratory simulations such as those performed by Kuenen (1960), which suggested rounding of grains by wind abrasion is 100–1000 times faster than rounding by fluvial abrasion. It has also been suggested that dune sand grains are relatively well rounded because wind action selectively transports more spherical, rounded grains (MacCarthy, 1935). However, more recent studies have indicated that most quartz dune sand grains are not well rounded, the exceptions being cases where the sands have been recycled from older sedimentary units (Goudie & Watson 1981, Goudie *et al.* 1987).

Most of the dune sand grains in the Simpson Desert of Australia are sub-angular to angular, with no noticeable rounding being accomplished in the present dune environment (Folk 1978). Folk suggested that this may be because the Simpson Desert dunes are partially fixed by vegetation and the grains have not been blown great distances from their fluvial source sediments.

Goudie & Watson (1981) examined fine and very fine sand grains in 108 dune sand samples from different parts of the world and also found that well rounded grains are relatively rare (about 8% of the grains examined). In dune sands from the Thar and California the predominant shape was sub-angular, although the sands from other areas were found to be predominantly sub-rounded. Only samples from Tunisia showed a predominance of rounded grains (Table 3.6). In two of the Namib Desert samples, Goudie & Watson (1981) found a clear increase in angularity with decreasing grain size. Considering the 108 sample data set as a whole, the very fine sand (3.5 ϕ) fractions were

Table 3.6 Values of mean roundness ($M\rho$) and mean percentage of grains in the rounded and well rounded classes of Powers (1953) shown for the 2.5ϕ and 3.5ϕ-size size fractions of desert dune sands from different parts of the world. (After Goudie & Watson 1981).

Desert	Sample size	2.5ϕ fraction		3.5ϕ fraction	
		$M\rho$	% rounded and well rounded grains	$M\rho$	% rounded and well rounded grains
Thar (Pakistan)	8	2.72	2.00	2.77	3.88
Thar (India)	20	2.98	2.95	2.87	2.55
Bahrain	12	3.51	12.25	3.40	16.00
Tunisia	7	4.01	50.00	3.19	7.00
Namib	20	3.53	15.30	3.30	19.90
Kalahari	8	3.21	2.00	3.16	3.25
California	3	2.89	6.33	2.64	2.67
Saudi Arabia	1	3.74	32.00	2.96	8.00
Mexico	1	3.64	22.00	3.46	16.00
Mean		3.19	9.64	3.04	7.97

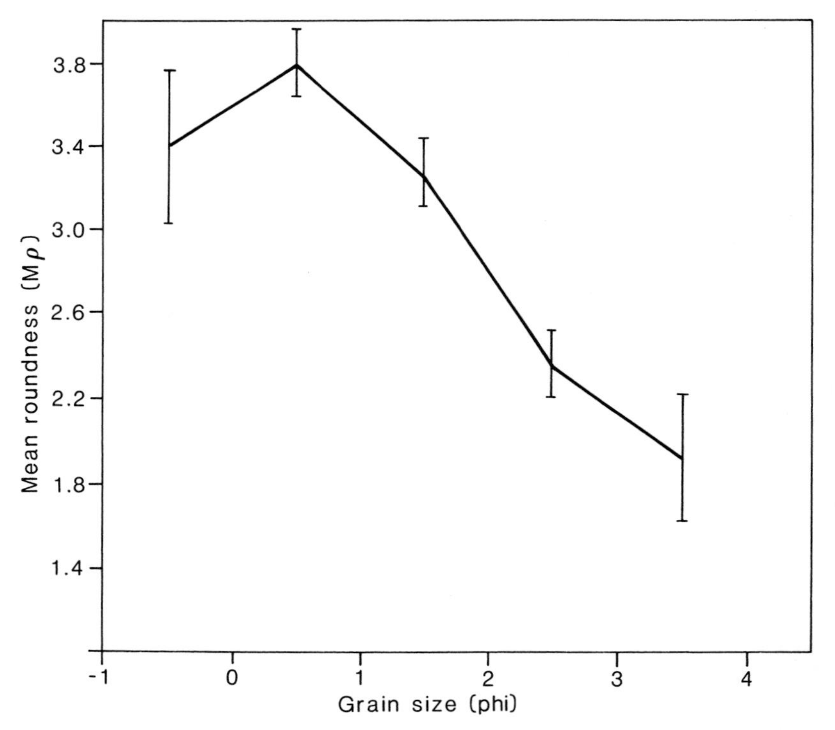

Figure 3.21 Relationship between mean roundness ($M\rho$) and grain size in aeolian sands from Kuwait. (After Khalaf & Gharib 1985).

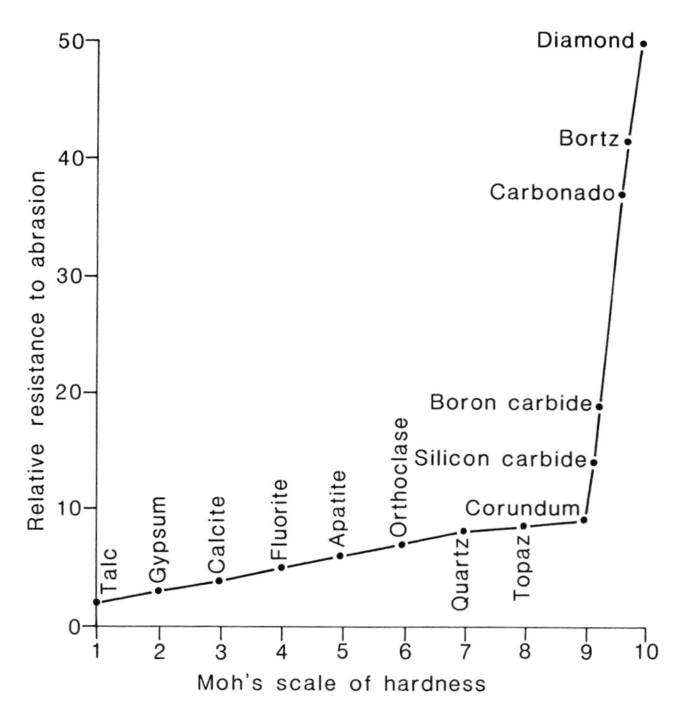

Figure 3.22 Relationship between Mohs' scale of hardness and relative resistance to abrasion of different minerals. (Modified after Dana & Harlbut 1959).

found to be systematically more angular than the fine sand (2.5 ϕ) fractions. Khalaf & Gharib (1985) and Ashour (1985) also found that mean roundness increases with increasing grain size in aeolian sediments from Kuwait and Qatar, respectively (Fig. 3.21). Khalaf & Gharib's data also show that the largest sand grains (larger than 1 mm) are less well rounded than the 0.5–1.0 mm diameter grains, possibly because the coarser grains are transported mainly by creep and experience less abrasion than finer grains which travel in saltation (Thomas 1987b).

The variable shape of grains in different size fractions often reflects differences in mineral composition, provenance, and geological history. Aeolian abrasion simulations by Marsland & Woodruff (1937) showed that the rate of rounding of different minerals is dependent on their hardness. The relative susceptibility to rounding by abrasion was found to be gypsum > calcite > apatite > magnetite > orthoclase > garnet > quartz. The relationship between Mohs' scale of hardness and resistance to abrasion is shown in Figure 3.22. In dune sands which contain mixtures of quartz and calcium carbonate grains, the latter are often more spherical and better rounded owing to their lower resistance to abrasion (Fig. 3.23).

Quartz grains recycled from older aeolian sandstones are likely to be better rounded than first-cycle grains derived from plutonic or metamorphic rocks. Similarly, first-cycle grains which have travelled only a limited distance from their source, owing either to low regional wind energy or to topographic constraints, are less rounded than first-cycle grains which have been transported great distances or which have been moved repeatedly backwards and forwards over a long period of time. In the Wahiba Sands of Oman, for example, much of the dune sand appears to have experienced a long history of cyclic deposition and reworking, with mixing of sands derived from different sources (Goudie *et al.* 1987). Most aeolian sand deposits consist of a mixture of grains derived from different sources, and although some grains may be well rounded others are often relatively angular.

There has been considerable debate concerning the importance of shape sorting during aeolian sand transport. Free (1911) was one of the first to suggest that grains of low sphericity would be blown by the wind more readily than spherical grains because they present a larger projected surface area (and therefore larger drag) to the wind in relation to their volume. However, MacCarthy (1935) found that coastal dunes in the eastern United States had a higher sphericity than the neighbouring beach sands and he interpreted this in terms of selective shape sorting by the wind. MacCarthy & Huddle (1938) subsequently reproduced such sorting in laboratory experiments and concluded that more spherical grains bounce higher following impact with the bed. Similar conclusions were reached by White & Schulz (1977), who noted that the saltation paths of spheres are higher than those of natural grains. Mattox (1955) concluded from field and laboratory experiments that aeolian shape sorting is likely to be unimportant in coastal areas where the transport distance between beach and dune is small. However, where transport distances are larger, he concluded that the wind should selectively transport less spherical grains. Williams (1964) and Willetts *et al.* (1982) found in laboratory experiments that spherical particles are transported more slowly than angular particles at low wind speeds, but that under strong wind conditions the reverse is true. Field evidence is equally contradictory. Coastal dune sands in several parts of the United States and Brazil have been found to show a higher degree of sphericity and roundness than their parent beach sands (Beal & Shepard 1956, Bigarella 1972). On the coast of Padre Island, Texas, more rounded, spherical grains become noticeably concentrated in coastal dunes after only a few feet of transport (Mazzullo *et al.* 1986). On the other hand, Stapor *et al.* (1983) reported than fewer spherical grains are preferentially transported inland in the coastal dunefield at Hout Bay, near Cape Town, South Africa.

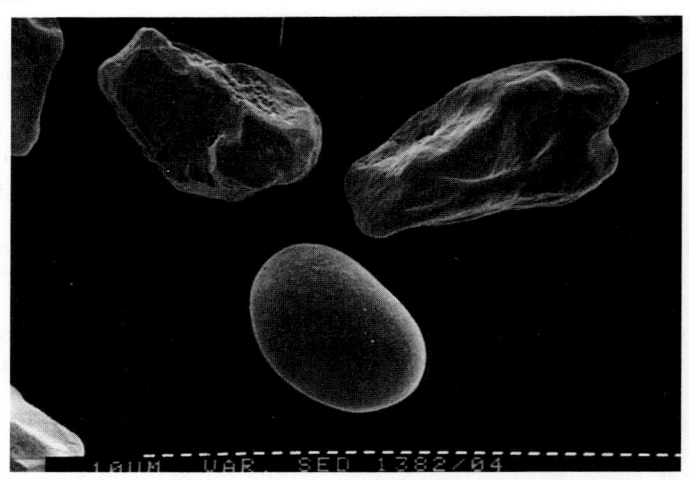

Figure 3.23 SEM micrograph showing a well-**rounded** calcite grain and more angular quartz grains from the Wahiba Sand Sea, Oman (sample collected by A. Warren). Scale bars = 10μm.

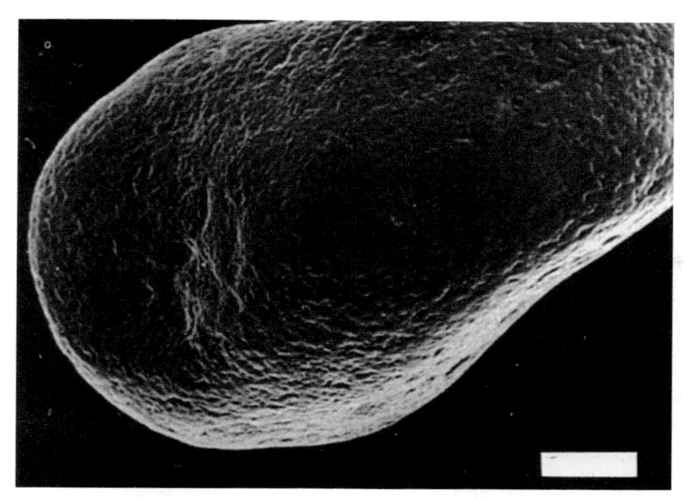

Figure 3.24 Scanning electron micrograph showing a well-rounded aeolian quartz sand grain with surface texture dominated by 'upturned plates', collected from the Sabha Sand Sea, south-central Libya (photograph by D. H. Krinsley). Scale bar = 30 μm.

3.6 Surface textures of aeolian sands

Many investigators have sought to obtain information about the origin and transport history of quartz sand grains by examining their surface textures using a scanning electron microscope. Surface textures of other minerals, particularly heavy minerals (Lin *et al.* 1974, Setlow 1978), have also been studied to a lesser extent.

Five types of textural features have been reported to be characteristic of aeolian quartz grains from modern hot deserts (Krinsley & Trusty 1985):

(a) General rounding of edges, regardless of whether the grains have high or low sphericity.
(b) 'Upturned plates' (Krinsley & Doornkamp 1973, Krinsley & McCoy 1978) which cover a high proportion of the surface of grains larger than 300–400 μm (Fig. 3.24). These plates appear as more or less parallel ridges ranging in width from 0.5 to 10 μm and have been interpreted to result from breakage of quartz along cleavage planes in the quartz lattice. Krinsley & Wellendorf (1980) suggested, on the basis of laboratory experimental evidence, that the spacing and size of upturned plates might be broadly related to wind energy. In nature the plates are often rounded to varying degrees as chemical processes. The presence of these modified plates gives rise to a 'frosted' appearance when the grains are viewed under a reflected light microscope (Margolis & Krinsley 1971).
(c) Equidimensional or elongate depressions, 20–250 μm in size, which occur predominantly on larger grains and which are caused by the development of conchoidal fractures during collisions. They are believed to develop as a result of direct impacts rather than glancing blows during saltation.
(d) Smooth surfaces which mainly occur on smaller grains (90–300 μm diameter), resulting from solution and precipitation of silica. These surfaces appear to be little affected by abrasion, possibly because small grains have less momentum and expend less energy when they collide during saltation.

(e) Arcuate, circular, or polygonal fractures which are mostly found on smaller (90–150 μm) grains. These probably have several different origins, including the development of fractures during direct impacts, salt weathering, and chemical weathering.

Quartz grains from coastal dune sands examined by Margolis & Krinsley (1971) were also found to have upturned plates on larger grains, although in fewer numbers than on warm desert grains. Equidimensional or elongate depressions are relatively rare. Smooth surfaces and elongate and arcuate cracks were also found to be less frequent on smaller coastal dune grains.

Periglacial aeolian sand also displays upturned plates, but only as patches on the grain surfaces. Equidimensional or elongate depressions are more frequently found on periglacial grains than coastal sand but are less abundant than on warm desert grains. Smooth surfaces on smaller grains, and arcuate, circular, or polygonal cracks are found relatively infrequently on periglacial aeolian grains. These differences may be attributable to the shorter aeolian transport history and shorter period of chemical weathering experienced in the coastal and periglacial environments studied by Krinsley and co-workers. In cases where coastal dune sands have experienced a long period of

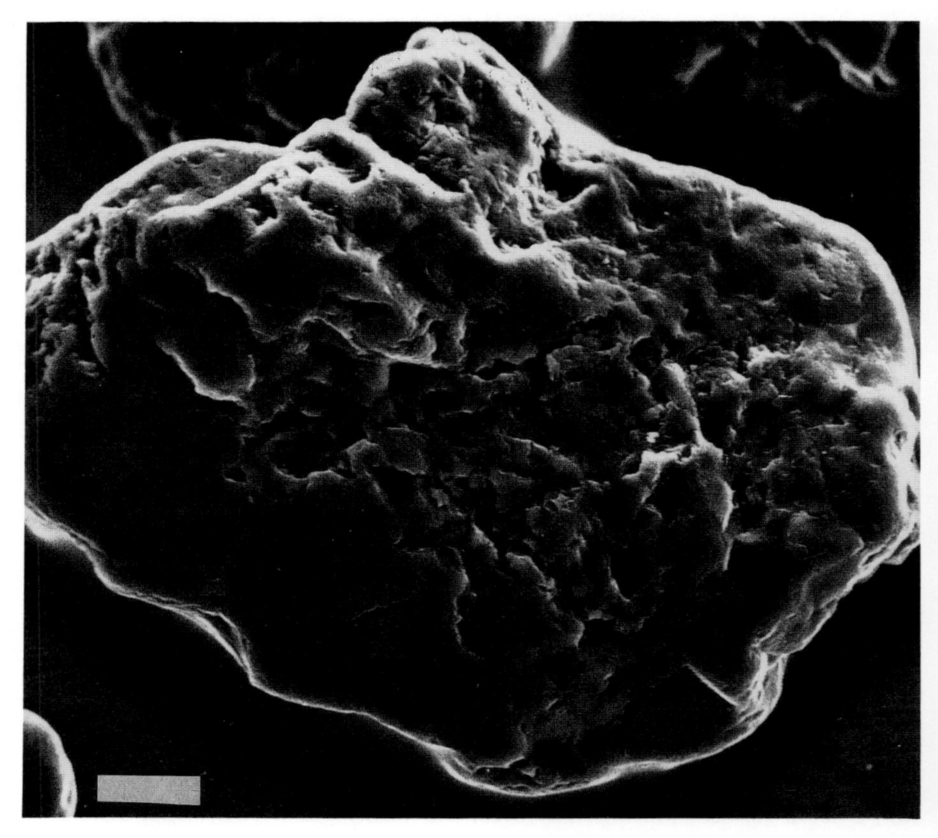

Figure 3.25 Scanning electron micrograph showing a chemically weathered quartz grain from an active dune at Cape Flattery, North Queensland. These dune sands have experienced several phases of podsolic weathering and aeolian reworking. Scale bar = 30 μm.

Figure 3.26 Scanning electron micrograph of a quartz sand grain from a seif dune in Sinai. The grain has a thin coating of amorphous silica and iron (III) oxyhydroxide. Scale bars = 10 μm.

weathering, involving several stages of aeolian reworking and re-deposition, as on the coast of eastern Australia, the grain surface textures are dominated by chemical features which include oriented etch pits, deep arcuate etch lines, and angular breakage features (Pye 1983a) (Fig. 3.25). The nature of the features on individual quartz grains depends on whether the grain is monocrystalline or polycrystalline, strained or unstrained, and on its previous weathering, transport, and diagenetic history (Little *et al.* 1978, Pye 1983a).

Chemical textures are also dominant on many desert aeolian sand grains which have experienced lengthy periods of weathering in the near-surface environment (Fig. 3.26). Such grains often have coatings of iron oxides, silica, or mixtures of non-crystalline alumino-silicate material and carbonate (Pye & Tsoar 1987). Folk (1978) described waxy-looking coatings on Simpson Desert dune sand grains as 'turtle-skin silica coats'. In pre-Quaternary dune sands the primary aeolian textures are normally completely obliterated by diagenetic texture, although a few cases of their preservation have been described (Krinsley *et al.* 1976).

A number of laboratory experimental investigations have examined the changes in grain shape and surface texture which occur as a result of aeolian abrasion (Nieter & Krinsley 1976, Kaldi *et al.* 1978, Lindé & Mycielska-Dowgiallo 1980, Krinsley & Wellendorf 1980, Wellendorf & Krinsley 1980, Whalley & Marshall 1986, Lindé 1987, Whalley *et al.* 1987). These studies have all shown that the main effect of abrasion is to produce rounding of edges and corners. Very angular grain projections are initially removed by edge chipping, which involved brittle fracture mechanisms. This is subsequently followed by smoothing of edges and corners. 'Upturned plates' appear in the later stages on parts of the grains which show positive curvature. The glancing nature of impacts between spinning grains in saltation is believed to play an important role in the rounding process (Whalley & Marshall 1986).

Table 3.7 Major types and sources of sand.

A Inorganic sand grains released by breakdown of igneous, metamorphic, and sedimentary rocks

- (i) rock fragments
- (ii) quartz (monocrystalline and polycrystalline)
- (iii) feldspars
- (iv) heavy minerals (oxides and silicates)
- (v) layer silicates (mainly micas)

B Inorganic grains released by breakdown of carbonate rocks

- (i) calcite (limestone and chalk fragments)
- (ii) dolomite (from limestone and dolomites)

C Inorganic grains formed in near-surface environments by chemical and physical processes

- (i) gypsum crystals
- (ii) clay pellets (mixtures of different clays and other minerals)
- (iii) carbonate ooids (mainly aragonite and high-Mg calcite)
- (iv) pyroclastic grains (glass shards, feldspars, amphiboles, pyroxenes)

D Biogenic skeletal carbonate grains and shell fragments (foram tests, algal and coralline debris, echinoderm plates, mollusc and gastropod shells, etc).

- (i) aragonite
- (ii) high-Mg calcite
- (iii) low-Mg calcite
- (iv) polymineralic carbonate

3.7 Porosity and permeablity of aeolian sands

Medium, well sorted dune sands may have porosities as high as 45%, depending on the sphericity of the grains and the packing arrangement. Most aeolian sands do not consist of spherical grains and have packing arrangements which are intermediate between the cubic and orthorhombic end-members shown in Figure 3.12. Typical porosities of uncemented aeolian sands lie in the range 34–40% (Kolbuszewski *et al.* 1950), although extremes of 25–50% have been recorded. Experiments have shown that there is an inverse relationship between the average wind velocity and the porosity of the deposited sand (Kolbuszewski 1953). Owing to their frequently finer grain size and better sorting, coastal dune sand bodies often have slightly higher porosities and permeabilities than do inland dune sands or interdune deposits (Pryor 1973).

Poorly sorted interdune deposits often form permeability barriers, especially where they have experienced early diagenetic cementation by carbonates or evaporites. Significant permeability differences between different aeolian stratification types have also been reported in aeolian sandstones (Goggin *et al.* 1986, Weber 1987, Chandler *et al.* 1989).

3.8 Sources and mineral composition of aeolian dune sand

Sand grains can be divided into four broad groups based on their mode of origin and source (Table 3.7). Quartz and silicate grains released by weathering and erosion of crustal rocks are by far the most important globally, but locally sands may consist

Table 3.8 Mineral proportions in the abundant types of plutonic igneous rocks. (Data from Wedepohl 1969, p. 248).

Mineral	Volume % of					
	granite	grandiorite	quartz diorite	diorite	gabbro	upper crust average
plagioclase	30	46	53	63	56	41
quartz	27	21	22	2	-	21
K-feldspar	35	15	6	3	-	21
amphibole	1	13	12	12	1	6
biotite	5	3	5	5	-	4
orthopyroxene	-	-	-	3	16	2
clinopyroxene	-	-	-	8	16	2
olivine	-	-	-	-	5	0.6
magnetite, ilmenite	2	2	2	3	4	2
apatite	0.5	0.5	0.5	0.8	0.6	0.5

largely or wholly of grains formed by breakdown of carbonate rocks or recent biogenic skeletal debris, or by near-surface chemical and physical processes.

3.8.1 *Weathering and erosion of crustal rocks*

Most quartz and feldspar grains of sand size found in modern sediments were originally derived from plutonic igneous or metamorphic rocks. Quartz and potassium feldspar are especially abundant in plutonic rocks of granitic or granodiorite composition, whereas these minerals are virtually absent in basic plutonic rocks such as gabbro (Table 3.8). Granitic igneous rocks are estimated to be about five times more abundant in the earth's crust than basic igneous rocks.

Feniak (1944) measured the size of minerals in thin sections of more than 200 massive plutonic rocks and found that the modal size of feldspar crystals is 1–2 mm (very coarse sand), whereas that of quartz grains is 0.5–1 mm (coarse sand). However, many quartz and feldspar grains in igneous rocks are fractured, so that individual crystals break into smaller pieces when released by weathering (Moss 1966, 1972).

Blatt (1967) found that weathering of granite in the arid southwestern United States lead to the release of sub-equal amounts of polycrystalline and monocrystalline quartz grains. The average size of the polycrystalline grains (1 mm) was found to be larger than that of the monocrystalline grains (0.5 mm). An average of 80–90% of the grains were found to be plastically deformed and exhibited undulatory extinction in thin section. In addition, many of the polycrystalline grains showed evidence of suturing at the intercrystalline boundaries. Several other studies have shown that polycrystalline quartz grains and polymineralic lithic fragments experience more rapid breakdown during transport than do monocrystalline quartz grains (e.g. Basu 1976).

Metamorphic rocks such as gneiss and migmatites, which may represent metamorphosed granites or siliceous sedimentary rocks, are also important sources of quartz and feldspar sand. In the Kora area of semi-arid Kenya, for example, regolith developed on gneiss and migmatites consists of 47–97% sand (Pye *et al.* 1985).

The proportion of sand present in the weathering products derived from any granitoid rock is dependent on the relative effectiveness of mechanical disintegration and chemical composition as weathering processes (Pye 1985a). In humid climates, the chemical decomposition of feldspars, biotite, and amphiboles is relatively rapid, with the result that the weathering products have a lower sand/silt plus clay ratio than in more arid climates. Relief and rate of surface erosion are also an important control on the composition of weathering products (Basu 1985).

Volcanic rocks also act as important sources of sand-sized particles in some areas. For example, the Great Sand Dunes dunefield in Colorado is composed mostly of volcanic rock fragments (51.7%) and quartz (27.8%). Much of the material has been transported from the San Juan Mountains by fluvial processes (Johnson 1967, Andrews 1981).

Sandstones and sandy conglomerates provide another important source of sand grains. The 'average' sandstone contains about 65% quartz (Blatt 1970), compared with about 27% quartz in the 'average' granite. Feldspar constitutes about 12% of the average sandstone, compared with about 65% in the average granite. This reflects the fact that feldspar is less durable than quartz both during mechanical abrasion and chemical weathering (Fig. 3.22). The less resistant nature of feldspars is due partly to their lower hardness, partly to their well developed cleavage and occurrence of twinning, and partly to the fact the feldspars are sometimes weakened by hydrothermal alteration before being released from the parent rock.

The grains in many sedimentary rocks have experienced several cycles of weathering, erosion, and deposition. As a general rule, the greater the number of episodes of recycling, the higher is the percentage of quartz in the resulting sandstone.

Holm (1960) and Whitney *et al.* (1983) concluded that much of the sand in the An Nafud sand sea of Saudi Arabia is derived from poorly cemented Palaeozoic sandstones which outcrop close by. The carbonate cement is extensively leached in outcrop and the weathered sandstone crumbles easily to loose sand.

Sand-sized fragments of limestone and dolomite are common close to areas where these rocks are undergoing predominantly physical weathering. Production of clastic debris from carbonate rocks is favoured where the rocks have been shattered by tectonic movements, by arid climatic conditions, and in cold climates where frost weathering is important. Owing to the lower hardness of calcite ($H=3$ on Mohs' scale) and dolomite ($H=4$), sand grains composed of these minerals should be abraded more rapidly than quartz ($H=7$) during transport. However, limestone and dolomite grains comprise more than 30% of some major sand sea deposits, such as those of the Wahiba Sands in Oman (Allison 1988). Calcrete and gypcrete fragments also comprise >5% of some arid region aeolian sands (e.g. Khalaf 1989).

3.8.2 Formation of sand-size particles in the near-surface environment

3.8.2.1 Gypsum sands Dunes consisting largely of abraded gypsum crystals occur on the margins of some playas which are subject to periodic wind scouring. Important examples include the White Sands dunefield adjacent to Lake Lucero in New Mexico (McKee & Moiola 1975) and the gypsum–oolite dunes close to the former shores of Lake Bonneville in Utah and Nevada (Jones 1938). The gypsum crystals are formed by chemical precipitation in lake or playa-bottom muds below the water level (Eardley & Stringham 1952). When the water level falls sufficiently the crystals are removed by the wind and deposited as dunes downwind of the lake. Initially most of the gypsum crystals are tabular or lozenge-shaped, but owing to their relative softness they become rounded by abrasion during aeolian transport. Jones (1938) reported that most of the abraded gypsum crystals in the Great Salt Lake Desert dunes were 63–250 μm in size, with a median diameter of about 155 μm (Fig. 3.27).

3.8.2.2 Clay pellets Wind erosion of dry playa sediments can also result in the formation of sand-size clay pellets which locally may be sufficiently abundant to form dunes. Active formation of clay pellets and clay dunes has been reported from the Texas Gulf Coast (Coffey 1909, Huffman & Price 1949, Price 1958), Algeria (Boulaine 1954, 1956) and West Africa (Tricart 1954). Fossil clay dunes occur in Australia (Stephens & Crocker 1946, Bettenay 1962, Bowler 1973), South America (Dangavs 1979) and southern Africa (Goudie & Thomas 1986). The formation of clay pellets is favoured in seasonally dry climates where playas experience cyclical wetting and drying. In Texas, Huffman & Price (1949) observed that movement of clay pellets begins in March and ceases in November when the winter rains cause waterlogging of the mud flats. During the dry season, cracking of the surface mud, combined with salt crystallization and hydration, causes fragments of the mud crust to be loosened and entrained by the wind. The existence of saline mud appears to be a requirement for large-scale clay pellet formation (Bowler 1973), since saline mud is hygroscopic and experiences frequent surface blistering, often on a diurnal basis. The mineral composition of the mud appears to be of minor importance. Individual pellets often consist of a mixture of minerals including quartz, kaolinite, illite, smectite and mixed-layer clay, chlorite, feldspar, and carbonates.

The eroded mud clasts become progressively more rounded during transport by creep or saltation (Fig. 3.28). After a few hundred metres of aeolian transport, the pellets establish a moderately well sorted, unimodal size distribution (Fig. 3.29).

In Texas the pellets are deposited on low transverse dune ridges downwind of the source areas. Elsewhere, as in parts of Australia, Argentina, and southern Africa, the aeolian clay accumulations form crescentic ridges, termed *lunettes* (Hills 1940). The shape of the lunettes is often controlled by wave action on the shoreline at the downwind end of the adjacent lake at times of high water level.

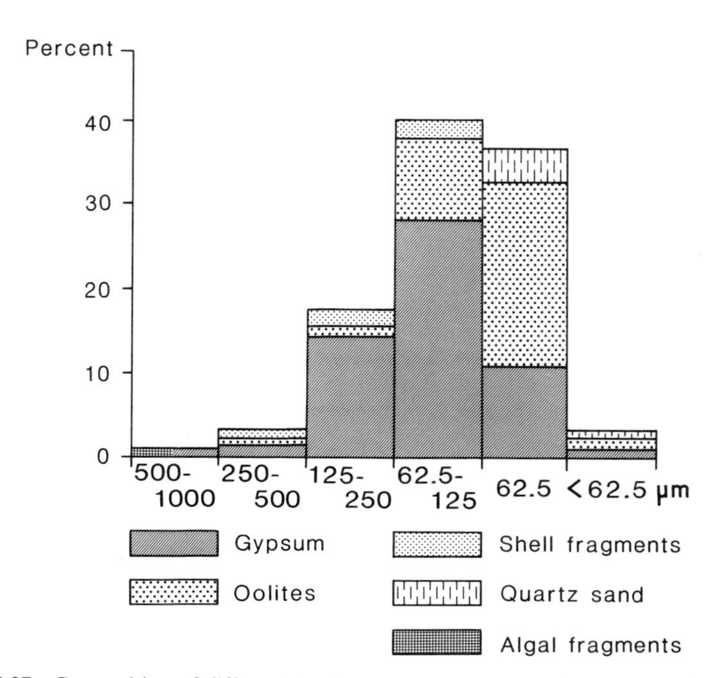

Figure 3.27 Composition of different size fractions in gypsum–oolite dunes of Great Salt Lake Desert, Utah. (Modified after Jones 1938).

Figure 3.28 Scanning electron micrograph showing sand-size clay pellets from the shores of a saline lake, Argentina (sample collected by A. T. Grove). Scale bars = 100 μm.

Transverse ridges and lunettes composed of aeolian clay rarely exceed 30 m in height but can be several hundred metres in width. The windward and leeward slopes rarely exceed 15°, and slip faces normally do not develop. The dunes accrete layer by layer and are characterized by near-parallel bedding. Once deposited on the dune, the pellets are wetted by rain and the clay regains its plasticity. Individual pellets eventually merge to form a more-or-less homogeneous mass.

3.8.2.3 Volcaniclastic sands In areas of recent volcanic activity, both coastal and inland dunes may contain large amounts of volcanic material. Many of the beaches in the Hawaiian Islands, for example, consist of 'black sands' containing large numbers of basaltic fragments, mafic minerals, and volcanic glass shards (Moberley *et al.* 1965). This mineralogy is reflected in that of the adjacent coastal dunes. Inland, in the Ka'u Desert,

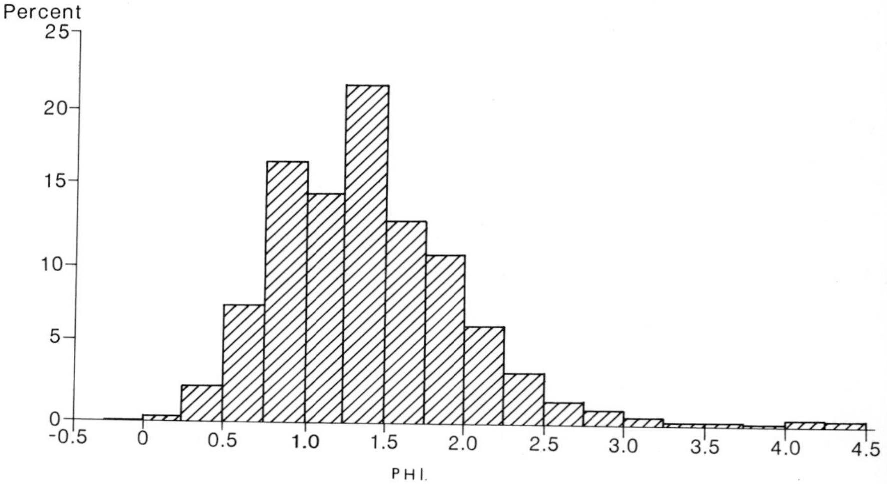

Figure 3.29 Grain size–frequency histogram of pellets from the surface of a clay dune, Argentina (sample collected by A. T. Grove).

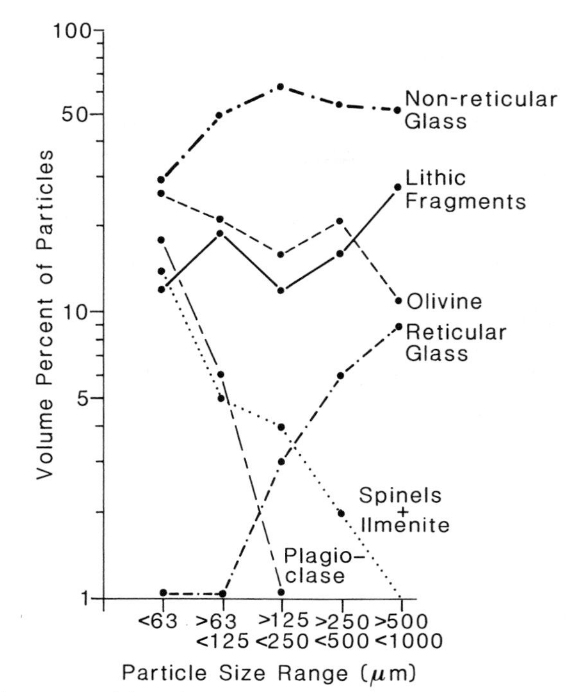

Figure 3.30 Phase composition of volcaniclastic dune sands as a function of particle size, Ka'U Desert, Hawaii. (After Gooding 1982).

on the southwest flank of Kilauea volcano, Hawaii, aeolian reworking of volcanic ash from the Keanakakoi Formation has formed dunes of similar mineralogy (Gooding 1982). Plagioclase, ilmenite, and magnetite were found by Gooding to be concentrated in the finer sand fractions, whereas olivine, lithic fragments, and non-reticulate glass were of similar abundance in all size fractions. Reticular glass was the only constituent found to increase in relative abundance with increasing grain size (Fig. 3.30).

3.8.2.4 Carbonate ooids and peloids Ooids are spherical or ovoid, well rounded carbonate grains consisting of a detrital nucleus and a concentrically laminated cortex (Fig. 3.31). The nucleus may be a fragment of carbonate of siliceous material. The cortex is usually composed of aragonite or high-magnesium calcite, sometimes with enclosed layers of organic matter. Most ooids are 0.1–1.5 mm in diameter. Some, such as those from the Great Salt Lake, Utah, display a radial aragonite microstructure (Kahle 1974).

The manner in which ooids grow has been much debated (Davies *et al.* 1978, Ferguson *et al.* 1978), but it is clear they can form both in lacustrine and shallow marine environments where the bottom sediments are periodically agitated.

Pelloids are spherical or ovoid-shaped carbonate grains which typically have a dark, structureless appearance in thin section. They are rich in organic matter and may contain included fragments of carbonate or siliceous material. Typically the size ranges from 0.1 to 3 mm in maximum dimension. Some represent faecal pellets or eroded fragments of mud which have been hardened by interstitial precipitation of calcium carbonate. Others are abraded fragments of shell or other debris which have been

Figure 3.31 Carbonate ooids from a littoral dune in the Bahamas. Scale bars = 10 μm.

micritized by algae and other boring organisms. Like ooids, peloids form both in lacustrine and marine environments (Bathurst 1975).

Jones (1938) reported that dune sands close to the Great Salt Lake in Utah contained significant numbers of ooids in addition to abraded gypsum crystals (Fig. 3.27).

3.8.3 Formation of biogenic carbonate sand

Many living organisms secrete calcareous hardparts composed of aragonite or high- or low-magnesium calcite. Among the more important sources of biogenic carbonate grains are calcareous algae, foraminiferal tests, echinoderm spines and plates, coral skeletons, and the shells of various organisms such as molluscs, pelecypods, ostracods, and gastropods. In addition, some soft-bodied organisms such as worms are important formers of calcified tubes.

Skeletal carbonate debris is broken down into smaller particles by both physical and biological processes. Waves and currents both cause breakage and interparticle abrasion, but in some environments various grazing and boring organisms are equally, if not more, important. The debris produced may have a wide variety of morphologies; shell fragments often form curved, platy grains.

Biogenic carbonate sand is produced mainly in shallow marine and lacustrine environments which have abundant nutrients and high rates of organic activity. It assumes greatest relative importance where rates of terrigenous sedimentation are low. Such areas are found particularly in arid climatic zones where fluvial discharge is limited, and on shallow shelves some distance away from the continental shoreline. Large accumulations of carbonate dune sand, much of it cemented to form aeolianite (Ch. 8), are found in Bermuda, along the shores of the Mediterranean, Natal, Western and South Australia, the Persian Gulf, and on many oceanic islands including Hawaii (McKee & Ward 1983, Gardner 1983b). Biogenic carbonate sand is not a major component of most desert sand seas, although in some instances it is significant. For example, marine foraminifera derived from the Indian Ocean are present in considerable numbers in the aeolianites (miliolites) of the Thar Desert, Pakistan (Goudie & Sperling 1977), and the Wahiba Sand Sea, Oman (Goudie *et al.* 1987, Allison 1988).

4 Mechanics of aeolian sand transport

4.1 Particle entrainment

4.1.1 Forces exerted on static grains by the wind

Wind flowing over a sand grain at rest on a horizontal surface exerts two types of force on the grain: (a) a drag force acting horizontally in the direction of the flow, and (b) a lift force acting vertically upwards. Opposing these aerodynamic forces are inertial forces, the most important of which is the grain's weight, which acts directly opposite to the lift force. Cohesive forces, which are attractive forces between neighbouring grains, and adhesive forces, which operate between grains and other surfaces, must also be taken into consideration for fine grains.

The drag force is the aggregate of the skin friction drag and the pressure drag (see Section 2.4.4). The latter results from positive pressure on the upwind face of the grain and negative pressure on its downwind side (Figs 2.21 and 4.1). The skin friction drag is the viscous stress acting tangentially to the grain surface. The total drag force (F_d) acting on the grain is given by

$$F_d \propto \tau_0 A \propto \rho u_*^2 A \qquad (4.1)$$

where τ_0 is the surface shear stress (Eqn 2.18), A is the grain's largest projected area and u_* is the friction velocity. For spherical particles of diameter d, $A = d^2\pi/4$. Hence the drag force on the sphere is given by

$$F_d = \beta \rho u_*^2 (\pi d^2/4) \qquad (4.2)$$

where β is a coefficient that depends partly on the ratio of the momentary velocities of turbulent fluctuations to the average wind velocity (Eqn 2.11), partly on the proportion of drag per unit area experienced by the grain due to its relative position amongst other grains on the bed, and partly on the height at which the drag force acts (Bagnold 1941, p. 86).

The lift force has been omitted from some theoretical and experimental considerations on the assumption that it is insignificant (Shields 1936, White 1940, Bagnold 1941, p. 32, Kalinske 1947). However, lift force is inherent to the Bernoulli effect and consequent aerodynamic thrust (Jeffreys 1929) (Eqn 2.26). Several authors have observed sand grains to rise almost vertically from the bed during wind tunnel studies (Chepil 1945a, White *et al.* 1976). The lift force arises because of the high wind velocity gradient near the bed. The flow velocity on the underside of a grain at rest on the bed is zero but on the upper side the flow velocity is positive. Hence there is a high static pressure under the grain and much lower pressure above it. The grain will be raised from the bed if the force resulting from the static pressure difference exceeds the inertial force due to the grain's weight, W, which is given by

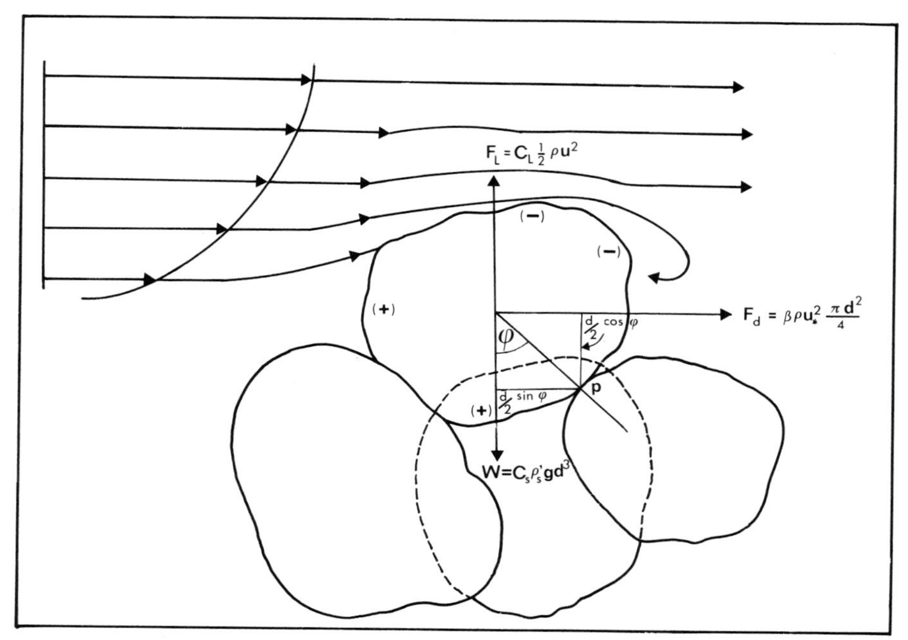

Figure 4.1 Schematic diagram showing the forces exerted on a static sand grain by the wind. On the upper left are the wind profile and the streamlines ensuing from it; (+) indicates relatively high pressure and (−) relatively low pressure on the grain surface. P is the pivot point about which the two moments $(d/2)\sin\varphi$ and $(d/2)\cos\varphi$ are calculated. For explanation of the three forces acting on the grain, see text.

$$W = C_s \rho_s' g d^3 \tag{4.3}$$

where ρ_s' is the density of the immersed grain ($\rho_s' = \rho_s - \rho$, ρ_s being the grain density and ρ the fluid density) and C_s is a shape coefficient such that $C_s d^3$ is the volume of the grain (for a sphere $C_s = 0.524$).

Experiments have shown that during turbulent flow there are large instantaneous variations in flow velociety and pressure which may produce a short-term lift force sufficient to raise grains from the surface (Einstein & El Samni 1949, Bisal & Nielsen 1962). The average lift force measured as pressure difference between the top and bottom of a hemisphere was found by Einstein & El Samni (1949) to be

$$F_L = \Delta p A = (C_L \rho U^2 A)/2 \tag{4.4}$$

where F_L is the lift force, Δp is the pressure difference between top and bottom of the hemisphere, C_L is the coefficient of lift [Chepil (1958b) amended the value of C_L of Einstein & El Samni (1949) to $C_L = 0.0624$], and U is the fluid velocity measured at 0.35 grain diameters from the theoretical surface that is represented by the roughness length (z_0).

Chepil (1958b) also found that the ratio F_L/F_D is constant for any size of roughness element and friction velocity within the designated range of Reynolds number:

$$F_L = cF_d \tag{4.5}$$

where c is a coefficient equal to 0.85 for near-spherical grains and wind velocity within the range required to move soil particles. In another experiment (Chepil 1961), c was found to have an average value of 0.74 for spheres with a diameter of 0.3 cm and 5.1 cm resting on the ground. In these experiments the drag force was always found to be greater than the lift force. Chepil also observed that the greater the surface roughness (and the grain size), and the greater the friction velocity, the higher is the significance of the lift force. It should be noted, however, that the particle Reynolds numbers in the experiments performed by Chepil and Einstein & El Samni are far larger than those typical of aeolian sand grains, and thus extrapolation of the results may not be strictly valid.

During aeolian sand transport, additional lift force may be generated by the rolling motion of grains which would further accelerate the flow of air moving over the top of the grains (Chepil 1945a). Strong lift forces may also be associated with small- and medium-scale vortices (Greeley *et al.* 1981).

4.1.2 *Threshold of grain movement*

When the wind velocity over a loose sand surface is slowly increased, a critical point is reached where some grains begin to move. This critical wind velocity is known as the *fluid threshold velocity* (Bagnold 1941, p. 33). Bagnold observed in wind tunnel experiments that the initiation of grain movement most often took place by rolling. Similar observations were reported by Malina (1941) and Chepil (1959). However, Bisal & Nielsen (1962) observed that at the very beginning of movement some grains began to vibrate backwards and forwards before leaving the surface almost vertically, as if ejected. Only a few grains were seen to roll along the surface before bouncing into the air. Similar observations were reported by Lyles & Krauss (1971), who examined the possibility that the particle vibration frequency was related to the frequency of turbulent fluctuations in the flow. They found that the peak turbulence frequency was 2.3 ± 0.7 Hz whereas the mean particle vibration frequency was 1.8 ± 0.3 Hz, the difference being attributed to a large density difference between the air and the grains.

The initiation of grain motion can be more fully understood by examining the forces acting on individual grains. Consider a flat surface covered by loose sand of uniform size. Grains in the uppermost layer of the bed are free to move upwards but their horizontal movement is constrained by adjacent grains. The point of contact between neighbouring grains acts as a pivot (p in Fig. 4.1) around which rotational movement takes place when the lift and drag forces exceed the inertial forces. The effectiveness of combined drag and lift forces in producing rotation about the pivot is measured by the product of the forces and their moments [$(d/2)\cos\phi$ and $(d/2)\sin\phi$ in Fig. 4.1]:

$$F_{\mathrm{d}}(d/2)\cos\phi = (W - F_{\mathrm{L}})(d/2)\sin\phi \qquad (4.6)$$

where ϕ is the angle at the centre of gravity of the grain between p and W (Fig. 4.1) and d is the grain diameter.

According to White (1940), the angle ϕ should be similar to the angle of internal friction which defines the angle at which sliding failure begins in granular materials (see Section 4.2.7). Grain rotation is easier when the friction angle is small. Spherical grains in a rhombohedral packing arrangement (Fig. 3.12), which approximates that found in some aeolian sands, would have a friction angle of 30°.

In wind tunnel experiments using loose, dry soil particles, Chepil (1959) found that the drag force acts slightly above the centre of gravity of the grains, reducing the friction angle to about 24°.

If the grains in Figure 4.1. are assumed to be spherical, substitution of Equations 4.2, 4.4, and 4.5 in Equation 4.6 gives

$$\beta\tau_{0(c)}(\pi/4)d^2\cos\phi = (\pi/6)\,\rho'_s g d^3\sin\phi - c\beta\tau_{0(c)}(\pi/4)d^2\sin\phi \tag{4.7}$$

where $\tau_{0(c)}$ is the critical surface shear stress at the threshold of grain movement. Equation 4.7 can also be expressed as

$$\frac{\tau_{o(c)}}{\rho'_s g d} = \frac{2}{3}\left(\frac{\sin\phi}{\beta(\cos\phi + c\,\sin\phi)}\right) \tag{4.8}$$

The threshold shear stress on a flat surface is determined by several factors, including the density, size distribution, packing, and shape of the grains. Since most natural sediments contain a range of particle sizes and shapes, it is difficult to define a single value of $\tau_{0(c)}$, which, indeed, may be better regarded as a statistical phenomenon (Chepil 1945b, Zingg 1953a, Nickling 1988).

Equation 4.8 shows that the threshold surface shear stress is directly proportional to the immersed density of the grains and the grain diameter. The packing of the grains is reflected by the angle ϕ and the grain shape and the sorting by β.

In the case of upward-sloping surfaces, the critical shear stress necessary to initiate grain movement is higher than that on a flat surface, whereas the reverse is true on a downward-sloping surface (Howard, 1977).

Sokolow (1894), Owens (1908), Jeffreys (1929), and Shields (1936) were among the first to study the initiation of sediment entrainment by water and air. Jeffreys (1934) defined the condition for a particle to move off the surface as

$$U = \propto (\rho'_s g d/\rho)^{1/2} \tag{4.9}$$

where U is the velocity of the fluid over the surface and $\propto$ is a constant that depends on the grain shape.

Shields (1936) developed a dimensionless coefficient which expresses the ratio of the applied tangential force to the force resisting grain movement:

$$\theta_t = \frac{\tau_{o(c)}}{\rho'_s g d} = f\left(\frac{u_{*_t} d}{\nu}\right) \tag{4.10}$$

where θ_t is known as the Shields threshold criterion (Miller *et al.* 1977). The term on the right-hand side of Equation 4.10 denotes the friction Reynolds number (Eqn 2.23) where u_{*_t} is the threshold friction velocity.

Bagnold (1941) defined the threshold friction velocity (u_{*_t}) in a form derived from the Shields criterion (Eqn 4.10):

$$U_{*_t} = A\left(\frac{\rho'_s g d}{\rho}\right)^{\frac{1}{2}} \tag{4.11}$$

where $A = (\theta_t)^{1/2}$. By substituting u_{*_t} for u_* in Equation 2.20, the threshold mean velociety (U_t), measured at any height z, can be found:

$$U_t = \frac{1}{k}A\left(\frac{\rho'_s g d}{\rho}\right)^{\frac{1}{2}}\ln(z/z_0) \tag{4.12}$$

Field and wind tunnel studies by several subsequent workers have generally verified Bagnold's findings (Chepil 1941, 1945b, 1958a, 1959, Zingg 1953a, Greeley *et al.* 1974a, Svasek & Terwindt 1974, Iversen *et al.* 1976a, White 1979, Iversen & White 1982). Some of these later results are compared with Bagnold's in Figure 4.2, which presents a modified Shields diagram relationship between Re_{*_t} (the particle friction Reynolds number at the threshold) and A. The curves all show a marked 'turn-up' at $Re_{*_t} < 1$, where A increases rapidly. For $Re_{*_t} > 1$, A becomes asymptotic, with an ultimate value which ranges between 0.1 and 0.118 according to different authors.

The relationship between u_{*_t} and grain diameter for quartz grains in air, determined in four separate sets of experiments, is shown in Figure 4.3. Some of the variation between the curves may be explained by the fact that different definitions of threshold have been used (Sagan & Bagnold 1975), and by differences in the size distributions of the sediments used (Nickling 1988).

All of the curves in Figure 4.3 show that particles in the size range 70–125 μm are most easily entrained. Bagnold (1937b) found in wind tunnel experiments that it was impossible to entrain silt-size particles of Portland cement with a wind velocity of 22 m s^{-1}, measured at a height of 10 cm. Similarly, Chepil (1941) found that silt particles smaller than 0.05 mm were very resistant to wind erosion and did not move at velocities up to 16.5 m s^{-1}, measured at a height of 15 cm.

A similar upturn in the threshold velocity curve for finer sizes in water was noted by Hjulstrom (1935), who emphasized the importance of interparticle cohesion. However, Bagnold (1941, p. 89, 1960) considered that cohesion is not the only cause of the high resistance of fine particles to entrainment. Bagnold pointed out that the forces exerted by fully turbulent flow on individual grains on the bed depend on the small-scale nature of the flow over the surface of the individual grains. This is indicated by the particle friction Reynolds number (Eqn 2.23). When $Re_* > 3.5$, the grain behaves as an isolated

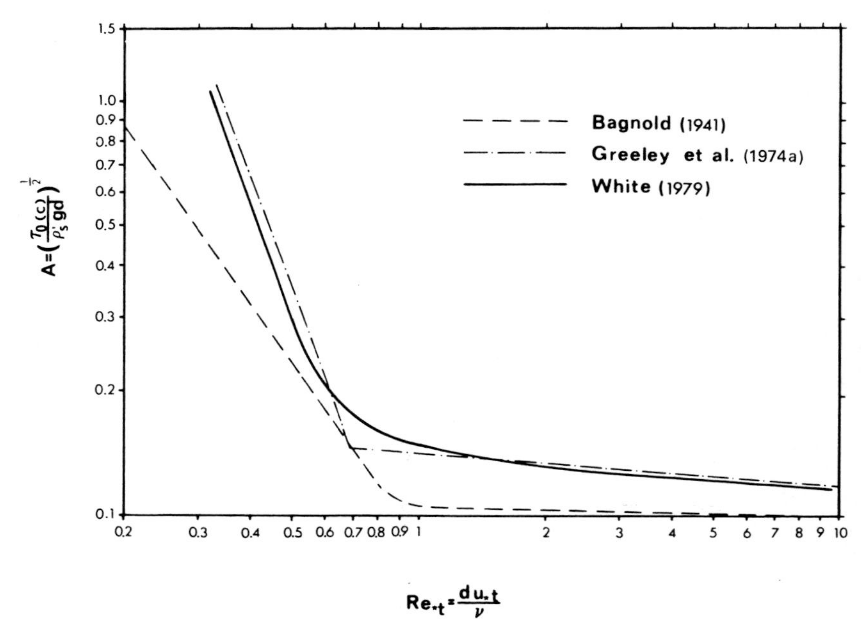

Figure 4.2 Shields diagram showing the relationship between Re_{*_t} and A determined in experiments by different authors.

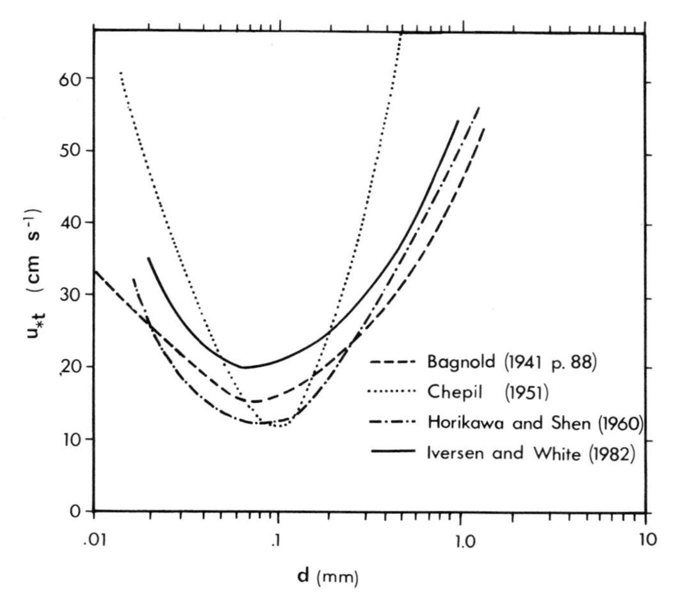

Figure 4.3 Threshold friction velocity (u_{*t}) curves for quartz grains of different diameter (d). Iversen & White's (1982) curve takes into account variations in Re_{*t} and cohesive forces. Chepil's (1951) curve relates to an equivalent diameter ($= \rho_s d/2.65$) of soil that contains a mixture of different size fractions.

obstacle in the path of the fluid and the surface can be regarded as aerodynamically 'rough' (Bagnold 1941, p. 89) (Fig. 2.17b). As the flow velocity increases and the thickness of the laminar sub-layer decreases (Eqn 2.22), those grains which protrude through it will be subject to increased drag and will be set in motion. When $Re_* < 3.5$, the surface can be regarded as aerodynamically 'smooth' (Fig. 2.17a) and the drag is distributed evenly over the whole of the bed surface.

However, there is strong evidence that cohesive forces also play a significant role in determining the threshold for fine particles. The cohesive forces include Van der Waals forces, electrostatic charges, and surface tension imparted by moisture films (Corn 1966). According to Sagan & Bagnold (1975), the 'turn-up' of the threshold curve for small particle sizes in air can be better explained in terms of cohesion rather than Reynolds number effects. However, because cohesion loses part of its effectiveness in water, they suggest that the early experimental data of Hjulstrom (1935) are in error. The idea that cohesive forces control the threshold velocity of small grains has been elaborated by Iversen *et al.* (1976a) and Miller & Komar (1977). Data obtained from low-pressure wind tunnel experiments which allowed separation of Reynolds number and cohesion effects have convincingly demonstrated the greater importance of the latter (Iversen & White 1982).

It has also been shown in wind tunnel experiments using materials of different densities that threshold curves are displaced to the left for denser materials and to the right for lighter materials compared with the curves shown in Figure 4.2 (Iversen *et al.* 1976a, Iversen & White 1982).

In summary, the threshold friction velocity for quartz sand grains larger than 0.25 mm in air can be written as:

$$u_{*_t} = 146d^{1/2} \tag{4.13}$$

Equation 4.13 is not valid for $Re_* < 1$ owing to the interparticle cohesive forces which are important for grains smaller than 0.1 mm.

4.1.3 *Impact threshold*

Bagnold (1937a) first observed that, when sand grains are introduced into the airflow at the upwind end of a wind tunnel, they initiate movement of other grains on the bed which can be sustained at a wind velocity below the fluid threshold velocity. This lower threshold velocity was termed the *impact threshold velocity* by Bagnold (1941). The velocity difference between the fluid and impact thresholds (Fig. 4.4) is due to the kinetic energy of the grains in motion. As discussed later in this chapter, sand grains larger than 0.1 mm move chiefly by a series of jumps (saltation), in which the impact of one grain on another provides a source of energy additional to the wind drag.

The impact threshold for grains larger than 0.25 mm can be calculated using Equations 4.11 and 4.12 by substituting $A = 0.08$ (Bagnold 1941, p. 94). Chepil (1945b) found that for grains larger than 0.1 mm, the coefficient A for impact threshold has a value of 0.085. Grains smaller than 0.06 mm are transported mainly in suspension because the settling velocities of such grains are lower than the vertical velocity component of

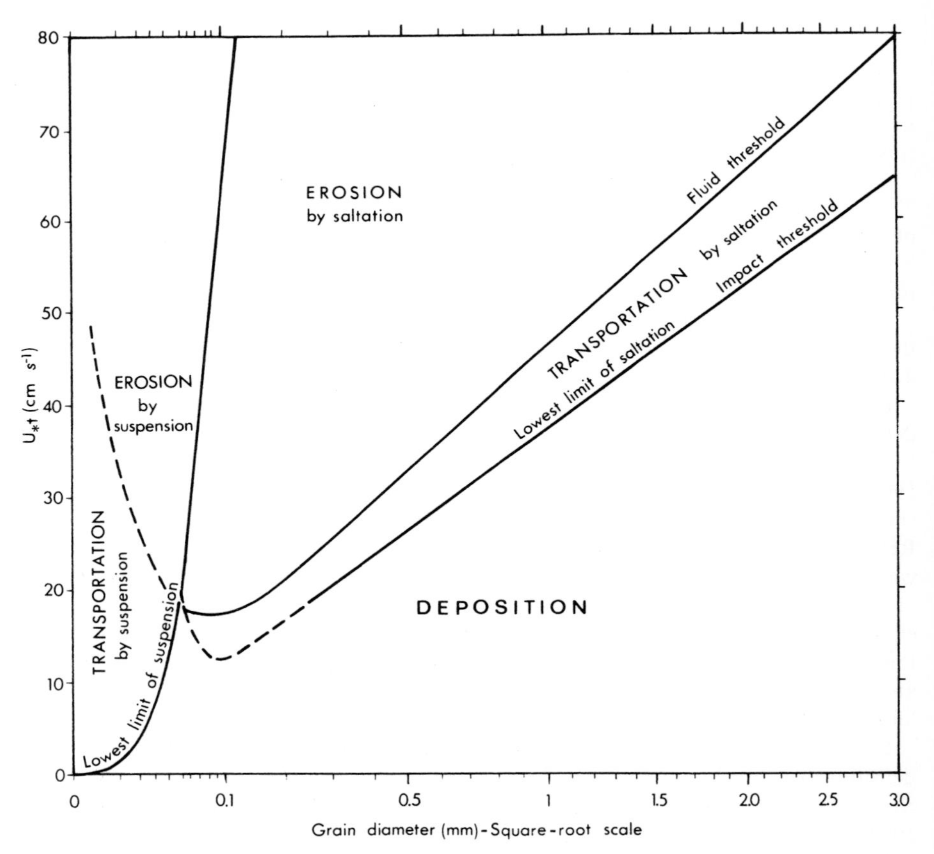

Figure 4.4 Variation of the fluid threshold velocity and the impact threshold velocity with grain size. The distinctions between the saltation and suspension modes of transport, and between erosion, transportation, and deposition, are also shown. (Data partly from Bagnold 1941, p. 88, and Chepil 1945b).

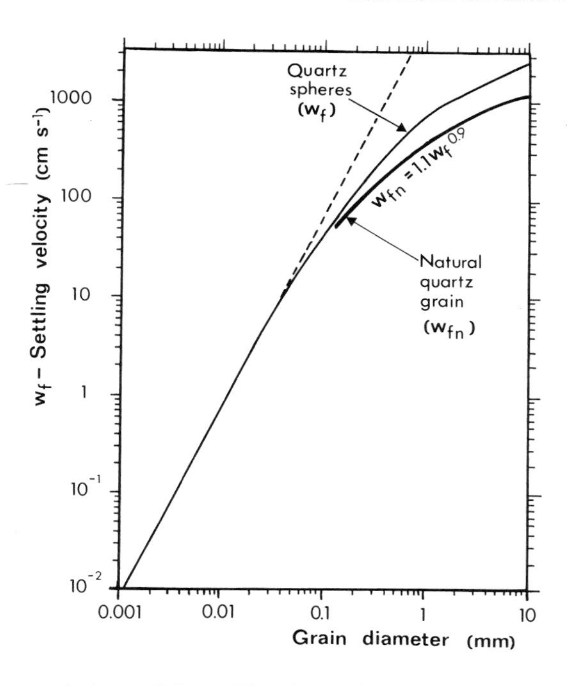

Figure 4.5 Calculated settling velocities (w_f) of quartz spheres (after von Engelhardt 1977) and measured settling velocities (w_{fn}) of natural quartz grains in air (after Cui *et al.* 1983). The dashed line shows the deviation from Stokes' law for particles larger than 0.04 mm.

turbulent airflow (Figs 4.4 and 4.5). Grains smaller than 0.06 mm do not produce significant impacts with other grains on the bed, and the fluid and impact threshold curves merge in the size range 0.06–0.1 mm. It should be noted, however, that entrainment of sediments composed of grains finer than 0.06 mm is often caused by ballistic impacts of saltating grains larger than 0.1 mm (Gillette *et al.* 1974).

Hjulstrom (1935, 1939) first used the threshold velocity curve to define domains dominated by erosion, transportation, and deposition by running water. In a similar way, the fluid threshold curve shown in Figure 4.4 can be regarded as the lowest erosion velocity curve, and the impact threshold curve can be regarded as the lowest transportation velocity curve.

4.1.4 Threshold velocities for poorly sorted sediments

Natural sediments and soils always contain a wide range of grain sizes. Hence, as noted above, there is no single fluid threshold velocity but rather a range of values related to the different size fractions in the mixture. Chepil (1945b) recognized maximal and minimal fluid threshold velocities necessary for entrainment of the largest and most erodible particles, respectively, in a soil.

Figure 4.6 compares the fluid threshold friction velocity curves for three mixtures, each containing a different range of particle sizes. The maximum equivalent diameter of the transported soil particles, plotted on the abscissa, is defined as $\gamma d/\rho_s$ (Chepil 1958a), where γ is the bulk density of the sand. Curve (a) in Figure 4.6 represents the case of a soil composed of erodible fractions with a limited range of sizes in which the ratio of the minimum to maximum equivalent diameter varies as $1{:}\sqrt{2}$. The value of the coefficient A (Eqn 4.11) for curve (a) for particles larger than 0.1 mm is about 0.1. Curve (b) represents the case of soil with a wider range of erodible particle sizes ranging from fine dust to a maximum equivalent diameter of 2 mm. The threshold velocity of this mixture is lower than that required to initiate movement of a soil containing only grains of the same maximum equivalent diameter because once the most erodible grains

(0.07–0.15 mm) begin to saltate they can initiate movement of larger grains as they impact on the bed. The appropriate value of A in case (b) is 0.085. Curve (c) represents a soil containing 15% non-erodible clods ranging up to 25 mm in diameter. In this case the threshold velocity required to move any given size is increased owing to the increased surface roughness caused by the non-erodible clods.

In nature, initial entrainment of sand grains results almost entirely from fluid drag and lift forces acting on a few of the smaller and more exposed grains. At wind velocities just above the fluid threshold, these grains dislodge other grains through impact, but the number of additional grains entrained in this way is limited because the grains which move first have relatively low momentum and lose further energy owing to friction with the bed. However, as the wind velocity increases further above the fluid threshold the grains gain greater momentum, which is transferred to other grains on the bed through impacts. Each impacting grain may then cause ejection of several further grains, producing a cascade effect (Iversen 1985, Nickling 1988; see Section 4.2.3). Wind tunnel studies have indicated that this transition from very limited movement to large-scale cascading saltation can take place over a very small wind velocity range (Nickling 1988).

4.1.5 *Effect of bedslope on threshold velocity*

The threshold of sand transport on dunes is influenced by the slope of the surface, being raised on positive gradients and lowered on negative gradients. Theoretical analyses of the effect of bedslope on threshold have been presented by several workers, but experimental verification is limited. Howard (1977) proposed the following relationship which predicts the threshold shear velocity on a sloping surface:

$$u_{*_t} = F^2 d\,[(\tan^2 \propto \cos^2 \theta - \sin^2 \chi \sin^2 \theta)^{1/2} - \cos\chi \sin \theta] \qquad (4.14)$$

Where $F = \beta(g\rho'_s/\rho)^{1/2}$, β is a dimensionless constant with a value of 0.31, d is the grain diameter, $\propto$ is the angle of internal friction of the sediment, θ is the bedslope angle, and χ is the angle between the local wind direction and the direction normal to the maximum bedslope. The threshold velocity, u_t, is similarly given by (Howard *et al.* 1978)

$$u_t = E(F/k)d^{1/2}\,[(\tan^2 \propto \cos^2 \theta - \sin^2 \chi \sin^2 \theta)^{1/2} - \cos \chi \sin\theta]^{1/2} \qquad (4.15)$$

Where E is a constant and K is the von Kármán constant (0.4).

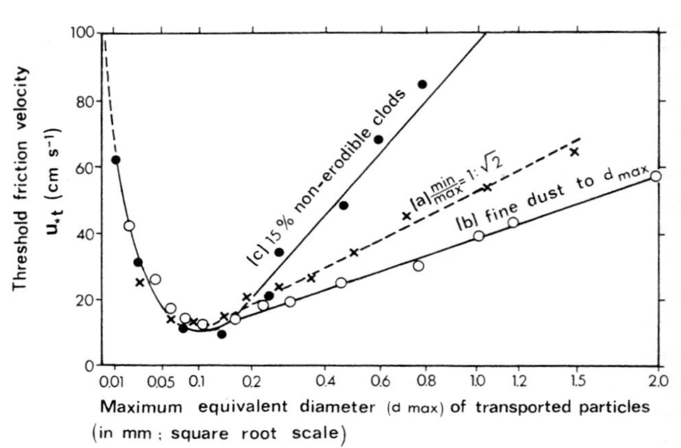

Figure 4.6 Relationship between the fluid threshold friction velocity and the maximum equivalent diameter of transported soil particles. For explanation, see text. (Adapted from Chepil 1958a).

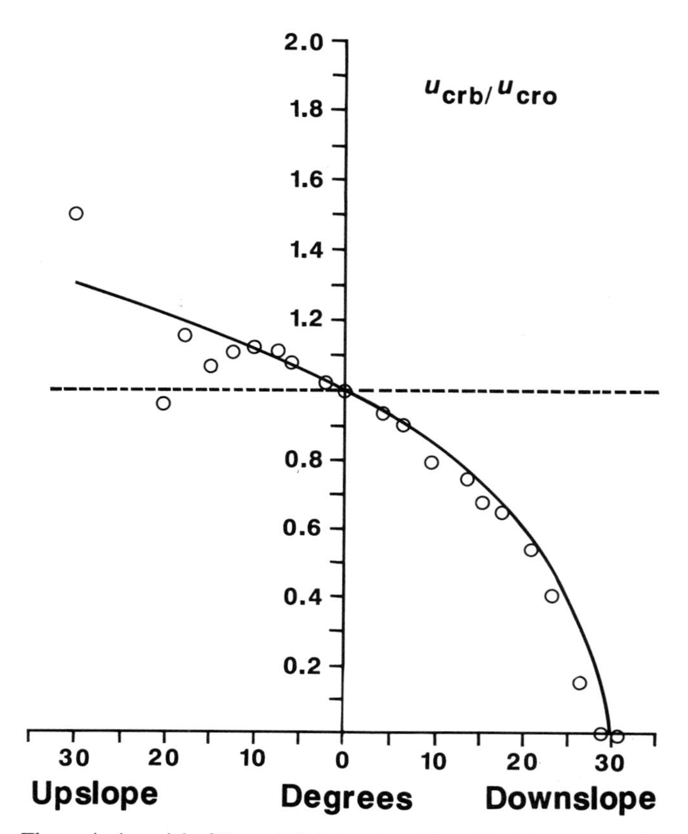

Figure 4.7 Theoretical model of Dyer (1986) for the effect of bedslope on threshold (solid line), compared with the experimental determinations on Saharan dunes (circles) made by Hardisty & Whitehouse (1988).

A simpler relationship for the effect of bedslope on threshold velocity, given by Dyer (1986), is

$$u_{*_t} = [(\tan \propto \; - \tan \theta / \tan \propto) \cos \theta]^{1/2} \tag{4.16}$$

This equation predicts that the threshold inceases slightly with increasing slope angle for positive gradients, and that the effect of bedslope angle on the threshold is much more pronounced for negative gradients. Experiments carried out on Saharan dunes by Hardisty and Whitehouse (1988), using a simple wind tunnel, showed close agreement with the relationship predicted by Eqn 4.16 (Fig. 4.7).

4.1.6 Effect of moisture content and cementing agents on threshold velocity

As discussed above, cohesive forces are significant for grains smaller than 0.1 mm, but larger grains are considered to be cohesionless unless affected by moisture derived from precipitation, groundwater, or tides. After wetting, moisture is retained by sand as a surface film, particularly at points of grain contact. The cohesion thus produced is the result of the tensile force between the water molecules and the sand grains (Chepil 1956, Bisal & Hsieh 1966).

Belly (1964) and Johnson (1965) demonstrated that for moisture contents of 0.05–4%, the relationship between fluid threshold velocity and moisture content is logarithmic, of the form

$$u_{*_{tw}} = u_{*_t}(1.8 + 0.6 \log W) \tag{4.17}$$

where W is the moisture content (%) and $u_{*_{tw}}$ is the fluid threshold velocity for wet sand. The wind tunnel used by Belly was unable to generate wind shears high enough to mobilize sand when the moisture content exceeded 4%. A limiting moisture content of approximately 4% was also found in studies by Azizov (1977) and Logie (1982).

'Dry' sand normally contains 0.2–0.6% moisture due to atmospheric humidity (Belly 1964, Hotta *et al.* 1985, Tsoar & Zohar 1985). Belly found that humidity has only a small effect on threshold velocity, although Knotternus (1980) concluded that it is significant, particularly if a small amount of organic material is present.

Measurements on a natural beach in The Netherlands by Svasek & Terwindt (1974) indicated fluid threshold velocities higher than those found by Belly (1964) for a given moisture content, although their data show a large amount of scatter. These authors observed that, once sand movement starts in one location, the impacts of incoming grains may overcome the effects of moisture tension in areas downwind which may have a higher moisture content. Further, there are difficulties in measuring the instantaneous moisture content of a thin surface layer of wet sand.

Experimental and field investigations in Japan (Horikawa *et al.* 1982, Hotta *et al.* 1985) suggested a linear, rather than a logarithmic, relationship between fluid threshold velocity and moisture content which is valid for grains in the size range 0.2–0.8 mm:

$$u_{*_{tw}} = u_{*_t} + 7.5W \tag{4.18}$$

The relationship developed by Hotta *et al.* (1985) is compared with that of Belly (1964) in Figure 4.8.

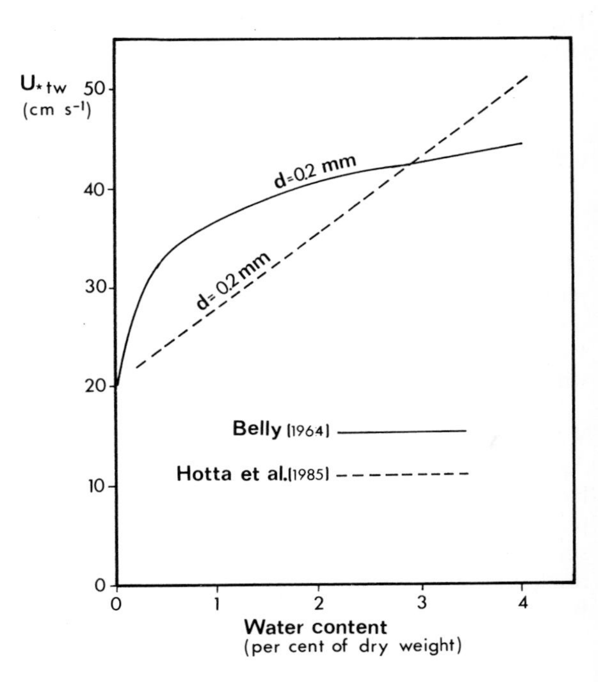

Figure 4.8 Variation of the threshold friction velocity of wet sand (u_{*tw}) with moisture content (W) for 0.2 mm diameter sand, as determined by Belly (1964) and Hotta *et al.* (1985).

Figure 4.9 A piece of salt-cemented crust (salcrete) formed by evaporation of salt spray on the upper part of a beach in northern California.

When evaporation rates are high the sand surface will dry out rapidly, lowering the threshold velocity and Equation 4.18 has to be modified accordingly (Hotta *et al.* 1985):

$$u_{*_{tw}} = u_{*_t} + 7.5W\,I_w \tag{4.19}$$

where I_w is an appropriate function of the evaporation rate and takes a value ranging from 0 to 1.

The presence of salts in relatively low concentrations can significantly raise the fluid threshold velocity for dry sand by acting as cement at points of grain contact (Pye 1980a, Nickling & Ecclestone 1981, Nickling 1984) (Fig 4.9). In wind tunnel studies using different concentrations of NaCl and KCl, Nickling & Ecclestone (1981) found a relationship between salt content and u_{*_t} which can be incorporated in Eqn 4.11 to give the modified relationship

$$u_{*_t} = A(0.97 \exp 0.1031S)\,[(\rho'_s/\rho)gd]^{1/2} \tag{4.20}$$

where s is the salt content in mg per gram of soil.

In addition to cementation by salts, sand grains may also be bound together by clay skins, fungal hyphae, algae, and lichens (Chen *et al.* 1980, Gillette *et al.* 1980, 1982, van den Ancker *et al.* 1985).

4.1.7 Effects of non-erodible roughness elements and vegetation on particle entrainment

Several studies have shown that the fluid threshold is affected by the presence of non-erodible roughness elements such as pebbles and crop stubble (Bagnold 1941, p. 173, Chepil 1950, Chepil & Woodruff 1963, Bisal & Ferguson 1970, Lyles & Allison 1976, Lyles 1977). As the roughness increases, so does the surface drag exerted on the wind. Experiments by Logie (1982) demonstrated that low densities of roughness elements (pebbles and glass spheres) actually lower the threshold velocity by promoting local flow acceleration and scouring, whereas high densities of roughness elements raise the threshold. For any size of roughness element there is a critical cover density, referred to as the *inversion point*, where the effect changes from accelerated erosion to protection. The value of the inversion point is also affected by the shape of the roughness elements (Logie 1982).

Buckley (1987) used the results of wind tunnel experiments to develop the following equation which defines the influence of vegetation cover on threshold velocity:

$$\bar{u}_{vt} = \bar{u}_t/(1 - kC) \quad (4.21)$$

where $\bar{u}_{vt}$ is the threshold velocity on vegetated loose sand measured at height z, $\bar{u}_t$ is the threshold velocity on bare loose sand, k is a constant dependent on plant shape (for small erect or spreading herbaceous dune plants $K = 0.018$, and for small rounded stemless plants $K = 0.046$), and C is the percentage plant cover (up to 17%).

It is often difficult to apply Equation 4.21 to field situations where the distribution of plants is uneven and where individual plants may cause local acceleration and deceleration of the wind. However, in general, no sand movement takes place when the vegetation cover exceeds 30% (Ash & Wasson 1983). The effects of vegetation in changing the wind velocity profile and reducing the friction velocity have been demonstrated by Olson (1958a), Chepil & Woodruff (1963) and Bressolier & Thomas (1977).

4.2 Transport of particles by the wind

4.2.1 Aeolian transport modes

Different modes of particle transport are defined by wind velocity and grain size. Grains that move very close to the bed are known as *bedload*. This mode consists of *saltation*, in which grains move forward by a series of jumps, and *surface traction*, in which grains roll or slide along the surface, due either to direct fluid drag or the impact of saltating grains. The surface traction load is also sometimes referred to as the *contact load*, since the grains do not lose contact with the surface. Other terms used to describe the forward movement of grains which do not lose contact with the bed, or do so only for very short periods, include *surface creep* (e.g. Greeley & Iversen 1985, p. 293) and *reptation*, a term derived from the Latin *reptare* = to crawl (Haff, cited by Anderson 1987b).

A second major transport mode is *suspension*, in which particles are lifted from the surface and carried large distances by the flow without regaining contact with the bed. Turbulent airflow is able to keep a grain in suspension when the vertical fluctuating velocity component of the flow (w', Eqn 2.11) exceeds the settling velocity of the grain (w_f). The calculated settling velocities of different sizes of quartz spheres, together with the measured settling velocities of natural quartz grains in air, are shown in Figure 4.5. In a neutrally stratified atmosphere, in which buoyancy effects due to thermal differences are unimportant (Sutton 1934, von Kármán 1937), the distribution of the vertical fluctuating velocity components near the ground is normally distributed, with a mean of zero and a standard deviation, $\sqrt{\overline{w'}^2}$, that is equal to Au_*, where A is a constant. The average value of A falls within the range 0.7–1.4 with an approximate mean value of 1.0; therefore, $\sqrt{\overline{w'}^2}/u_* \approx 1.0$ (Lumley & Panofsky 1964, p. 134, Bagnold 1973, Pasquill 1974, p. 77). The ratio w_f/u_* provides a measure of a grain's susceptibility to be carried in suspension (Francis 1973, Tsoar & Pye 1987, Fig. 4.10).

The demarcation line of $w_f/u_* = 1$ is arbitrary. There is no sharp division between bedload and the suspended load but rather a gradual transition (Nickling 1983). Pure bedload transport occurs when the vertical velocity components associated with turbulence have no significant effect on particle trajectories. This happens when $w_f/u_* >> 1$. Pure suspension occurs when the particle settling velocity is very small relative to the friction velocity ($w_f/u_* << 1$). Where w_f/u_* assumes a value close to 1, particles move in modified saltation (Hunt & Nalpanis 1985, Nalpanis 1985), in which they display random trajectories through the flow transitional between saltation and suspension. The arbitrary boundary between pure and modified saltation, as determined theoretically by Nalpanis (1985), is shown in Figure 4.10. It corresponds approximately to the value of $w_f/u_* = 1.25$, which was regarded by Bagnold (1973) as the point at which a solid particle becomes liable to suspension. The upper limit of pure suspension can be taken as $w_f/u_* \approx 0.7$ (Gillette *et al.* 1974).

The sedimentological significance of the distinction between bedload and suspended

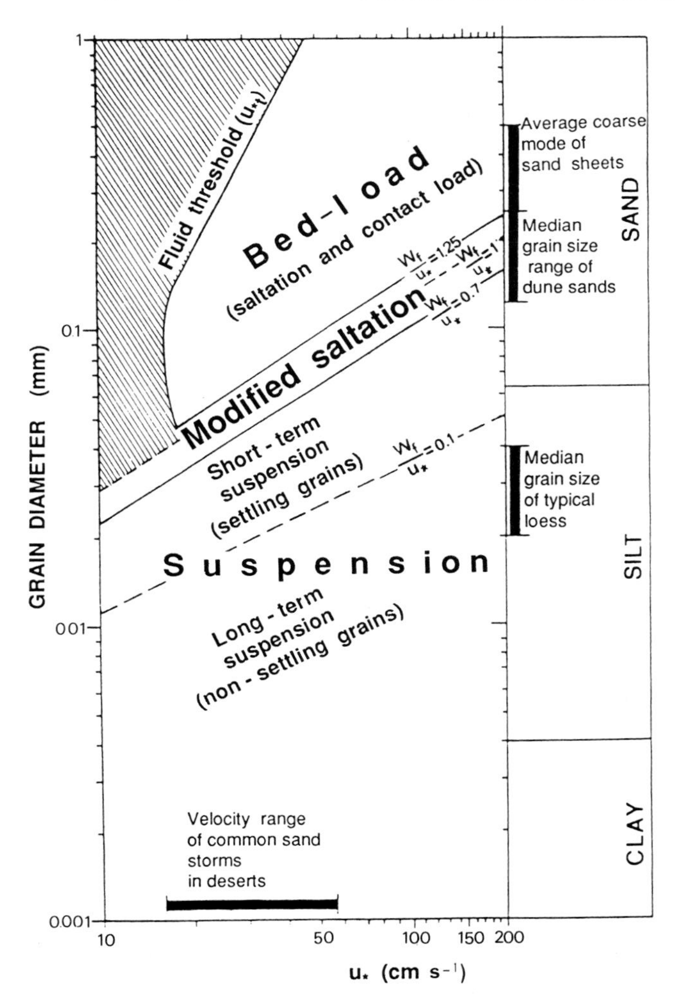

Figure 4.10 Modes of transport of quartz spheres at different wind shear velocities. (After Tsoar & Pye 1987).

load lies in the distance particles are carried by the wind. Grains 0.1–0.3 mm in diameter, which saltate most easily during typical wind storms, form sand dunes whereas particles smaller than 0.1 mm, which are transported in suspension, are carried larger distances and are ultimately deposited as loess (Fig. 4.10), and grains larger than 0.3 mm move mainly by rolling and tend to become concentrated in residual sand sheets. Wind action is relatively effective in separating coarse sand, medium – fine sand and silt fractions, although aeolian sediments consisting of mixtures of these sizes are found in some transitional environments, as discussed in Chapter 3. Perfect separation of different sizes of particles does not occur since aeolian sediment transport is a stochastic process in which the trajectories of individual grains are affected to varying degrees by random turbulent fluctuations of the wind, and also by considerable natural variability in the nature of grain – bed collisions (Ungar & Haff 1987, Anderson 1987a).

4.2.2 Suspension

As noted above, suspension transport occurs when the vertical velocity fluctuations associated with air turbulence are greater than the grain's settling velocity. The settling velocities of quartz spheres in the size range 0.001–0.05 mm can be calculated according to Stokes' Law (Green & Lane 1964, p. 67):

$$w_f = Kd^2 \tag{4.22}$$

where d is the diameter and K is given by $\rho_s g/18\mu$ (where ρ_s is the grain density, g is the acceleration due to gravity and μ is the dynamic viscosity of air). For quartz spheres, K is taken to be $8.1 \times 10^5\,cm^{-1}\,s^{-1}$ in air at sea level.

During typical sandstorms, when values of u_* range between 0.18 and $0.6\,m\,s^{-1}$, the corresponding maximum size of particles which can be transported in suspension ranges from 0.04 to 0.06 mm diameter (Fig. 4.10).

In order to remain suspended in the atmosphere for a considerable period of time, particles must experience a high ratio of upward to downward movements. When $w_f/u_* = 0.4$, the ratio is 0.5 (Gillette 1979, 1981), and there is a low probability that grains will remain suspended for a long period. For long-term suspension a w_f/u_* ratio of <0.1 is required, which corresponds to a maximum particle size of 0.015–0.02 mm during typical windstorms (Fig. 4.10) (Gillette 1979, 1981). Particles smaller than this size are referred to as *non-settling grains* and larger silt grains as *settling grains* (Tsoar & Pye 1987). Typical loess deposits are composed mainly of settling grains which are transported in short-term suspension, whereas the fine dust carried large distances and deposited in the oceans is composed mainly of non-settling grains transported in long-term suspension (Tsoar & Pye 1987, Pye 1987).

The preceding discussion provides only a broad outline of grain transport in suspension, since this topic essentially lies outside the scope of this book. More detailed models of suspension transport are discussed by Anderson & Hallet (1986) and Anderson (1987b).

4.2.3 Saltation

The term saltation (from the Latin *saltare* = to leap) was introduced for the first time by McGee (1908) to describe the jumping movement of grains transported along the bed by running water. Joly (1904) and Owens (1927) were among the first to describe this phenomenon during wind transport. The characteristic ballistic trajectory of saltating grains in air (Fig. 4.11) was demonstrated photographically by Bagnold (1936) and subsequently by Chepil (1945a) and Zingg (1953b).

The nature of saltation has been extensively investigated using a range of wind tunnel,

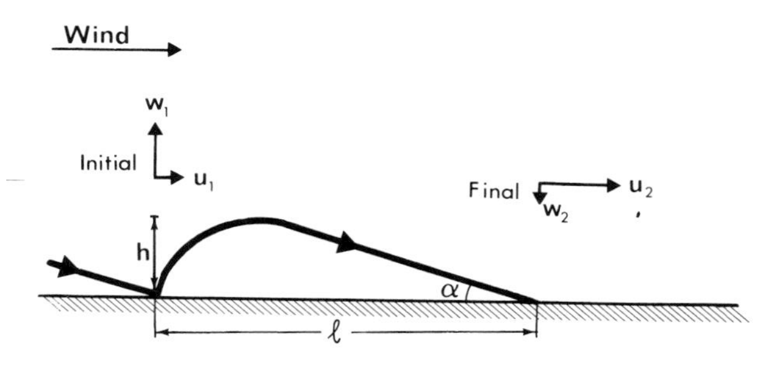

Figure 4.11 Characteristic path of a saltating grain; h and l are the maximum jump height and jump lengths, respectively, α is the impact angle, w_1 and w_2 are the initial and final upwind velocities, and u_1 and u_2 are the initial and final horizontal velocities of the grain. (After Bagnold 1937b).

numerical modelling, and field approaches (Bagnold 1936, 1941, Owen 1964, 1980, Tsuchiya 1970, White & Schulz 1977, Gerety & Slingerland 1983, Jensen *et al.* 1984, Horikawa *et al.* 1984, Rumpel 1985, Anderson & Hallet 1986, Willetts & Rice 1986a,b, 1988, 1989, Jensen & Sorensen 1986, Ungar & Haff 1987, Werner & Haff, 1988, Anderson & Haff 1988). These studies have shown that the nature and rate of saltation are influenced by several factors, including grain size and shape, the wind velocity (represented by u_*), and the nature of the surface over which saltation takes place.

Owing to the large difference in density between air and transported quartz grains ($\rho_s/\rho = 2150$), the settling velocities of quartz grains in air are 60–80 times higher than in water. Grains transported in air require a fluid velocity 29 times greater than that required to transport the same size of grains in water. The impact force of grains in air is also much greater than that of grains transported in water (Iversen *et al.* 1987). Consequently, in water, few grains are dislodged from the bed by the impact of saltating grains, whereas this is the principal mechanism responsible for grain movement during aeolian sand transport. The effectiveness of saltation in air reflects the operation of a positive feedback process in which a few grains which are initially entrained by aerodynamic drag give rise to a chain reaction in which each impacting grain causes the ejection of several other grains from the bed. However, the presence of saltating grains has the effect of modifying the wind velocity profile near the bed, thereby acting as a self-regulating mechanism.

The trajectory of a saltating grain in air depends on whether the grain is entrained by direct fluid lift/drag or by the impact of other saltating grains, or whether the grain is already saltating and rebounds off the bed in a series of jumps known as *successive saltation* (Tsuchiya 1970, Rumpel 1985). Bagnold (1935b, 1936), Chepil (1945a), and Bisal & Nielsen (1962) observed that many grains rise almost vertically from the bed, but more recent studies have reported mean ascent angles of between 34° and 50° with respect to the horizontal (Tsuchiya 1970, White & Schulz 1977, Nalpanis 1985). Grains which rebound from the surface during saltation have a lower mean angle of ascent (21–33°, the higher values being for coarser grains) than grains which are ejected from the bed by 'splashing' during impact (52–54°; Willetts & Rice 1985a). This may be partly explained by the significant forward momentum possessed by rebounding grains at the time of impact, and partly by the fact that ejected grains have a lower ascent velocity and therefore spend longer near the bed where the lift force is most significant (Anderson & Hallet 1986). On the other hand, the horizontal drag force increases with height (Chepil 1961). Figure 4.12 shows, two-dimensionally, the pattern of pressure distribution and the resultant force exerted on the surface of an 8 mm diameter sphere subject to a friction velocity of 98 cm s^{-1}. It shows that the lift force vanishes at a height of about 2.5 cm while the drag force continues to increase with height in proportion to the increase in wind velocity.

Additional lift force may be generated if the grains spin while moving along their

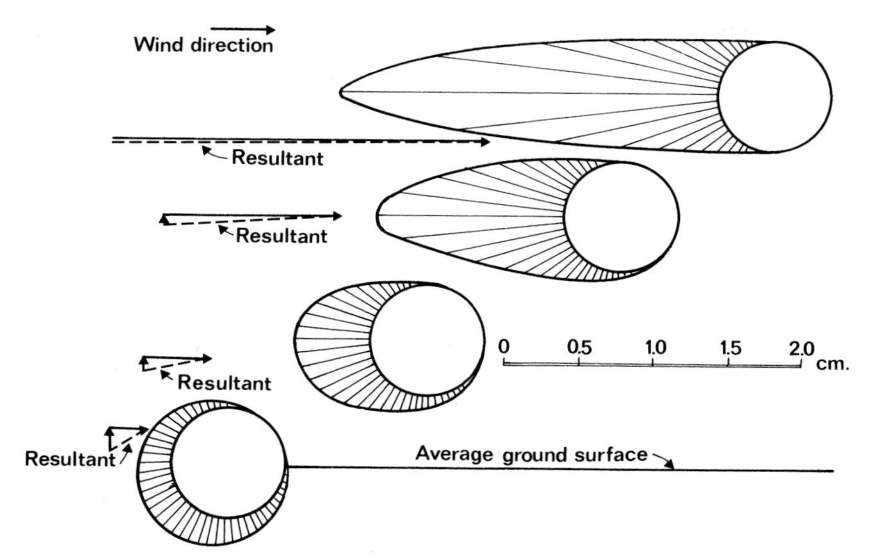

Figure 4.12 Pattern of pressure around a sphere, 8 mm in diameter, subjected to a friction velocity of 98 cm s^{-1} at various heights above the ground. Also shown is the resultant force acting on the grain at different heights. (After Chepil 1961).

trajectory. Bagnold (1936) and Bisal & Nielsen (1962) reported rotation of only a few grains, but Chepil (1945a) observed that about 75% of grains spin at a rate of 200–1000 r.p.s. Particle spin was also emphasized by White & Schulz (1977), who estimated the spinning rate to be 115–500 r.p.s. White (1982) reported a mean particle spin rate of 350–400 r.p.s.

As a grain spins, the air streamlines around the grain become asymmetric. On the underside of the grain (point Q in Figure 4.13), the grain and the air adjacent to it move against the wind direction. On the upper side (point P in Figure 4.13) they move in the same direction as the wind flow. In accordance with the Bernoulli equation (Eqn 2.26), this causes differences in pressure between P and Q (and also between R and S). Lift force is induced perpendicular to the flow direction and to the axis of grain rotation. This type of lift, known as the *Magnus effect*, is significant for sand grains larger than 0.1 mm in diameter (White & Schulz 1977, White 1985).

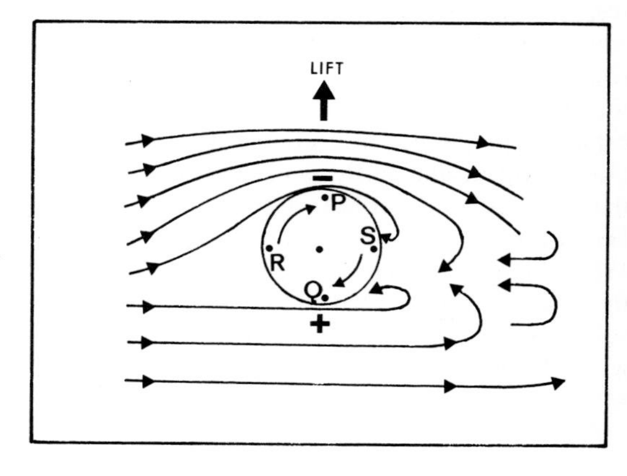

Figure 4.13 Effect of grain spin in producing additional lift (Magnus effect). In addition to streamline asymmetry and the resulting pressure difference on the front and back of the grain (cf. Fig. 2.21), there is also asymmetry above and below the grain. Velocity is high near P, where the airflow is reinforced by the spinning motion of the grain, and low near Q, where the grain's rotation opposes the airflow.

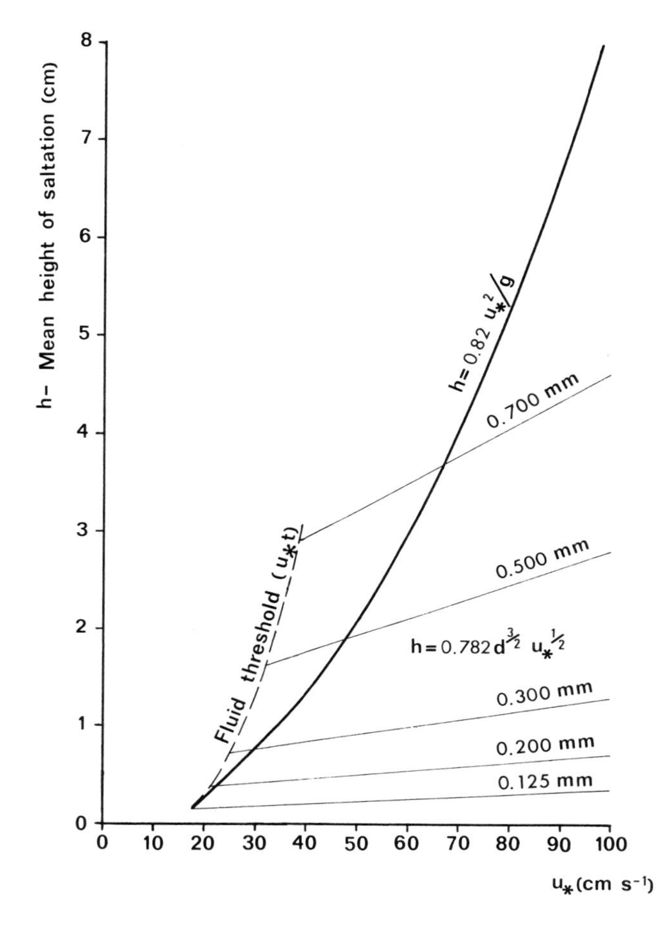

Figure 4.14 Mean ascent height of saltating sand as a function of friction velocity (bold line), according to Owen's (1980) theoretical analysis, and as a function of grain size and friction velocity for uniform sand (thin lines), according to the empirical analysis of Zingg (1953a).

A rebounding grain starts to spin after it strikes the surface a glancing blow. The spinning rate is greatest immediately after impact and decreases as the grain progresses along its trajectory. A comparison of observed and calculated theoretical trajectories showed good agreement when allowance is made for a Magnus effect equivalent to spin rates of 100–500 r.p.s. (White & Schulz 1977, White 1982). However, these results have been questioned by Jensen & Sørensen (1983) and Hunt and Nalpanis (1985).

The two forces oppose the lift forces exerted on the grains: the drag force (Eqn. 2.29) and the downward acting force due to the grain's weight (Eqn 4.3). Because air has a much lower density than water, the weight force is much more significant in air. In the case of a 1 mm diameter spherical grain, lifted with a vertical velocity (w_1) of $10\,\mathrm{cm\,s^{-1}}$, the weight force will be about 1300 times greater than the drag force (Middleton & Southard 1978). In the absence of drag and lift forces, a rebounding grain would rise to a height of $w_1^2/2g$ if all the kinetic energy could be converted into potential energy. However, the maximum ascent during saltation will be less than $w_1^2/2g$, depending on grain size, the forward velocity of the grain, and the angle of ascent (Anderson & Hallet 1986). The smaller the grain, the greater is the effect of the drag force, the lower is the rate of ascent, and the greater is the horizontal acceleration. When $w_1 = 70\,\mathrm{cm\,s^{-1}}$ and $u_* = 20\,\mathrm{cm\,s^{-1}}$, a grain of 0.2 mm diameter attains a maximum ascent rate of $0.7w_1^2/2g$, but a grain of 0.1 mm diameter ascends at a maximum rate of $0.53w_1^2/2g$ (Nalpanis

1985). The relationship between the mean ascent height (h), friction velocity (u_*), and grain size (d), for several well sorted sand size ranges, was found by Zingg (1953a) to be

$$h = 0.782 d^{3/2} u_*^{1/2} \tag{4.23}$$

in which h is expressed in cm, d in mm and u_* in cm s^{-1} (Fig. 4.14).

Observations during sandstorms of moderate intensity also confirmed that large grains bounce higher than small grains (Chepil & Milne 1939, Bagnold 1960, Sharp 1964).

Wind tunnel studies have indicated a general decrease in mean size of saltating grains with height above the surface, at least up to a height of 2 cm (Gerety & Slingerland 1983, Williams 1964). The size of grains at any given height becomes larger with increasing u_*, since there is a tendency for coarse grains to bounce higher under strong winds. However, most of the grains near the top of the saltation layer are small, probably because small grains are lifted to a greater height by turbulent eddies (i.e. these grains are transported in modified saltation). In field studies, De Ploey (1980) and Draga (1983) observed a decrease in the size of saltating grains with height, even though large numbers of coarse grains were found to bounce more than 50 cm above the surface.

The saltation layer does not have a clearly defined upper limit, but its maximum height

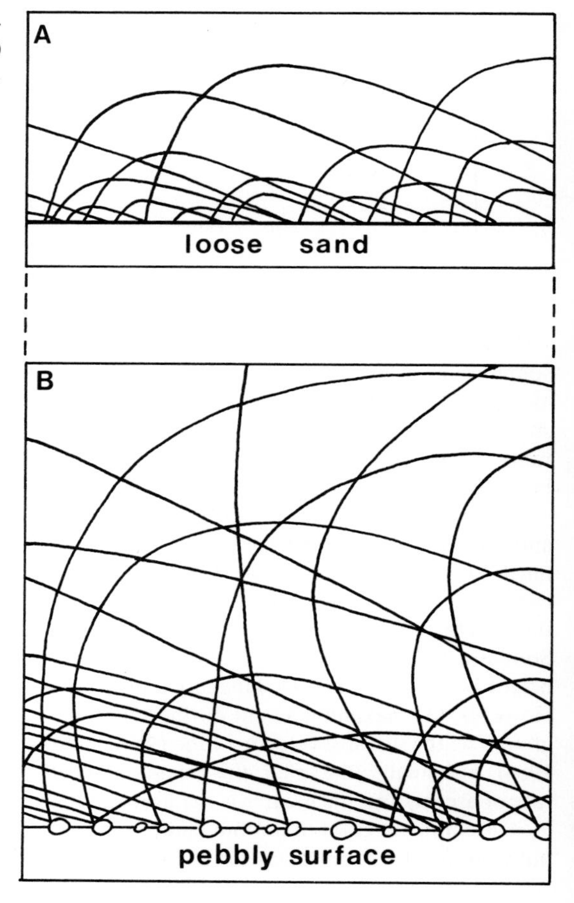

Figure 4.15 Schematic representation of saltation trajectories over (a) a loose sand and (b) a pebbly surface. (After Bagnold 1941, p. 36).

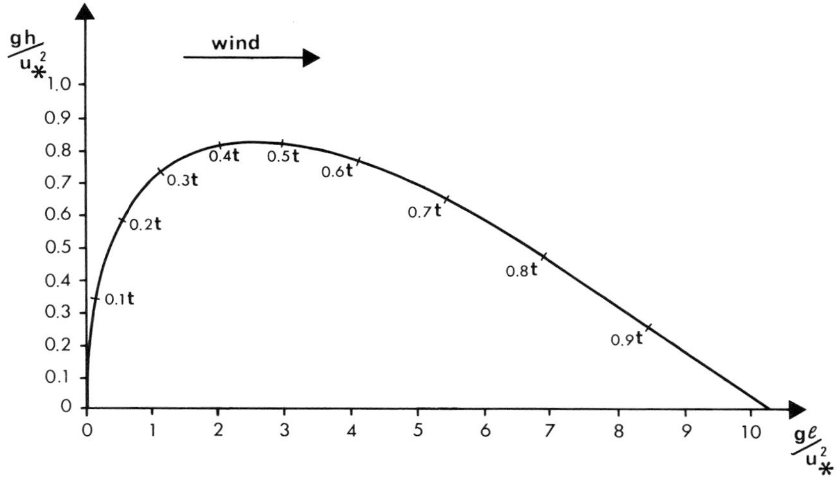

Figure 4.16 Depiction of the mean saltation trajectory after Owen (1980); h is the maximum jump height, l is the horizontal distance between the lift-off point and the impact point, and t is the time taken to complete the trajectory. The coordinates are non-dimensional and the ordinate scale is exaggerated by a factor of five.

is approximately ten times the mean saltation height (which is generally below 1 cm in wind tunnel studies). Field measurements of loam soil particles in saltation, with wind velocities of 7–10 m s^{-1} measured at a height of 30 cm, indicated a mean saltation height of about 5 cm (Chepil & Milne 1939). On hard desert surfaces such as rock pavements and gravel fans, grains of any given size bounce higher than on loose sand (Bagnold 1941, p. 36, Sharp 1964) (Fig. 4.15). The maximum saltation height on such hard surfaces may exceed 3 m, with a mean saltation height of more than 20 cm.

The distribution of grain flux with height during saltation is highly skewed, with the highest concentration of particles being found close to the bed.

According to theoretical calculations by Owen (1980), the mean saltation height is given by $0.82u_*^2/g$ and the mean saltation path length by $10.3u_*^2/g$ (Fig. 4.16). The mean horizontal distance travelled during saltation was predicted by Owen to be approximately 12 times the mean maximum jump height. This ratio is higher than the values measured by Bagnold (1936) and Chepil (1945a) and those computed by Tsuchiya (1970), but lower than that predicted by White & Schulz (1977). The height and length of the saltation trajectory are dependent on grain shape and also on grain size and friction velocity. Both Williams (1964) and Willetts (1983) observed that grains of low sphericity move in flatter, longer trajectories than spherical grains. The flight time of particles from take-off to touch-down is of the order of 0.1–0.2 s (Anderson & Hallet 1986). It takes about half this time for the particle to reach the vertex of its trajectory.

The initial ascent velocity (w_1) of the characteristic grain is proportional to friction velocity since the characteristic grain is considered to be ejected from the surface by the mean force of an impacting grain whose final velocity is controlled by u_* (Bagnold 1936, 1937b, Owen 1964). Thus $w_1 = Bu_*$, where B, the *impact coefficient*, was found to be 0.8 for a grain with a typical size of 0.25 mm (Bagnold 1936). More recent studies have suggested higher values of about 2 for the impact coefficient (White & Schulz 1977, Nalpanis 1985).

After reaching its vertex, a grain starts to fall at an accelerating velocity. Computations by Hunt & Nalpanis (1985) show that at the vertex of the trajectory sand particles (ca 0.2 mm diameter) have horizontal velocities which are about half the mean wind velocity at the same height. The drag force of the wind induces horizontal acceleration as the grain descends. Since the average drag force is greater than the force of gravity, the grain hits the surface at an angle of 6–20° (average 14°). This angle decreases as the grain size decreases (Bagnold 1936, Chepil 1961, Sharp 1963, White & Schulz 1977, Nalpanis 1985, Willetts & Rice 1985a). As the impact angle decreases, less energy is expended during collision with the bed, and the impacting grain will ricochet with a high proportion of its energy retained. The mean ascent velocity of rebounding fine dune sand (0.15–0.25 mm) was found by Willetts & Rice (1985a) to be 240 cm s^{-1} (about 3–5 times their lift-off velocities), while ejected grains of the same size attained a mean ascent velocity of only 31 cm s^{-1}.

Because they are accelerated by the wind, grains land with more energy than they had when they left the surface. Particles which leave the surface with a wide range of vertical velocities impact on the surface with 3–5 times their initial velocities or 10–20 times their initial kinetic energy (Anderson & Hallet 1986). Part of this energy is dissipated through inelastic deformation and frictional rotation of grains on the bed. During such impacts several new grains may be ejected into the flow. Sometimes the impacting grain buries itself in the bed, but more frequently it bounces off the surface with varying degrees of energy loss.

Mitha *et al.* (1986) observed that, when rebound occurs, several other grains are usually ejected with low energy from a region approximately ten grain diameters across, centred just ahead of the impact site. Up to ten grains may be ejected by each impact, each with a mean ejection speed of less than 10% of the impact speed (Willetts & Rice 1986a).

As discussed in Section 4.2.4, the saltating grains extract momentum from the wind, thereby modifying the wind velocity profile near the bed and reducing the friction velocity below the fluid threshold. Sand transport is then maintained wholly by the impact of saltating grains (Bagnold 1973, Anderson 1987b, Ungar & Haff 1987, Werner 1988). Eventually a condition of equilibrium, known as *steady-state saltation*, is attained between the near-surface wind velocities and the grains in saltation.

Individual grains do not move in continuous saltation, even when there is no change in mean wind strength. After rebounding several times, a saltating grain may come to rest before being mobilized once again by the impact of another saltating grain. Wind tunnel experiments have suggested that the mean forward velocity of movement decreases proportionately with increasing grain size.

In laboratory wind tunnel experiments, Barndorff-Nielsen *et al.* (1985b) found that sand grains in the 0.28–0.48 mm size range had a mean forward velocity of 0.6–1.8 cm s^{-1}, including rest periods but excluding time spent buried under ripples. The mean forward velocity increases, although not to a great extent, with increasing friction velocity (Fig. 4.17) (White & Schulz 1977, Gerety 1984, Barndorff-Nielsen *et al.* 1985b, Willetts & Rice 1985a).

4.2.4 Wind velocity profile during saltation

As sand grains accelerate from rest they extract momentum from the wind, thereby changing the velocity profile near the ground. Wind profile measurements by Bagnold (1936) over a wet surface of uniform sand, and repeated when the sand dried out and became mobile, are shown in Figure 4.18. Whereas the wind velocity gradients over the stable sand surface converge at a point (z_0) just above the ground surface, those measured during active sand movement converge at a higher focal point, z', 0.2–0.4 cm above the surface. This effect of moving sand on the near-bed wind velocity profile has been confirmed by many later workers (Chepil & Milne 1941, Chepil 1945b, Horikawa &

Shen 1960, Belly 1964, Hsu 1973, 1974, Vugts & Cannemeijer 1981a,b, Ungar & Haff, 1987).

The wind profiles measured by Bagnold (Fig. 4.18) showed marked kinks up to a height of 3 cm, which reflect deviations from the logarithmic velocity profile law (Eqn 2.20). The width and height of the kink increase with friction velocity (see also Zingg 1953a, Gerety & Slingerland 1983). Bagnold (1941, p. 63) believed that these kinks correspond to the mean saltation height of uniformly sized grains. For sands with non-uniform grain size, the kink in the velocity curves is not clearly seen since different sizes have different characteristic trajectories. Some parts of the observed wind velocity profiles show negative deviations from the logarithmic velocity profile (Bagnold 1936, Horikawa & Shen 1960, Kawamura 1964) while others show positive deviations (Belly 1964, Walker 1981, Gerety & Slingerland 1983, Ungar & Haff 1987). Bagnold's (1936) data show a negative kink (i.e. lower than expected velocities) at a height of 1 cm and a postive deviation (i.e. higher velocities) at a height of less than 0.5 cm. Bagnold hypothesized that the negative deviations correspond approximately to the saltation trajectory height, whereas the positive deviations reflect speeding up of the wind by the accelerating grains just before they hit the bed.

By disregarding the kinks and fitting straight lines to the velocity profile data through

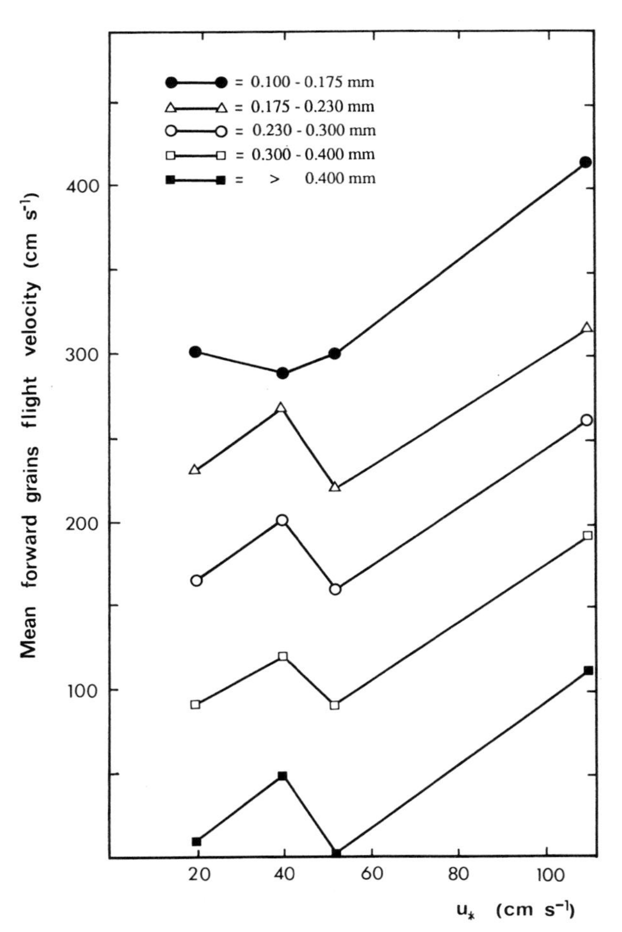

Figure 4.17 Variation in estimated mean forward grain velocities with u_* for different sizes of quartz grains. (Data of Gerety 1984).

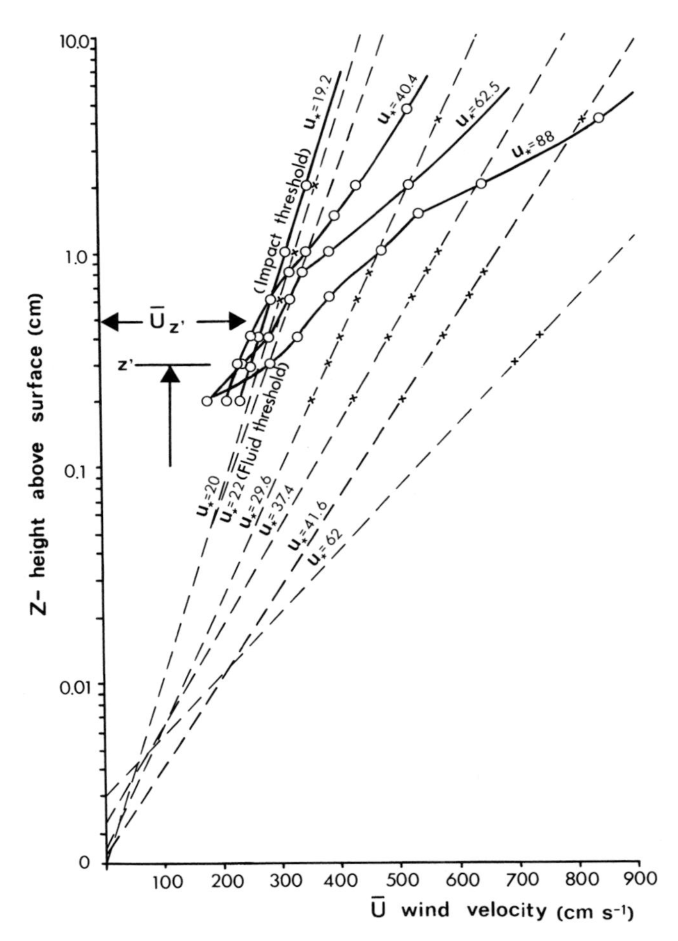

Figure 4.18 Wind velocity profiles measured by Bagnold (1936) over a bed of 0.25 mm uniform sand with (solid lines) and without (dashed lines) sand movement.

the focal point, Bagnold (1936) was able to rewrite the logarithmic wind velocity equation for flow over stable surfaces (Eqn 2.20) in a form applicable to sand surfaces on which saltation is taking place:

$$\bar{U} = (1/k)u_*\ln (z/z') + \bar{u}_{z'} \qquad (4.24)$$

where z' is the average height of the focal point and $\bar{u}_z{}'$ is the mean wind velocity at height z' and is also the threshold velocity at that height. These two variables were found to be unchanged for a given grain size regardless of changes in the wind velocity.

Zingg (1953a) found that, except for the largest sand size used, the focal point is related to the grain diameter (d) by

$$z' = 10d \qquad (4.25)$$

and

$$\bar{u}_{z'} = 8889d \qquad (4.26)$$

Like Bagnold, he found that wind velocity profiles over moving sand surfaces do not form a straight line when plotted with a logarithmic height scale.

A comparison of measured drag, obtained from the determination of pressure drop in the wind tunnel, with calculated drag estimated from the slope of the straight velocity profile lines (Fig. 4.18, Eqn 4.25), showed close agreement (Bagnold 1936).

The wind velocity below z' actually becomes smaller as u_* increases (Bagnold 1941, p. 60). This was interpreted by Chepil & Woodruff (1963) as being due to a higher concentration of saltating grains at higher values of u_*. This explanation agrees with Bagnold's (1941, p. 32, 1973) argument that the total applied shear stress is transmitted to the grains resting on the surface by the impact of wind-accelerated saltating grains and not directly by the airflow. Accordingly, the consistent velocity $\bar{u}_{z'}$, is the speed below the fluid threshold velocity which is achieved when the sandflow achieves a steady state (Bagnold 1973).

Gerety (1984, 1985) questioned the validity of determining u_* for mobile sand surfaces by fitting logarthimic law regression lines to velocity data above some assumed focal height which is considered to represent a fixed roughness. Gerety reviewed the experimental wind velocity profile data from several different studies and concluded that two distinct flow zones can be identified: an inner two-phase flow zone near the bed, 2–3 cm thick, containing a high concentration of saltating grains, in which the wind profile deviates from that predicted by the logarithmic equation (Eqn 2.20), and an outer, essentially grain-free, zone which obeys the logarithmic law. There is commonly a gradual curved transition (kink) between the two parts of the velocity profiles. Gerety also pointed out that, in the data obtained by herself, by Zingg (1953a) and by Chiu (1972), the wind velocity gradients near the bed are not steep and there is no well defined single focal height.

The saltation layer can be regarded as an aerodynamic roughness to the flow, so Equation 4.24 can be written as

$$U = (1/k)u_*\ln(z/h) + \bar{U}_h \qquad (4.27)$$

where h is the height of the saltation layer, proportional to $u_*^2/2g$ (Section 4.2.3), and does not have a direct dependence on grain size, and $\bar{U}_h$ is the mean wind velocity at the top of the saltation layer. Accordingly, the logarithmic velocity profile can be written as (Owen 1964)

$$\bar{U}/u_* = (1/k)\ln(2gz/u_*^2) + D' \qquad (4.28)$$

where D' is a constant that determines the ratio $\bar{U}/u_*$ at the height of the saltation layer ($2gz/u_*^2 = 1$). Owen (1964) plotted data for uniform sand [from Bagnold (1936) and Chepil (1945a,b)] and non-uniform soil [from Zingg (1953a)] to find $D' = 9.7$. Within the saltation layer, where $2gz/u_*^2 < 0.25$, there appears to be a tendency for the mean wind velocity to become constant. However, the parameter $u_*^2/2g$ is not a good predictor of the height of the saltation layer, although it does seem to work as a roughness height. It is not easily defined for natural sand showing a wide range of grain size and trajectory heights (Gerety 1984).

4.2.5 Contact load (surface creep)

The nature of the contact load is poorly understood as most wind tunnel studies have focused on saltation. Several investigators have indicated that at the onset of motion grains first roll along the suface (Bagnold 1941, p. 32, Chepil 1945a, Maegley 1976,

Seppala & Lindé 1978). In the field, Carroll (1939) observed coarse grains to roll along the surface or proceed by intermittent jerky movements for short distances.

The moment that saltation becomes fully developed, the residual wind stress on the surface becomes relatively small and insufficient to maintain contact movement of the coarsest grains directly by wind drag. Such grains are, however, pushed forward by the impact of saltating grains, a process referred to by Bagnold (1941, p. 33) as surface creep. A high-velocity saltating grain can impel, through impact, a grain that is six times its diameter and more than 200 times its own weight (Bagnold 1941, p. 35).

The demarcation between saltation and contact load is arbitrary. Grains in the size range 0.1–0.5 mm saltate most readily during typical sandstorms, whereas 0.5–2.0 mm grains move mainly by creep. Particles larger than about 2 mm are not moved except under extremely high wind velocities (Folk 1971a). Some grains smaller than 0.5 mm may, however, move by creep before beginning to saltate (Nickling 1983, Gerety 1984, Willetts & Rice 1985a). Following impact by a saltating grain, some of the emergent grains have very low energies and rise only a short distance from the surface, or are simply displaced laterally without losing contact with the surface (Mitha *et al.* 1986). The term reptation includes the various small-scale transitional movements of such grains.

The difference in the rate of movement between grains in saltation and grains in surface creep is of the order of 200–400 times. The proportion of surface creep is independent of wind velocity but varies as a function of grain size (Horikawa & Shen 1960). Measurements on sand dunes and in the wind tunnel (Bagnold 1938a, Willetts & Rice 1985b) indicate that surface creep represents about one quarter of the total transport. A much lower proportion was reported by Nickling (1978, 1983), whereas a higher proportion was indicated by Anderson (1987b).

4.2.6 Sand transport rate

Several authors have developed mathematical models aimed at predicting the mass transport of windblown sand. Most of these models are based on theoretical considerations but incorporate empirical coefficients (O'Brien & Rindlaub 1936, Bagnold 1936, Chepil 1945a,b, Kawamura 1964, Zingg 1953a, Kadib 1965, Hsu 1971a,b, Lettau & Lettau 1978, White 1979). In all of these equations the sand transport rate is represented by q, which has units of mass per unit width per unit time ($\mathrm{kg\,m^{-1}\,s^{-1}}$).

One of the most frequently cited equations is that formulated by Bagnold (1936) based on a consideration of the relationship between the wind velocity and the rate of sand movement. According to Bagnold, a sand grain lifted from the surface has a horizontal velocity of u_1. After travelling a distance l, it strikes the ground with a horizontal velocity u_2. The momentum extracted from the wind over the distance l is

$$m(u_2 - u_1)/l \tag{4.29}$$

where m is the mass of one grain. As u_1 is very small relative to u_2, as an approximation u_1 can be neglected and the rate of momentum drawn from the wind by a mass q_s of sand in saltation (per unit width in unit time) is then given by

$$q_s = u_2/l \tag{4.30}$$

This expression measures the resisting force per unit area, which is the surface shear stress (Eqn 2.18); hence

$$q_s u_2/l = \rho u_*^2 \tag{4.31}$$

The ratio l/u^2 was found to equate approximately with the time taken for the grain's ascent, which is w_1/g (w_1 is the initial velocity of rise; Fig. 4.11) when no drag is present (Bagnold 1936). It was assumed that $w_1 = Bu_*$ (Section 4.2.3), so that

$$q_s = B(\rho/g)u_*^3 \tag{4.32}$$

The total sand transport q consists of the saltation transport q_s plus the creep transport. Bagnold assumed that the latter comprises one quarter of the total transport load, hence

$$q = (4/3)B(\rho/g)u_*^3 \tag{4.33}$$

The impact coefficient, B (found to be 0.8 for uniform sand with an average diameter of 0.25 mm), increases with increasing grain size. When the sand has a wider range of sizes, considerable bouncing occurs, thereby increasing the impact effect and giving higher values for B. For grain sizes found in typical aeolian dune sands (0.1–1.0 mm), Bagnold (1941, p. 67) found that q varies approximately as the square root of the grain diameter (Bagnold 1941, p. 67). As Equation 4.33 is valid only for uniform sand with an average diameter of 0.25 mm, the total sand transport corresponding to other grain sizes (within the range 0.1–1.0 mm) can be given by the expression (Bagnold 1937b)

$$q = C(d/D)^{1/2}(\rho/g)u_*^3 \tag{4.34}$$

where D is a standard grain diameter of 0.025 cm, d is the mean grain diameter of the sand in question, and C is a constant (related to B) with values (Bagnold 1941, p. 67) of 1.5 for nearly uniform sand, 1.8 for naturally graded sand found on sand dunes, 2.8 for poorly sorted sand with a wide range of grain sizes, and 3.5 for a pebbly surface.

The variation in the coefficient C indicates that the sand transport rate is higher over a surface of poorly sorted sand or pebbles than over a surface of uniform sand. This is because sand saltates more readily over a hard surface or a surface containing larger particles.

A similar equation was proposed by Zingg (1953a), who carried out experiments using a much wider range of particle sizes than Bagnold:

$$q = C(d/D)^{3/4}(\rho/g)u_*^3 \tag{4.35}$$

where $C = 0.83$. Owen (1964) concluded that Zingg's power function of 3/4 is more appropriate for larger grain sizes whereas the value of 1/2 used by Bagnold (Eqn 4.34) is appropriate for finer sands.

Equation 4.34 suffers from the limitation that it predicts unrealistic transport rates when u_* is below threshold (Belly 1964). To correct this, Bagnold (1954b, 1956) suggested a modified equation which includes a threshold term:

$$q = \propto C(d/D)^{1/2}(\rho/g)(\bar{u}/\bar{u}_t)^3 \tag{4.36}$$

where $\propto$ is a constant equal to $[0.174/\ln(z/z')]^3$, $\bar{u}$ is the mean wind velocity at height z', and $\bar{u}_t$ is the threshold velocity at height z'.

Kawamura (1964) also developed a transport equation which takes into account the threshold velocity:

$$q = K(\rho/g)(u_* - u_{*_t})(u_* + u_{*_t})^2 \tag{4.37}$$

where K is a constant (indicated by wind tunnel experiments to have a value of 2.78). Field measurements showed that K ranges from 2.3 to 3.1 for beach sand with a size range of 0.1–0.8 mm and a median diameter of 0.3 mm (Horikawa *et al.* 1984, 1986). For a wet sand surface with 3–4% water content, K ranges from 1.8 to 2.5, reducing the amount of blown sand to about 80% of that from a dry sand surface (Horikawa *et al.* 1984).

Equation 4.37 can be modified when u_{*_t} is replaced by $u_{*_{tw}}$ in Equation 4.18 (Hotta *et al.* 1985).

Lettau & Lettau (1978) also suggested a refinement of Equation 4.37 which incorporates a threshold term:

$$q = C_1(d/D)^{1/2}(\rho/g)u_*^2(u_* - u_{*_t}) \tag{4.38}$$

where the constant $C_1 = 4.2$.

The mass flux of saltating grains cannot increase indefinitely with increasing wind velocity, since a greater concentration of grains in the saltation layer lowers the residual shear stress transmitted to the particles (Owen 1964). Therefore, a steady-state transport rate requires that the resistance to flow exerted by saltating grains should always be such that the wind velocity near the base of the saltation layer should remain constant and never exceed the impact threshold value (Bagnold 1973).

The ratio between friction velocity (measured above the saltation layer, Eqn 4.28) and the threshold friction velocity, known as the *transport stage* (Bagnold 1973), can define the concentration of sand in the saltation layer. Analysis by Maegley (1976) yielded

$$\rho_m/\rho \approx 1 - (u_{*_t}/u_*)^2 \tag{4.39}$$

where ρ_m is the mass concentration of sand in the saltation layer.

When the friction velocity is approximately three times the threshold value, the transport stage approaches zero and the concentration of the grains in the saltation layer reaches saturation, defined by a ratio of 0.9. Gerety & Slingerland (1983) reported that, for heterogeneous sand with a mean size of about 0.18 mm, saturation was achieved at $u_* = 60\,\text{cm s}^{-1}$.

In Equation 4.34, q depends solely on friction velocity and not on the transport stage of the saltation layer. Owen (1964) introduced the variation of q due to the transport stage by examining it in the light of the experimental data obtained by Bagnold (1936) and Zingg (1953a):

$$q = [0.25 + (w_f/3u_*)]\,[1 - (u_{*_t}/u_*)^2]\,(\rho/g)u_*^3 \tag{4.40}$$

The effect of grain size is taken into account in the above Equation 4.40 through the grain settling velocity (w_f).

Since the transport stage varies as the square of u_{*_t}/u_*, it follows that, when the friction velocity becomes large, the transport stage approaches zero and q is dominated mainly by u_*^3 (Eqn 4.40). At friction velocities not much above the threshold value, the dependence of q on the transport stage is predominant.

A different method for computing rate of sand transport was developed by Hsu (1971b, 1973, 1974), who related it to the third power of the friction Froude number (Fr):

$$q = KFr^3 = K[u_*/(gd)^{1/2}]^3 \tag{4.41}$$

where K is a dimensional sand transport coefficient with the same dimensions as q. K is

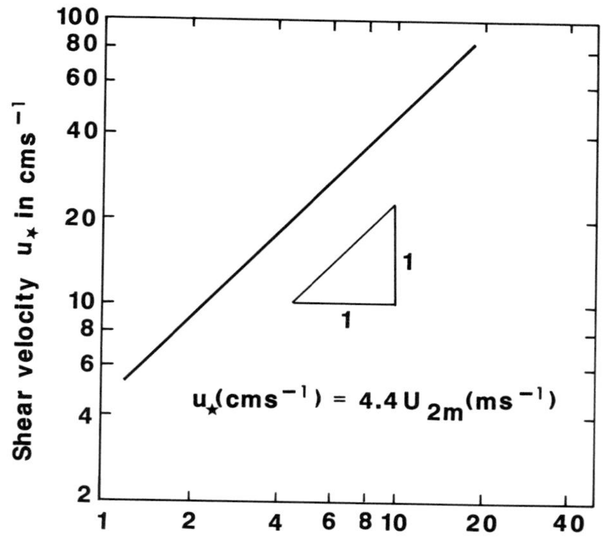

Figure 4.19 Relationship between friction velocity and the wind velocity at a height of 2 m, based on field data from several different parts of the world. (After Hsu 1974).

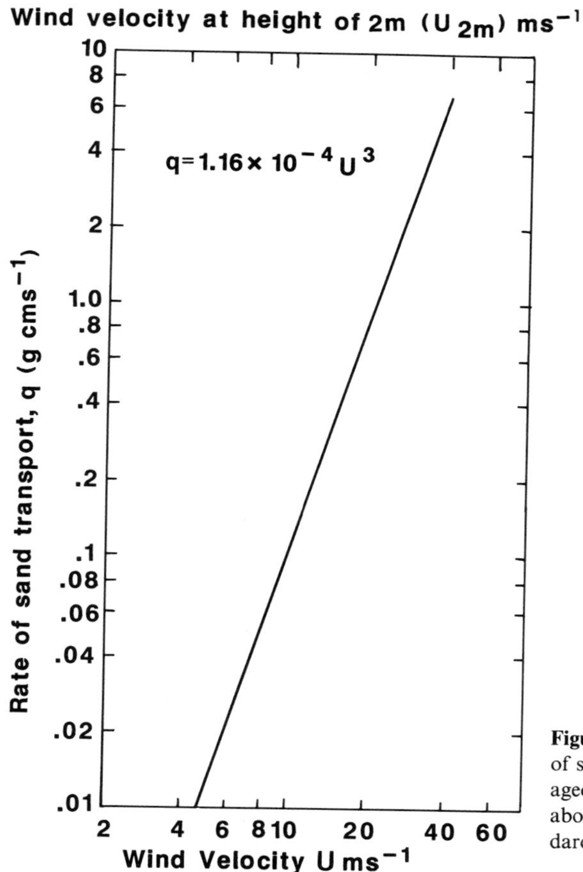

Figure 4.20 Relationship between rate of sand transport and the hourly averaged wind velocity at a height 2–10 m above the ground, for sand with a standard particle size of 0.25 mm. (After Hsu 1974).

related to the grain size as $K \times 10^4 = \exp(4.97d) - 0.47$ (where d is in mm and K is in g cm^{-1} s^{-1}). For sand with a mean diameter of 0.25 mm, $K = 2.17 \times 10^{-4}$ (Hsu 1974, p. 1621).

Based on field data from several areas, Hsu (1974) derived a relationship between u_* and the wind velocity at a height of 2 m under conditions of active sand transport (Fig. 4.19). Hsu showed that, according to the logarithmic wind velocity profile law modified for conditions of active sand transport,

$$u_* = 0.044\bar{U}_{2\,\mathrm{m}} \tag{4.42}$$

and

$$u_* = 0.037\bar{U}_{10\,\mathrm{m}} \tag{4.43}$$

where $\bar{U}_{2\,\mathrm{m}}$ and $\bar{U}_{10\,\mathrm{m}}$ are the averaged hourly wind velocities at 2 and 10 m, respectively.

For application to wind velocity data from standard meteorological stations, Hsu proposed that the average of Equations 4.42 and 4.43 should be used:

$$u_*\,(\mathrm{cm\,s^{-1}}) = 4.0\bar{U}\,(\mathrm{m\,s^{-1}}) \tag{4.44}$$

where $\bar{U}$ is the mean hourly wind velocity in the direction of transport. Substituting values of g, K and d in Equation 4.44, the sand transport rate (for sand with a mean diameter of 0.25 mm) is given by

$$q = 1.16 \times 10^{-4}\bar{U}^3 \tag{4.45}$$

The resulting relationship is shown in Figure 4.20.

A further equation which predicts the sand transport rate was proposed by Kadib (1965) who, basing his approach on work by Einstein (1950), defined the sand transport rate in terms of the intensity of sediment transport (ϕ) through the expression

$$q = \phi(\rho_s' g)(\rho_s'/\rho)^{1/2}(gd^3)^{1/2} \tag{4.46}$$

The intensity of sediment transport (ϕ) is related to the flow intensity by

$$A^*\phi/(1 + A^*\phi) = F(\psi^* B^* - 1/\eta_0) \tag{4.47}$$

where A^* and B^* are constants with values of 43.5 and 0.143, respectively, $F(x)$ is the normal distribution integral between the limits of infinity and $(\psi^* B^* - 1/\eta_0)$, ψ^* is a measure of the flow intensity, given by $\psi^* = \zeta\psi/I$ [where ζ is related to the depth of laminar sublayer, I is a measure of the disturbance to the bed caused by impacting grains, and the dimensionless parameter ψ is given by $\psi = (\rho'_s/\rho)gd/u_*^2$], and η_0 is the normalized standard deviation of the turbulent lift force, with a value of 0.5.

Based on wind tunnel studies aimed at simulating aeolian transport on both Earth and Mars, White (1979) proposed the following universal sand transport equation:

$$q = 2.61u_*^3(1 - u_{*_t}/u_*)\,(1 + u_{*_t}^2/u_*^2)\rho/g \tag{4.48}$$

Sarre (1987) compared several sand transport equations and showed that they predict widely differing results for the same grain size and friction velocity (Fig. 4.21). For sand with a mean size of 0.2 mm and u_* values of 40–60 cm s^{-1}, the highest transport rates are predicted by Kawamura's equation and the lowest by Zingg's equation.

Sand transport rates in the field are often influenced by the presence of vegetation. Some sand movement can take place with plant covers of up to 40% (Ash & Wasson 1983, Buckley 1987). A modified version of Equation 4.34 was developed empirically by Wasson & Nanninga (1986). Buckley (1987) also developed a modified equation based on experimental work:

$$q = B[\bar{U}(1 - kC) - \bar{u}_{z'}]^3 \tag{4.49}$$

where $B = \propto C(d/D)^{1/2}(\rho/g)$, taken from Equation 4.34, $\propto = [k/\ln(z/z')]^3$, taken from Equation 4.24, $\bar{U}$ is the average wind velocity measured at a height of 0.5 m (valid up to 15 m s^{-1}), $\bar{u}_{z'}$ is the threshold velocity at height z', C is the plant cover in per cent (up to 17%), and k is a constant dependent on plant shape (see Eqn. 4.21). When $C = 0$, Equation 4.49 becomes the summation of Equations 4.24 and 4.34.

All of the sand transport equations referred to above relate to sand transport on a flat surface, which is rarely found in nature. According to Bagnold (1956, p. 294), the transport rate on an inclined surface, q_1, can be expressed as

$$q_1 = q/[\cos\theta(\tan\propto + \tan\theta)] \tag{4.50}$$

where $\propto$ is the angle of internal friction of the sand and θ is the bedslope. This equation predicts that bedslope has only a small effect on the sand transport rate at slope angles up to 30°. However, wind tunnel experiments on Saharan dunes (Hardisty & Whitehouse 1988) suggested that the sand transport rate is much more strongly dependent on surface slope (Fig. 4.22).

Laboratory wind tunnel experiments by Williams (1964) showed that at low wind

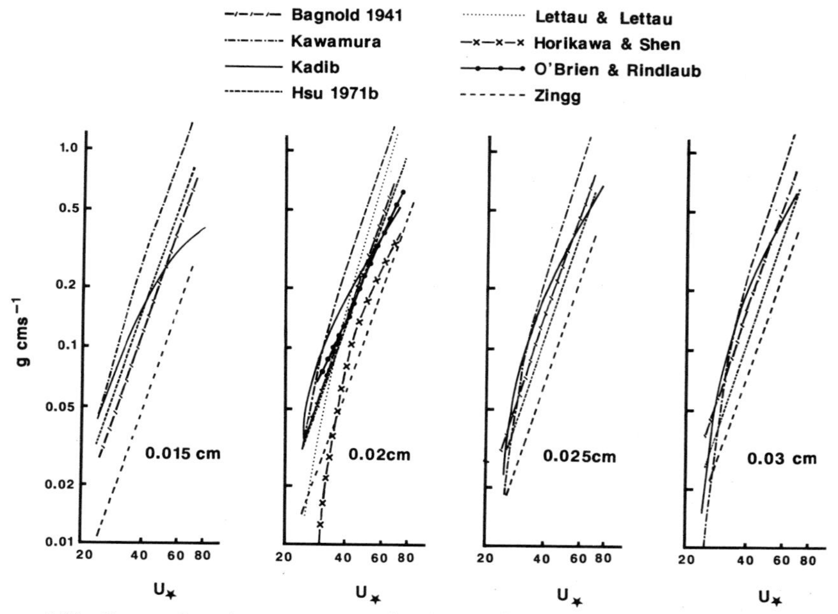

Figure 4.21 Rates of sand movement predicted by different transport equations for sand of different sizes. (After Sarre 1987).

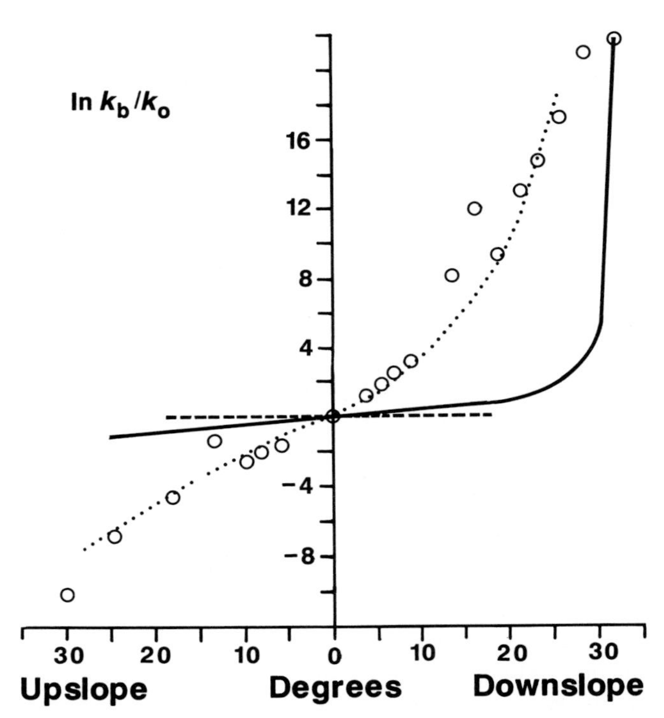

Figure 4.22 The effect of bedslope on sand transport rate according to the theoretical model of Bagnold (1956) (solid line) and the field experimental data (circles and dotted line) of Hardisty & Whitehouse (1988). k_b/k_o is the ratio of the transport rate on a sloping surface to that on a flat surface.

shear velocities the transport rate increases with decreasing grain sphericity, whereas the reverse was found at high shear velocities. In the velocity range 55–95 cm s^{-1}, grain shape was found to have little influence. Based on these experimental results, Williams (1964) proposed the following average relationship for a wide range of grain sizes:

$$q = a' u_*^{b'} \rho / g \tag{4.51}$$

where a' and b' were found to have values of 0.1702 and 3.422, respectively.

Willetts *et al.* (1982) and Willetts (1983) obtained broadly similar results in their laboratory wind tunnel studies, and found that the exponent b' in Equation 4.51 increases with increasing grain sphericity, with values ranging from 2.15 to 4.05.

Comparisons between predicted sand transport rates and those measured in the field have shown only moderate agreement. Berg (1983) found that predicted rates calculated using the Bagnold (1936) and Hsu (1971b) equations were an order of magnitude higher than those measured during a tracer study. The Kadib (1965) equation provided a closer correspondence with observed transport rates for medium sand (mean diameter 0.65 mm) but underestimated the observed rates for coarse sand (mean diameter 1 mm). Sarre (1988) found that the sand transport rates measured on a North Devon beach showed the closest agreement with the rates predicted by the White (1979) equation. The correlation was found to be particularly close at friction velocities above 0.28 m s$-^{1}$. In this study, moisture contents of up to 14% in the top millimetre of beach sand were found to have no significant effect on the sand transport rate. However, at higher levels the effect of moisture became increasingly important. With $u_* = 64\,\mathrm{cm\,s^{-1}}$ and moisture content of 16%, the transport rate was reduced to 67% of the dry sand value, whereas with 22% moisture content the sand transport was reduced almost to zero.

Differences between observed and predicted sand transport rates can be accounted for by several different factors. Sand transport is difficult to measure accurately in the field, since all sand trap designs have an effect on the windflow around them (see Section 10.2), and tracer and survey methods give only broad approximations. The natural pattern of sand transport also varies locally and on short time scales, reflecting the occurrence of turbulent bursting phenomena in the near-surface wind field, variations due to surface microtopography, vegetation, non-erodible roughness elements, surface crusting, and small-scale atmospheric instabilities arising from temperature differences between the sand surface and the overlying air (Borowka 1980).

Although some intermittent sand transport may take place owing to turbulent gusts when u_* lies below the threshold, the accuracy of predicted sand transport rates is apparently not improved if short-term variations in wind gustiness are taken into account (Lee 1987).

In most cases the sand transport equations provide estimates of the maximum rate of sand movement which is likely under specific meteorological and topographical conditions. However, in rare circumstances they may underestimate the actual transport rate (Reid 1985, Jensen *et al.* 1984, Horikawa *et al.* 1984, Watson 1989, Sarre 1989).

4.2.7 *Avalanching of sand on dune slip faces*

In most dunefields, sand transport also takes place by avalanching on dune lee-side slip faces.

The tangential force (F) acting on a grain on a slope depends on the slope angle (θ) and the weight (W) of the grain (Fig. 4.23):

$$F = W \sin\theta \tag{4.52}$$

This force is opposed by the force of intergranular friction. The static friction force (F_s) is proportional to the grain's pressure (normal stress, N) on the surface:

$$F_s = f_s N \tag{4.53}$$

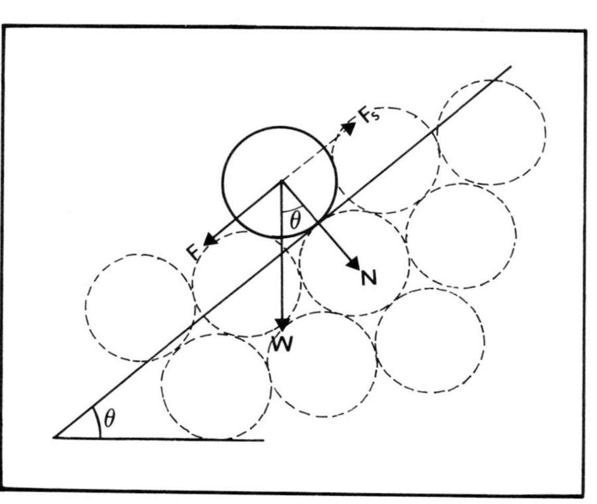

Figure 4.23 Forces acting on static, cohesionless grains forming an angle θ relative to the horizontal. F is the tangential force, F_s is the static friction force, W is the weight of the grain, and N is its normal pressure.

where f_s is the static friction coefficient for grains at rest. Equation 4.53, known as the *Coulomb frictional equation*, is similar to Equation 2.9 in which shear stress increases as shearing strain (normal stress) increases. Avalanching occurs when the tangential force (F) exceeds the static friction force (F_s). When the sand grains are just about to slide, the magnitude of the tangential force (shear stress) is

$$F = \tan\theta N = F_s \tag{4.54}$$

According to Equations 4.53 and 4.54, the tangent of the slope angle in the condition of incipient sliding (θ_s) is equal to the static pressure coefficient:

$$f_s = \tan\theta_s \tag{4.55}$$

The angle θ_s, referred to as the *angle of internal friction*, is the angle at which grains begin to slide. The angle is affected by the shape and surface textural characteristics of the grains (see Chapter 3). Generally, it increases with departure of the grains from a spherical form (Carrigy 1970) and with increasingly rough surface texture.

As soon as avalanching starts, the friction force takes on a characteristic value, F_k, known as the *kinetic friction force*:

$$F_k = f_k N \tag{4.56}$$

where f_k is the *kinetic friction coefficient*.

The tangential force required to initiate avalanching is greater than that required to maintain sliding (Rabinowicz 1965). Accordingly, $f_s > f_k$. The kinetic friction coefficient is given by

$$f_k = \tan\theta_k \tag{4.57}$$

where θ_k is known as the *angle of repose*. It represents a condition of balance between intergrain kinetic friction and the force of gravity (weight) acting to pull them downslope (Van Burkalow 1945).

Although the above explanation is satisfactory for a single grain on a slope, it does not work satisfactorily for a mass of grains in loose contact. Before movement of such a mass can take place, the packing arrangement must be disturbed (Jenkin 1931, 1933). This involves an expansion of the whole grain mass, in which process, known as *dilation*, energy is expended (Bagnold 1956).

The angle of repose is smaller than the angle of internal friction, and it represents the angle at which sliding ceases. Slopes which maintain the angle of repose, known as slip faces, are ubiquitous on the lee side of sand dunes. Laboratory experiments have shown that the angle of repose for medium-fine sands varies from 30.5 to 35.45° (Jenkin 1933), but is typically 32–34° (Allen 1970a, Carrigy 1970).

Failure on dune slip faces occurs when deposition of grains blown over the dune crest (*grainfall*) causes oversteepening of the upper part of the slope (Fig. 4.24). Failure of the oversteepened slope commonly takes place by the formation of a series of avalanche 'tongues', each 10–20 cm across (Fig. 4.25). This process of grain transport is sometimes referred to as *grainflow*.

Anderson (1988) has shown that grainfall deposition in the lee of aeolian dunes reaches a maximum up to a few tens of centimetres from the brink. Beyond the point of maximum deposition the rate of deposition declines exponentially. A *pivot point* can be identified on the lee slope above which the slope is subject to oversteepening by grainfall,

Figure 4.24 Photograph showing the movement of sand over the brink of a transgressive dune on the Oregon coast. The slip face (to the right) has an average slope of 33°, whereas the gentler windward slope to the left has a maximum slope of 18°, decreasing to less than 5° near the crest (which in this case coincides with the brink). A plume of sand grains transported to the brink in saltation can be seen falling over the top of the slip face.

Figure 4.25 Avalanche tongues on the middle part of the slip face shown in Figure 4.24.

with periodic adjustment by grainflow, and below which the slope steadily accumulates, partly by grainfall but mainly by grainflow from upslope. Under typical wind conditions the pivot point is located approximately 2 m downslope from the brink of a 5 m high dune and 3 m downslope from the brink of a 10 m high dune (Anderson 1988).

A common feature observed in lee slope grainflow deposits is reverse grading, in which larger grains become concentrated towards the top of the flow (Sallenger 1979). Two hypotheses have been proposed to explain reverse grading. The *dispersive stress* hypothesis (Bagnold 1954a) points out that dispersive stress is greatest close to the shear plane and that large grains can exert a higher stress than small grains. Larger grains therefore move upwards through the flow to equalize the stress gradient. According to the second hypothesis, small grains simply filter through the gaps between larger grains until they come to rest near the shear plane. This process is referred to as *kinetic filtering* (Middleton 1970). Experiments performed using grains of equal size but different density have demonstrated that the first hypothesis is valid, since the dispersive stress also depends on grain density. However, the occurrence of kinetic filtering during grain flows has not been disproved.

Because coarse grains move towards the top of the flow, they show a tendency to roll faster and further than smaller grains, which are confined by neighbouring grains closer to the shear plane. This process is an important factor contributing to the concentration of coarser grains in the basal layers of aeolian dune deposits (Pye 1982b).

In some situations the resistance to grain movement is increased owing to intergranular cohesion. Thus the tangential force required to initiate motion is:

$$F = f_s N + C \tag{4.58}$$

where C represents the cohesion force.

Cohesion may be due to moisture (Nickling 1978), salt (Land 1964, Pye 1980b, Nickling 1984), or electrostatic forces (Corn 1966, Greeley & Leach 1978, Iversen *et al.* 1976a). When moisture or considerable amounts of fine particles are added to sand, the cohesive forces become dominant in Equation 4.58. For this reason, dune sands which contain a few per cent of fines can form slopes which are almost vertical. In deserts, where the surface sand is usually dry, the slip faces of active dunes are normally maintained in the range 32–34°. In coastal dunes, however, slip faces may have slightly steeper angles owing to the effect of moisture and salt (Land 1964, Hunter *et al.* 1983).

5
The formation of sand seas and dunefields

5.1 Definition of sand seas and dunefields

Deposits of aeolian sand cover approximately 6% of the global land surface area, of which about 97% occurs in large arid zone sand seas. On average, about 20% of the world's arid zones are covered by aeolian sand, although the proportion varies from as little as 2% in North America to more than 30% in Australia and >45% in Central Asia (Mabbutt 1977, Lancaster & Hallward, 1984).

The term *sand sea* conveys a general impression of a large sand-covered area, but a clear distinction between sand seas, dunefields, and sand sheets has not always been made. Wilson (1973, p. 78) proposed that the term *erg* be used to describe 'an area where wind-lain sand deposits cover at least 20% of the ground, and which is large enough to contain draas' (the third-order aeolian bedforms recognized by Wilson – see Chapter 6). 'Erg' is an arabic word used by local people in the northwest Sahara to describe areas of wind-deposited sand of virtually any size. In practice, most sedimentologists and geomorphologists have considered that ergs or sand seas (the names are often used interchangeably) must cover a minimum area of 125 km^2 (Fryberger & Ahlbrandt 1979, Thomas 1989b). Smaller areas are defined as dunefields or, if they contain no significant dune bedforms, sandsheets.

5.2 Global distribution of sand seas

The location of the world's major active sand seas is shown in Figure 5.1a. Many of these sand seas have large areas of stabilized dunes on their margins, which in many cases were active around the time of the last glacial maximum (Sarnthein 1978, Sarnthein *et al.* 1981). The distribution of known active sand seas at 18 000 yr BP is shown in Figure 5.1b.

Most of the major northern hemisphere sand seas are concentrated in the sub-tropical desert belt which extends across North Africa and the Arabian sub-continent into Iran and Pakistan, or in the mid-latitude desert basins of Central Asia. Smaller dunefields occur in the Southwestern and Midwestern United States.

The Sahara Desert contains the largest number of sand seas with an area >12 000 km^2 (Table 5.1). The Saharan sand seas occur in two broad belts which run north and south of the Tibesti and Hoggar highlands and which converge in the west (Fig. 5.2a). The largest of the Saharan ergs is the Erg Chech in southern Algeria, which has an area of 319 000 km^2. Many of the Saharan sand seas occupy structural basins (Wilson 1971), although a significant minority do not (Mainguet 1978). Two large ergs occur in southern Africa, the Namib (Lancaster 1989c) and the Kalahari (Lancaster 1989d).

Almost a third of the Arabian sub-continent is covered by sandy deserts (Holm 1960), the largest being the Rub Al'Khali, which has an area of more than 560 000 km^2 (Fig. 5.2b).

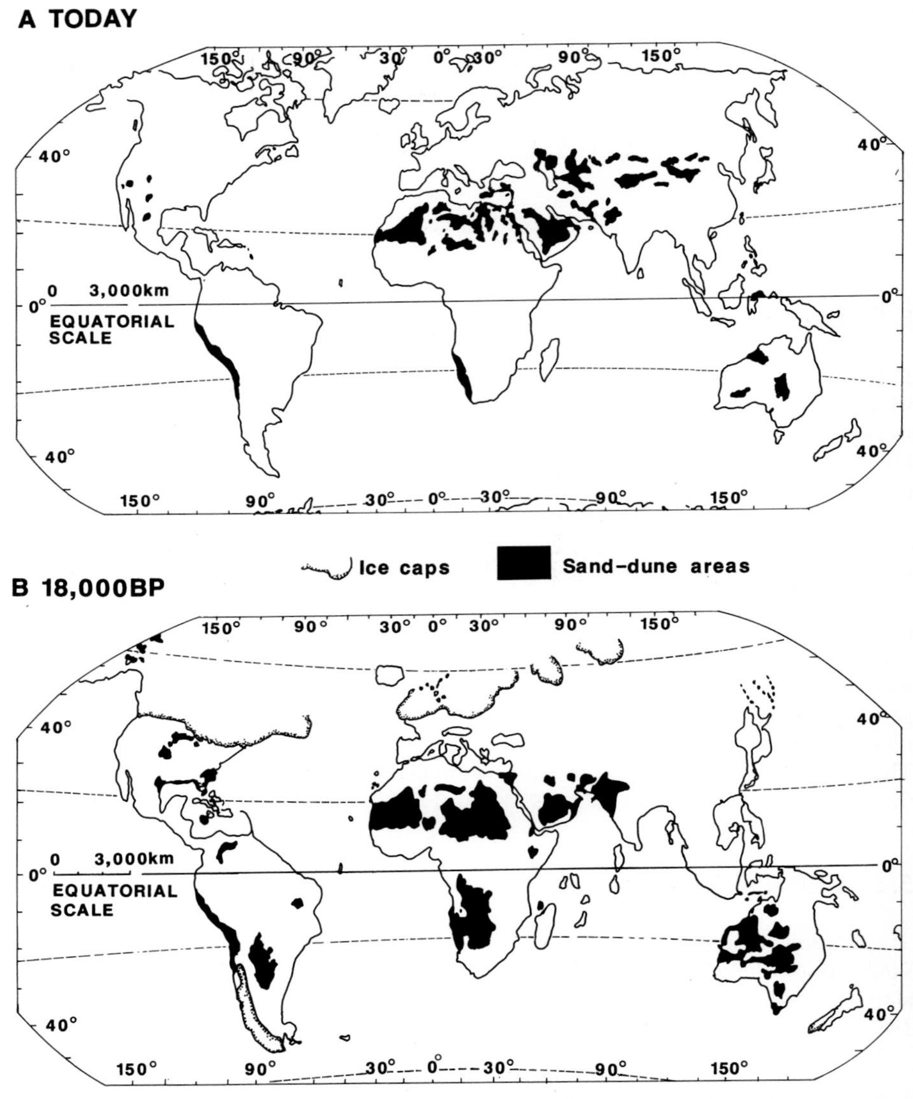

Figure 5.1 Global distribution of major ergs (A) at the present day and (B) at the last glacial maximum (18 000 yr BP). Modified after Sarnthein (1978) and Goudie (1983b). Little information is available about the extent of the Central Asian and Chinese sand seas during the late Pleistocene.

The southern Rub Al'Khali occupies a huge basin bounded by the Hijaz Plateau to the west, the Hadramaut Highlands to the south, and the Oman Mountains to the east. The desert covers a large alluvial fan complex of post-Pliocene age which grades gently towards the northeast.

In Central Asia, several large sand seas occur in deep basins which are bounded by high mountain ranges. The largest sand sea of this type is the Takla Makan in Xinjiang Province, western China. A number of large sand seas also occur on the northwestern side of the mountains in the southern Soviet Union. These include the Kara-Kum and

Table 5.1 Ergs in the world larger than 12 000 km². (After Wilson 1973).

No.	Name	Area(km²)	A*	SC†	BO‡	No.	Name	Area(km²)	A*	SC†	BO‡
	North Africa										
1	Abu Moharik		L			31	Ramlat Wahibah	16 000	L		D
2	Great Sand Sea	105 000	L		D	32	Ramlat Sabatayn	14 000	L		
3	Sudanese Qoz	240 000	F		D	33	Al Nefud	72 000	L	Q	D
4	Erg Rebiana	65 000	L			34	'Nafud complex'	25 000	L	O	D
5	Erg Calanscio	62 000	L								
6	Edeyen Murzuq	61 000	L	Q	D		Asia				
7	Edeyen Ubari	62 000	L		D	35	Thal Desert	18 000	F		
8	Issaouane-N-Irarraren	38 500	L	O	D	36	Thar Desert	214 000	F		
9	Erg Oriental	192 000	L	O	D	37	Ryn Peski	24 000	L		
10	Erg Occidental	103 000	L	O	D	38	Peski Kara-Kum	380 000	L		D
11	Erg er Raoui	11 000	L	O	D	39	Peski Kyzyl-Kum	276 000	L		D
12	Erg Iguidi	68 000	L	O	D	40	Peski Priaralskye	56 000	L		D
13	Erg Chech-Adrar	319 000	L	O	D	41	Peski Muyunkum	38 000	L		D
14	North Mauretanian Erg	85 000	L	O	D	42	Peski Sary Isnikotrav	65 000	L		D
15	South Mauretanian Erg	65 000	F	Q	D	43	Peski Dzosotin	47 000	L		
16	Trarza and Cayor Erg	57 000	F		D	44	Takla Makan	247 000	L		
17	Ouarane, Aouker,	206 000	L/F	O	D	45	East Takla Makan	14 000	L		
	Aklé, etc.					46	South Ala Shan	65 000	L		
18	El Mréyé	63 000	L	Q	d(D)	47	North Ala Shan	44 000	L		
19	Erg Tombuctou	66 000	L	O	D	48	South-east Ala Shan	14 000	L		
20	Erg Azouad	69 000	F	O	D	49	East Ala Shan	12 000	L		
21	Erg Gourma	43 000	F	O	D	50	West Ala Shan	27 000	L		
22	West Azouak	35 000	F	O	D	51	Ordos	17 000	L		
23	East Azouak	34 000	L/F	O	D	52	'Peski Lop Nor'	18 000	L		
24	Erg Bilma-Ténéré	155 000	L	Q	d						
25	Erg Foch	13 000	L	O	d		Australia				
26	Erg Djourab	45 000	L	O	d	53	Victoria Desert	300 000	F		d
27	Erg Kanem	294 000	F		D	54	Great Sandy-	630 000	F		d
							Gibson Desert				
	Arabia					55	Simpson Desert	300 000	L/F		d(D)
28	Rub al Khali	560 000	L	Q	D	56	'Northern Desert'	81 000	F		d
29	Al Dahana	51 000	L	O	D						
30	Al Jafura	57 000	L	O	D		South Africa				
						57	Namib Desert	32 000	L		D
						58	Kalahari Desert		F		

Numbers refer to the figures indicated. Gaps in the table are due to absence of data.
* Activity: L = active erg; F = fixed erg.
† Sand cover: O = open 20–80%, Q = quasi-closed 80–100%.
‡ Bed-form order: D = draas predominant; d = dunes predominant; d(D) = dunes predominant, some draas.

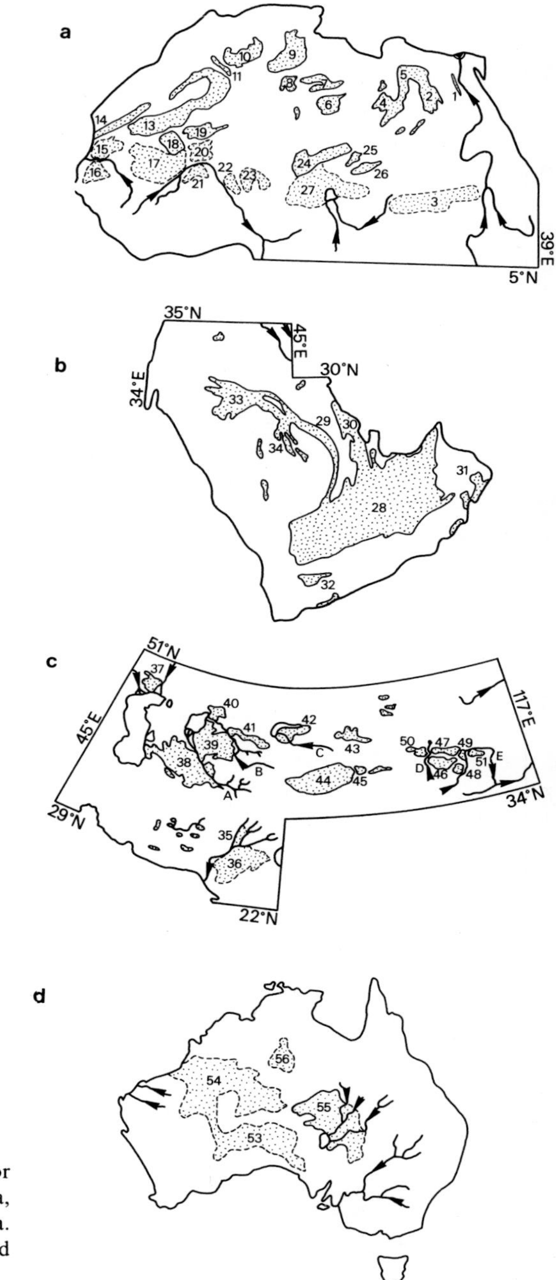

Figure 5.2 Distribution of major ergs in (a) the Sahara, (b) Arabia, (c) Central Asia, and (d) Australia. (After Wilson, 1973). The numbered ergs are listed in Table 5.1.

Kyzyl Kum to the east of the Caspian Sea and Aral Sea, respectively (Fig. 5.2c).

In the southern hemisphere large areas of windblown sand cover much of Central Australia and southwest Africa. Smaller dunefields occur near the equator in Peru and in Argentina. Fossil sand seas of late Pleistocene age have also been identified in parts of the Orinoco and Amazon River basins (Tricart 1974).

Australia has four major desert sand seas, the largest occupying the Great Sandy Desert and part of the Gibson Desert (Fig. 5.2d). In southern Africa there are two sand seas occupying areas of $>12\,000\,km^2$, the Namib, and the Kalahari.

Few coastal dune complexes are large enough to be classified as sand seas on the basis of size. Some of the largest coastal dune formations are found on the coasts of eastern and southern Australia, Natal, and Oregon. The coastal dunefield at Cape Bedford and Cape Flattery in North Queensland, for example, has an area of about $600\,km^2$ (Pye 1982a), compared with a dune-covered area of $630\,000\,km^2$ in the Great Sandy–Gibson Desert. Desert sand seas with an area of more than $32\,000\,km^2$ contain over 85% of the global total of aeolian sand (Wilson 1970, 1973).

A clear distinction between coastal and desert dunes is sometimes difficult to make, since sand supplied from coastal sources may be formed into dunes which migrate tens of kilometres inland and merge with dunes composed of sand derived from inland sources. Examples where this has occurred include the coastal deserts of Namibia (Lancaster 1982a, Lancaster & Ollier 1983), Peru (Finkel 1959), and Oman (Goudie *et al.* 1987). In other instances, continental sand seas have prograded seawards under the influence of offshore winds, thereby supplying sand to the coastal zone. Fryberger *et al.* (1983) described such an example on the shores of the Arabian Gulf near Dhahran, Saudi Arabia.

5.3 Factors controlling the distribution and magnitude of sand seas

There are three basic requirements for the formation of large sand seas and dunefields: (a) availability of a large supply of sand; (b) sufficient wind energy to transport the sand or rework it *in situ*; and (c) suitable topographic and climatic conditions maintained over a long period which allow accumulation of a large thickness of sand.

5.3.1 Sand sources and dunefield development

The magnitude of sand supply is dependent on the nature of the lithologies which outcrop in an area, by weathering and denudation rates, and by the effectiveness of other sediment transport processes which sort and transport sand to sites where it becomes exposed to wind action.

In a simple situation, unvegetated sandy regolith formed by the weathering of sandstones or other rocks in an arid climate may be reworked by the wind to form dunes more or less *in situ* (case 1 in Fig. 5.3). Examples are provided by the dunes of the Navajo Country (Hack 1941) and the Killpecker Dunes of Wyoming (Ahlbrandt 1974). However, more complex transport event sequences are involved in the formation of a majority of inland dune complexes (cases 2 and 3 in Fig. 5.3). Fluvial processes often play a key role in presorting and concentrating the products of weathering before aeolian transport takes place (Smith 1982). The Great Sand Dunes of Colorado, for example, are formed of sand which has been transported into the San Luis Valley by rivers flowing from the San Juan and Sangre de Cristo Mountains (Fig. 5.4). The immediate source of the aeolian sand is an area of abandoned levees and dry oxbow lakes on the northern side of the Rio Grande (Johnson 1967). The predominant direction of aeolian transport is northeasterly under the influence of strong southwesterly winds which predominate for much of the year.

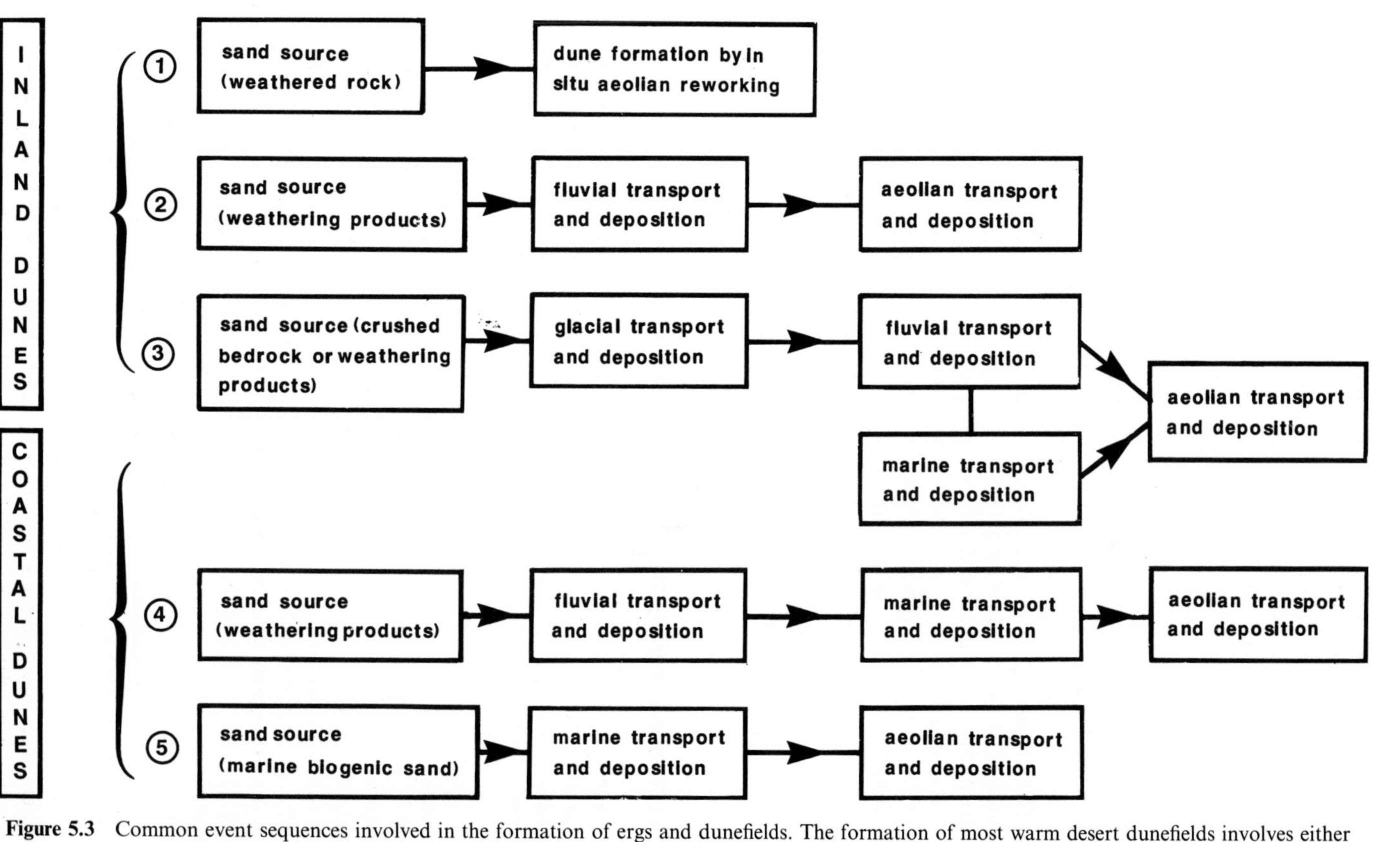

Figure 5.3 Common event sequences involved in the formation of ergs and dunefields. The formation of most warm desert dunefields involves either event sequence 1 or event sequence 2. Cold climate dunefield formation may involve event sequence 2 or event sequence 3.

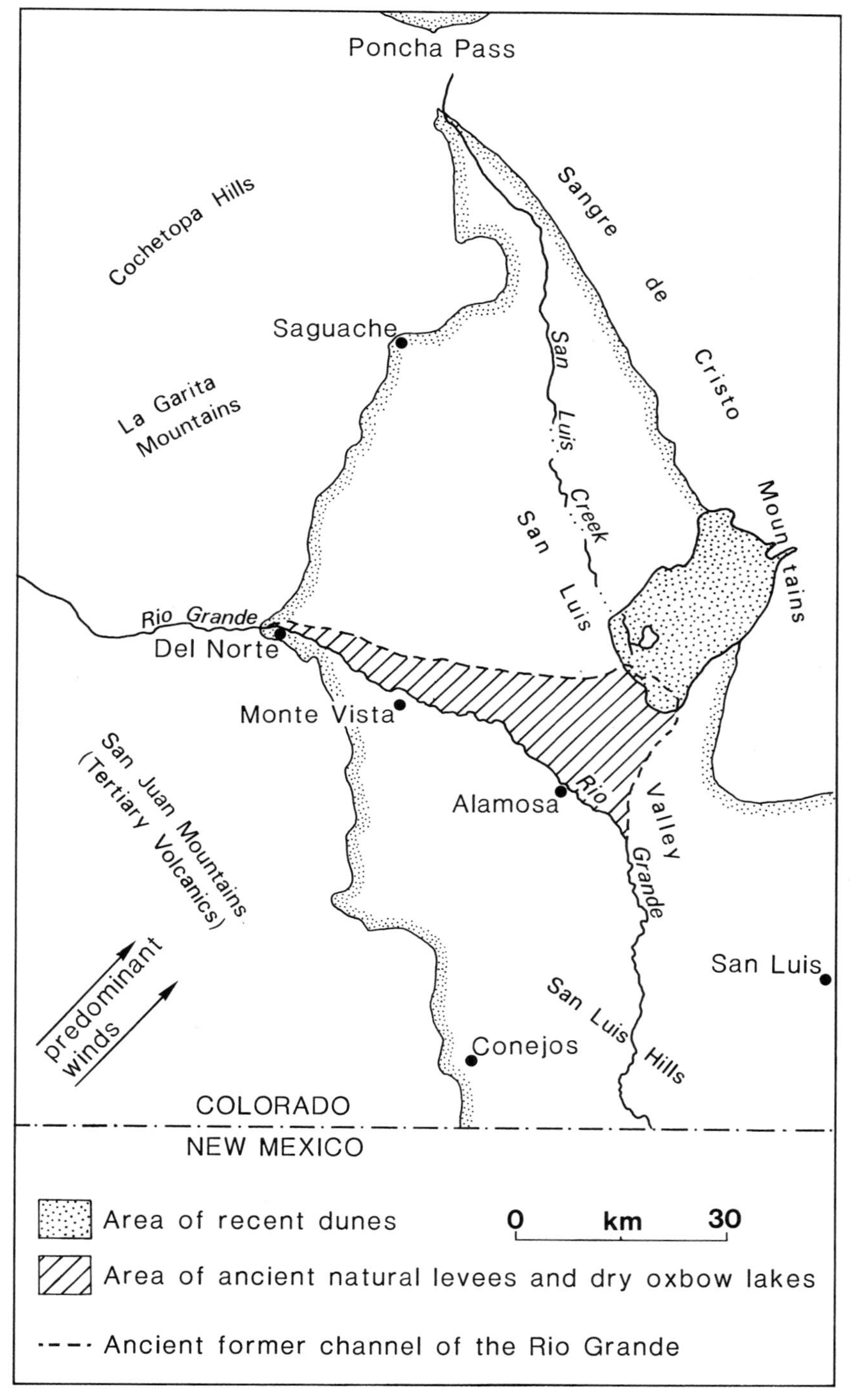

Figure 5.4 Geomorphic setting of the Great Sand Dunes, Colorado. (Modified after Johnson 1967).

Many other inland dune complexes, including the Simpson–Strzelecki desert of Central Australia (Fig. 5.5) (Wasson 1983a,b, Wopfner & Twidale 1988) and the southern deserts of Iraq (Al-Janabi *et al.* 1988), have been formed by deflation of exposed fluvial and lacustrine sediments. In the Kalahari, dunes of late Pleistocene to Recent age have formed partly by reworking of the surface sediments of the Kalahari Beds, which are weakly consolidated fluvio-aeolian sediments that have been accumulating in a slowly subsiding intra-cratonic basin since Jurassic times (Thomas 1984, 1988b).

Long-distance fluvial transport of sand is involved in some instances. For example, the desert dune complexes of northern Sinai consist of sand transported from East Africa by the River Nile. A considerable proportion of this sand has also been

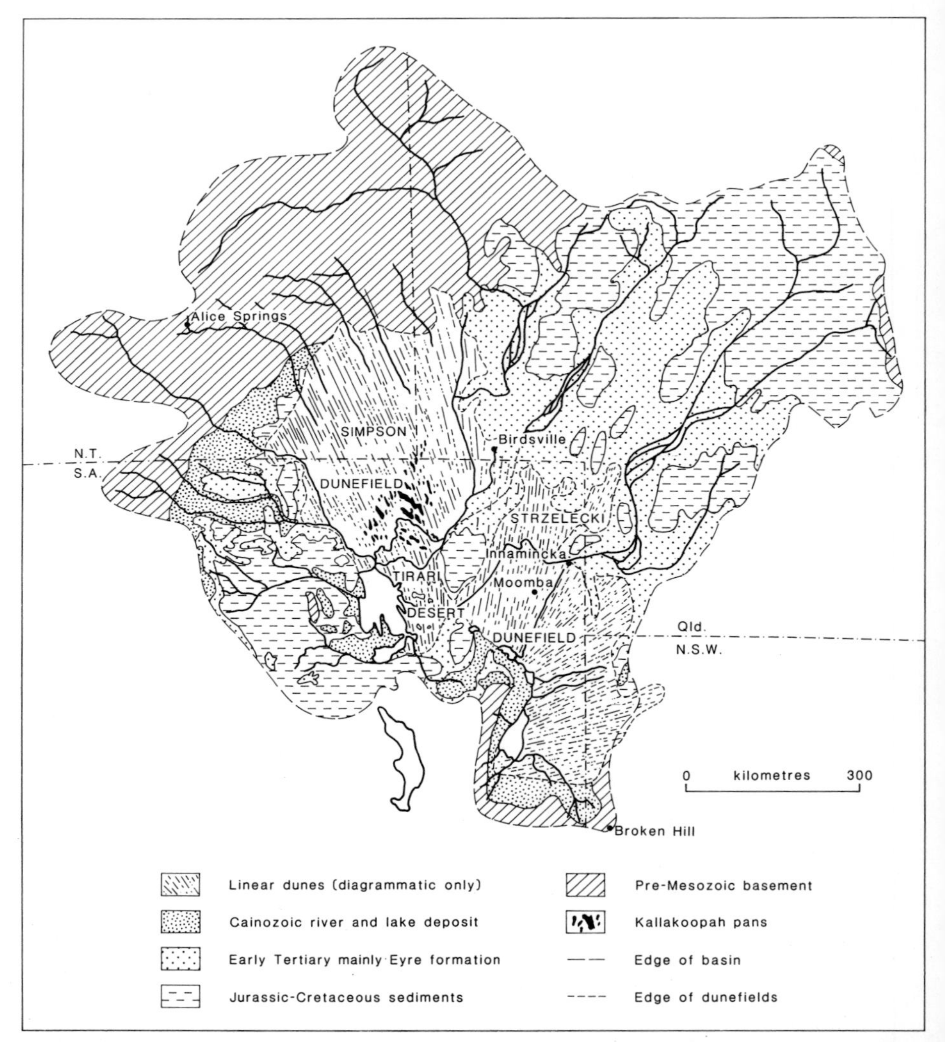

Figure 5.5 Major geomorphological and sedimentary features of the Lake Eyre Basin, a basin of internal drainage in which late Cenozoic fluvial sediments have been extensively reworked by the wind to form linear dunes. (Modified after Wasson 1983a).

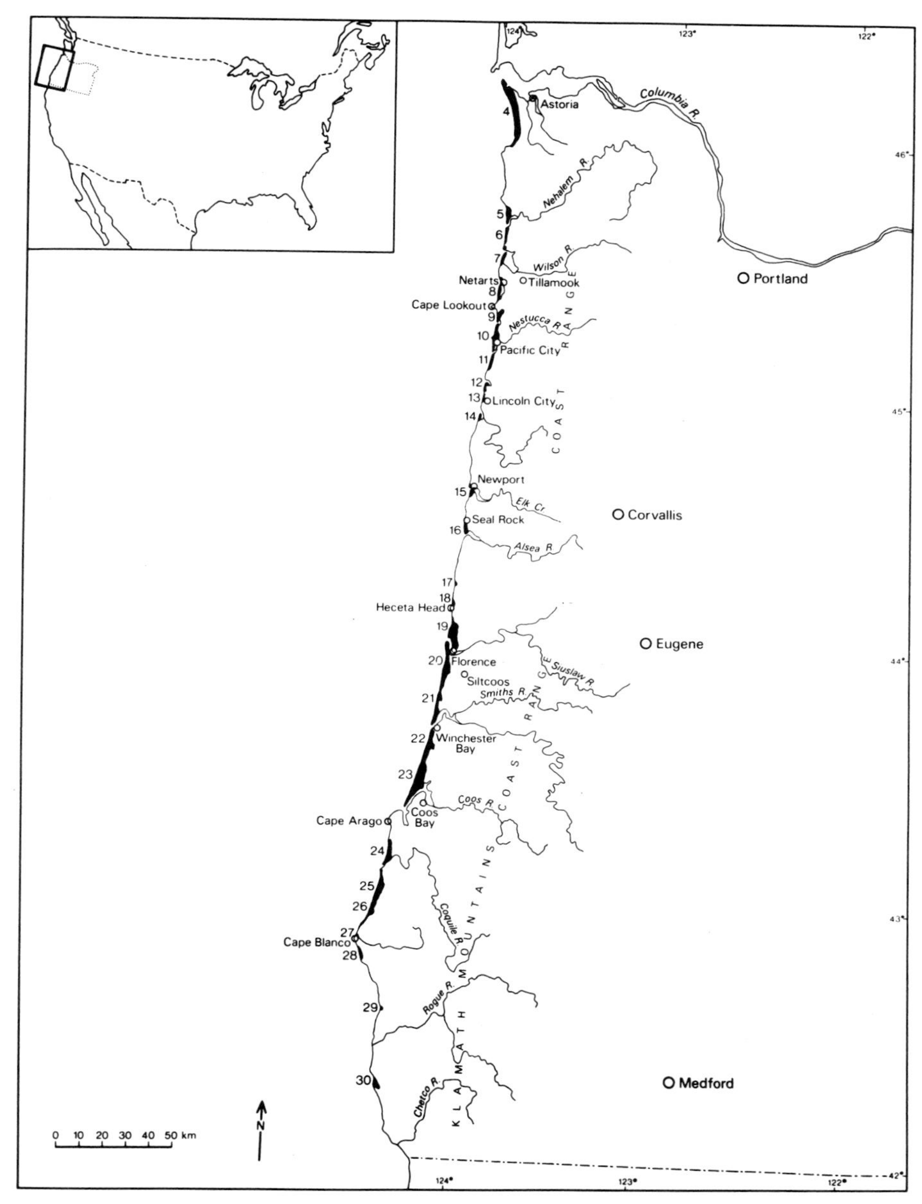

Figure 5.6 Distribution of coastal dunefields on the Oregon coast in relation to major river mouths. Dunefields are numbered after Cooper (1958).

transported along the Sinai coast from the Nile Delta by marine processes, before being blown inland (Tsoar 1978).

The formation of coastal dune complexes normally involves several stages of transport involving fluvial, glacial, marine, and aeolian processes (cases 4 and 5 in Fig. 5.3). Cooper (1958) introduced the concept of a *receptive shore* to which well sorted sand is supplied by marine processes before being blown inland by the wind. The sand supplied to the shore may be derived from relict deposits on the continental shelf or be supplied from rivers. The location of a receptive shore is usually determined by the regional topography and the prevailing wind and wave conditions. Headlands which protrude seawards from the general line of the coast often create receptive shores by acting as barriers to the longshore movement of sand, thereby forming a closed *sediment compartment* (Davies 1974). On the coast of eastern Australia, the largest dune complexes occur on the southern side of headlands which have intercepted the flow of northward-drifted littoral sand over a long period (Thom 1978, Pye 1983b). Many of the Oregon coastal dunefields also show a relationship with headlands and the mouths of rivers which have supplied sand (Cooper 1958) (Fig. 5.6).

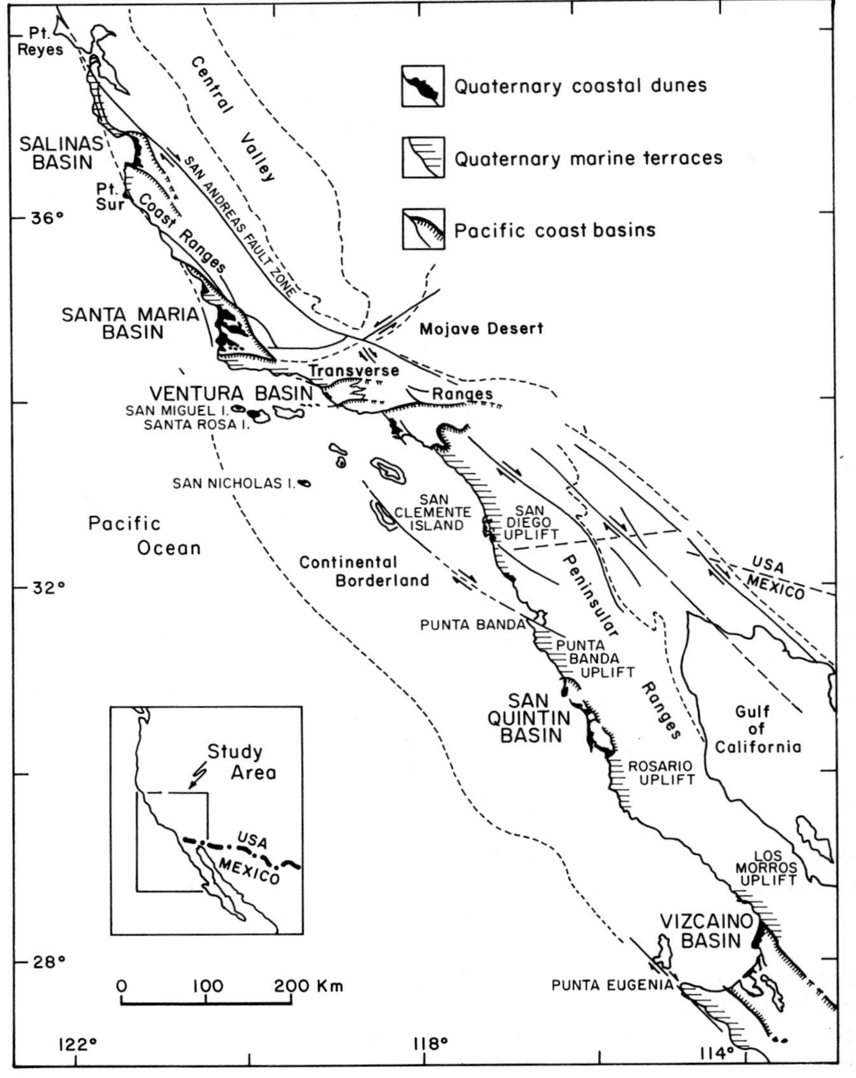

Figure 5.7 Relationship of Quaternary coastal dunefields to marine terraces and structural basins along the California coast. (After Orme & Tchakerian 1986).

Sections of the coast which form tectonic basins often act as long-term sinks for littoral sand, whereas tectonic highs rarely provide favourable sites for large-scale sand accumulation and preservation. A relationship between the location of major coastal dune complexes and small tectonic basins is clearly seen on the coast of California (Fig. 5.7) (Orme & Tchakerian 1986).

5.3.2 Relationship between sand deposits and climate

As a general principle, aeolian processes are areally more important in arid areas where vegetation cover is sparse and the soil moisture content is low (Marzolf 1988). For this reason, the largest active sand seas occur in areas which receive <250 mm of annual rainfall. However, dunes can form in any climatic regime where bare sand is exposed and where the wind is strong enough to entrain sand. Jennings (1964, 1965) noted that coastal dunes appear to be less well developed in humid tropical climates compared with temperate latitudes, but subsequent studies demonstrated that there is no general lack of dunes in humid tropical areas (Swan 1979, Pye 1982a, 1983b). Coastal wind energy is, however, lower in equatorial latitudes than in the trade wind belt and in the zone of mid-latitude westerlies (Fig. 2.5) (see also Pye 1985b), and consequently there is a lower potential for aeolian sand transport. However, high rainfall and humidity are of minor importance as limiting factors in dune development, and the absence of dunes on many tropical shores can be explained mainly by the poorly sorted nature of beach sediments and low wind energy resulting from the coastal orientation relative to the prevailing wind direction (Pye 1983b, 1985b). Large coastal dunes occur in many areas, both tropical and extra-tropical, which receive an annual rainfall of >2000 mm, including parts of North Queensland (Pye 1982a) and Oregon (Cooper 1958, Hunter *et al.* 1983). In these areas the greatest amount of sand transport occurs during the wetter months of the year, when strong winds are most frequent (Pye 1980a). Because strong winds are frequent in exposed coastal areas, there is no rainfall limit for the occurrence of coastal dunes.

In inland desert regions, where wind energy is considerably lower than on the coasts (Fig. 2.5), an annual rainfall of 250 mm often provides an upper limit for active dunes. In some areas, such as the northwest Sahara (Sarnthein & Diester-Haas 1977), the rainfall limit for active dunes may be as low as 25–50 mm per year (Table 5.2). However, a clear distinction between active and fossil dunes is not always easy to make. In southern Africa, for example, the present limit of dune activity has been equated with the 100 or 150 mm isohyets (Lancaster 1981c). Extensive areas of vegetated and partially vegetated dunes which occur beyond these limits have generally been regarded as fossil, having formed during more arid periods of the Pleistocene (Grove 1958, 1969, Lancaster 1981c, 1989d). Recent work, however, has demonstrated the importance of vegetation in the development of some linear dunes (Ash & Wasson 1983, Tsoar & Møller 1986), and Thomas (1988a) has suggested that the dunes in the southwest Kalahari are partially active.

Variations in temperature with latitude affect aeolian processes in two distinct ways. First, since the density of air varies inversely with temperature (Table 2.1), and because the drag force is proportional to air density (Eqn 4.1), it has been suggested that winds at high latitudes should entrain and transport sediment more effectively than at low latitudes (Smith 1965, p. 160). Using the Bagnold entrainment equation (Eqn 4.11), Selby *et al.* (1974) calculated that a granule of 3 mm diameter can be carried to a height of 2 m by a wind of velocity 36.05 m s^{-1} when the air temperature is -70 °C, but requires a velocity of 45.42 m s^{-1} at a temperature of $+50$ °C (Fig. 5.8). This factor, combined with the higher frequency of strong winds at high latitudes (Fig. 2.5), may explain the

Table 5.2 Rainfall limits of active and fossil dunes (mm).

Location	A Present rainfall limit for active dunes (mm)	B Present rainfall limit of fossil dunes (mm)	C Distance between A and B (km)	Source
West Africa	150	750–1000	600	Grove (1958)
N. Kalahari	150	500–700	1200	Lancaster (1981c)
Zimbabwe	300	500	400	Flint and Bond (1968)
N.W. India	175–200	850	350	Goudie *et al.* (1973)
Australia	200	1000	800	Glassford and Killigrew (1976)
N.E. Brazil	-	600	-	Tricart (1974)
Venezuela	-	1400	-	Tricart (1974)
Arizona	250	300–380	-	Hack (1941)

relatively common occurrence of granule ripples and coarse sand sheets in extra-glacial parts of the Antarctic and the Arctic (Pissart *et al.* 1977, Selby *et al.* 1974, Good & Bryant 1985).

The second effect of temperature is related to the fact that threshold velocities are raised and aeolian entrainment is suppressed on surfaces which experience seasonal freezing or are covered for part of the year by snow (McKenna-Neumann & Gilbert 1986). During the summer melt season, sand in the upper active layer also remains wet at shallow depth (Good & Bryant 1985). However, this factor is not in itself sufficient to prevent the formation of sand dunes at high latitudes (Calkin & Rutford 1974, Selby *et al.*

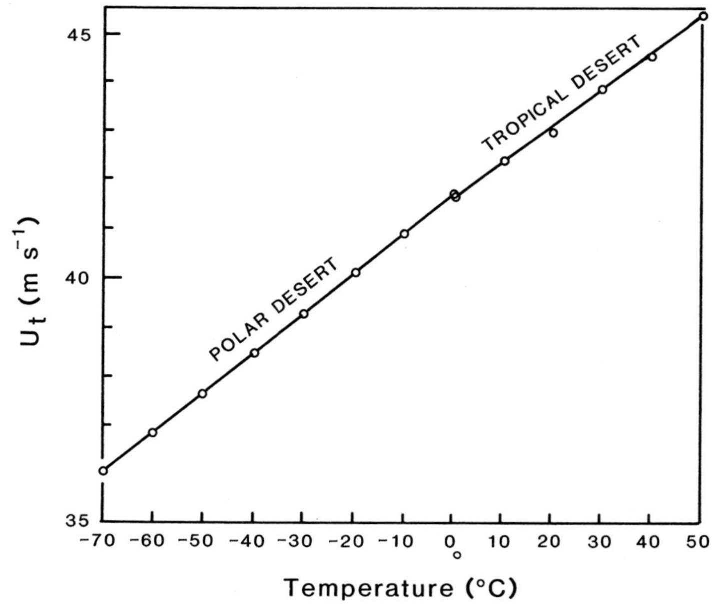

Figure 5.8 Threshold wind velocity required to lift a 2 mm granule to a height of 2 m at various temperatures. (After Selby *et al.* 1974).

1974, Carter 1981, Koster & Dijkmans 1988, McKenna-Neumann 1989, Williams *et al.* 1987).

5.3.3 Time required for the development of ergs and dunefields

If rates of sand supply are high, or if strong winds are able to rework large areas of bare sandy deposits *in situ*, very extensive dunefields can form within a few tens or hundreds of years. However, the formation of very thick, extensive sand sea deposits takes at least several thousand years, and requires geological and climatic conditions which change only slowly.

The mean thickness of sand in modern ergs is considerably less than that found in the geological record. Wilson (1973) calculated that the mean sand thickness in a number of Saharan ergs ranged from 21 to 43 m, whereas that of the Simpson Desert erg is only about 1 m. In the central Namib Desert, where some individual dunes approach 100 m in height, the maximum mean sand thickness is only 30 m, whereas on the fringes of the Namib the mean sand thickness is less than 10 m (Lancaster 1988c). Some modern aeolian sand sheets, such as the Selima Sand Sheet of southern Egypt, consist of only a few centimetres of blown sand overlying fossil soils and fluvial sediments (Haynes 1982, Breed *et al.* 1987). This compares with mean thicknesses of up to several hundred metres found in some Mesozoic and Palaeozoic ergs [e.g. see papers in Kocurek (1988a)]. The greater thickness of aeolian sequences in the geological record compared with modern examples may be explained by a combination of the following factors: (a) there has been preferential preservation of those ancient sequences which grew vertically over long periods in slowly subsiding basins or more rapidly in rift-valley settings; (b) some thick sequences may represent multiple 'stacked' ergs which migrated laterally before coming to rest at the accumulation site; (c) aeolian processes may have been more effective in the geological past, especially before the development of land plants.

The age structure of the present sand deserts is not well documented, partly because of the difficulties of dating sand deposits older than 100 000 years. However, stratigraphical and radiocarbon dating evidence indicates that the present dune forms are late Pleistocene or Holocene in age, but in some areas these dunes overlie older sand formations. In the Namib, for example, predominantly arid conditions have prevailed since Mesozoic times (Ward *et al.* 1983). The present sand sea probably began to form in Pliocene times, but is widely underlain by aeolian, fluvial, and playa sediments of early to middle Tertiary age (Ward 1988).

5.4 Development of sand seas in relation to topography

Sand seas can be regarded as static or dynamic depending on whether they show a shift in position over time. Static sand seas occur mainly in one of two situations (Fig. 5.9): in topographic depressions or upwind or downwind of major topographic obstacles. Sand seas found in topographic depressions can be formed in two distinct ways, by aeolian reworking of sediments which were transported to the basin by fluvial processes (Fig. 5.9a), or by accumulation of sand transported to the basin by wind (Fig. 5.9b). Static sand seas which develop either upwind or downwind of major topographic obstacles do so due to local deceleration or convergence of the regional wind flow (Fig. 5.9c). Dynamic sand seas occur mainly in flat terrain where erg migration in the downwind direction is allowed to proceed unhindered (Fig. 5.9d).

Wilson (1973) maintained that virtually all ergs are confined to basins and terminate at any pronounced break of slope. The margins of some ergs, such as those of the Issaouane-n-Irarraren in Algeria, closely follow topographic contours for several

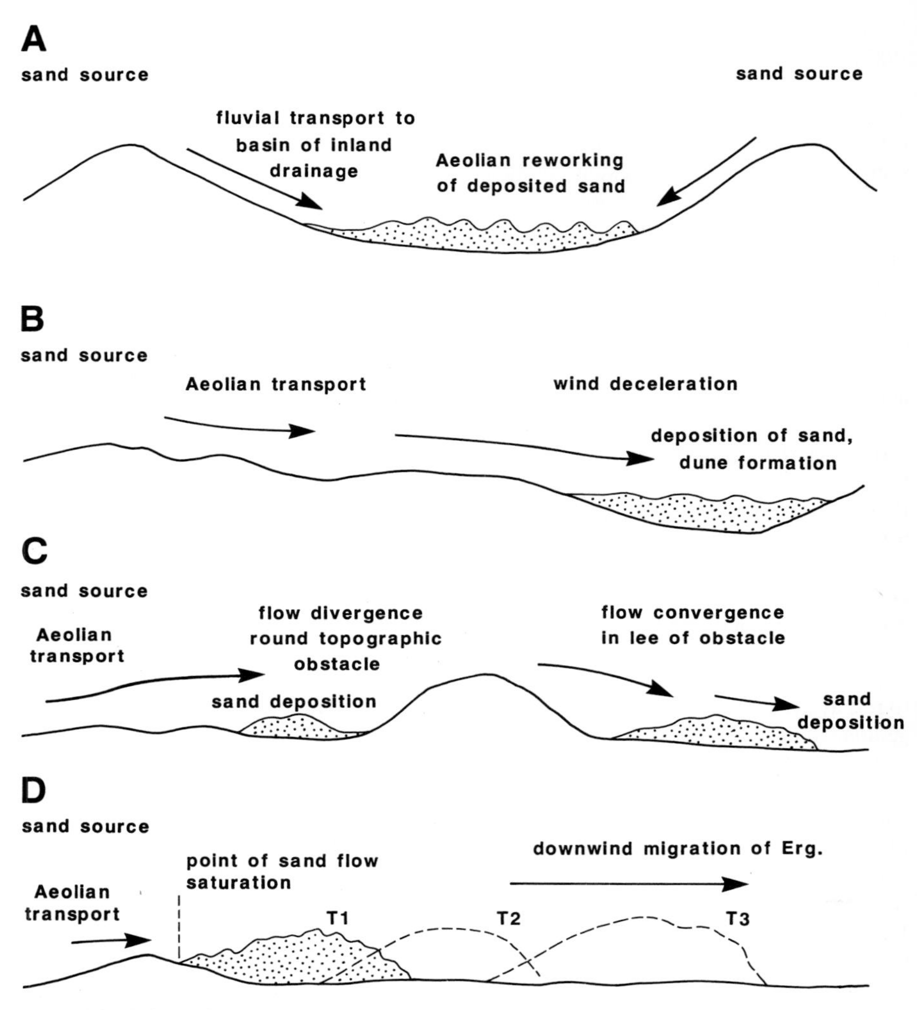

Figure 5.9 Schematic diagram showing topographic situations in which static (A – C) and dynamic (D) ergs may form.

hundred kilometres. However, within individual basins, the location of ergs is almost independent of relief (Wilson 1973). Where sand is blown large distances by the wind, topographic highlands such as mountain ranges act as barriers to the flow, leading to sand accumulation. Such a situation occurs in eastern Algeria, where the Tinhrert Plateau and the Tademait Plateau extend roughly perpendicularly to the south to southeastward regional drift of sand (Fryberger & Ahlbrandt 1979) (Fig. 5.10). The Great Sand Dunes dunefield in Colorado has also formed where the sand drift from southwest to northeast is blocked by the Sangre de Cristo Mountains (Fig. 5.4). At Ferris Dunefield, Wyoming, sand has accumulated mainly on the upwind side of the Ferris and Seminoe Mountain Ranges (Fig. 5.11) (Gaylord & Dawson 1987). The convergence of airflow north of Table Mountain causes a two- to three-fold acceleration towards Windy Gap. On passing through the Gap the concentrated flow ascends into a standing wave pattern, analogous to a hydraulic jump (Fig. 5.12). Below the zone of

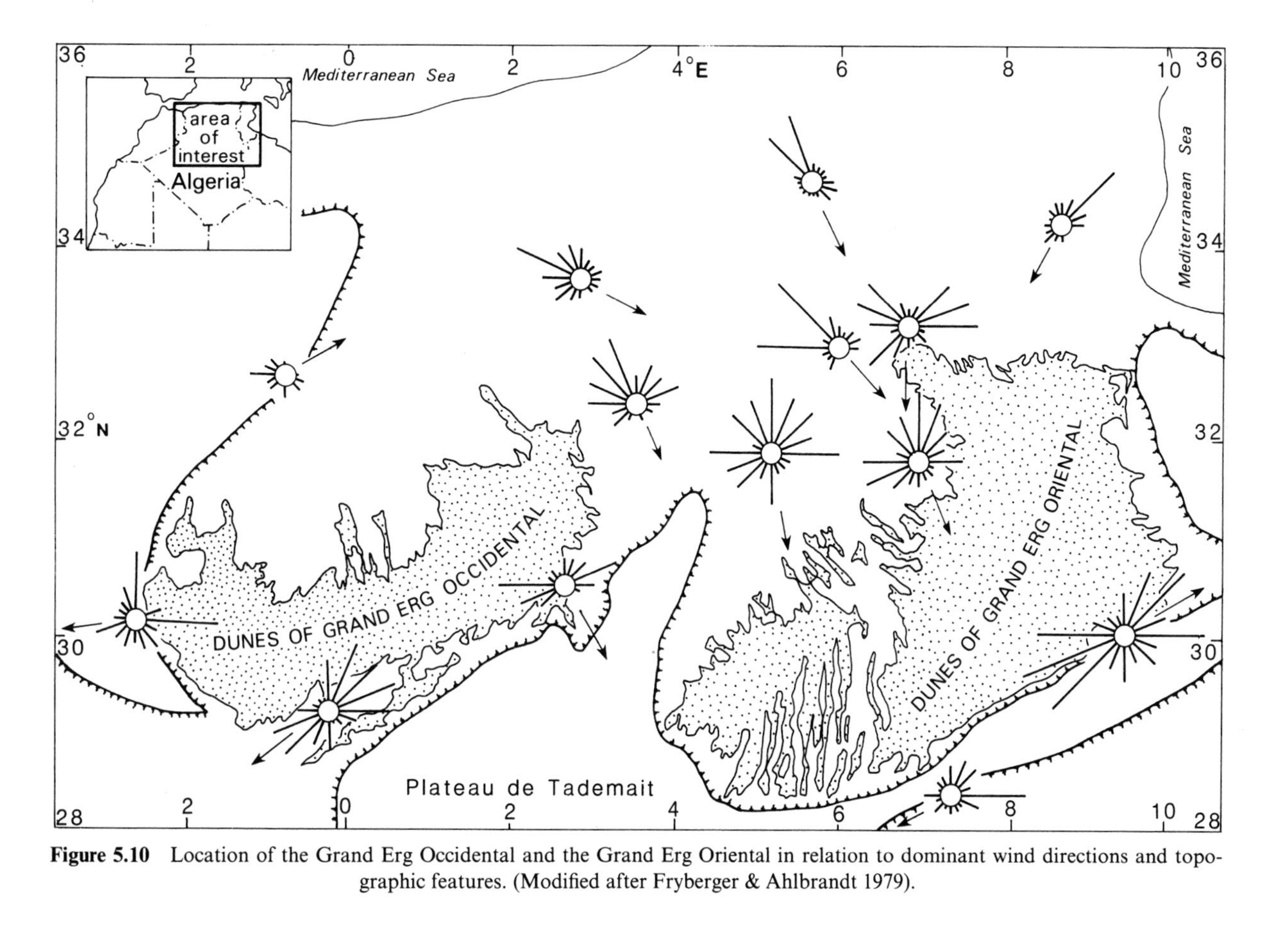

Figure 5.10 Location of the Grand Erg Occidental and the Grand Erg Oriental in relation to dominant wind directions and topographic features. (Modified after Fryberger & Ahlbrandt 1979).

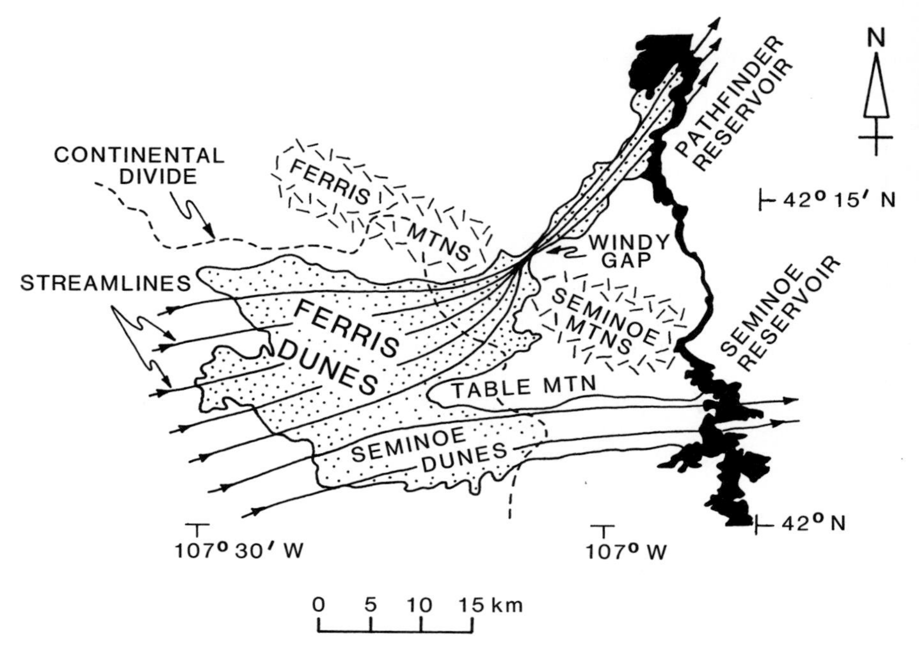

Figure 5.11 Location map of Ferris Dunefield, Wyoming, showing relationship of typical streamlines (100 m above the ground) to major topographic features. Note splitting of airflow by Table Mountain and confluence of flow through Windy Gap. Prevailing winds are from the west-southwest. (After Gaylord & Dawson 1987).

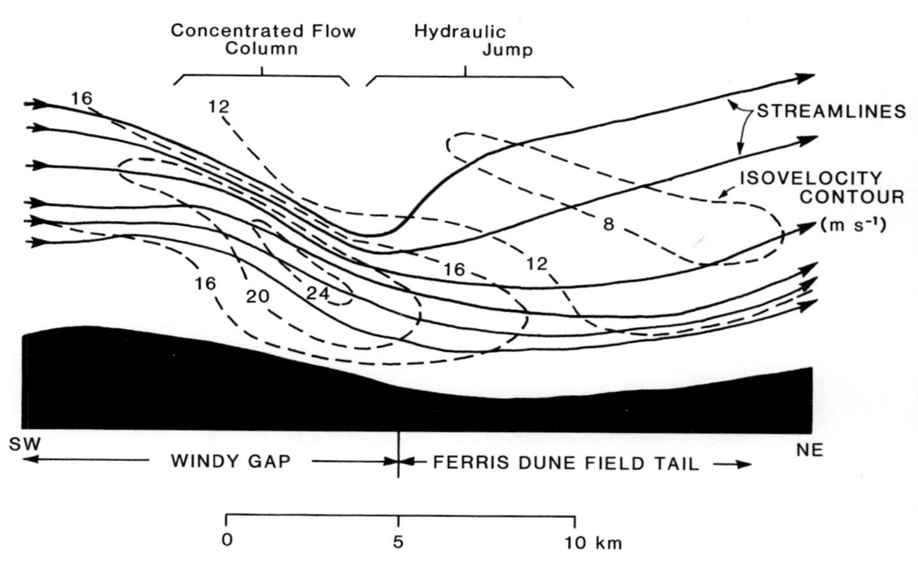

Figure 5.12 Vertical profile showing compression of streamlines and acceleration of the wind over Windy Gap, with streamline divergence and flow deceleration over the Ferris Dunefield Tail, Wyoming. Vertical exaggeration is about 6.5 ×. (Modified after Gaylord & Dawson 1987).

diverging and decelerating flow is a second elongated area of aeolian sand accumulation (Fig. 5.11). Highly turbulent, chaotic winds in the zone immediately below the hydraulic jump have created an elongate deflation hollow characterized by complexly mixed, poorly sorted aeolian deposits.

5.5 Wind regime and regional sandflow paths

The direction and rate of aeolian sand transport are strongly governed by the wind regime, i.e. the velocity distribution and directional variability. Regional sandflow resultants can be determined approximately using one of three methods: (a) from wind velocity data using a sand transport rate equation such as that proposed by Bagnold (1941, p. 67) (see Section 4.2.6); (b) by analysis of sandstorm duration and direction records (e.g. Dubief 1952); and (c) from the size and orientation of aeolian bedforms and sandstreaks. Using the first method resultants can be calculated for any time period and grain size, provided that adequate wind records are available. Unfortunately, this is often not the case in remote desert regions.

Using a combination of these methods, Wilson (1971) constructed a composite regional sandflow map for the Sahara (Fig. 5.13), which shows a clear divide across the central Sahara that runs into a complex sandflow 'gyre' located in eastern Libya. On the south side of this divide the sandflow is predominantly in a southwesterly direction; on the north side sandflow is predominantly in a northeasterly or easterly direction.

Based on the sandflow concept, Wilson (1971) developed a theoretical model to account for the development of ergs with and without bedforms. Central to these models is the relationship between the potential sandflow rate, $\bar{Q}\sigma$, which for any given grain size is governed by the wind velocity, and the actual mean sandflow rate, $\bar{Q}$. According to Wilson, if a wind stream crosses from a bare rock surface, where it is carrying no sand, to a sandy surface, it will gradually erode the bed until the flow becomes saturated (i.e.

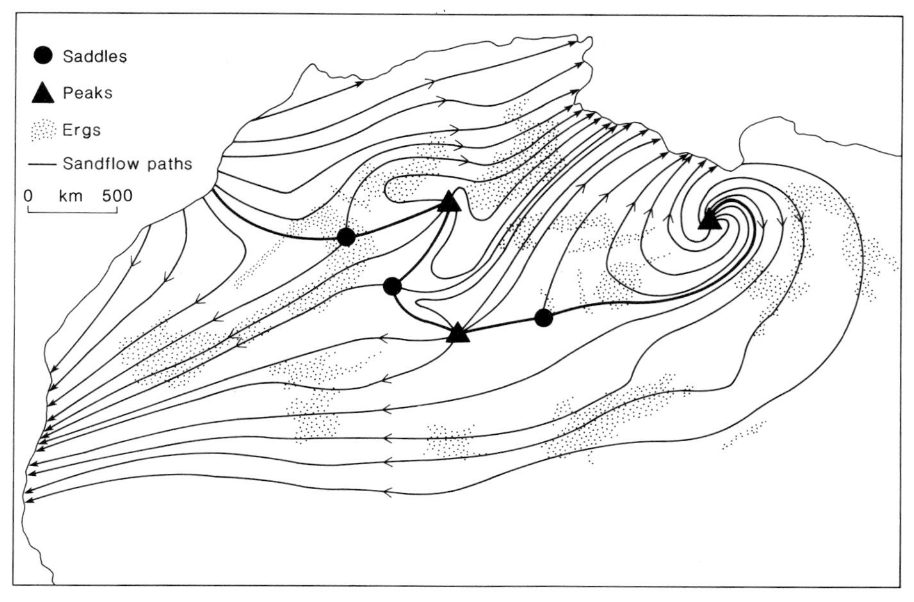

Figure 5.13 Sandflow map of the Sahara. (Simplified after Wilson 1971).

$\bar{Q}\sigma = \bar{Q}$) at some distance downwind from the roughness boundary. For an erg without bedforms, deposition of sand cannot occur unless the sandflow is both saturated and decelerating or converging. Consequently, the sand will continue to move downwind until such conditions are encountered. It follows that ergs cannot form at sandflow peaks which are points of sandflow divergence; they can only form in areas of convergence, i.e. within sandflow 'centres' or on either side of a 'saddle'.

Wilson (1971) pointed out that this simple model has to be modified for natural ergs whose surfaces are covered by bedforms. Sedimentation on such bedforms is controlled not only by the regional wind system but also by secondary flows associated with the bedforms. Such bedforms can survive in equilibrium with undersaturated sandflows, but if the degree of saturation is decreased there must be a point at which the dunes can no longer survive because they lose sand faster than they can trap it. Wilson termed this the *metasaturation point*. Although a sandflow may not be fully saturated, and therefore is unable to maintain a complete sand cover, under conditions of metasaturated sandflow it may be possible for individual dunes to grow with bare ground between them, even without convergence or deceleration of the regional sandflow. Bedforms at the upwind end of the incipient erg remove sand from the airflow crossing them so that the sandflow downwind is depleted. However, once the bedforms close to the upwind edge are fully grown they no longer deplete the sandflow. As a result, a front of equilibrium advances downwind, leaving fully grown bedforms behind it. The metasaturation zone advances ahead of the advancing equilibrium front until sustained divergence or acceleration of the regional sandflow prevents further extension, thereby fixing the downwind margin of the erg. Any further growth of the erg can then occur only by thickening of the saturation zone.

This basic model is to some extent complicated by the fact that the aeolian bedforms may themselves move downwind over time. Given sufficient time, entire ergs composed of migrating bedforms might also move downwind. Whether this happens will depend on the nature and mobility of the bedforms, which are in turn determined by the sand availability, wind regime, and vegetation cover.

The maintenance of the sand supply from upwind is a key factor which determines whether the discontinuous deposits at the upwind erg margin are maintained. An exhaustion of the sand supply will mean that the deposits at the erg margin will be eroded (Wilson 1971).

Partial support for Wilson's (1971) ideas about the formation of ergs has been provided by studies using satellite imagery. These have shown that there is often a distinct sequence of bedform types from the margin to the centre of many modern ergs (Breed & Grow 1979). Small barchanoid or transverse dunes, zibars, and sand sheets are the most common forms found on erg margins, while erg centres consist largely of complex megadunes. Some modern sand seas show clear evidence of migration which has produced a distinctive spatial association of sedimentary facies (e.g. Fryberger *et al.* 1983). Further evidence of long-term erg migration is provided by vertical sedimentary sequences in the geological record (Clemmensen & Abrahamsen 1983, Porter 1986). Palaeozoic and Mesozoic erg sequences often show three distinct sedimentary units which, from the bottom to top, have been termed *fore-erg*, *central erg*, and *back-erg* facies, respectively (Fig. 5.14) (Porter 1986). However, Sweet *et al.* (1988) have shown that the direction of erg migration need not be parallel to the resultant sand transport direction. In the case of the Algodones dunefield, California, migration has apparently occurred in an easterly direction, oblique to the resultant sand flow direction (S24°E), owing to a localized secondary airflow generated by interaction between the regional winds and the dunefield (Sweet *et al.* 1988).

As discussed in Section 4.2.6, a number of equations can be used to calculate potential

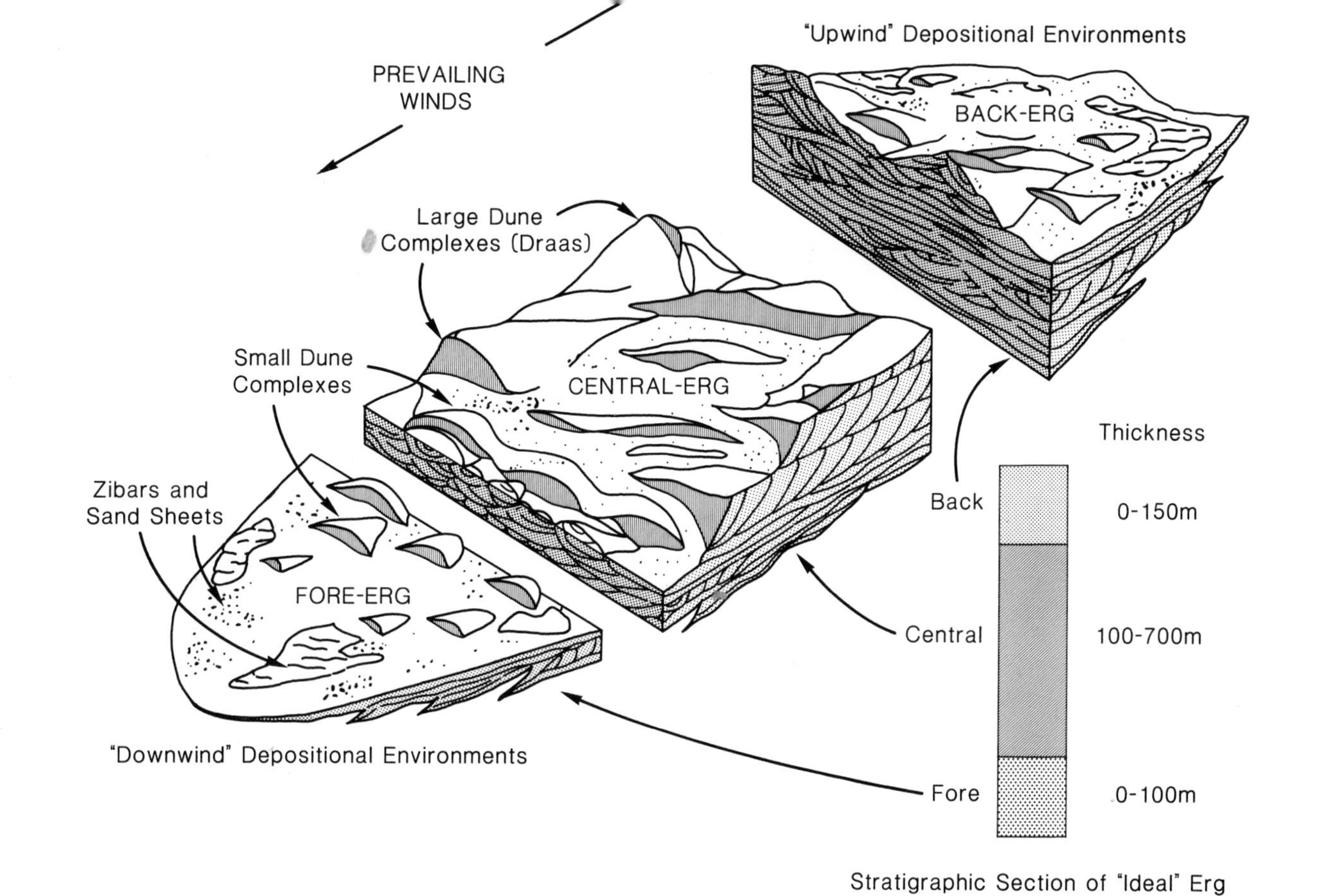

Figure 5.14 Conceptual model of a migrating erg, showing the stratigraphic sedimentary sequence of the 'ideal' erg. (After Porter 1986).

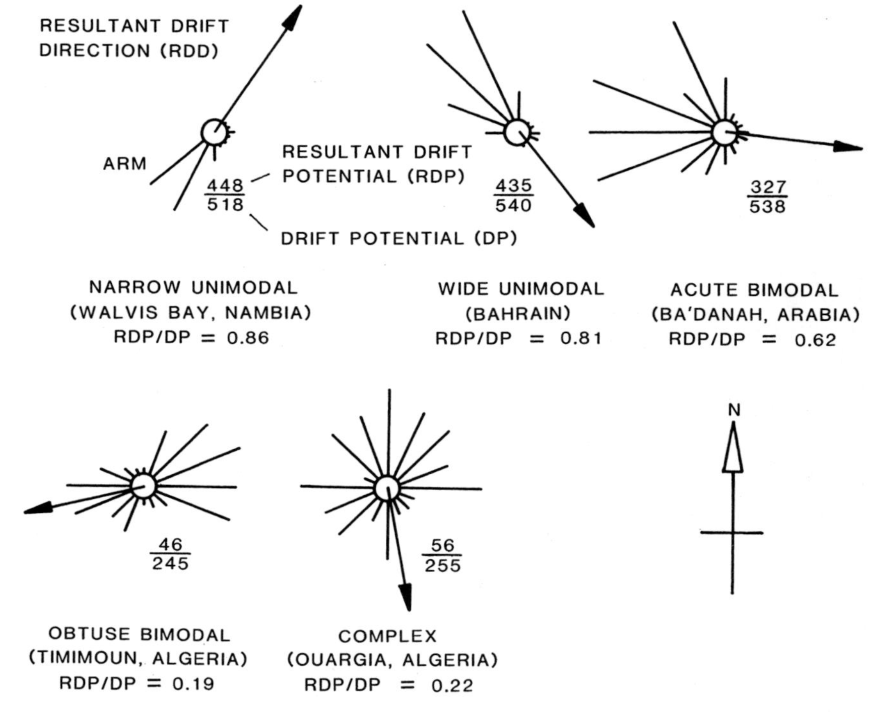

Figure 5.15 Examples of annual sandflow regimes. (After Fryberger & Dean 1979).

sand transport rates. By combining these equations with observed wind frequency and direction data, indices of regional sand drift potential can be calculated. Fryberger & Dean (1979) used a derivation of the sand transport equation developed by Lettau & Lettau (1978) (Eqn 4.38) to calculate *drift potentials* (DP), which are expressed numerically in vector units. Fryberger and Dean's index of drift potential is given by

$$Q_{\mathrm{p}} \propto U^2[U - u_{\mathrm{t(i)}}]t \qquad (5.1)$$

where Q_{p} is a proportionate amount of sand drift, U is average wind velocity measured at a height of 10 m, $u_{\mathrm{t(i)}}$ is the impact threshold wind velocity, and t is the time the wind blew as a percentage of the total record.

A number of contrasting drift potential roses are shown in Figure 5.15. The direction of the vector resultant of drift potentials for sixteen points of the compass is defined as the *resultant drift direction* (RDD), while the magnitude of the vector resultant is defined as the *resultant drift potential* (RDP) [see Fryberger & Dean (1979, pp. 146–7) for computation details]. An index of the directional variability of the wind is given by the ratio RDP/DP; the greater the directional variability of the effective sand-transporting winds, the lower is the RDP/DP ratio.

Average drift potentials for thirteen desert regions, calculated by Fryberger & Dean (1979), are listed in Table 5.3. Deserts such as the Thar and Takla Makan, which have relatively low drift potentials, are located near the centres of semi-permanent high- or low-pressure cells, whereas deserts with relatively high drift potentials, such as those of North Africa and Saudi Arabia, lie on the margins of such cells and are influenced to a greater extent by tradewind circulations or mid-latitude depressions (Fryberger & Ahlbrandt 1979).

Table 5.3 Average annual drift potentials for 13 desert regions based on data from selected stations. (After Fryberger & Dean 1979).

Desert region	Number of stations	Average annual drift potential (in vector units)
High-energy wind environments		
Northern deserts, Saudi Arabia and Kuwait	10	489
Northwestern Libya*	7	431
Intermediate-energy wind environments		
Simpson Desert, Australia	1	391
Western Mauritania	10	384
Peski Karakumy and Peski Kyzylkum, USSR	15	366
Erg Oriental and Erg Occidental, Algeria	21	293
Namib Desert, South Africa	5	237
Rub' al Khali, Saudi Arabia	1	201
Low-energy wind environments		
Kalahari Desert, South Africa	7	191
Sahelian zone, Niger River, Mali	8	139
Gobi Desert*, Peoples Republic of China	5	127
Thar Desert*, India	7	82
Takla Makan Desert*, Peoples Republic of China	11	81

* *DP*s estimated.

However, there are important variations in sand drift potential within individual deserts. Transfer of sand from areas of high energy to areas of low energy has been documented in the Jafurah Sand Sea of Saudi Arabia (Fryberger *et al.* 1984) (Fig. 5.16) and in the Namib Desert (Lancaster 1985b). It should also be stressed that Fryberger & Dean's parameters, like other indices of sandflow based on Bagnold's equation (e.g. Lancaster 1985b), provide a measure of potential rather than actual sandflow. The latter is governed to a large extent by the distribution of sand sources and vegetation cover, and also by wind velocity and direction.

Analysis of satellite images has shown that regional sandflow paths are often well defined, reflecting a close interaction between topography and surface winds (Mainguet 1978, Mainguet & Cossus 1980, Mainguet & Chemin 1983). Currents of sand-laden air are channelled between and around highland areas, but sand deposition leading to erg formation is not restricted to topographic depressions. Mainguet *et al.* (1984) observed that in the Sahara large sand accumulations may form in several different situations anywhere where there is deceleration of the regional winds. This can occur in several different situations: (a) where a large topographic obstacle lies transverse to the sand stream; (b) where the sand stream divides to flow around an obstacle; (c) where two or more sand streams converge; and (d) where a sand stream moves into an area of wetter climate and thicker vegetation cover. Examples of sand accumulation in areas of flow divergence and flow convergence are found upwind and downwind, respectively, of the Eglab Massif (Fig. 5.17).

Mainguet (1978) identified four main sand streams which she suggested transport sand over distances of thousands of kilometres in the Sahara (Fig. 5.18):

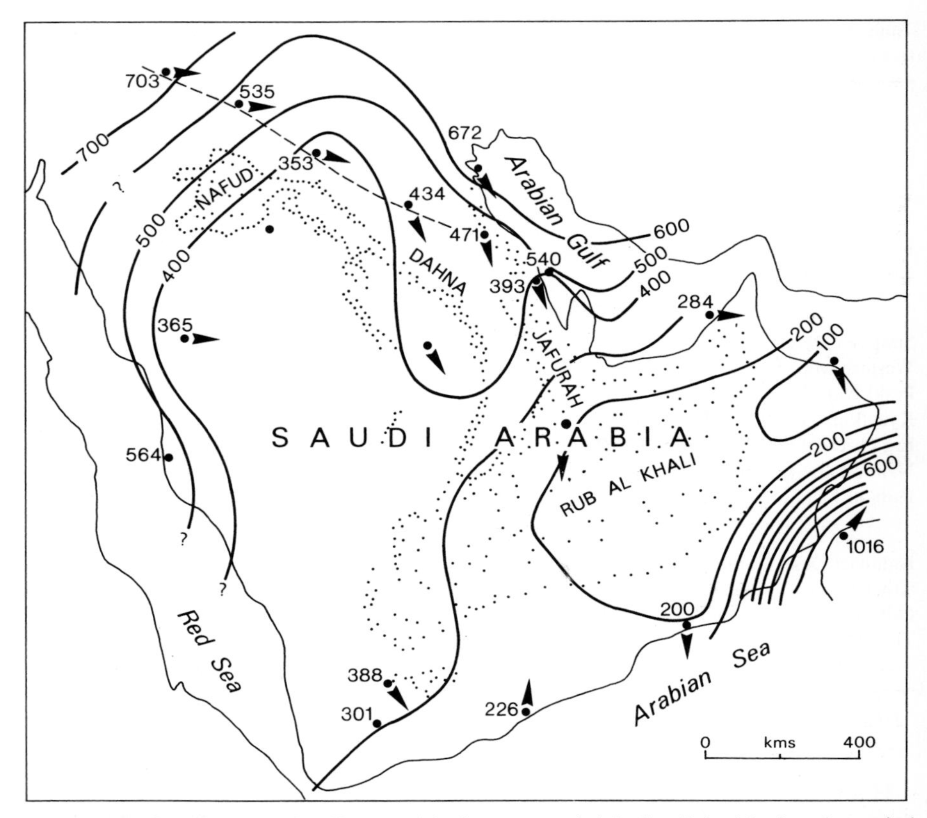

Figure 5.16 Contour map of drift potentials (in vector units) in Saudi Arabia, based on wind records from the National Climatic Center, Asheville, North Carolina. (After Fryberger *et al.* 1984).

(a) An eastern sand stream which starts at about latitude 29°N and sweeps through Egypt before dividing upwind of Tibesti. The two branches converge downwind (southwest) of Tibesti in the area of the Erg of Fachi Bilma, Niger.

(b) A central sand stream which starts between Gebel el Assaouad and the Hamada el Homra, which it bypasses to the south before sweeping southwestwards along the southern border of the Erg of Mourzouk. It then passes between the Hoggar and the Air Mountains and splits into two branches, one of which is deflected north around the Adrar des Iforas towards the Erg Chech, while the other continues towards the Mauritanian ergs.

(c) A western sand stream which is formed by the coalescence of two tributary flows, one of which originates in the southern part of the Great Eastern Erg and the other south of the Great Western Erg. The two sand streams converge south of Kreb en Naga and continue for a further 1500 km before reaching the Atlantic coast between 16 and 20°N.

(d) An Atlantic coastal sand stream which originates near Cap Juby in Mauritania and runs southwards almost parallel to the coast before entering the sea at the same latitude as the western sand stream.

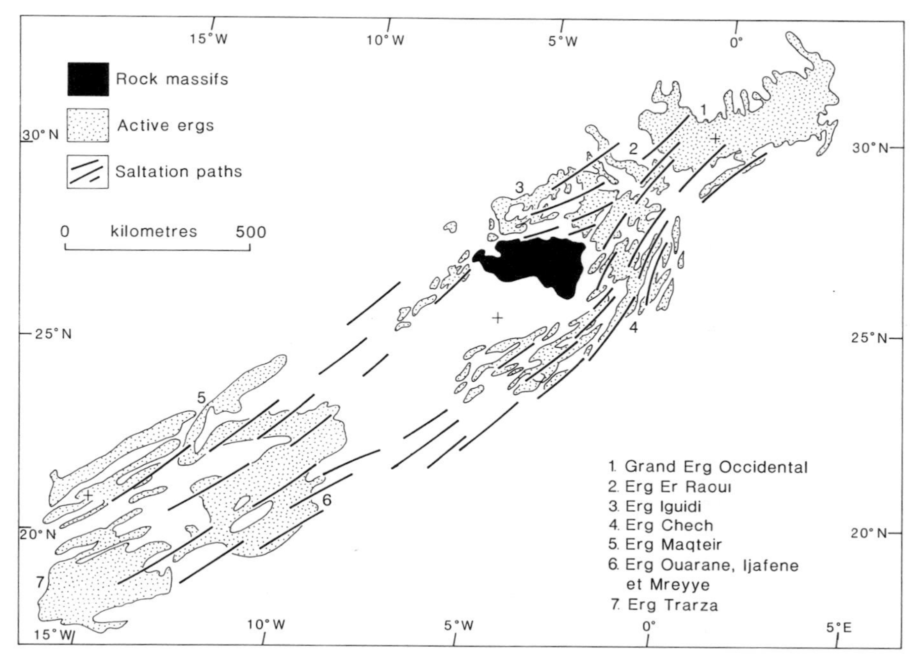

Figure 5.17 Formation of ergs at points where sandstreams divide and converge around the Eglab Massif, Algeria. (Modified after Mainguet *et al.* 1984).

5.6 Evolution of ergs in response to climatic changes

Work carried out since the late 1950s has revealed that many of the present-day ergs were very much more extensive during earlier periods of the Quaternary (Fig. 5.1b). Large areas of the African continent are covered by fossil dunes which are now degraded, cultivated, or forested. In West Africa, the limit of active dune formation at times in the late Pleistocene moved southwards more than 600 km from its present position (Grove 1958, 1969, Prescott & White 1960, Grove & Warren 1968, White 1971, Sarnthein 1978, Talbot 1984) (Table 5.2). Similar enlarged ergs of late Pleistocene age have been identified in the Kalahari (Flint & Bond 1968, Heine 1982, Lancaster 1981c, 1989d, Thomas 1984, Thomas & Goudie 1984), northwest India (Allchin *et al.* 1978, Goudie *et al.* 1973, Wasson *et al.* 1983), and Australia (Bowler *et al.* 1976, Wyrwoll & Milton 1976). In South America, small ergs occupied parts of the Sao Francisco and Orinoco catchments in the late Pleistocene (Tricart 1974). Aeolian activity was also more extensive in the Great Plains and the Carolinas of North America (Price 1944, 1958, Wells 1983, Carver & Brook 1989).

Radiometric dating and other evidence indicates that the now fossilized dunes in many of these areas were last active between 20 000 and 13 000 years ago, with maximum aeolian activity occurring around the time of the last glacial maximum (Sarnthein 1978; Sarnthein *et al.* 1981; Bowler, 1978). An exception is provided by the Southwestern United States, where conditions were wetter and aeolian activity was suppressed at this time compared with the Holocene.

Conversely, around the time of the mid-Holocene climatic optimum (about 6000 yr ago), the area of active dunes in many areas was less extensive than at present (Sarnthein

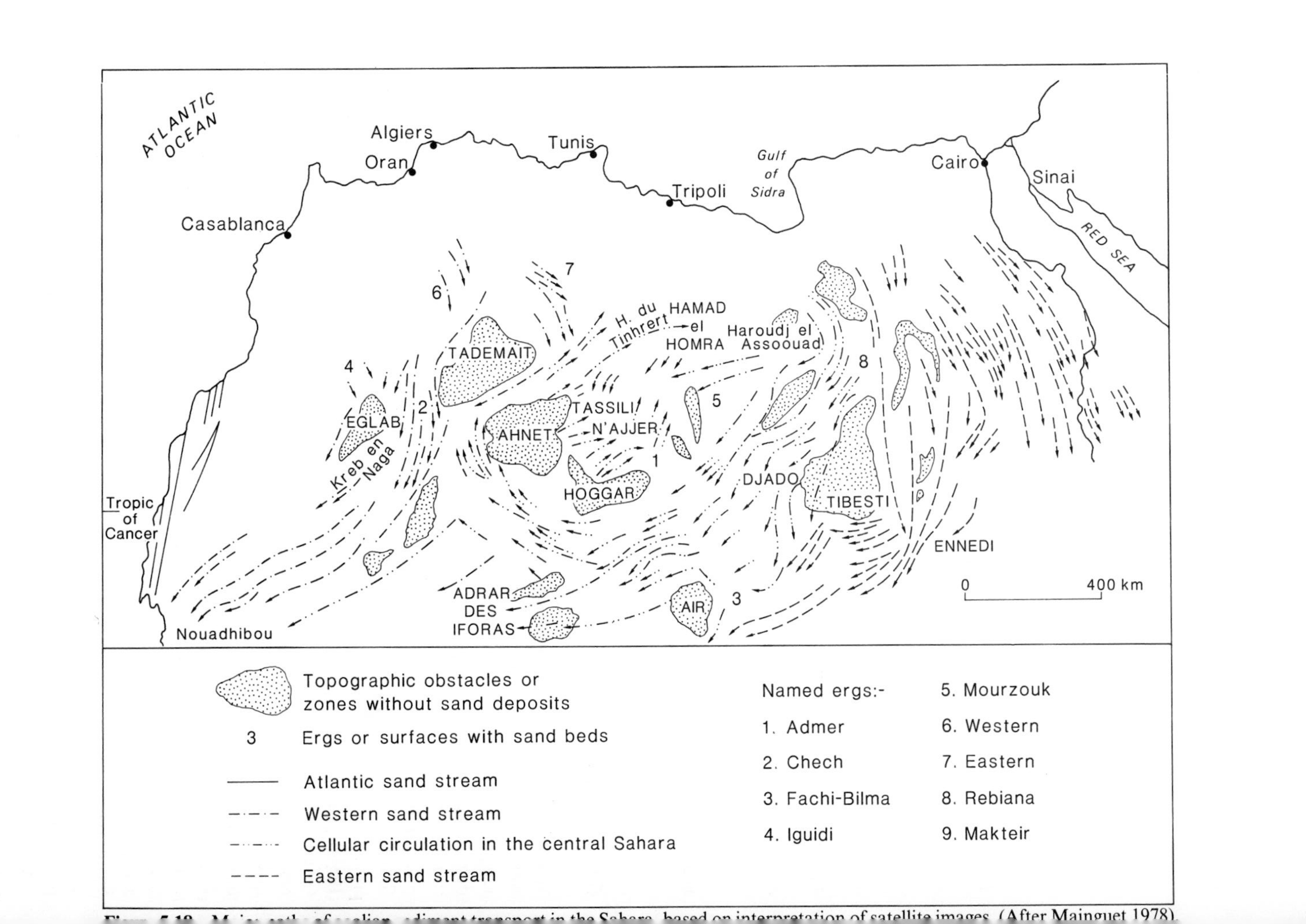

Figure 5.18 Major paths of aeolian sediment transport in the Sahara, based on interpretation of satellite images (After Mainguet 1978).

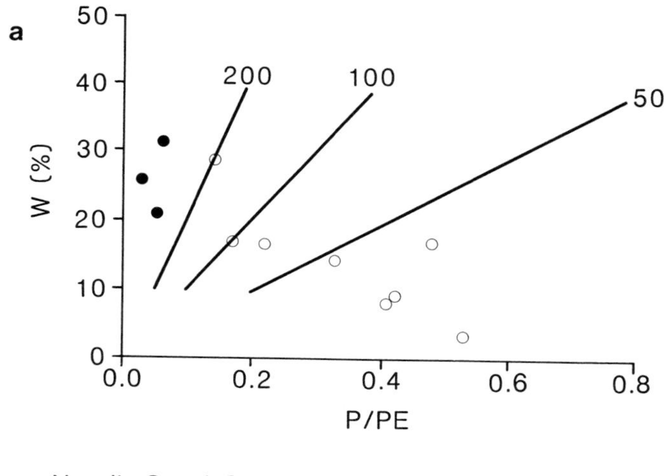

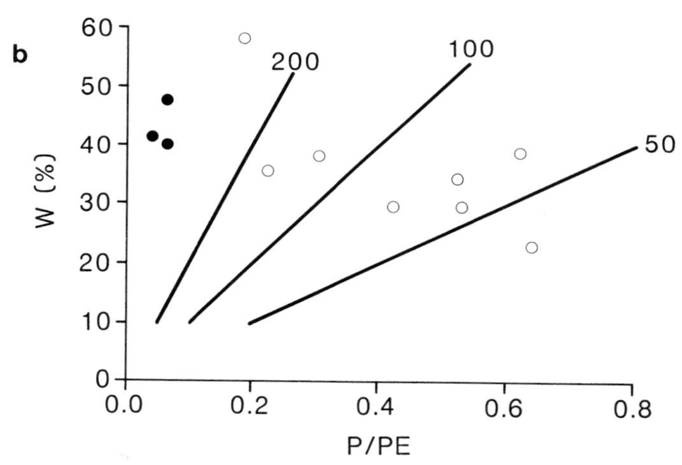

Figure 5.19 Relationship between the percentage of time the wind is above threshold velocity for sand transport (W) and effective precipitation (P/PE) for arid areas in southern Africa (a) today and at (b) 18 000 yr BP. (After Lancaster 1988c).

1978). Much of the Sahara experienced wetter conditions during the early to mid-Holocene, as indicated by high lake levels and a variety of floral, faunal, and archaeological evidence. The timing and magnitude of the changes in climatic conditions varied between different areas, but dunes in most areas were stabilized for varying periods and experienced weathering, pedogenesis and partial reworking by fluvial processes (Rognon & Williams 1977, Talbot & Williams 1979, Talbot 1984). Similar, although not simultaneous, changes in rainfall affected the sand deserts of Saudi Arabia (Whitney *et al.* 1983) and southern Africa, although it is unlikely that the dunes in the hyper-arid part of the central Namib were ever completely stabilized (Lancaster 1988c).

Fluctuations in environmental conditions before the last glacial maximum are also indicated by a large amount of stratigraphical, sedimentological, and botanical evidence, although the timing of such changes is poorly documented beyond the limits of the radiocarbon timescale. However, there is strong evidence that several desert areas experienced wetter conditions immediately prior to the last glacial maximum (Rognon & Williams 1977, Bowler *et al.* 1976, Bowler 1978).

The increased aeolian activity around the time of the last glacial maximum probably resulted from combined changes in wind regime, temperature, and rainfall, since all three factors influence the moisture balance and vegetation cover within an area. The relative importance of changes in each of these three factors is difficult to determine. However, evidence provided by dust records in ocean cores demonstrates that some wind systems, such as the trade winds, were certainly stronger during glacial times due to intensified latitudinal temperature and pressure gradients (Parkin & Shackleton 1973, Parkin & Padgham 1975). It is unlikely, however, that changes in wind intensity alone can account for the greater extent of active dunes in the Last Glacial period. Even taking into account increases in wind strength, Talbot (1984) estimated that a reduction in rainfall of 25–50% would have been required to account for dune activation in the Sahel between 13 000 and 20 000 yr BP.

A dune *mobility index*, M, can be calculated which relates wind energy and effective precipitation (Talbot 1984, Lancaster 1988c):

$$M = W/(P/PE) \tag{5.2}$$

where W is the percentage of time the wind is blowing above the threshold for sand transport (taken to be 4.5 m s^{-1}), P is the annual rainfall, and PE is the annual potential evapotranspiration (Thornthwaite 1931).

Based on field observations in the Sahel (Talbot 1984) and southwest Africa (Lancaster 1988c), critical values of M have been identified for different degrees of dune activity. For fully active dunes it is suggested that values of M must exceed 200; between $M = 100$ and 200 dune plinths and interdune areas are stabilized; between $M = 50$ and 100 only dune crests are active; and for $M < 50$ the dunes are entirely stabilized (Lancaster 1988c) (Fig. 5.19).

5.7 Effect of sea-level changes on coastal dunefields

Coastal dune activity is influenced not only by changes in wind strength, rainfall, and evaporation rates, but also by changes in sea level and rates of marine sediment supply. The relationship between sea level changes and the development of coastal dunes has been much debated, and it is evident from data collected in many different areas that dunes can form during high and low sea level stands and marine transgressions and regressions.

Bretz (1960) concluded that the carbonate dunes of Bermuda, which in many places can be traced laterally into beach and nearshore marine sands at, or slightly above, modern sea level, were formed mainly during high sea level stands (Fig. 5.20a). Bretz argued that the present dunes could not have originated on, and migrated large distances across, the Bermuda Platform during times of low sea level, since the carbonate dunes rapidly become cemented during subaerial diagenesis.

An opposite conclusion was reached by Sayles (1931) in the context of Bermuda and by many other workers elsewhere (Wright 1963, Coetzee 1975/6a, Hobday 1977). According to these authors, dunes now exposed along the shoreline originally formed on the continental shelf during glacial low sea level stands. According to this model, during

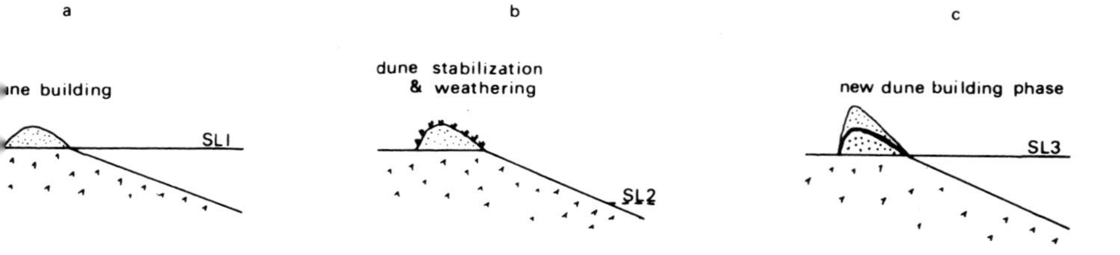

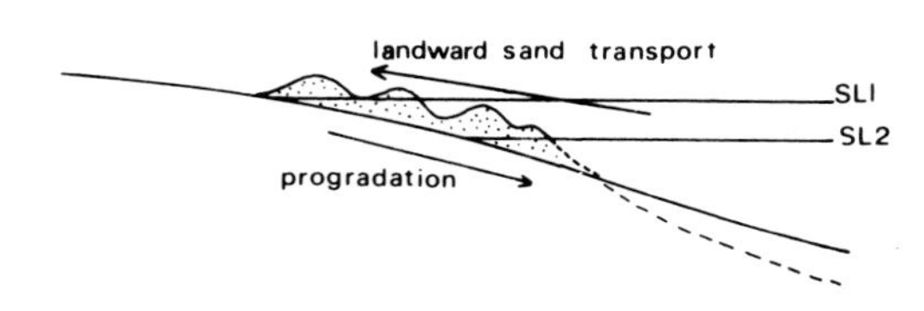

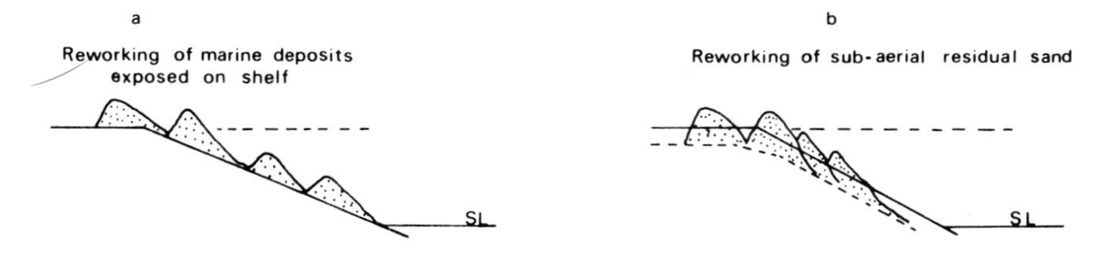

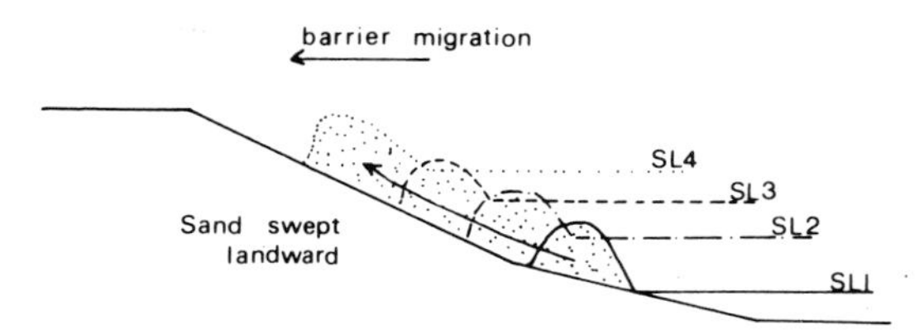

Figure 5.20 Four alternative models of coastal dune development in response to sea level rise. For explanation, see text. (Modified after Pye 1984).

transgressions some dunes were submerged while others advanced onto higher ground where, deprived of their sand supply, they became stabilized (Fig. 5.20c).

A third model suggests that most dune formation takes place while the sea level is falling during the transition from interglacial to glacial conditions (Fig. 5.20b). Following this argument, a falling sea level lowers the wave base and increases the continental shelf area over which landward sand transport can take place (Schofield 1975).

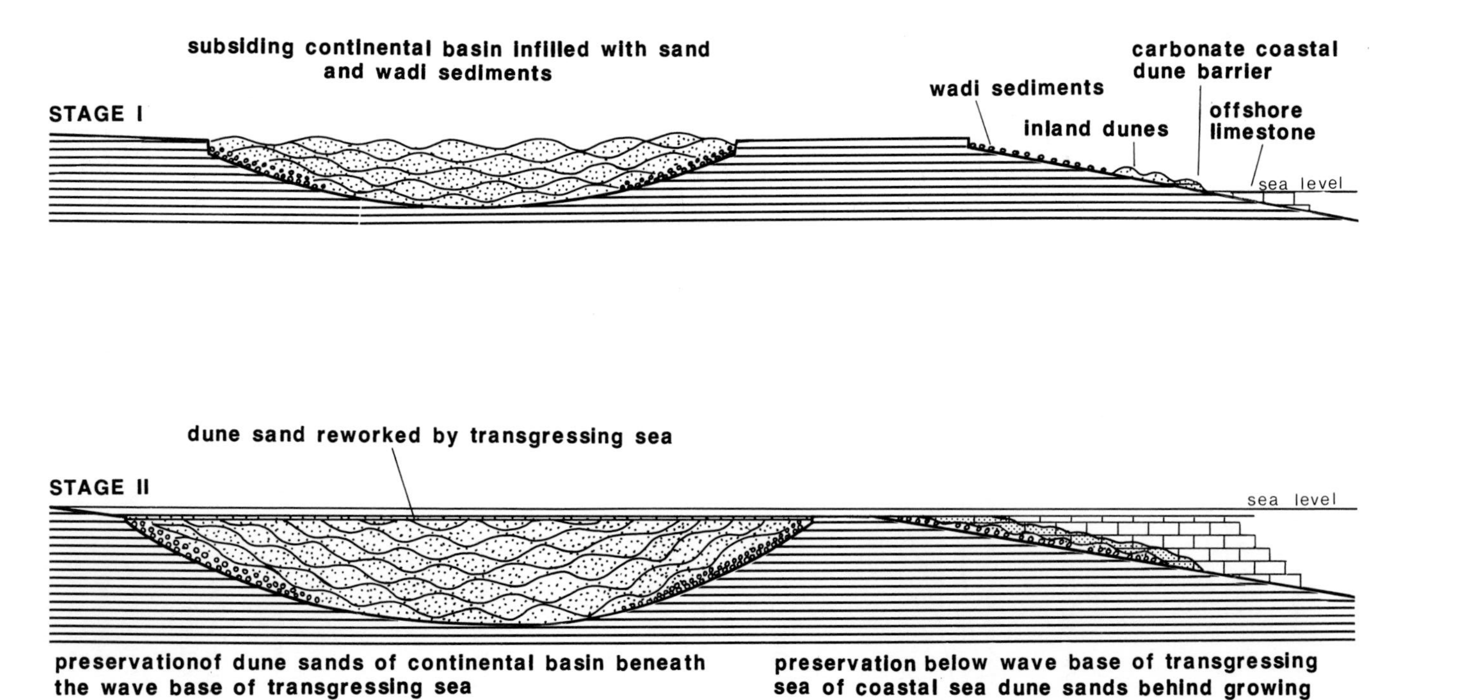

Figure 5.21 Schematic cross-sections to illustrate how unconsolidated dune sands may be preserved beneath the wave base of a transgressing sea. (After Glennie 1970, p. 9).

Conversely, a rising sea level raises the wave base and reduces the shoreward movement of sand.

A fourth model (Fig. 5.20d), originally proposed by Cooper (1958) in relation to dune development on the Oregon coast, suggests that marine transgressions are responsible for initiating episodes of transgressive coastal dune development, whereas regressions lead to shoreline progradation and beach ridge construction. Rising sea levels cause shoreface erosion and offshore movement of sand (Bruun 1962), but in areas of high wind energy the destruction of foredune vegetation may allow large amounts of sand to be blown landwards as transgressive dunes (Thom 1978). Detailed morphostratigraphic studies, supported by radiocarbon dating, have indicated that such a sequence of events occurred on many exposed parts of the eastern Australian coast during the post-glacial marine transgression (Pye 1984, Pye and Bowman 1984). During the later part of the Holocene, when the sea level had not varied in this area by more than ±1 m, smaller scale episodes of transgressive dune activity may be related to fluctuations in wind and storm wave climate, or to local aboriginal burning (Thom 1978). Here, as in many other areas (e.g. Filion 1984), it is difficult to separate the effects of changes in sea level, climate, and human disturbance with any degree of certainty. In suitable topographic situations, coastal dune sand bodies may be partially or completely drowned during a period of rising sea level. Examples have been described from the northeast coast of Australia (Pye & Rhodes 1985) and Baja California (Fryberger *et al.* 1990).

5.8 Effect of sea-level changes on continental dunefields

Where regional sand streams blow offshore, the margins of desert ergs may extend across the continental shelf, especially during periods of falling sea level. Along the coast of West Africa, for example, Sarnthein & Diester-Haas (1977) described an accretionary slope composed of aeolian sand turbidites which formed during the Last Glacial period of low sea level. During the ensuing marine transgression, the windblown deposits were largely reworked as liquefied sand flows and high-density turbidity currents.

Rapid sea level rise may submerge parts of a coastal erg with varying degrees of dune destruction and marine reworking (Glennie & Buller 1983, Eschner & Kocurek 1986, 1988). The degree to which the dune topography is preserved during a marine transgression is largely determined by the rate of sea level rise, the wave energy regime of the transgressing sea, and the extent to which the dunes have undergone early diagenetic cementation (Chan & Kocurek 1988). If the dunes have accumulated in a slowly subsiding basin separated from the coast by a topographic high which eventually is overtopped by the rising sea, only the top of the aeolian sediment sequence is likely to be reworked (Glennie 1970) (Fig. 5.21). Other things being equal, dune morphology is likely to be destroyed and the sands extensively reworked if the rate of sea level rise is slow and the storm wave energy of the transgressing sea is high. Carbonate dune structures are less likely to be destroyed because they are often rapidly cemented during subaerial exposure. Submerged aeolianite ridges have been identified on many continental shelves, including those of Bermuda, Israel, Natal, and Mozambique (Almagor 1979, Hobday & Orme 1975).

6
Aeolian bedforms

6.1 Types of aeolian sand accumulation and bedform terminology

Based on field and air photograph measurements, Wilson (1972a) recognized a hierarchy of aeolian bedforms consisting of four components: two types of ripples (aerodynamic ripples and impact ripples), dunes, and draas (Table 6.1). *Draa* is a North African term for a large sand hill (Capot-Rey 1945, Price 1950). The spacing of the three highest orders of bedform was attributed by Wilson to different scales of atmospheric instability. Wilson also suggested that there is a relationship between bedform spacing and grain size, with the larger, more widely spaced bedforms consisting of coarser sand. The explanation offered for this relationship was that mobilization of larger grains requires higher shear velocities associated with larger scale atmospheric flows. However, this *granulometric control hypothesis* has not been supported by subsequent work. Wasson & Hyde (1983a) showed that draas cannot always be distinguished from dunes on the basis of grain size, and there is continuum of scale between the two. More recently, Havholm & Kocurek (1988) proposed that draa should be used as a purely morphological term for any aeolian bedform with smaller superimposed dunes. This definition includes both complex and compound forms in the sense used by McKee (1979b). In most cases, draas are distinctly larger than simple dunes, but exceptions can be found (Havholm & Kocurek 1986). In our view it is preferable to use the term *megadune* to describe very large aeolian bedforms. Megadunes can be simple, complex, or compound (see Section 6.3.1).

Table 6.1 Wilson's hierarchy of aeolian bedforms. (After Wilson 1972a).

Order	Name	Wavelength (m)	Height (m)	Origin
1	draas	300–5500	20–450	aerodynamic instability
2	dunes	3–600	0.1–100	aerodynamic instability
3	aero-dynamic ripples	0.015–0.25	0.002–0.05	aerodynamic instability
4	Impact ripples	0.05–2.0	0.0005–0.1	impact mechanism

Figure 6.1 Wind ripples (foreground) and megaripples (background) in Rice Valley, California.

6.2 Ripples

6.2.1 The general nature of sand ripples

Two main types of wind ripples can be recognized on the basis of size: (a) normal ripples with wavelengths of < 1 cm up to 25 cm, and (b) larger ripples, termed *ridges* by Bagnold (1941, p. 149) but more widely known as *megaripples* (Greeley & Peterfreund 1981, Greeley & Iversen 1985, p. 154, Tsoar 1990a) (Fig. 6.1), which can have wavelengths of up to 20 m and heights of up to 1 m (Bagnold 1941, p. 155, Wilson 1972a). Megaripples are often composed of coarse sand and, occasionally, pebbles (Newell & Boyd 1955, Weir 1962, Smith 1965, Sakamoto-Arnold 1981). In the latter case they are referred to as granule megaripples to distinguish them from sand megaripples.

Any ripple profile can be divided into four elements: stoss slope, crest, lee slope, and trough. In the case of aeolian ripples the maximum inclination of the stoss slope ranges from 8 to 10°, whereas that of the lee slope ranges from 20 to 30° (Sharp 1963).

There are three main classes of ripples (Fig. 6.2): (a) asymmetric wind ripples, (b) asymmetric aqueous current ripples, and (c) symmetrical wave ripples (Twenhofel 1950, p. 568, Tanner 1967). A dimensionless indicator, the *ripple index*, *RI*, defined as the ratio of the ripple wavelength (L) to ripple height (h), can often be used to distinguish the different types of ripple. Wind ripples typically have $RI > 10$–15, whereas for water ripples $RI < 10$–15 (Cornish 1897, Bucher 1919, Bagnold 1941, p. 152, Sharp 1963, Tanner 1967, Ellwood *et al.* 1975, Brugmans 1983). The aeolian ripple index varies inversely with grain size and directly with wind velocity (Sharp 1963, Walker & Southard 1982). Although the ripple index is not a decisive parameter (Goldsmith 1973), ripple geometry can help to identify aeolian paleoenvironments and directions of sediment transport in ancient sandstones (McKee 1945, 1979b).

A measure of the lateral continuity of ripples is provided by the ratio of the mean crest length to the mean wavelength; this ratio was termed the *horizontal form index* by Allen

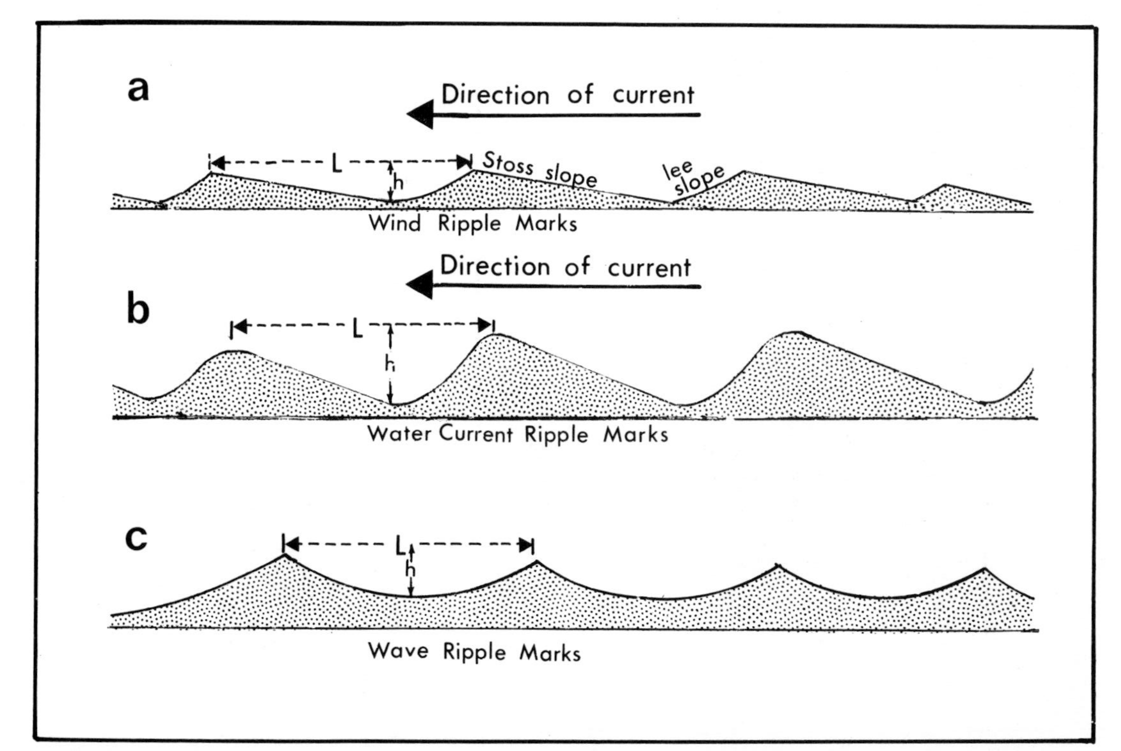

Figure 6.2 Profiles of three types of ripples formed by (a) wind, (b) water currents, and (c) waves. L = ripple wavelength; h = ripple height. (After Twenhofel 1950, p. 568).

(1963) and the *continuity index* by Tanner (1967). Normal aeolian ripples, whose sinuous crests run in a direction transverse to the local wind direction, typically have horizontal form indices of 10–100.

6.2.2 *Effect of wind velocity and grain size on aeolian ripple development*

Normal aeolian ripples form only in sediments of fine sand or coarser grade. As discussed below, the formation of such ripples depends on the impact of saltating fine and medium sand grains and the resulting creep of coarser grains. Ripples can form in beds of very fine sand and silt, but they are discontinuous and display a characteristic linguoid morphology (Greeley & Iversen 1985, p. 155). The geometry of these ripples is determined by local variations in the surface shear stress rather than by ballistic impacts. They are therefore referred to as *fluid drag ripples* (Bagnold 1941, p. 166) or *aerodynamic ripples* (Wilson 1972a) to distinguish them from *impact ripples* or *ballistic ripples*. The upper grain size limit for the formation of aeolian impact ripples is limited only by the wind velocity.

The mean size of sand comprising ripples is usually coarser than that of the underlying sand body as a whole (Sharp 1963, Tsoar 1990a). Within individual ripples, the mean grain size is coarsest at the ripple crest (Tsoar 1990a). Within-ripple grain size differences are more pronounced when the parent sand is poorly sorted.

At wind velocities just above the fluid threshold, some grains move forward by creep, but only a relatively small number of grains enter saltation. These grains have relatively short trajectories and have relatively low energy when they strike the bed. Under such conditions ripples do not develop. At moderate wind velocities the fine and medium sand grains saltate readily and induce the forward creep of coarser grains when they strike the bed. However, at high wind velocities the coarse grains also start to saltate, causing the ripples to lengthen and flatten out (i.e. *RI* progressively increases). At a certain critical wind velocity the ripples disappear and a planar surface is formed (Bagnold 1937a). This type of surface is analogous to a subaqueous plane bed of the upper flow regime (Simons *et al.* 1965). There is an abrupt change at this stage from $RI = 80$ or 100 to $RI = \infty$ (Walker 1981).

In wind tunnel studies using uniform, well sorted sand, ripples have been observed to disappear when u_* reaches 65–95 cm s^{-1}. When the sand is poorly sorted, the critical velocity increases (Bagnold 1941, p. 151, Walker 1981). As a general rule the critical velocity is about three to four times the fluid threshold velocity.

Ripples are good indicators of local wind direction as their crests are orientated perpendicular to the wind, with the steeper lee slope on the downwind side. Slight fluctuations in wind direction are not reflected by changes in the ripple alignment. A mean flow divergence of $> 20°$ is required to initiate a new set of ripples (Sharp 1963).

Deflection of sand transport over a sloping surface causes a deviation in the alignment of the ripple crests with respect to the regional wind direction (Howard 1977). On dune lee slopes, for example, ripples often move across the slope under the influence of vortices created by flow separation at the dune crest.

Field and wind tunnel observations have shown that winds of 12–14 m s^{-1} produce visible undulations on a smooth surface of loose sand in less than 1 min. Within 2–3 min the surface is transformed into a series of transverse ripple marks which become fully developed after 10–15 mins (Cornish 1914, p. 79, Sharp 1963, Seppala & Lindé 1978, Walker 1981, Brugmans 1983, Rubin & Hunter 1987). The development of megaripples takes much longer. Bagnold (1941, p. 156) found that it took 2 h to form ripples with a wavelength of 18 cm in the wind tunnel. However, Bagnold's view that giant pebble megaripples with a wavelength of 20 m and height of 60 cm develop over a period of decades or centuries has been challenged by other authors, who suggest that the process may take only a few weeks (Sharp 1963), or even hours with very high wind velocities (Sakamoto-Arnold 1981). The rate of movement of individual sand ripples has been observed to vary between 0.9 and 8.1 cm min^{-1} under wind velocities ranging between 7.2 and 13.4 m s^{-1} (Cornish 1914, p. 82, Sharp 1963). The relationship between wind velocity and rate of ripple migration in this range is approximately linear.

Ripples are found in areas of net deflation, net deposition, and in places where sand transport occurs but where there is no net erosion or deposition. Ripples which migrate while net deposition takes place are referred to as *climbing ripples* (Allen 1968, pp. 100–108). The angle of climb may be either positive on upward-sloping surfaces or negative on downward-sloping surfaces.

6.2.3 *Models of ripple formation*

Ripples have attracted the attention of researchers for many years (Rae 1884, Joly 1904, King 1916) and the literature dealing with their formation is voluminous [see Hogbom (1923) for a review of earlier work]. However, a fully satisfactory model of ripple development is still lacking, since a full explanation of their formation and movement requires the application of non-linear dynamics, with its attendant difficulties (Werner *et al.* 1986).

Many early hypotheses of ripple formation (e.g. Cornish 1897) referred to the effect of aerodynamic forces acting on sand grains in a way similar to the mechanism proposed for ripple formation in flowing water. They ignored the fact that, because of the large density difference between air and sand, aerodynamic forces in air are less effective than their counterparts in water.

Several of the aerodynamic hypotheses employed the *Helmholtz theorem*, which predicts the occurrence of wave-like oscillations at the interface of two media of different densities flowing with different velocities. An equation that predicts the bedform wavelength (L) when the height of the wave is very small relative to its length was given by Hogbom (1923):

$$L = \left(\frac{2\pi}{g}\right)\left(\frac{u^2}{\rho_s/\rho}\right) \tag{6.1}$$

where u is the wind velocity, ρ_s is the density of the lower layer, ρ is the density of the upper layer (air density), and g is the acceleration due to gravity.

Hogbom took surface loose sand to be the lower layer, and by substituting in Equation 6.1 obtained predicted wavelengths consistent with those of wind ripples. Von Kármán (1947) considered it more appropriate to regard the saltation layer as the lower heavy fluid stratum. This idea was subsequently adopted by Cooper (1958, p. 34), Folk (1976a) and Brugmans (1983). Brugmans suggested that fluctuations in wind velocity and surface shear stress, or in the sizes of impacting grains generated by the oscillating flow, can explain the alternation of fine-grained ripple troughs and coarse-grained ripple crests. Using values of 0.5–0.7 g cm^{-3} for the density of the lower layer of air with saltating grains, Brugmans obtained predicted ripple wavelengths in accordance with those observed in the field.

Other investigators have emphasized the role of saltation which causes coarser grains to creep along the surface. Joly (1904) suggested that ripples are initiated by small increases in bed roughness and that their height gradually increases as grains accumulate by saltation and rolling. Eventually the ripple reaches a height which is in equilibrium with the wind. At this stage the grains are removed from the ripple crest as fast as they arrive. Joly concluded that the wavelength and height of ripples are interdependent because, for a given wind velocity, the grains have a characteristic path length.

Bagnold (1937a, 1941, p. 144) developed this idea into the ballistic theory of ripple formation. According to this model, any chance irregularity in the sand surface will be enhanced, since more saltating grains strike the windward side of the irregularity than the leeward side (Fig. 6.3). As a result, more grains are ejected into saltation, and move forward by surface creep, on the windward slope. According to Bagnold (1941, p. 148), the relative impact intensity (I_β), which represents the propelling force, is given by

$$I_\beta = 1 - (\tan \beta / \tan \alpha) \quad (6.2)$$

where α is the impact angle (relative to the horizontal) of descending grains in saltation and β is the angle between the sand surface and the horizontal, taken as positive for the lee slope and negative for the windward slope (Fig. 6.3). When $\beta > \alpha$ the saltating grains will never hit the surface and $I_\beta = 0$.

If a surface irregularity has windward and leeward slopes of 4° and the impact angle is taken to be 14°, grains on the windward slope will be pushed forward with an intensity that is about 1.8 times that on the lee slope (Eqn 6.2). Consequently, the windward and leeward slope angles will change.

Once irregularities are initiated on the sand surface, a greater number of ejections occur from the windward than from the leeward slopes. According to Bagnold's theory, the pattern of impacts and forward creep movements is repeated at regular intervals governed by the characteristic saltation path length (which should increase with wind velocity and grain size).

The limitation of ripple height was explained by Bagnold in the following way. Over the crests the wind velocity increases with height at a greater rate than it does over the troughs. As the ripple grows vertically it will eventually attain a critical height where grains which reach the crest are immediately swept off again. The critical height is dependent on grain size, since stronger winds are required to move larger grains. Bagnold (1941, p. 152) suggested that for typical fine dune sands a steady-state ripple profile is attained when $RI = 30$–70. From then on the ripples migrate downwind by erosion of the windward slope and deposition on the lee slope, but do not change shape as long as the wind velocity and direction remain constant.

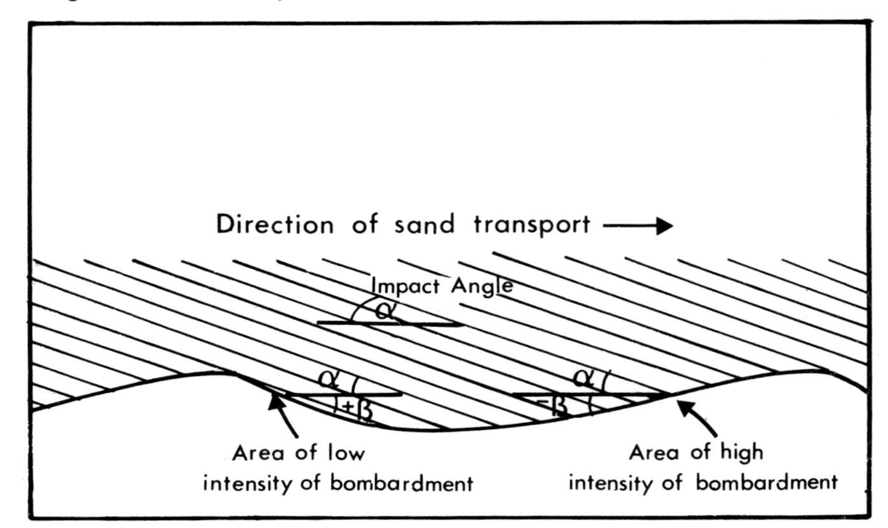

Figure 6.3 Differential intensity of saltation impact on the windward and lee slopes of a ripple. (After Bagnold 1941, p.146).

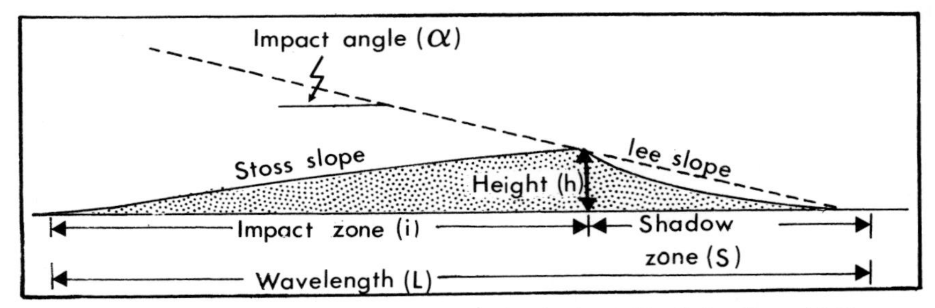

Figure 6.4 Parameters used to describe the ripple profile. (After Sharp 1963).

In his investigation of ripples on the Kelso Dunes in California, Sharp (1963) observed that the ripple spacing increases with time, even though the wind velocity remains constant, and therefore is unlikely to be controlled by a characteristic saltation path length as conceived by Bagnold. Sharp divided the ripple wavelength (L) into two parts, an impact zone (i) and a shadow zone (s) (Fig. 6.4). The length of s depends directly on the ripple height (h) and inversely on the impact angle (α). Since h increases directly with grain size, so does s; α varies inversely with wind velocity, but s should vary directly with wind velocity. However, h varies inversely with wind velocity, at least for higher velocities, so the exact dependence of s on velocity is not clear.

The length of i varies directly with h and inversely with the inclination of the windward slope. Sharp claimed that, for a steady-state condition, the latter varies inversely with α and grain size and directly with the energy of the impacting grains. Therefore, i is expected to increase with increasing grain size, which also brings about an increase in h. An increase of wind velocity decreases α and increases the energy of the impacting grains, so i decreases with velocity.

It can be concluded that if the grain size increases, so do i and s, which together comprise the ripple wavelength (L). Field observations (Sharp 1963) and wind tunnel experiments (Bagnold 1936, Walker 1981) showed that L increases with velocity. Therefore, the increase in s with velocity should exceed the decrease in i. Although the mean saltation path length is dependent on the wind velocity and the size of the saltating grains, it is independent of the size of the creeping grains forming the ripple. Hence there need be no relationship between the ripple wavelength and a 'characteristic saltation path length' as suggested by Bagnold (Folk 1977b, Walker 1981, Anderson & Hallet 1986).

It has been found empirically that, for any given value of u_*, the ripple wavelength and height decrease with decreasing grain size in the range 0.78–0.32 mm, but then start to increase again for finer sizes down to 0.2 mm (Walker 1981). Ripple wavelength was also observed to increase as the sand sorting became poorer (Walker 1981, Walker & Southard 1982). Poorly sorted or bimodal sands containing very coarse grains form higher ripples because the coarsest grains which accumulate at the crest are too large to be removed except by the strongest winds (Bagnold 1941, p. 156, Tsoar 1990a). As more and more coarse grains accumulate near the crest, the dimensions of the ripple will gradually increase, eventually forming a megaripple. However, under conditions of very strong winds even the coarsest grains can be moved and hence the megaripple pattern is replaced by an almost flat surface (Bagnold 1941, p. 157, Wilcoxon 1962, Walker 1981).

As pointed out by Bagnold (1941, p. 155), the essential difference between ripples and megaripples lies in the relative magnitudes of the wind strength and the size of the crest grains. In the case of ripples, the wind is strong enough to remove the topmost crest grains whenever the crest height reaches a certain limiting height. In the case of megaripples the wind is not sufficiently strong, relative to the size of the crest grains, to achieve this. Bagnold suggested that the conditions necessary for the growth of megaripples are (a) availability of sufficient coarse grains which have a diameter 3–7 times larger than the mean diameter of grains in saltation, (b) a constant supply of fine sand in saltation to sustain forward movement of the coarse grains by creep, and (c) wind velocity below the threshold to remove coarse grains from the megaripple crest.

The wavelength of megaripples may increase indefinitely, although at a progressively slower rate, as long as the sand supply is maintained. Bagnold (1941, p. 156) considered that the dimensions of a megaripple should vary as the square root of its age, and that very large megaripples seen in the field must therefore have taken decades or centuries to form. However, other workers have concluded that granule megaripples can form in a much shorter time (Sharp 1963, Sakamoto-Arnold 1981). It is not clear, however, whether in these instances u_* was above the threshold to entrain individual granules by direct fluid drag.

Ellwood *et al.* (1975) concluded that Bagnold's concept of the characteristic saltation path length of saltating grains can be related to the development of all ripples including megaripples. The abrupt change in wavelength between ripples and megaripples which is often seen in the field (Fig. 6.1) was attributed by Ellwood *et al.* (1975) to local differences in sand size, particularly the proportion of coarse grains present. They suggested that when the content of coarse grains exceeds a certain critical proportion, saltation occurs almost entirely by ricochet, whereas when the bed contains a lower proportion of coarse grains ejection plays a more important role. Since ejection produces less efficient saltation than ricochet, a sudden change in sand grain size may

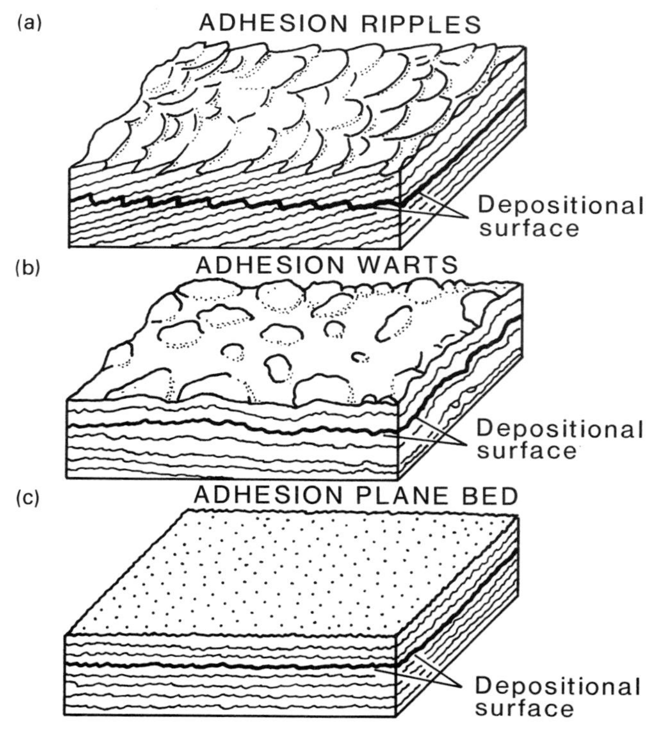

Figure 6.5 Block diagrams illustrating the plan view shape and cross-sectional structure of the three main types of adhesion structures. (After Kocurek & Fielder 1982).

produce an equally sudden change in the mean saltation path length and consequently in ripple wavelength.

Anderson (1987b) has suggested that while ripple (including megaripple) wavelengths are affected by grain trajectory lengths, they do not correspond with a 'characteristic' or mean saltation path length. According to Anderson, ripple spacing is a function of the probability distribution of the total trajectory population, in which low-energy reptating grains outnumber higher energy saltating grains by about nine to one. Model predictions suggest that the ripple crest spacing should be approximately six times the mean reptation path length.

6.2.4 Adhesion ripples

When saltating dry sand blows across a wet or damp surface, some of the grains become trapped by surface tension. Van Straaten (1953) first described the resulting structures and referred to them as 'anti-ripplets'. However, Reineck (1955) was the first to document their formation in detail, both experimentally and in nature. He recognized two distinct forms, *adhesion ripples* (*haftrippeln*) and *adhesion warts* (*haftwarzen*). Hunter (1969) described the formation of adhesion ripples, which he termed *aeolian microridges*, on modern beaches and identified the first possible ancient example. Hunter (1980) recognized an additional adhesion structure which he termed 'quasi-planar adhesion stratification'. Similar structures were recognized by Kocurek & Fielder (1982) and termed *adhesion plane bedforms*.

Adhesion ripples are small, sub-parallel ridges perpendicular to the wind (Fig. 6.5a). They typically have wavelengths of less than 1 cm and heights of 0.3–3 mm. Values of the horizontal form index are generally < 3. The ripple crests are often slightly convex in the upwind direction. In cross section the upwind slopes are commonly much steeper than the downwind slopes, indicating that the features grow slowly upwind by accretion of saltating sand (Hunter 1973). Adhesion ripples climb over the deposits of their upwind neighbours to generate climbing adhesion ripple structures.

Adhesion ripples have been observed on marine beaches and on desert interdune flats and playas. Growth of the structures can only take place as long as moisture is drawn to the surface by capillary action. Once the surface has accreted to the point where this can no longer take place, sand then saltates across the surface with little disturbance of the structure.

Adhesion warts are distinguished from adhesion ripples by their irregularity and open-arched nature (Fig. 6.5b). They have a more random distribution than adhesion ripples and their formation appears to be favoured by a rapidly shifting wind direction which encourages vertical rather than lateral growth (Reineck 1955, Olsen *et al.* 1989).

An adhesion plane bed is generally smooth, with irregularities not much larger than grain roughness. Adhesion laminations are typically 1–2 mm thick and often display a crinkly appearance due to spatial irregularities in the accretion process (Fig. 6.5c) (Kocurek & Fielder 1982). They commonly form where sand input to a damp surface is predominantly by grainfall rather than saltation.

6.3 Sand dunes

6.3.1 Classification of sand dunes and other aeolian sand accumulations

A sand dune can be defined simply as a hill or ridge of sand piled up by the wind. The maximum linear dimension of individual dunes ranges from less than 1 m to several tens of kilometres, while the height ranges from a few tens of centimetres to more than 150 m. Very large dunes, on which smaller dunes may be superimposed, are referred to as

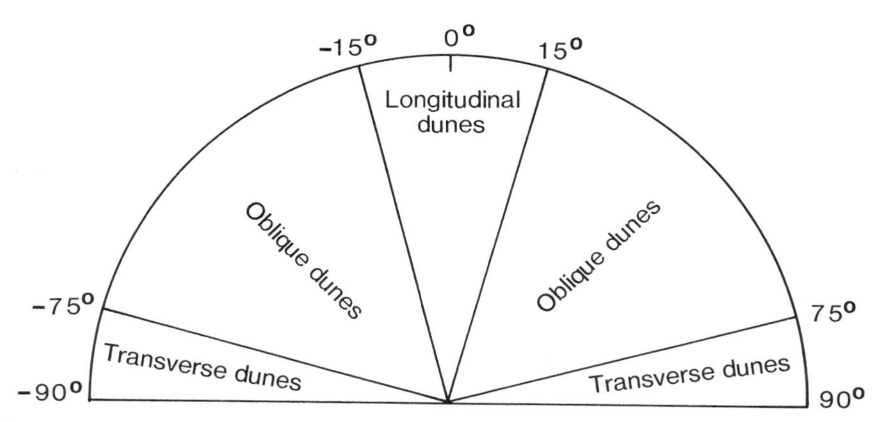

Figure 6.6 Morphodynamic classification of relatively straight-crested dunes in terms of angle between average dune trend and long-term sand transport resultant. (After Hunter *et al.* 1983).

megadunes or draas. Dunes may also be linked together to form dune chains or dune networks.

A useful distinction can be made between *simple*, *compound*, and *complex* dunes (or megadunes) (McKee 1979a). Simple dunes consist of individual dune forms which are spatially separate from their neighbours. Compound dunes consist of two or more dunes of the same type which have coalesced or are superimposed. Complex dunes consist of two or more different types of simple dunes which have coalesced or are superimposed.

Many attempts have been made to classify dunes based on a combination of shape, number and orientation of slip-faces relative to the prevailing wind or resultant sand drift direction, and degree of form mobility (e.g. Aufrère 1931, 1933, Bryan 1932, Melton 1940, Hack 1941, Smith 1946, 1953, 1963, Kuhlman 1960, Holm 1968, Mainguet & Callot 1974, Mabbutt 1977, p. 225, McKee 1979b, 1983, McKee & Breed 1974, McKee *et al.* 1977, Breed & Grow 1979, Hunter *et al.* 1983, Mainguet 1983, 1984b, Wasson & Hyde 1983a, El Baz 1986, Thomas 1989b, p. 242). A considerable number of local names and geometric or organic analogies have been used to describe dunes [see Stone (1967) and Breed & Grow (1979, pp. 284–96)], and the use of different names for basically similar features in different parts of the world has resulted in some confusion. As discussed in more detail below, the relationship between wind regime, slip face orientation, and direction of dune movement (or extension) is complex, and descriptive terms such as 'longitudinal' and 'transverse', which carry genetic implications, are frequently used inappropriately. Hunter *et al.* (1983) suggested that the term *longitudinal dune* should be applied only where the orientation of the long axis of the dune deviates by less than 15° from the resultant sand transport direction, while a *transverse dune* should have long axes which are within 15° of being normal to the resultant sand transport direction. Dunes whose long axes show a larger deviation from the resultant sand transport direction are referred to as *oblique dunes* (Fig. 6.6). In practice, it is often difficult to classify dunes accurately in this way because long-term wind data are commonly not available from the dune areas, and local sand transport directions may deviate significantly from those predicted using data from the nearest weather station owing to the effects of topography and secondary circulations in the atmosphere.

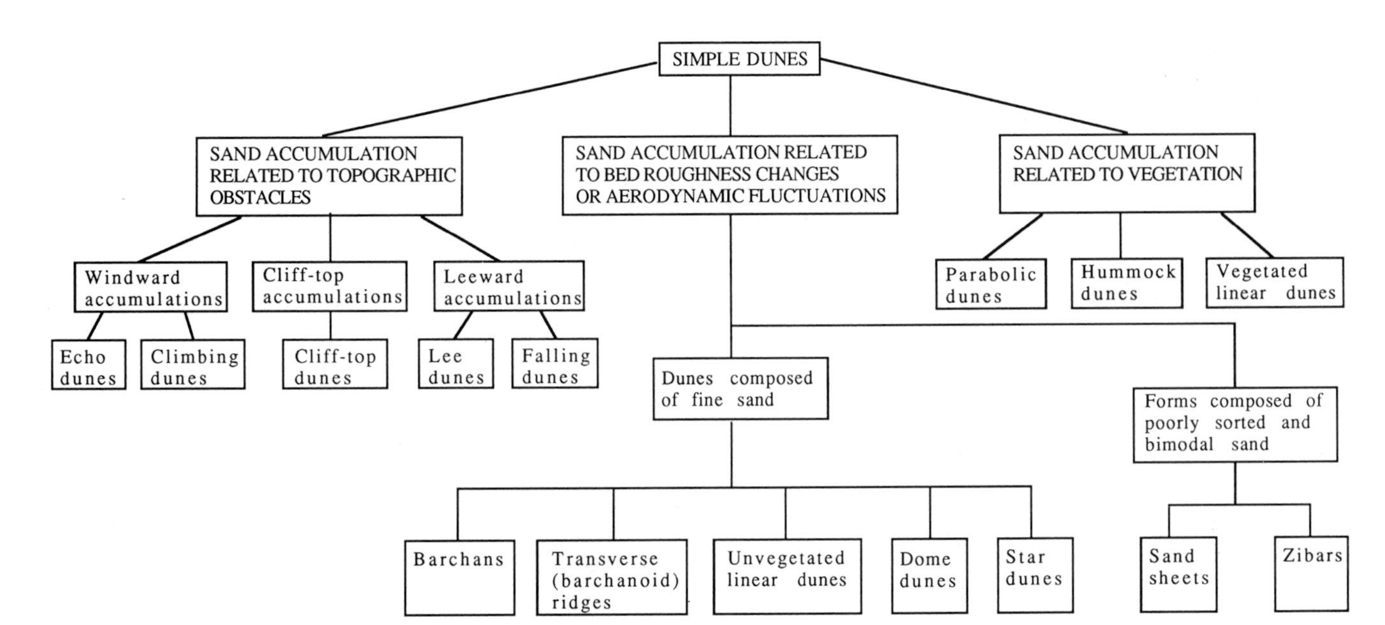

Figure 6.7 Classification of major dune types.

Many dune ridges show dynamic behaviour typical of both transverse and longitudinal dunes, and Carson & MacLean (1985a,b, 1986) suggested the term 'hybrid' to describe such dunes. This term has not, however, been widely accepted and is considered inappropriate by some authors (Hunter *et al.* 1985).

In the classification system shown in Figure 6.7, simple dunes are divided into three basic groups: (a) those whose development is related to topographic obstacles, (b) those which can be regarded as self-accumulated (autogenic dunes), and (c) those whose development is strongly influenced by vegetation (phytogenic dunes). The first category is divided into windward accumulations, which include *climbing dunes* and *echo dunes*, leeward accumulations, which include *lee dunes* and *falling dunes*, and *cliff-top dunes*, which may form in a zone of flow separation just downwind from the crest of an escarpment. Self-accumulated simple dunes include *barchans*, *transverse* (*barchanoid*) *ridges*, *unvegetated linear dunes* (*seif dunes*), *dome dunes*, and *star dunes*. Dunes formed by accumulation of sand related to the presence of vegetation include *parabolic dunes*, *vegetated linear dunes*, and *coppice* or *hummock dunes*.

Linear dunes (vegetated and unvegetated) are by far the most common dune type found in deserts, followed by transverse dunes. However, there is considerable variation between different regions (Tables 6.2 and 6.3). In humid region coastal dunefields hummock dunes and parabolic dunes are the most common types, whereas in arid and semi-arid region coastal dunefields barchans and transverse barchanoid ridges are dominant (Inman *et al.* 1966, Pye 1984, Illenberger 1988).

6.3.2 Dune accumulation influenced by topographic obstacles

6.3.2.1 Lee dunes Topographic obstacles such as boulders, escarpments, and hills induce zones of airflow acceleration, deceleration, and enhanced turbulence (Gaylord & Dawson 1987). Consequently, there will be either erosion or accumulation of sand, or both simultaneously in different places. The resulting dunes are static, i.e. they do not advance or elongate once they have attained a steady state.

Bagnold (1941, pp. 189–90) used the term *sand shadow* to describe a tapering accumulation of sand formed in the lee of an obstacle where the wind velocity is locally reduced. Allen (1982, pp. 197–200) used the less specific term *current shadow* for such features in both aeolian and aqueous environments. However, the terms lee dune or *shadow dune* are more widely used in the aeolian literature.

Some shadow dunes evolve from initially horseshoe-shaped sand accumulations which have been termed *current crescents* (Allen 1982, pp. 189–91) (Fig. 6.8a–d). As the airflow is deflected around and over the obstacle, a horseshoe vortex is created (Fig. 6.9), and sand is initially deposited both in front of the obstacle and on either side as two tapering wings (Figs 6.8a & 6.10) (see also Greeley *et al.* 1974b). The two wings eventually coalesce as the arms of the horseshoe vortex gradually transfer sand towards the centreline of the obstacle, and over time the shadow dune becomes higher, longer, and narrower (Fig. 6.8d).

Small shadow dunes commonly form in the lee of boulders and clumps of vegetation (Kadar 1934, Beheiry 1967, Goldsmith 1973). The size and shape of the lee dune are related to those of the obstacle. Large linear dunes in the lee of mountains have been described from the Libyan Desert (Kadar 1934), southern California (Smith 1978, 1982), Peru (Bosworth 1922, p. 303, Grolier *et al.* 1974, Howard 1985) and Chad (Mainguet & Callot 1978, p. 113).

Lee dunes are best developed under a nearly unidirectional wind regime. They therefore break up downwind, at a distance where the topographic obstacle is no longer effective, into individual barchan dunes which are the preferred form in open areas of

Table 6.2 Sand sea characteristics. (Data of Fryberger & Goudie 1981, Lancaster *et al.* 1987).

	Sahara				Southern Africa			Asia			N. America
	West ergs	South ergs	North ergs	NE ergs*	Namib	SW Kalahari	Arabian ergs	Thar	Takla Makan	Ala Shan	Gran Desierto
area (thousand km^2)	207	447	306	161.5	34	100	743	214	261	109	5.5
dunes (% of area)	54.7	52.5	64.1	56.3	54.6	86.5	70.1	68.3	66.4	32.2	70.0
sand sheets (% of area)	45.3	47.5	35.9	39.3	45.4	13.6	23.2	31.8	33.6	67.8	30.0
uncertain (% of area)	–	–	–	4.5	–	–	6.7	–	–	–	–

* Excluding Egyptian and Libyan ergs.

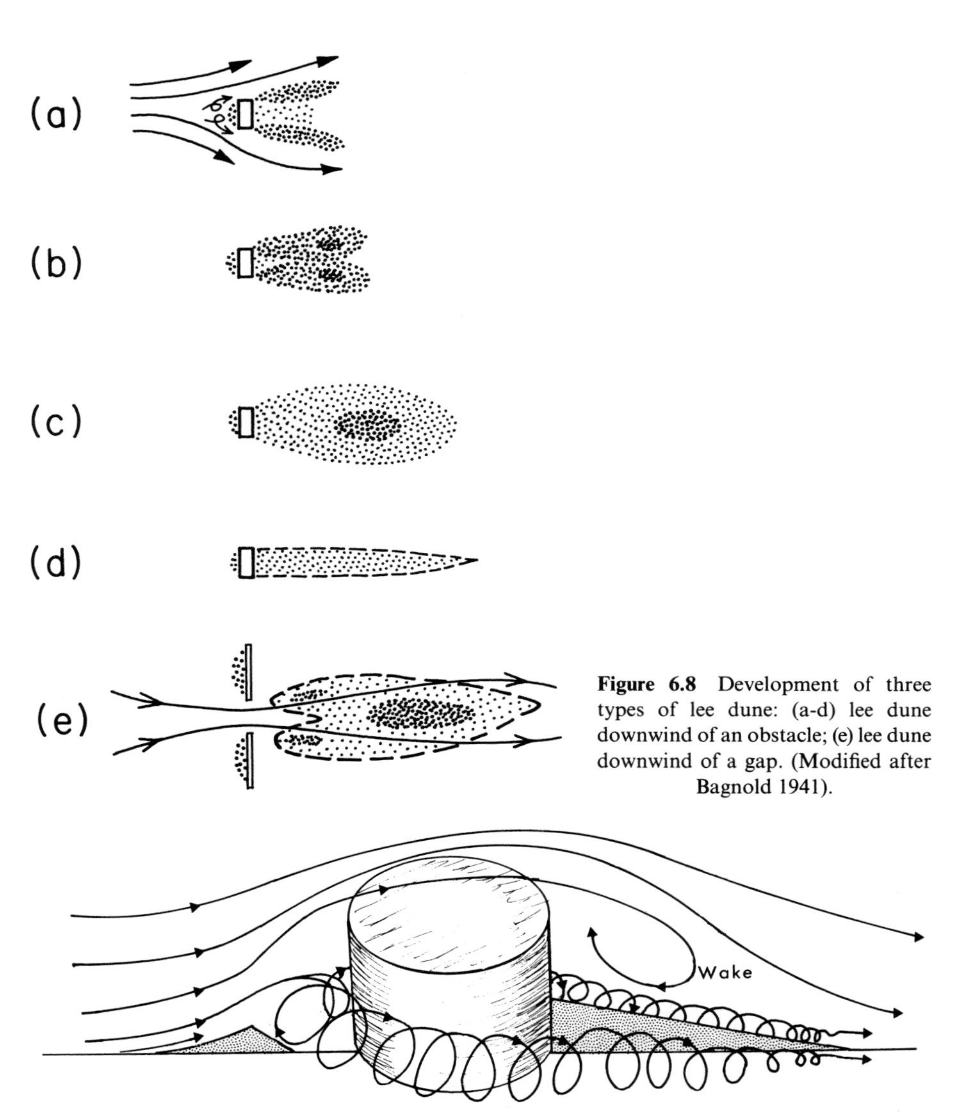

Figure 6.8 Development of three types of lee dune: (a-d) lee dune downwind of an obstacle; (e) lee dune downwind of a gap. (Modified after Bagnold 1941).

Figure 6.9 Schematic diagram showing the development of a horseshoe vortex around an obstacle. Sand accumulations are stippled. (Modified after Ash & Wasson 1983, Tsoar 1983b, Greeley 1986).

unidirectional wind (Fig. 6.11) (Bagnold 1941, p. 194, Smith 1954, Grolier *et al.* 1974, Howard 1985).

Accumulation of sand can also occur in the lee of a gap between two obstacles (Fig. 6.8e). The sandflow is accelerated and funnelled through the gap but fans out and decelerates on the lee side, leading to sand deposition (Bagnold 1941, p. 192).

Sand accumulations are also sometimes found in the lee of cliffs. Elongated lee dunes form where irregularities in the cliff line initiate convergence of the streamlines as they

Figure 6.10 Development of paired lee accumulations and windward sand ramp adjacent to an obstacle. Wind direction is from left to right. This represents stage (a) shown in Figure 6.8.

pass over the brink (Fig. 6.11). If the cliff brinkline is straight there is a general reduction in flow velocity, leading to the formation of falling dunes along the entire cliff foot and associated talus slopes (Evans 1962).

6.3.2.2 Echo dunes Wind tunnel experiments by Tsoar (1983b) showed that the horizontal near-surface wind velocity in front of a vertical obstacle starts to drop at a distance of $d/h = 3.3$ (where d is the horizontal distance upwind from the obstacle and h is the height of the obstacle) and reaches a minimum at $d/h = 0.75$, where the two opposite flow directions meet (Fig. 6.9). From this point the wind velocity increases to a maximum (about half the undisturbed wind velocity) at $d/h = 0.275$ and falls again to zero near the base of the obstacle (where $d/h = 0$). In these experiments sand accumulated between $d/h = 2$ and 0.3, forming an echo dune with a crestline at $d/h = 0.5$. No accumulation occurred between $d/h = 0$ and 0.3 due to erosion induced by the reverse flow of the horseshoe vortex.

Echo dunes commonly form in front of cliffs (Fig. 6.12) but can only be maintained in the long term if sand is moved laterally along the cliff line by vortices located between the cliff and the echo dune. The sand may ultimately escape through drainage channels or other breaks in the cliff.

When the echo dune is small, the forward wind velocity on the echo dune crest is lower than that of the reverse flow of the vortex in front of the bluff body. Consequently, the echo dune grows in height by sand deposition until a condition of steady state is achieved. Tsoar's (1983b) experiments showed that this stage is reached when the height of the echo dune is about one third of that of the obstacle. Sand arriving at the crest is then moved into the trough between the dune and the cliff.

If the escarpment is initially not sufficiently steep, a vortex with strong reverse flow will not develop and sand is able to climb the escarpment as a climbing dune (Fig. 6.13). Simulation work has shown that initial slope angles of less than 60° bring about the formation of climbing dunes (Tsoar 1983b).

6.3.2.3 Cliff-top dunes A zone of reduced wind velocity is frequently observed just downwind of the crest of escarpments (Jackson 1976, Bowen & Lindley 1977, Marsh & Marsh 1987) (Fig. 2.25). Consequently, sand often accumulates at such sites forming cliff-top dunes (Fig. 6.13). Since sand can be blown up slopes as steep as 60°, the formation of cliff-top dunes does not necessarily require cliff recession or erosion of a sand ramp as suggested by Jennings (1967).

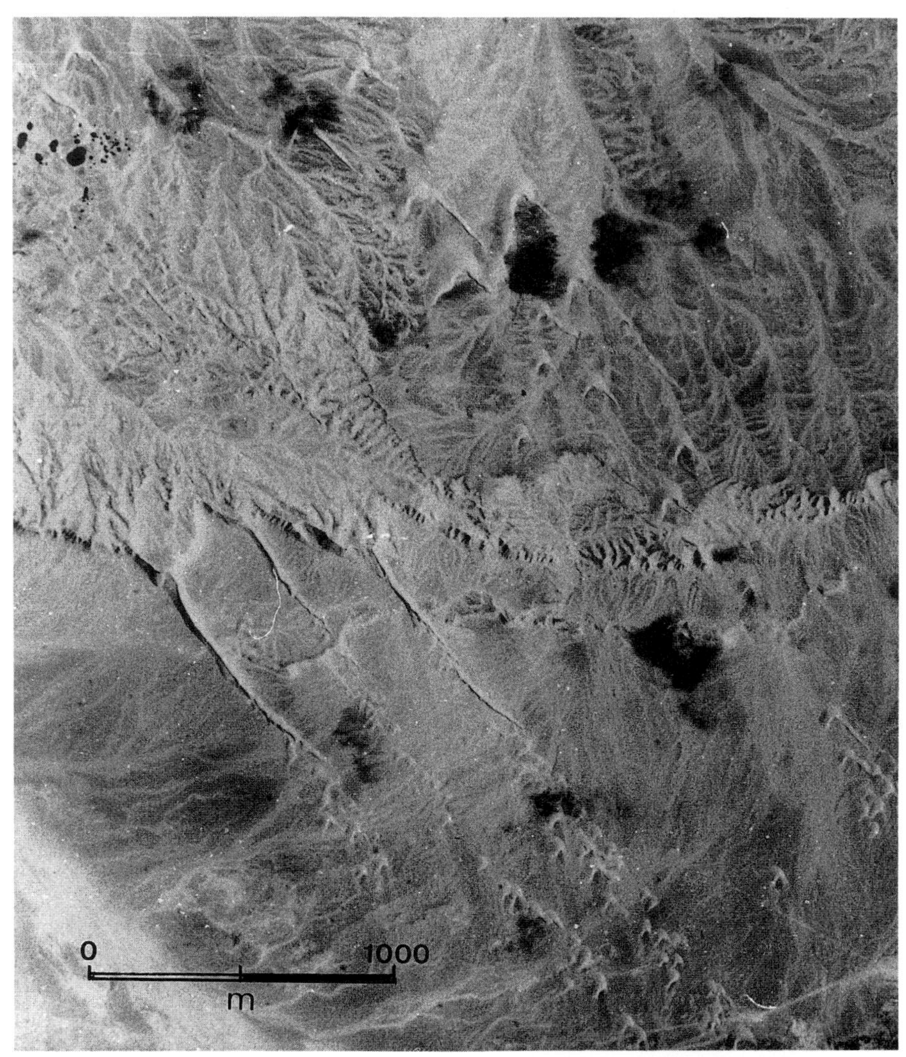

Figure 6.11 Air photograph showing lee dunes formed downwind of an escarpment in northern Sinai. Note how the lee dune breaks up downwind into barchans.

Figure 6.12 Echo dunes developed in front of a cuesta in northern Arizona. The extremities of the dune develop into climbing dunes where drainage channels cut the cliff.

Figure 6.13 Climbing and cliff-top dunes, southern Negev.

6.3.3 Formation of self-accumulated dunes

6.3.3.1 Dune initiation Wind-blown sand has a propensity for self-accumulation into dunes in the absence of topographic obstacles. Bagnold (1937, 1941, pp. 169–71) suggested that this is due to the fact that saltating grains bounce off a hard desert surface (pebbles or bedrock) more effectively than over a bed of loose sand (Eqn 4.34). The sand transport rate over a random sand patch is, therefore, relatively lower than that over its surroundings, leading to accretion of the sand patch and extension of its upwind margin.

This process is effective only if there is a constant supply of sand from upwind and under conditions of strong winds which can transport the sand over a rough pebble surface and deposit it on the sand patch. Under gentle wind conditions the sand is trapped by the pebble surface while the sand patch is eroded and extends downwind (Bagnold 1937).

An alternative explanation, widely known as the *waveform theory*, suggests that sand becomes concentrated into dunes owing to the existence of wave-like motions or secondary circulations in the atmosphere. These secondary air motions, which may be intrinsic to the flow or generated by bed irregularities further upwind, cause variations in surface shear stress and may therefore be expected to generate spatial variations in sand transport rate (Wilson 1972a). This should produce alternating transverse or longitudinal zones of erosion and deposition, which lead to the formation of a regular series of ridges and troughs (Wilson 1972a, Folk 1976a, Warren 1979).

Bagnold (1941, pp. 176–9) also recognized that when sand drift takes place over a surface of uniform roughness the sand is often concentrated into longitudinal strips. He hypothesized that this is due to secondary rotational eddies which move parallel to the primary flow direction. The formation of longitudinal sand strips up to 2 m wide and 20 cm high is commonly observed on the surface of sand sheets and beaches (Fig. 6.14), and there is considerable experimental evidence for the existence of secondary longitudinal vortices in turbulent flows (Grass 1971, Weedman & Slingerland 1985, Allen 1985). However, there is no direct observational evidence that longitudinal sand strips evolve into dunes. Incipient barchan dunes without slip-faces are sometimes seen migrating across sand surfaces on which sand strips are developed (Fig. 6.15), without any apparent genetic association. In other instances sand accumulates on beaches and sand plains as low, static sand mounds (incipient dome dunes) rather than as sand strips. At present the aerodynamic processes responsible for these different forms of aeolian sand accumulation are not well understood.

Figure 6.14 Depositional sand strips formed on the upper foreshore, Saunton Sands, North Devon, UK.

Figure 6.15 Barchans without slip-faces forming on the beach at Formby Point, Merseyside, UK.

Few field experimental investigations of dune initiation in the absence of vegetation have been undertaken, and the question of why only some sand mounds accrete and evolve into mobile dunes with slip faces remains unanswered. Lettau & Lettau (1978, pp. 140–3) reported that experimental conical mounds of sand placed on a bare Peruvian Desert surface lost more than 60% of their volume within a few days, apparently because even the largest pile was too small to give rise to a lee-side vortex which would sweep sand back towards the dune and prevent the escape of loose grains downwind. Clearly, growth of a sand patch into a dune requires that more sand is supplied to and retained on the patch than escapes downwind. If the upwind supply of sand is restricted for any reason, the sand patch is likely to experience wastage.

6.3.3.2 Development of a steady-state dune profile Whatever the reasons for the initiation of a sand mound, its growth causes further perturbations in the wind flow which affect the pattern of surface shear stress and hence sand transport rates over the growing dune. The wind flow is accelerated up the windward slope towards the crest, the magnitude of the amplification factor being dependent on the shape of the mound (Fig. 6.16; Section 2.4.5). As the incipient dune grows higher, so the crest is exposed to stronger wind velocities and both the surface shear stress and the sand transport rate increase. The dune can grow vertically only as long as the rate of sand supply to the crest is not exceeded by the rate of sand removal. Eventually, a condition of steady state (Chorley & Kennedy 1971) is achieved in which the form of the dune is in balance with the sand transport over it (Allen 1968, p. 102, Howard *et al.* 1978, Walmsley & Howard 1985, Tsoar 1985). The precise nature of the equilibrium form will reflect a balance between sand supply, grain size, and wind energy factors. Any subsequent alteration in wind or sand supply will produce a change in the dune morphology which acts to regulate the effect of the change and brings the dune to a new state of balance between input and output (self-regulation by negative feedback). For a dune to maintain itself in a steady state, the rate of sand transport should steadily increase up the windward slope towards the crest. In this manner the wind at any point on the windward slope is able to carry all of the sand previously eroded and can erode the bed at the rate necessary to maintain a constant slope profile (Tsoar 1985). This compensation principle is similar to that of a graded stream in which the bed slope is adjusted to carry the sediment load. When a steady state is achieved, sand eroded from the windward side of the dune is deposited in equal volume on the lee side, and the dune advances without substantially changing its shape.

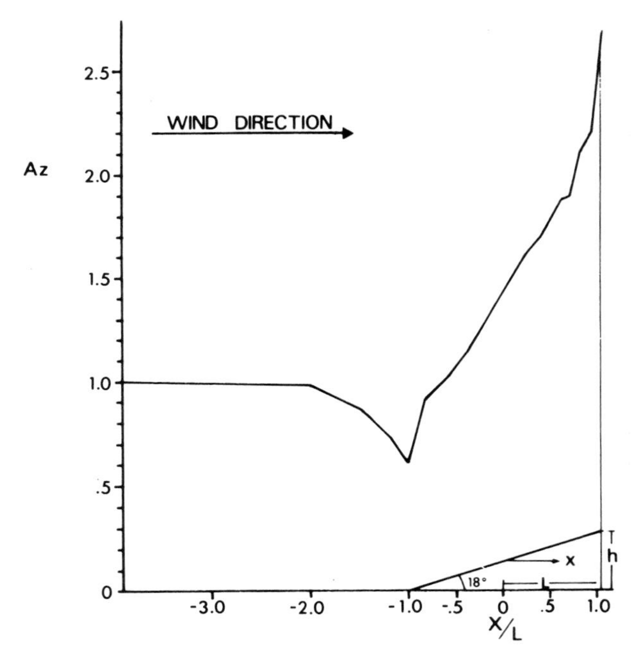

Figure 6.16 Rate of change of speed-up ratio (A_z) over (a) a uniform windward slope model in a wind tunnel ($h = 7\,\mathrm{cm}$, $u_1 = 216\,\mathrm{cm\,s^{-1}}$) and (b) a convex model ($h = 2\,\mathrm{cm}$, $u_1 = 223\,\mathrm{cm\,s^{-1}}$). Velocity measurements taken at a height of 0.6 cm. (After Tsoar 1985).

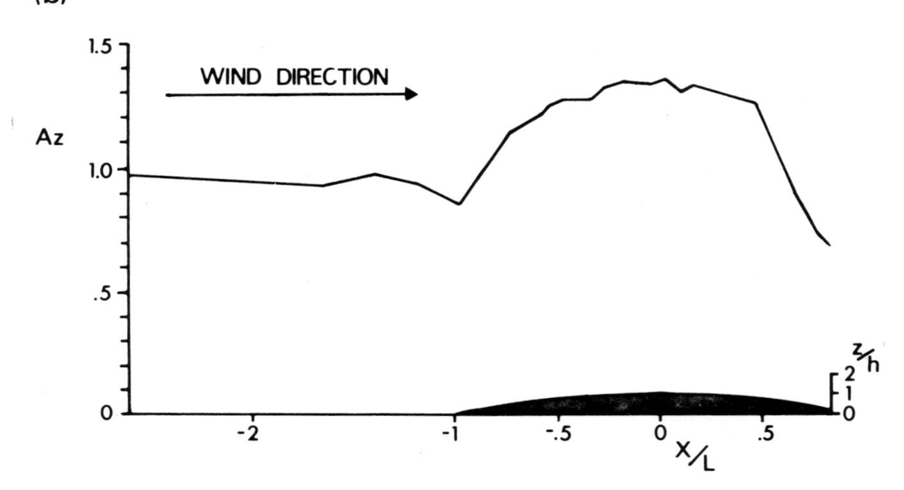

The rate of advance of a dune in a condition of steady state should be constant at all points on the dune. The rate of advance (c) can be derived from the equation presented by Bagnold (1941, p. 200):

$$c = [(\mathrm{d}q/\mathrm{d}x)/\gamma \tan\alpha] \tag{6.3}$$

where $\mathrm{d}q/\mathrm{d}x$ is the rate of sand removal or deposition per unit area, α is the local slope angle, and γ is the bulk density of sand.

According to Tsoar (1985) and Lancaster (1985a), the rate of erosion is defined by the wind velocity as specified by the rate of increase of the speed-up ratio, which is dependent on the shape of the slope (Fig. 6.16). However, there are a number of difficulties in relating steady-state dune profiles simply to the wind velocity (Watson 1987). First, rates of erosion and deposition across sand dunes are dependent on variations in shear stress rather than wind velocity (Hsu 1971b, Lai & Wu 1978, Howard & Walmsley 1985). The relationship between wind velocity and u_* is linear for uniform boundary layer conditions, but flow over a rough surface such as a dune distorts the logarithmic wind velocity profile and pressure distribution, thereby influencing the shear velocity (and hence the shear stress (τ), since $\tau = \rho u_*^2$). The maximum wind velocity recorded at some fixed height (e.g. 2 m) therefore does not necessarily coincide with the maximum surface shear stress.

Measurements over model dunes in a wind tunnel by Lai & Wu (1978) showed that, whereas the wind velocity increases to a maximum over the crest, maximum shear stress occurs on the steepest part of the windward slope. Shear stress was found to decrease on the higher convex slope elements. If flow separation occurs at the crest, shear stress at this point will fall to zero. On the other hand, rates of erosion on the steepest parts of the windward slope are limited by the higher critical shear stress required to move sand up a steep slope, whereas lower shear stresses are required to erode sand of the same size on the flatter slopes near the crest. Lai & Wu concluded that maximum erosion would coincide with maximum shear stress on the steepest part of the windward slope and maximum deposition would be at, or just in the lee of, the crest, upwind of the point of flow separation.

Field measurements have confirmed that wind profiles over dunes deviate from the logarithmic law. Mulligan (1985, 1987) found that, owing to the effects of flow amplification and compression of streamlines, surface stress as measured by u_* attained a maximum on the middle of the windward slope. Lancaster (1987) reported increasing deviations from the logarithmic wind profile law with increasing height on the windward slope of a 40 m high star dune in the Gran Desierto, Mexico. However, in this instance the maximum surface shear stress was found at the dune crest.

A second problem encountered in relating an 'equilibrium' dune profile to wind conditions arises from temporal fluctuations in wind strength and direction which continually bring about changes in the shape of the dune. During periods of gentle winds, u_* may be below the threshold for sand movement near the foot of the dune, but above it on the upper windward slope. Bagnold (1941, p. 198) pointed out that sand movement will begin at a point on the upper windward slope where the wind acceleration is at a maximum. In the case of flow over a rounded dune profile, the height of the dune crest may either remain constant or be lowered slightly, depending on the distribution of sand deposited on the lee slope. However, with flow over a sharp-crested dune, where flow separation occurs, accretion of the crest may take place (Lai & Wu 1978, Watson 1987), at least if the wind approaches the crest obliquely (Tsoar 1985), rather than erosion as suggested by Lancaster (1985a). During periods of strong winds u_* will be above the threshold for sand movement from the foot of the windward slope to the crest, and the dune is more likely to advance with little change in the steady-state profile, since at all points on the windward slope the dune is able to maintain a balance between the erosion rate and the surface slope.

Since fluctuations in wind strength and direction are typical of all environments, dunes are never likely to be in state of perfect equilibrium and therefore are best regarded as being in a quasi-steady state.

For a dune to maintain a quasi-steady-state profile in the medium term, the rate of sand input to the dune must be equal to the sand output. If the two are not in balance the dune will grow or shrink in size. Any change in dune dimensions resulting from a change in sand volume may have an effect on the flow over the dune, on the pattern of surface shear stress, and hence on the dune morphology. Dunes often increase in size if they incorporate sand from an underlying sandy surface as they migrate. Wasting of dunes can occur if no sand is supplied from upwind or by erosion of the substrate, but sand continues to be blown from the lee slope or flanks of a dune faster than the dune form as a whole can migrate. In the case of parabolic and vegetated linear dunes which extend over a non-sandy substrate, the dune height usually decreases as the dune migrates downwind, since progressively more of the available sand becomes fixed in the vegetated arms (Section 6.4.2).

6.3.3.3 Flow separation and the development of a dune slip-face As the airflow passes over the crest of an incipient dune, the compressed streamlines spread out and there is a reduction in mean flow velocity, with the result that less sand can be carried (Section

Figure 6.17 Oblique air photograph of a large barchan dune in the Salton Sea area, southern California. Note the asymmetry due to elongation of the right horn by winds blowing from the left side of the dune. The elongated horn is truncated by a fluvial channel. This photograph was taken in 1965; by 1985 the dune had virtually disappeared because very little of its sand escaped being washed away by the channel during floods. (Photograph by J. S. Shelton).

2.4.3). The rate of sand deposition is zero at the crest, reaches a maximum at some point on the lee slope, and declines downwind thereafter. According to Bagnold (1941, p. 201), the distance between the crest and the point of maximum deposition depends on the magnitude of the lag between the change in wind condition and the corresponding change in sand movement. Since this lag distance may be expected to remain constant for a given wind velocity, the point of maximum deposition should become closer to the dune crest as the dune grows in size. Consequently, the upper part of the lee slope should advance faster than the lower part, leading to an increase in the leeward slope angle. As the lee slope steepens, the wind experiences increasing difficulty in deflecting downwards sufficiently rapidly to follow the surface, eventually leading to flow separation (Bagnold 1941, p. 202) (Section 2.4.3). Sand accumulation is enhanced in this relatively sheltered region of low velocities, thereby accentuating the steepening process. Eventually the angle of repose is exceeded and a slip face develops (Section 4.2.7). Field observations indicate that slip faces can develop on dunes only 50 cm high, and that there is a minimum slip face height for very fine uniform sands of about 30 cm (cf. Bagnold 1941, p. 211). The minimum height of the slip face increases with increasing grain size. For this reason slip faces usually do not develop on low ridges composed of coarse sand (zibar).

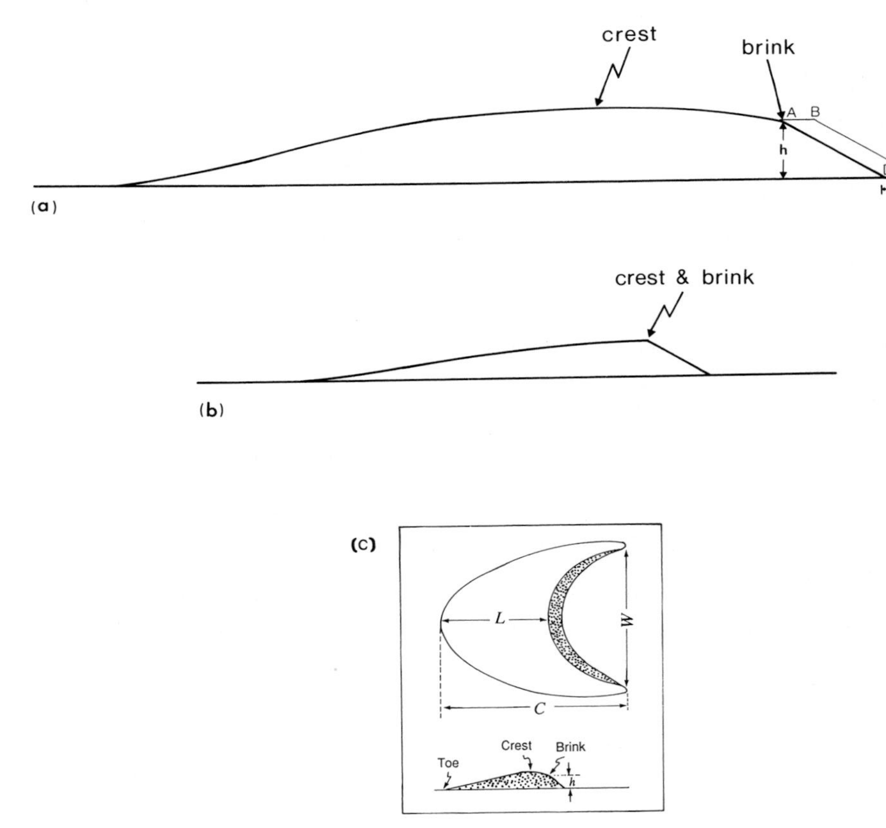

Figure 6.18 Terminology and parameters used to describe barchan morphology. (a) A barchan with a separate brink and crest; the parallelogram ABCD represents the area occupied by sand deposited on the slip face when the dune advances by a distance *c*; (b) a barchan with coincident brink and crest; (c) parameters used to characterize the plan morphology of a symmetrical barchan.

Figure 6.19 Oblique air photograph of a complex megabarchan on the Snake River plain, Idaho. (Photograph by J. S. Shelton).

6.3.4 Simple barchans and transverse barchanoid ridges

Barchans are isolated crescentic dunes whose horns point downwind (Figs 6.17 and 6.18). The windward slope is typically convex with an average maximum slope of 12° while the leeward slope is characterized by a slip face at 33–34°. Some barchan dunes have a separate crest and brink (Fig. 6.18a), whereas in other cases the two coincide (Fig. 6.18b). Small barchans tend to be flatter than larger forms and have a smaller angle between the windward flank and the desert floor (Hastenrath 1987).

Although barchans are relatively rare in the great sand seas of the world (Gautier 1935, p. 47, Jordan 1965, Higgins *et al.* 1974), they are one of the best known dune types and their genesis has been much discussed (Douglass 1909, Barclay 1917, Finkel 1959, Hastenrath 1967, 1978, 1987, Lettau & Lettau 1969, 1978, Rempel 1936, Long & Sharp 1964, Norris 1966, Inman *et al.* 1966, Gad-el-Hak *et al.* 1976, Howard *et al.* 1978, Haff & Presti 1984, Walmsley & Howard 1985, Wipperman & Gross 1986).

Small barchans form rapidly on beaches and in other areas where sand is blown over a relatively hard substrate (Walker & Matsukura 1979, Hummel & Kocurek 1984, Bourman 1986) (Fig. 6.15). These small dunes are ephemeral, being easily destroyed following a change in wind and sand supply conditions. At the other extreme compound mega-barchans which occur in some dunefields (Simons 1956, Norris 1966) (Fig. 6.19) may persist for hundreds of years.

Patches of sand migrating over a hard substrate develop a crescentic plan form even before they are high enough to develop a slip face (Fig. 6.15), because sand is transported more rapidly across and around the sides of the sand patch than across its centre. As more sand is trapped in the middle of the patch, it grows in height until it is sufficiently high to induce flow separation and form a slip face. The precise form of the dune is determined by the rate of sand supply, degree of sandflow saturation, and wind regime (Howard *et al.* 1978).

Barchans can migrate long distances downwind without experiencing major changes in size or shape (Norris 1966). In some instances they act almost as a closed system, in which sand is prevented from escaping from the dune by reverse flows associated with vortices on the leeward side of the dune. However, in most cases migrating barchans act as an open system in dynamic equilibrium, in which the input of sand from upwind is equal to the downwind losses from the horn tips. A change in wind conditions or sand supply will cause the barchan to change its size and probably its shape.

There is general agreement that barchans form in vegetation-free areas where sand supply is limited and the winds are almost unidirectional (Fryberger & Dean 1979, Breed & Grow 1979, Glennie 1983b, Wasson & Hyde 1983) (Fig. 6.20). Winds from a secondary direction may cause asymmetry of the dune by elongating one of the horns (Fig. 6.17) (Kerr & Nigra 1951, 1952, Clos-Arceduc 1967, 1969, p. 38, Mainguet 1984b, Tsoar 1984).

The rate of barchan advance is directly related to the rate of sand transport over the brink and inversely related to the brink height (Beadnell 1909, 1910, Rempel 1936, Bagnold 1941, p. 204, Finkel 1959, Long & Sharp 1964, Hastenrath 1967, 1978, Tsoar 1974, Embabi 1982, Hidore & Albokhair 1982, Haff & Presti 1984). Using geometric reasoning (Fig. 6.18a), Bagnold (1941, p. 204) showed that

$$c = q/\gamma h \tag{6.4}$$

where c is the rate of dune advance, q is the sand transport rate over the brink, γ is the bulk density of the sand, and h is the brink height.

Migration rates in excess of up to 30 m yr^{-1} have been recorded for small barchans, but 5–10 m yr^{-1} is more typical for larger dunes (Norris 1966, Inman *et al.* 1966, Lettau & Lettau 1969, Hastenrath 1967, 1987, Watson 1985).

Where sand supply increases, individual barchan dunes may link up to form a sinuous-crested ridge oriented perpendicular to the strongest wind or resultant sand

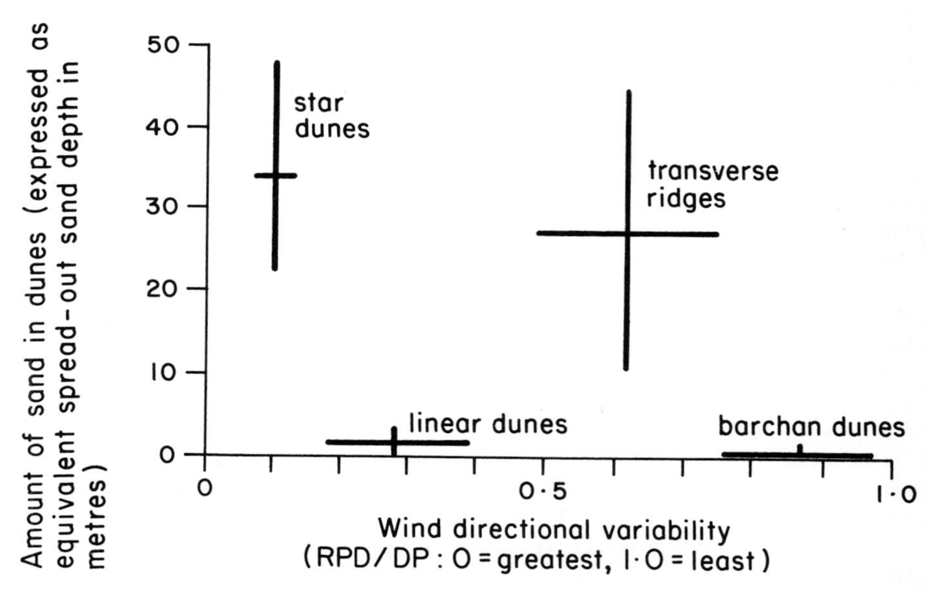

Figure 6.20 Occurrence of major dune types in relation to sand volume and wind directional variability. (After Wasson & Hyde 1983a).

Figure 6.21 Crescent-shaped transverse dunes along the coastal plain of the Negev. (Photograph by V. Goldsmith).

drift direction. Many transverse ridges display alternating *barchanoid* (downwind-facing) and *linguoid* (upwind-facing) elements viewed in plan and alternating peaks and saddles viewed in section (Cooper 1958, p. 31, Cooke & Warren 1973, p. 288). The barchanoid element of one ridge is frequently followed in the next ridge downwind by a linguoid element (Fig. 6.21) (Inman *et al.* 1966). Such transverse ridges often form the dominant component of complex dune networks which have been referred to as *gridiron* or *fishscale* patterns (Wilson 1972a,b), *aklé* (Monod 1958, Cooke & Warren 1973, p. 288, Illenberger 1988), or *reticulate networks* (Warren & Kay 1987). The longitudinal elements of these complex dune patterns are formed by sand ridges which extend downwind, most commonly from the lee side of the linguoid element (Cornish 1914, p. 45, Beadnell 1910, Besler 1980, p. 159, Illenberger 1988). These longitudinal ridges were termed *lee projections*, by Cooper (1958, p. 31), who interpreted them as erosional residuals formed by the convergence of right- and left-handed longitudinal vortices emanating from saddles in the transverse ridge crestline. The lee projections are thus analogous to the lee shadow dunes described in Section 6.3.2.1 (see also Reid 1985).

Not all transverse ridges display well developed barchanoid and linguoid elements, some being almost straight (Cooper 1967, Mainguet 1984b). The variable degree of expression of barchanoid character and superimposition of longitudinal elements has been interpreted as reflecting differences in the importance of two-dimensional wave flow versus three-dimensional vortex flow in the atmospheric boundary layer (Cooke & Warren 1973, pp. 289–91, Wilson 1970, 1972b). However, it is extremely difficult to separate cause and effect in bedform–flow interaction, and there is no direct proof that primary instabilities in the atmosphere initiate the observed forms.

Transverse and barchan dunes in a reversing wind regime change their profile dramatically as the slip face alternates from one side of the dune to the other (Cornish 1897, King 1918, Dann 1939, Bagnold 1941, p. 217). Dunes which change their slip face orientation in this way are known as *reversing dunes* (Lindsay 1973, McKee 1979a, 1982).

Figure 6.22 Oblique air photograph of complex transverse megadunes in the Algodones dunefield, California. (Photograph by J. S. Shelton).

Very large complex and compound barchan and transverse megadunes occur in some areas, including the Algodones dunefield of California (Fig. 6.22) and the Thar Desert of northwest India (Kar 1990).

6.3.5 Linear dunes

Linear dunes are characterized by their considerable length, relative straightness, parallelism, regular spacing, and low ratio of dune to interdune areas (Lancaster 1982b). As with other dune types, simple, complex, and compound varieties can be recognized. Simple linear dunes consist of a single narrow dune ridge with a straight or sinuous crest line which may be rounded or sharp in cross-section. Two major types are recognized (Tsoar 1989), unvegetated and vegetated. Because of the lack of vegetation, the former has a sharp crest which gives it its name *seif* (*sword* in Arabic). The second type is vegetated and has a more rounded crest. Seif dunes occur in many African sand seas, where they have often been referred to as *silk* dunes (pl. *slouk*) (Monod 1958, Mainguet *et al.* 1974). They are also found extensively in Sinai and other parts of the Middle East (Tsoar 1978). Vegetated linear dunes occur widely in Central Australia (Madigan 1936, 1946, Mabbutt 1968, Folk 1971a, Twidale 1972a, Buckley 1981), the Kalahari (Lewis 1936, Goudie 1970, Thomas 1986a, Lancaster 1981a, 1986), the Thar Desert of northwest India and Pakistan (Verstappen 1970, Kar 1987), southern Israel (Striem 1954, Tsoar & Møller 1986), Arizona, and California (Hack 1941, Breed & Breed 1979). This dune type is discussed more fully in Section 6.4.3.

Compound linear dunes consist of two or more closely spaced or overlapping dune ridges on the crest of a much larger plinth. The constituent ridges generally rise no more than 40 m above the plinth, which is 0.5–1.0 km wide (Table 6.3).

Table 6.3 Relative occurrence % of major dune types in selected ergs.(Data of Fryberger & Goudie 1981, Lancaster *et al.* 1987, Thomas 1989b).

	Sahara				Southern Africa			Asia			N. America	
	West ergs	South ergs	North ergs	NE ergs*	Namib	SW Kalahari	Arabian ergs	Thar	Takla Makan	Ala Shan	Gran Desierto	Mean
linear (total)	65	46	36	30	60	99	71	20	33	5	0	42.7
simple & compound	65	46	15	6	34	99	43	20	28	5	–	32.8
complex: transverse imposed	–	–	6	13	–	–	–	–	5	–	–	2.2
star imposed	–	–	15	11	26	–	28	–	–	–	–	7.3
transverse (total)	35	54	52	26	22	0	21	38	56	84	70	41.6
simple	1	8	1	–	22	–	1	13	6	27	40	10.8
compound	–	–	11	3	–	–	–	–	–	–	18	2.9
complex	34	46	40	23	–	–	20	24	50	57	8	27.5
star	0	0	12	43	18	0	8	0	0	9	33	11.1
parabolic	0	0	0	0	0	1	0	42	0	0	0	43.9
dome	0	0	0	1	0	0	0	0	11	3	0	1.3

* excluding Egyptian and Libyan ergs.

Complex linear dunes consist of very large ridges along which are distributed en echelon peaks which reach a height of 150–200 m in Arabia (Holm 1960) and 100–200 m in Namibia (McKee 1982, Lancaster 1982a, 1988a). Secondary dunes, usually oblique or transverse to the main trend, are often developed on their surfaces (Lancaster 1982a, Livingstone 1986, 1989a) (Fig. 6.23).

Wasson & Hyde (1983a) concluded that linear dunes in general occur principally in areas of limited sand availability and moderately variable wind regimes (Fig. 6.20), although their conclusions were heavily influenced by data from Central Australia. Fryberger & Dean (1979) also noted that linear dunes are found in areas with a wide range of wind energy and directional variability. According to these authors, they occur commonly in wide unimodal (winds from one broad directional sector) and bidirectional wind regimes, and sometimes occur in areas with complex wind regimes (two or more distinct modes). Many linear dunes extend parallel to the resultant sand drift direction, even in complex wind regimes (Fryberger & Dean 1979), and can thus be described as longitudinal according to the terminology of Hunter *et al.* (1983). Others can be considered as oblique dunes according to Hunter *et al.*'s classification since their long axis orientation deviates by more than 15° from the calculated resultant sand drift direction (e.g. Carson & MacLean 1986). However, some unvegetated linear dunes

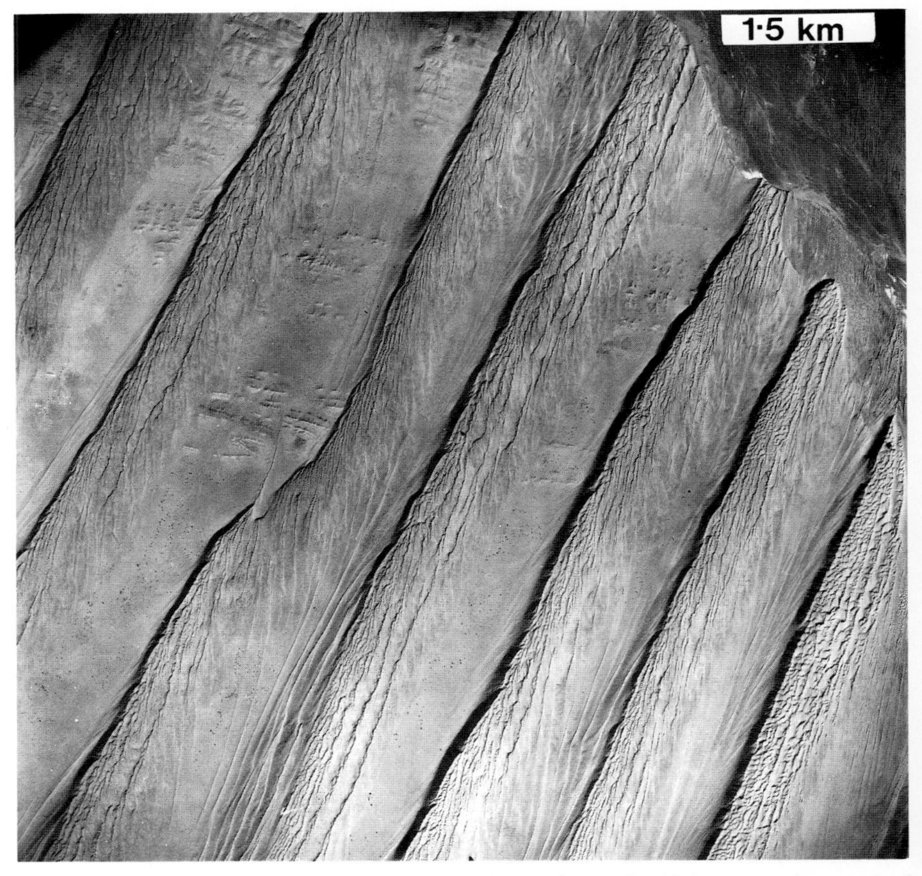

Figure 6.23 Complex asymmetric megadune with superimposed seif dunes, northern end of Wahiba Sands, Oman. (Courtesy of K. W. Glennie).

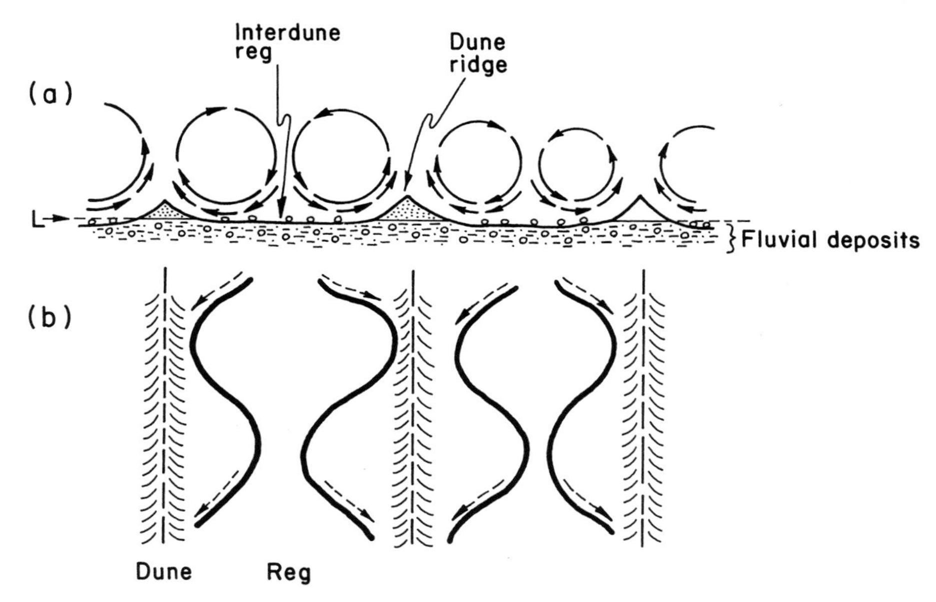

Figure 6.24 Model for the formation of longitudinal dunes by helicoidal flow. (a) Transverse section to primary wind flow direction; (b) plan view. Sand is blown from the interdune areas towards the dune ridges, leaving a bimodal lag deposit in the depressions. The dashed lines in (b) indicate the direction of surface sand transport. (Modified after Bagnold 1953a and Folk 1971a, b).

which have been described as oblique (e.g. Sneh 1982, 1988) are actually seifs formed in a bidirectional wind regime (Tsoar 1978).

Many authors have supported the theory that linear dunes develop owing to the presence of parallel helical vortices in unidirectional wind regimes (Bagnold 1953b, Hanna 1969, Cooke & Warren 1973, p. 295, Glennie 1970, pp. 89–95, 1987, Wilson 1972a, Folk 1971a,b, 1976a, Warren 1979), although little direct proof of a causal relationship has been provided (Leeder 1977, Warren 1984).

Bagnold (1953b) proposed that horizontal roll vortices would develop under conditions of strong geostrophic wind and strong thermal heating. He suggested that paired roll vortices moving parallel to the dominant wind direction might sweep sand out of the troughs onto the adjacent ridges (Fig. 6.24). As noted above, recent work has shown that seif dunes can form in bidirectional or complex wind regimes, and field studies have provided evidence that longitudinal helical vortices are not required (Tsoar 1978, 1982, 1983a, 1989, Livingstone 1986, 1988, 1989). There is no doubt that longitudinal roll vortices do occur in the lower atmosphere, and are probably responsible for some linear aeolian features such as sand strips, sand streets and erosional lineation (Hanna 1969, Wippermann 1969, Hastings 1971, Whitney 1978), but it has not been convincingly shown that they have a determining influence on the morphology and development of major linear dunes (Lancaster 1982b, Tsoar 1983a, 1989, Livingstone 1988, Wopfner & Twidale 1988).

6.3.5.1 Development of seif dunes Bagnold (1941, p. 224) noted that the detailed shape of seif dunes varies according to differences in the long period wind regime, but the essential feature common to all is a single continuous, often sinuous ridge which rises and falls at regular intervals to form a series of peaks and saddles. This gives the ridge the appearance of a chain of 'teardrops' when viewed from the side. Many seif dunes, such as those in northern Sinai (Tsoar 1978, 1983a) (Figs 6.25–6.27), are highly sinuous in plan, but some are relatively straight. In the latter case the peaks and saddles along the crest tend to be less well developed. All seif dunes have a sharp crest which shows seasonal or shorter term reversal of the slip-face. The height of the slip-face is typically one-third to half of the height of the dune as a whole. Active seifs are devoid of vegetation except along the parts of the basal plinth.

Bagnold (1941, p. 223) proposed a model to explain how seifs evolve from barchans by extension of one horn in a bidirectional wind regime. According to Bagnold's model,

Figure 6.25 Oblique aerial view of sinuous seif dunes, Sinai Desert.

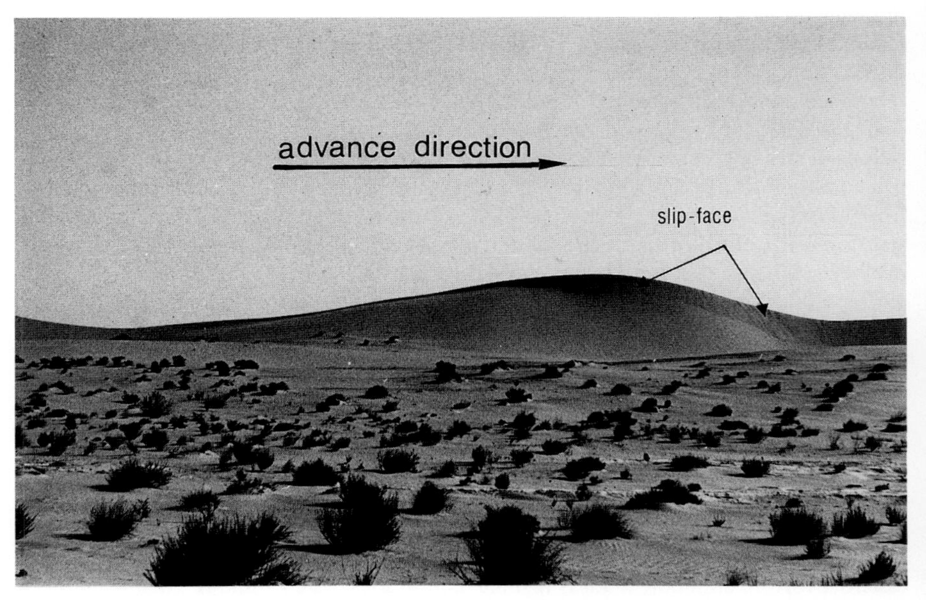

Figure 6.26 'Tear drop' form resulting from rhythmic development of peaks and saddles on a sinuous seif dune. (After Tsoar 1983a).

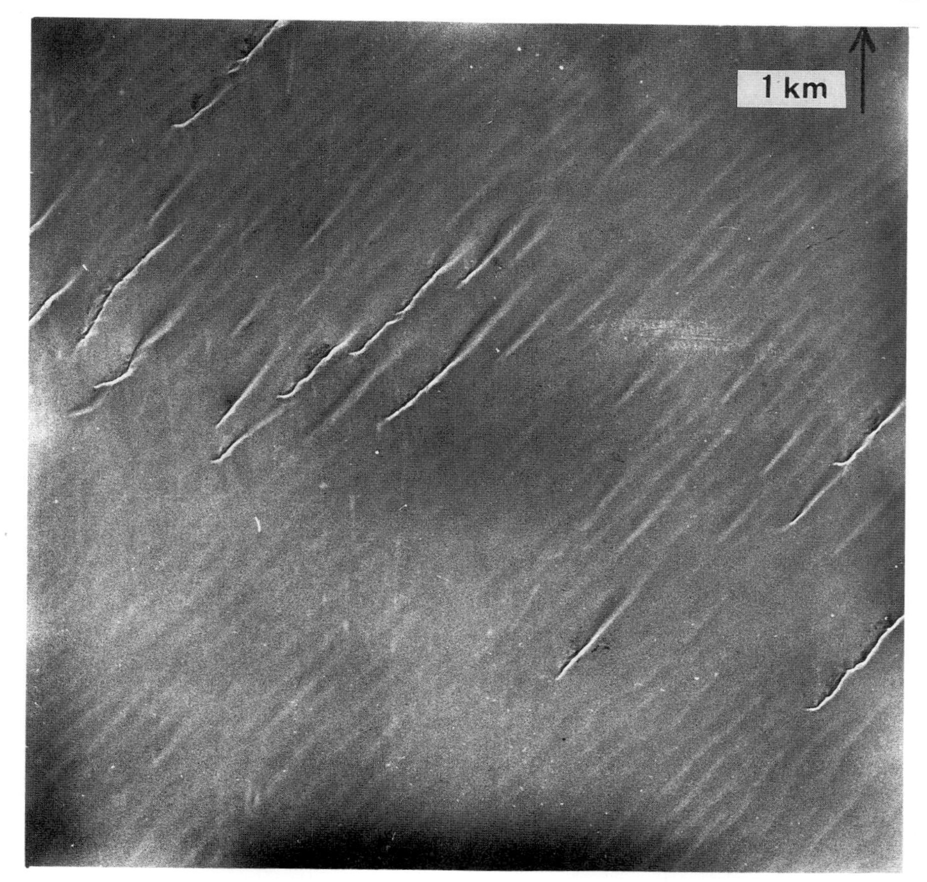

Figure 6.27 Vertical air photograph showing development of seif dunes from linear zibar, Ténére Desert, Niger. (Courtesy of A. Warren).

occasional strong storm winds from a secondary direction elongate the horn on the side from which the strong wind blows (Fig. 6.28a).

Tsoar (1984) proposed a modified version of this model which emphasizes that both strong winds from the primary direction and gentle winds from the secondary direction play a part in elongating the horn on the side opposite to the secondary gentle wind direction (Fig. 6.28b). Once a barchan begins to evolve into a seif dune, the longitudinal element elongates faster than the advance rate of the transverse element of the barchan. Examples of this mode of seif formation have been described from Sinai by Tsoar (1984) and from southern Africa by Lancaster (1980).

Less commonly, seif dunes originate from linear zibar (Warren 1972, Tsoar 1978, 1983a) (Fig. 6.27) or due to overgrazing on vegetated linear dunes (Tsoar & Møller 1986). In the former case the coarse-grained surface of the zibar appears to act as an efficient saltation conduit for sand blown from upwind, rather than providing the actual source of sand.

Bagnold (1941, p. 224) first proposed that seif dunes elongate downwind in a bidirectional wind regime. Subsequent detailed fieldwork on a 6–13 m high seif dune in northern Sinai (Tsoar 1978, 1983a, Tsoar & Yaalon 1983) has provided further evidence of this process. Using flow visualization and sand tracer techniques (Fig. 6.29), Tsoar showed that when the wind strikes the dune obliquely (a common situation) and undergoes flow separation at the crestline, part of the flow is diverted along the lee slope parallel to the crest (Fig. 6.30). Further down the lee slope, beyond the line of reattachment, the wind resumes its general direction. Further evidence of sand transport parallel to the crest on the upper lee slope is provided by the orientation of ripple crests (Fig. 6.31).

The phenomenon of lee flow diversion has also been observed during oblique flows over transverse dunes (Sharp 1963) and barchan dunes (McKee 1945, Knott 1979) and in laboratory experimental studies using different model shapes (Allen 1968, p. 161, Mulhearn & Bradley 1977, Jackson 1977, Tsoar *et al.* 1985, Barndorff-Nielsen 1986).

Field observations and wind tunnel experiments have shown that the degree of flow diversion along the lee slope is strongly dependent on the angle between the approach wind and the dune crest and on the cross-sectional profile of the dune perpendicular to the incident wind (Tsoar 1983a, Tsoar *et al.* 1985) (Figs 6.32 and 6.33). Since the crestline of a seif dune meanders, there are segments where the incident wind approaches the crest obliquely and others where it is almost transverse (Fig. 6.34). In northern Sinai the dominant summer winds approach the dunes from the northwest. Deflection of the flow

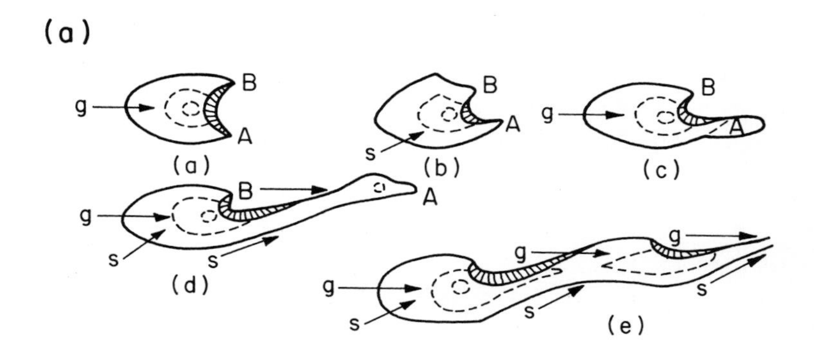

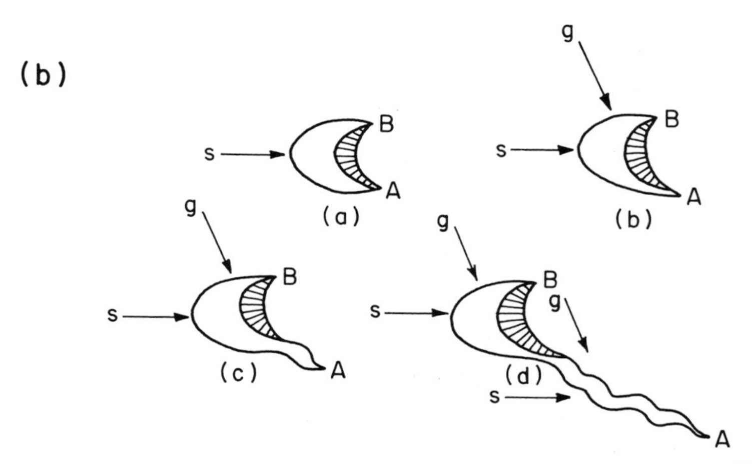

Figure 6.28 Models for the development of seif dunes from barchans, (a) as suggested by Bagnold (1941) and (b) as modified by Tsoar (1984); s indicates a strong wind and g a gentle wind.

Figure 6.29 Wind flow diversion of the lee slope of a seif dune demonstrated by smoke. Note the separation of flow at the crest, the reattachment on the lee slope and the diversion at the reattachment line.

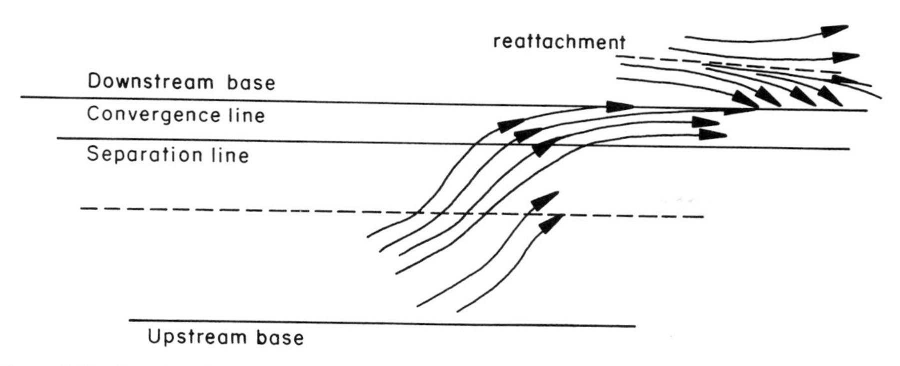

Figure 6.30 Results of flow visualization over a three-dimensional symmetrical rounded model in which the wind approaches the crest at an angle of 25°. (After Tsoar *et al.* 1985).

along the lee side is greatest where the wind approaches the crest at an angle of 40 ± 10°. Tsoar (1978, 1983a) also found that in areas of maximum deflection the diverted flow is stronger than the incident wind as measured at the crest (Fig. 6.32), resulting in erosion of this part of the lee slope as it is in summer (between A′ and B′ in Fig. 6.34). Winds approaching the crest in summer between B′ and C′ at > 40° are also deflected on the lee slope but speeds are much lower. Consequently, there is net accretion of the lee slope between B′ and C′. During the winter, when the dominant winds blow from southwest, erosion occurs on the lee slope between B and C where the wind approach angle is most oblique to the crest line. Accretion of sand occurs in the zone of minimal flow deflection between A and B (Fig. 6.34). Because of the lack of uniformity in the effect of the wind on the two sides of the seif, there are variations in the rate of sand erosion and

Figure 6.31 Ripples on the lee slope of a seif dune indicating transport parallel to the crestline. The incident wind direction is oblique to the crestline as demonstrated by two smoke candles (arrows).

deposition. Since the summer wind is more effective, the seif widens in the accretion areas and narrows in the erosion areas formed by the summer wind. Widening also leads to greater height, which accounts for the presence of peaks and saddles (Fig. 6.34). Tsoar (1983a) observed that the peaks and saddles migrate slowly downwind at a fairly steady average rate of 0.7 m per month. The elongation rate of the dune was 1.7 times the rate of displacement of the peaks and saddles, indicating that new peaks and saddles must be created near the downwind end as the dune extends. On this dune virtually all of the sand moved over the crest was moved along the dune by the secondary circulation cell which dominated a large part of the lee slope. Under these conditions little sand is lost from the dune, except at its end, which extends at a relatively rapid rate.

The model of seif dune extension by lee side-flow diversion may not be entirely applicable to very large unvegetated linear dunes where the back eddy caused by flow separation occupies only a small part of the lee slope (Livingstone 1986, 1988). Studies on a complex linear megadune in the Namib confirmed that sand transport parallel to the crest occurs with oblique incident winds, but only the uppermost part of the lee slope was found to be affected by a three-dimensional flow separation vortex (Livingstone 1986, 1988, 1989a). No lateral sequence of net erosion and depositional zones, as described by Tsoar, were identified, and sand was observed to escape from the lee slope and cross the interdune corridors from one dune to the next.

The apparent differences between the Sinai and Namib dunes may be partly attributable to differences in scale and morphology, and partly to differences in wind regime. Winds in the Namib blow seasonally from almost opposite directions (southwest and east), whereas in Sinai both the winter and summer winds blow from westerly quadrants. Along-dune transport may therefore be expected to be greater in the case of the Sinai dunes. The great height of the Namib linear dunes also suggests that vertical growth may

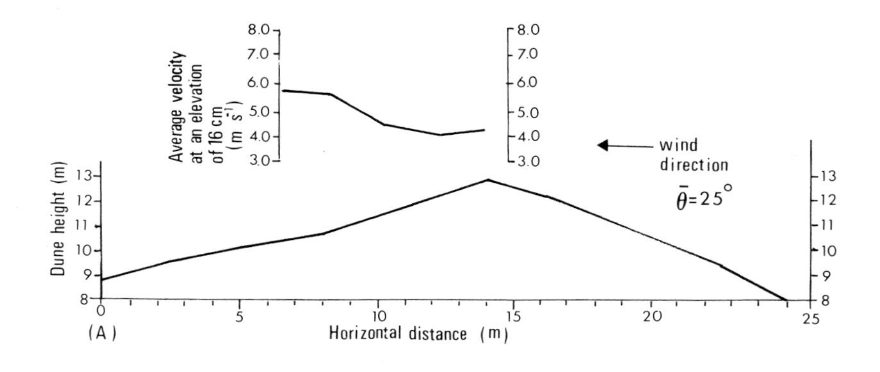

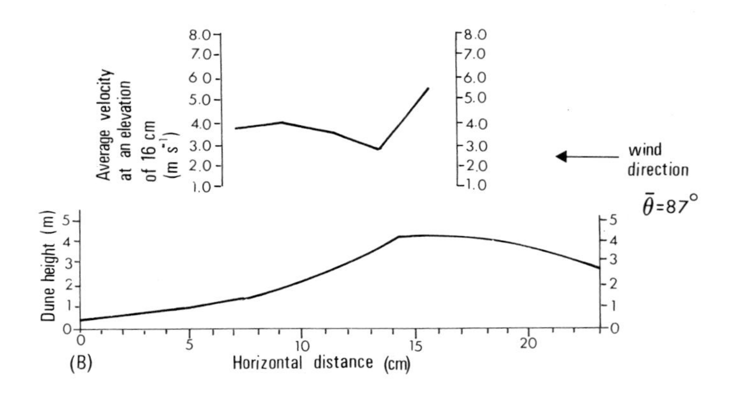

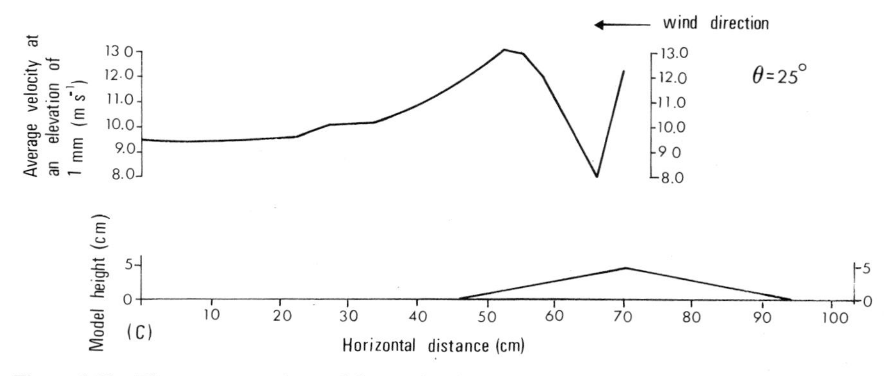

Figure 6.32 Three cross-sections of dunes showing changes in wind velocity at the surface on the lee slope. θ is the angle between the direction of wind approach and the crestline. A and B are based on field measurements and C on wind tunnel simulations. (After Tsoar 1983a, Tsoar *et al.* 1985).

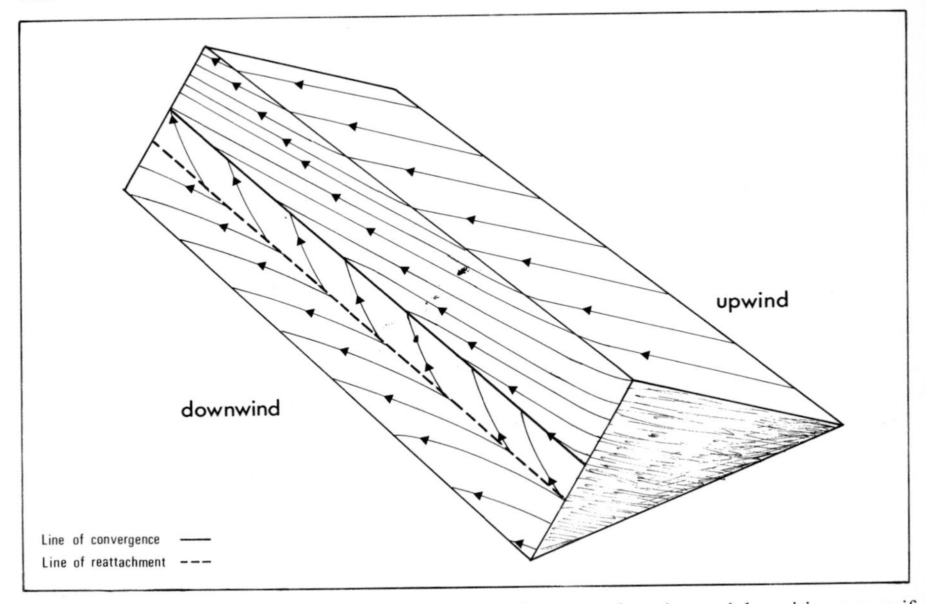

Figure 6.33 Generalized sand transport directions and pattern of erosion and deposition on a seif dune under the influence of winds which approach the crest obliquely, based on field observations (Tsoar 1978, 1983a, Tsoar & Yaalon 1983).

have been favoured at the expense of the rate of dune extension. However, it is still uncertain whether the Namib dunes are in equilibrium with present-day winds. It is possible that the gross morphology of these dunes was determined by stronger Pleistocene winds, which may have created much larger lee circulation cells, and that present-day winds have simply modified the crestal areas.

6.3.5.2 Oblique dunes According to the definition given by Hunter *et al.* (1983), oblique dunes have crestlines which form an angle of 15–75° with the long-term resultant sand transport direction (Fig. 6.6). Several authors have claimed to recognize oblique dunes in both modern and ancient aeolian deposits (Rubin & Hunter 1985, Sneh 1988, Clemmensen & Blakey 1989). While there are no theoretical reasons why oblique bedforms as defined by Hunter *et al.* should not exist, and there is some field experimental evidence for their formation (Rubin & Hunter 1987), there are grounds for questioning the value of the term as an aid to understanding of dune morphogenesis (see also Carson & MacLean 1985a). One fundamental difficulty in interpreting dunes which are apparently oblique arises from the frequent uncertainty as to whether such dunes are in equilibrium with present-day wind conditions. This is especially true in the case of complex and compound megadunes whose origins may extend back to the Pleistocene. There are many parts of the world where dunes are essentially fossil and are oblique to the present resultant sand transport direction deduced from modern wind data. Such areas include the Timbuktu region (Breed *et al.* 1979, p. 324) and the northwest Kalahari (Fryberger & Dean 1979, p. 164). A further fundamental problem concerns the fact that the wind data used to calculate resultant sand transport directions often relate to meteorological stations which are tens or even hundreds of kilometres away from the dunes, sometimes in completely different topographic settings. Finally, the question arises as to whether the resultant sand transport direction determines the gross morphology and orientation of dunes. Since there is strong evidence that in some cases it

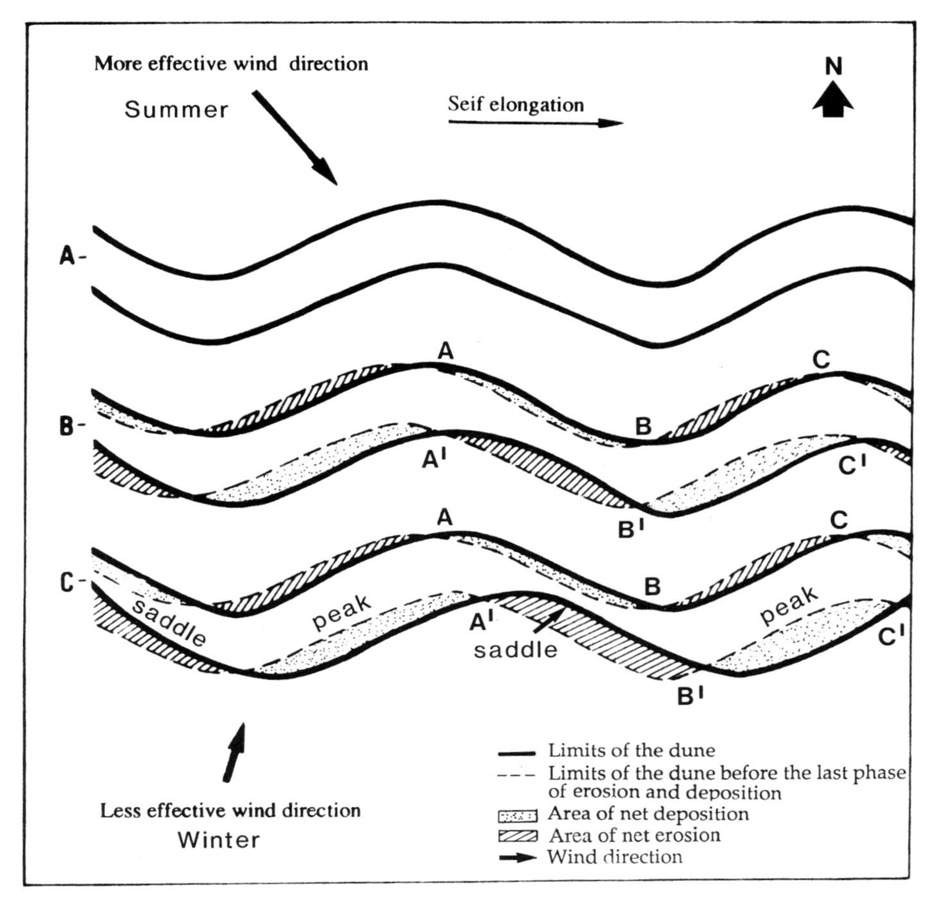

Figure 6.34 Schematic model showing the formation of peaks and saddles along a sinuous seif dune. (A) Limits of the seif dune having no peaks and saddles; (B, C) form changes after two successive phases of deposition and erosion.

does not, terms such as longitudinal, oblique, and transverse, defined relative to the theoretical resultant sand transport direction, have limited morphogenetic relevance.

Cooper (1958, pp. 49 and 53–5) first applied the term 'oblique' to certain Oregon coastal dune ridges which he considered lie at an angle to both the dominant winter southwesterly winds and the secondary summer northwesterly winds (a wind regime which is classified as obtuse–bimodal according to the scheme of Fryberger & Dean 1979, p. 149). Cooper considered that the ridge crests probably coincided with the long-term resultant sand transport direction, but provided no specific evidence.

Hunter *et al.* (1983) showed that the crestal orientation lies at an angle of between 15 and 75° to the resultant transport direction calculated using 1 year's wind data for the coastal station of Newport. However, since this station lies approximately 100 km north of the Umpqua South part of the Coos Bay dune sheet where they worked, and local wind conditions in the dunefields are modified to some extent by local topography, it is uncertain how representative the Newport wind data are of actual conditions on the dunes studied.

Hunter *et al.* (1983) showed that the gross morphology of these dune ridges is controlled by south-southwesterly winter storm winds, to which they are transverse. Moderate northwesterly summer winds modify the dune forms but not the dune trend. The orientation of the dune slip faces reverses during the summer, but the principal net effect of the summer winds is to transfer sand along the dune ridges towards the east, such that the dunes become larger in this direction (Carson & MacLean 1985a, Hunter *et al.* 1985). The internal structures of the dunes confirm northward migration of the ridges during wet (winter) conditions, while summer deposits are generally not preserved. We consider, therefore, that these ridges are better regarded as transverse forms which display minor reversing behaviour and secondary sand transport parallel to the crest which is reflected in lateral asymmetry of the dunes.

There is growing evidence that in bidirectional wind regimes where the two directional components are of unequal magnitude, dune ridges may show dynamic behaviour typical of both transverse and seif dunes (Rubin & Hunter 1985, Carson & MacLean 1986, Hesp *et al.* 1989, Rubin 1990). Some seif dunes move sideways but show predominant elongation, while some transverse dunes display net lateral transfer of sand parallel to the crest as the slip face advances. These dunes lie within a continuum between true transverse dunes and seifs. The term *hybrid* has been suggested by Carson & Maclean (1985a,b, 1986) to describe such dunes. However, our view, like that of Hunter *et al.* (1985), is that this term lacks precision. The hybrid dunes of the Williams River area in northern Saskatchewan described by Carson & MacLean (1986) occur in a bidirectional wind regime of seasonally opposing winds. Both of the opposing winds blow almost perpendicular to the dune axes but the secondary component is slightly oblique. Like the Oregon dunes studied by Hunter *et al.* (1983), these dunes are therefore essentially transverse dunes which display seasonal reversing behaviour. Net movement of the ridges is predominantly in the direction of the strongest seasonal wind, but owing to the influence of the secondary wind the ridges also show a degree of extending behaviour.

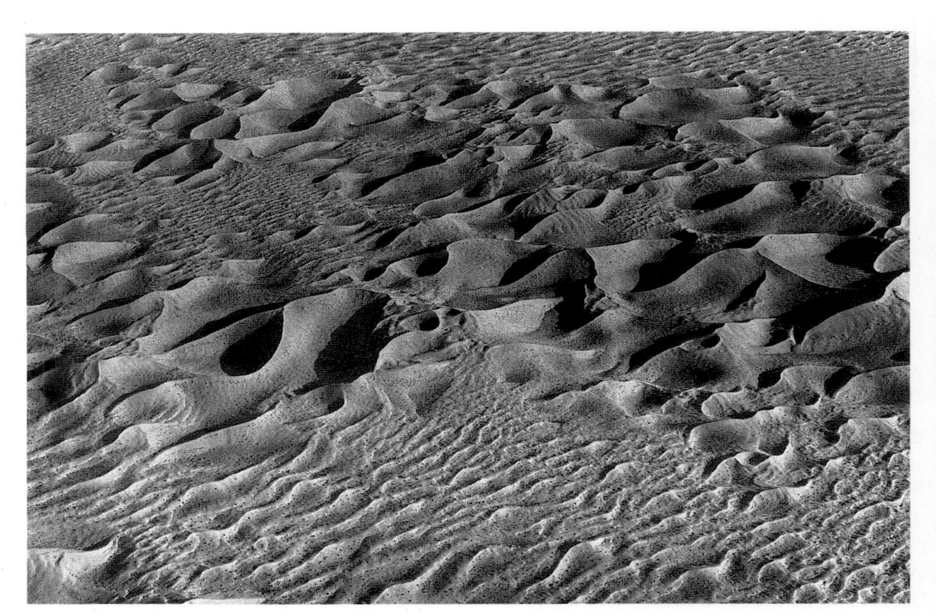

Figure 6.35 Oblique aerial view of star megadunes, Gran Desierto, Mexico. (Courtesy of D. Ball).

6.3.6 *Star dunes*

Star dunes are characterized by their large size, pyramidal morphology, and radiating sinuous arms (Lancaster 1989a,b) (Fig. 6.35). They occur as simple forms with three or more radial arms joined at a single summit, as compound forms with multiple peaks connected by cols, or as complex forms superimposed on linear megadunes (Breed & Grow 1979, McKee 1982, Walker 1986, p. 469, Nielson & Kocurek 1987). In the Sahara they are known as *demkhas*, *ghourds*, *rhourds*, or *oghourds* (Aufrère 1935, Capot-Rey 1945, Mainguet & Callot 1978). Other terms which have been used to describe them include *sand massifs* (Bagnold 1951), *pyramidal dunes* (Holm 1960), *stellate dunes* (Glennie 1970), *sand mountains* (Cooke & Warren 1973), and *horn* or *cone-shaped dunes* (Zhenda 1984).

Approximately 11% of all dunes are of the star type, constituting about 5% of aeolian depositional surfaces (Table 6.3) (Fryberger & Goudie 1981). Important occurrences of star dunes include the Grand Erg Oriental in Algeria, the Erg Fachi-Bilma in Niger, the southeastern Rub-al-Khali in Saudi Arabia, the Gran Desierto in Mexico, the Ala Shan Desert in China, and the Namib Desert (Capot-Rey 1945, Holm 1960, Mainguet & Callot 1978, Breed & Grow 1979, Breed *et al.* 1979, Lancaster 1983a, 1986, 1989b,c, Zhenda 1984, Lancaster *et al.* 1987, Walker *et al.* 1987). Small groups of star dunes are also common in the Basin and Range deserts of North America (Sharp 1966, Andrews 1981, Smith 1982, Nielson & Kocurek 1987). The only sand sea where star dunes cover a large part of the area is the Grand Erg Oriental, where approximately 40% of the dunes are of this type (Breed *et al.* 1979).

Star dunes typically have a mean width of 500–1000 m and a mean height of 50–150 m, although there are significant variations between different deserts (Table 6.4). Exceptional star dunes more than 300 m high have been reported from Namibia and the Ala Shan Desert. Breed & Grow (1979) identified two groups of star dune spacings: a large group with spacings of 1000–1400 m and a smaller group with spacings of 2000–4000 m. In general, the spacing of star dunes increases with increasing dune height (Lancaster 1989b). In some areas star dunes occur in complex star dune chains. In other areas there are transitional forms from complex linear dunes or complex transverse dunes to star dunes.

Star dunes typically occur in obtuse bimodal or complex (multimodal) wind regimes, whether they are of low, intermediate, or high energy (Fryberger & Dean 1979). The stellate form has generally been explained as a response to sand transporting winds which blow from different directions at different times of the year (Holm 1960, Glennie 1970, Cooke & Warren 1973, McKee 1982), but recent work has pointed to the importance of secondary circulations induced by the star dune form itself (Nielson & Kocurek 1987, Lancaster 1989a,b). Many areas in which star dunes are developed lie close to the poleward margins of desert regions, where the effects of seasonal changes in wind directions are more marked compared with the equatorial margins, where the wind regimes are dominated by trade-wind circulations (Lancaster 1989b).

Studies of surface airflow and sand transport over individual star dunes in the Namib, Gran Desierto, and Dumont Dunefield, California, have shown that the major arms of the dunes tend to be aligned transverse or slightly oblique to the two major directions of sand transport (Lancaster 1989a,b, Lancaster *et al.* 1987, Nielson & Kocurek 1987). The minor arms of the dunes are aligned parallel to these directions and transverse to the secondary wind direction. The major ridges of star dunes in the Rub-al-Khali also appear to be transverse to the dominant sand transport directions from the northwest and east-southeast (Lancaster 1989b), but the arms of the star dunes in the Grand Erg Oriental have a more complex pattern of alignments.

A close association between the occurrence of star dunes and topographic barriers

Table 6.4 Mean spacing, width, and height of (a) linear dunes and (b) star dunes in selected ergs. (Data from Hack 1941, Breed & Grow 1979, Wilson 1972a, Holm 1960, Lancaster 1982b, Lancaster *et al.* 1987, and Walker *et al.* 1987).

(a) *Linear dunes:* locality	Spacing (km)	Width (km)	Height (m)
Southern Africa	0.90	0.22	10–25
	0.70	0.29	5–20
	1.41	0.38	-
	0.15	0.04	2–10
	3.17	1.48	100–200
	2.20	0.88	50–160
	3.24	-	-
	1.93	0.94	-
Mauritania	3.28	-	-
	1.90	0.65	25–45

(b) *Star dunes* Locality	Spacing (km) Mean (range)	Width (km) Mean (range)	Height (m) Mean (range)
Namib	1.33	1.0	145
	(0.6–2.6)	(0.4–1.0)	(80–350)
Niger	1.0	0.61	
	(0.15–3.0)	(0.2–1.2)	
Grand Erg Oriental	2.07	0.95	117
	(0.8–6.7)	(0.4–3.0)	
SE Rub-al-Khali	2.06	0.84	(50–150)
	(0.97–2.86)	(0.5–1.3)	
Gran Desierto	2.98	2.09	
	(0.15–4.0)	(0.07–6.0)	
Dunes in clusters	0.31	0.18	80
(Gran Desierto)	(0.16–0.49)	(0.09–0.36)	(10–150)
Ala Shan	1.37	0.74	(200–300)
	(0.3–3.2)	(0.4–1.0)	

was noted by Breed & Grow (1979). The effect of major topographic features is probably both to increase the complexity of the regional windflow and to generate secondary flows through differential surface heating. Topographic features also create barriers to the sandflow, leading to accumulation of thick sand deposits which are associated with the occurrence of star dunes (Wasson & Hyde 1983a) (Fig. 6.20).

The processes responsible for the initiation and subsequent early development of star dunes are currently uncertain. Cornish (1914) suggested that star dunes are initiated at the centre of convection cells, while Clos-Arceduc (1966) put forward the view that they develop at the nodes of stationary waves in oscillating flows. Wilson (1972a) speculated that star dunes might develop at points where sandflow paths cross, while Mader & Yardley (1985) suggested star dunes may form by extension of other dune types into regions of complex wind regime. The latter hypothesis is the only one for which there is

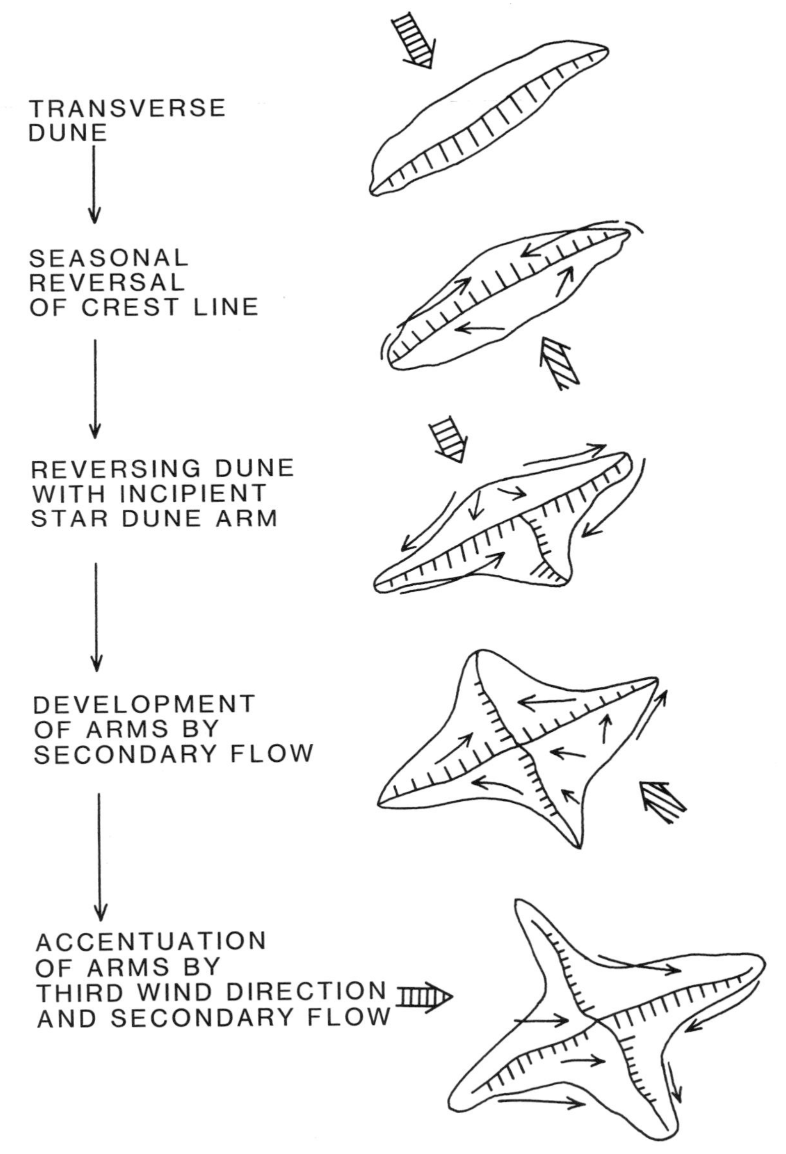

Figure 6.36 A model for star dune formation by development of secondary flow circulations as transverse dunes migrate into an area of multidirectional winds. (After Lancaster 1989b).

empirical support. Nielson & Kocurek (1987) observed that small star dunes in the Dumont Dunefield, California, form during winter periods of variable wind regime, but are modified into barchan forms during the summer period of unidirectional winds. This suggests there may be a minimum size for the survival of the star dune form.

Observations of surface wind velocity and sand transport on a 40 m high star dune in the Gran Desierto revealed a high degree of interaction between the airflow and the dune morphology in response to seasonal wind changes (Lancaster 1989a,b). These interactions concentrate sand deposition on the central parts of the dune, giving rise to its pyramidal shape, and also lead to some extension of the radiating arms. The major arms on this dune are oriented approximately transverse to summer southeast and winter north-northwest winds (Fig. 6.36). In periods of northerly or southerly winds, flow separation at the crestline of the arms generates a wide zone of leeside secondary flow which moves sand along the base of the avalanche face towards the central part of the dune. Spring westerly winds move sand obliquely up the southern and northern arms of the dune and outwards along the eastern arm. Large-scale flow separation and diversion are replaced by the development of strong helical eddies in the immediate lee of the main crestline which move sand towards zones of lower flow velocity at the end of the dune arms.

In the eastern part of the Gran Desierto, simple and compound barchanoid ridges migrate northwards and enter an area where northerly winds are more vigorous. Initially they develop reversing crests, and by the time they reach the northern margin of the sand sea large star forms have developed on their tops (Lancaster *et al.* 1987, Lancaster 1989a). In the northwestern part of the sand sea, transverse dunes migrate towards the southeast. As they do so they encounter progressively stronger southerly winds which reduce their rate of forward migration. Faster moving dunes from upwind eventually collide with them, adding to their bulk and encouraging the growth of high reversing and then star dunes.

The development of arms sub-parallel to the principal flow directions (perpendicular to the transverse dune axis) was suggested by Lancaster (1989a,b) to be caused by secondary leeside flows from the margins of the ridge towards its centre. These flows probably arise as a result of negative pressures in the wake region which is created as the primary flow passes over the ridge (Hunt *et al.* 1978). The two secondary flows converge near the centre of the ridge and are deflected downwind. Sand deposition at the point of convergence leads to extension of a linear ridge parallel to the primary flow direction. These ridges may therefore be considered analogous to the leeside projections seen on some transverse dunes, the difference being that in the case of star dunes they develop on both sides of the transverse ridge in a reversing wind regime. Vertical growth of the central part of the dune is favoured by long-term focusing of the deflected sand flows at this point. Simple star dunes of this type can thus be viewed as reversing transverse dunes with bidirectional lee projections.

In a complex multimodal wind regime, oblique flows to the crests of the arms are more frequent. It is not yet clear why, in some instances this leads to extension of the arms and to the development of lower, flatter forms, whereas in other cases sand is concentrated near the centre of the dune, leading the development of a high, steep form. Much work remains to be done to document the sand transport dynamics on these dunes.

6.3.7 Dome dunes

Dome dunes are relatively low, flat-crested forms, often without slip faces, which are circular or elliptical in plan. Fryberger & Goudie (1981) estimated that they comprise only 1.3% of the dunes in the major sand seas. They are absent in many deserts and are common only in some of the Chinese deserts [especially the Taklamakan (Zhenda 1984)] and parts of Saudi Arabia (Holm 1953).

Small coastal dome dunes form fairly frequently on beaches. These dunes are typically less than 1 m high and less than 14 m in diameter (McKee & Bigarella 1979, p. 99). McKee (1966) described simple desert dome dunes in the White Sands dunefield of New Mexico which are typically 150–200 m in diameter, 6–10 m high, round or oval in plan, and lack slip faces. Complex dome dunes in the Taklamakan are considerably larger, being 40–60 m high and 500–1000 m in diameter. Secondary dunes are extensively developed on their surfaces (Zhenda 1984). Some of the domes in the Takla Makan have linear 'tails' (Breed & Grow 1979, p. 281). Features of similar size which occur in northern Saudi Arabia were termed giant domes by Holm (1953). These dunes also have small barchanoid ridges developed on their surfaces.

In many dunefields, including the White Sands dunefield (McKee 1966) and the Killpecker dunefield of Wyoming (Ahlbrandt 1974), dome dunes occur close to the upwind margin of the dunefields, leading some authors to suggest that dome dunes develop where winds are sufficiently strong and unidirectional to effectively retard normal upward growth of dune crests (McKee & Bigarella 1979, p. 98, Breed & Grow 1979, p. 280). However, this hypothesis has not been adequately tested using field wind data. Glennie (1972, p. 1052) expressed an opposite view, although unsubstantiated by evidence, that sand is deposited in linear strips when wind velocities are high and as low oval mounds when the velocities are lower.

McKee (1966, p. 26) noted that the dome dune he studied at White Sands was composed of coarser, more poorly sorted sand than the other dune types further downwind. However, the possible significance of grain size in determining the morphology of dome dunes also remains a matter of conjecture.

Goldsmith *et al.* (1977) used the term *medaño* (Spanish for coastal sand hill) to describe the high, steep, isolated dunes without vegetation which occur in some coastal dunefields including Currituck Spit, Virginia/North Carolina, Coos Bay, Oregon, and the south end of Lake Michigan. This type of dune was considered by Goldsmith (1985) to be a distinctive feature of coastal areas, although when viewed from the air they bear some resemblance to dome dunes. Medaños form in a bimodal or polymodal wind regime which moves sand up towards the summit from several directions. Small slip faces which develop intermittently near the dune crest reverse their orientation in response to changes in wind direction.

6.4 Vegetated dunes

Although vegetation has long been recognized as a major factor controlling coastal dune morphology, its role in deserts has probably been underestimated (Thomas & Tsoar 1990). Hack (1941) noted the interaction of vegetation cover, wind strength, and sand supply as controlling factors which determine the formation of basic dune types, but Wasson & Hyde (1983a) regarded vegetation as a modifying factor (e.g. from transverse to parabolic dunes) rather than a primary determinant. Where desert dunes are currently partly vegetated, the dunes have often been regarded as partly or wholly relict (e.g. Twidale 1981). However, some perennial vegetation is to be expected in desert sand dune areas which receive some rainfall (see Ch. 9). Only in hyper-arid and overgrazed regions is vegetation entirely absent.

6.4.1 Hummock dunes

The term *hummock dune* is used here to describe any irregularly shaped mound of sand whose surface is wholly or partially vegetated. This definition is wider ranging than that used by some authors who have restricted the term to describe small (up to 3 m high and 1–8 m diameter) mounds of sand trapped by clumps of plants (Tinley 1985, Illenberger 1988). Our definition includes the complexes of irregular, overlapping vegetated *sandhills* found in many humid coastal dunefields, and the smaller, isolated features which have previously been referred to by different authors as *hedgehogs* (Ranwell 1972), *shadow dunes* (Hesp 1981), *coppice dunes* (Melton 1940, Lancaster 1989c, p. 42), *nebkhas*, and *rebdou* (Guilcher & Joly 1954, Cooke & Warren 1973, p. 317). The latter features occur both in deserts and in coastal dunefields.

Hummock dunes range in size up to 30 m high and 100 m across, but a height of < 10 m is more typical (Fig. 6.37). They vary considerably in plan profile and in cross section, depending on the nature of the colonizing vegetation and subsequent form modification due to wind scouring. In coastal dunefields they sometimes form a series of discontinuous chains parallel to the shoreline, but commonly they display a haphazard distribution.

The morphology of hummock dunes is strongly dependent on the shape, density, and growth characteristics of the associated vegetation. Some littoral stoloniferous species, such as *Ipomea-pes-caprae*, *Canavalia maritima*, and *Spinifex hirsutus*, which are common on shorelines around the Pacific, are able to extend laterally to cover large areas very quickly. They encourage sand deposition over a wide area, leading to the development of sand platforms or broad, low dune ridges. However, the seaward side of such ridges is often made steeper by periodic wave erosion.

In humid temperate areas, grasses such as *Agropyron junceiforme* and *Elymus arenarius* are the most common pioneer dune species. They have a more irregular distribution and a tussock-like growth form which tends to encourage the development of steep-sided embryo dunes (van Dieren 1934, Salisbury 1952, Ranwell 1972, p. 141). These species, together with *Ammophila* species, which tends to become dominant just inland from the shore, have a high capacity for vertical growth and can withstand high rates of sand sedimentation (Gemmell *et al.* 1953, Laing 1954, Olson 1958a,b, Ranwell 1958) (Fig. 6.38).

Ammophila and *Agropyron* species also have a high capacity for rapid stoloniferous extension, with the result that where they are dominant the dunes tend to form a flatter, more uniform surface than in areas where *Elymus arenarius* is the dominant pioneer species or where shrubs such as *Salix repens*, *Populus* and *Hippophae* are important constituents of the dune vegetation community (Fig. 6.39).

Figure 6.37 A hummock dune on the coastal plain of Israel.

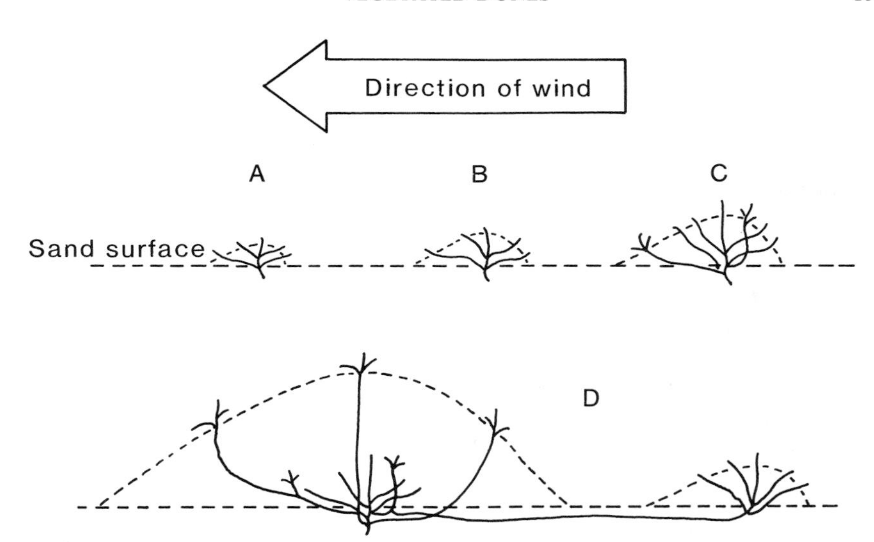

Figure 6.38 Stages in the development of 'embryo' dunes encouraged by growth of *Agropyron junceiforme*. (After Nicholson 1952, Ranwell 1972).

Figure 6.39 Irregular hummock dunes covered by *Ammophila arenaria* at Formby, Merseyside, UK. This type of vegetated dune morphology, punctuated by blowouts and associated sand mounds, is typical of coastal dunefields in many humid temperate parts of the world.

Experiments by Hesp (1981) demonstrated how pyramidal-shaped shadow dunes are formed in the lee of isolated *Festuca* or *Ammophila* tussocks (Fig. 6.40). Sand is deposited in a triangular wake region by eddies which flow from the edges of the plant towards the plant 'centreline' (Fig. 6.41). Near the ground the flow is diverted around the plant; towards the top, part of the flow is deflected over the plant, part is deflected around it, and part moves through the less dense upper foliage.

The height to which a shadow dune can be built is determined primarily by the width

Figure 6.40 Shadow dune developed in the lee of a clump of vegetation.

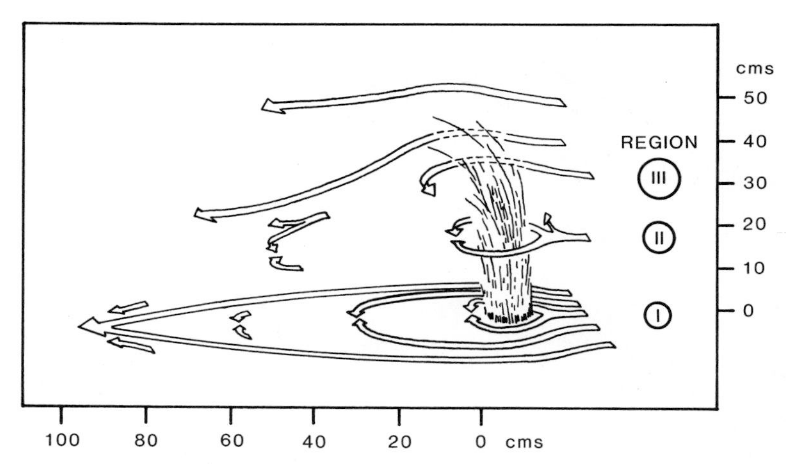

Figure 6.41 Three-dimensional diagram showing the major components of flow past a *Festuca* or *Ammophila* plant, based on observed smoke plume behaviour, cotton tuft traverses, and grain movement. The wake region shown is related to a free stream velocity of $7\,\mathrm{m\,s^{-1}}$. In region I symmetrically opposed horizontal reverse flows dominate within a triangular wake which narrows to leeward. The structure is similar in region II (not shown for clarity), with an added vertical upwind component and downwards downwind component. The latter is within a separation envelope formed in region III where flow moves through the less dense part of the plant. (After Hesp 1981).

Figure 6.42 A parabolic dune which has migrated only a short distance from its point of origin, Cape Flattery, North Queensland (view looking downwind).

of the plant and the angle of repose of the sand. The wider the base of the plant, the higher is the ridge that can be built before the angle of repose is attained. This relationship can be expressed as (Hesp 1981)

$$h = (w/2)\tan \theta \qquad (6.5)$$

where h is the frontal height of the dune, w is the plant width, and θ is the angle of repose.

The height of the shadow dune is greatest in the immediate lee of the plant, where the reverse flows are greatest, and decreases progressively with distance downwind. Hesp found that dune length decreases with increasing wind velocity, but for any given velocity the dune length increases (albeit irregularly) with plant width. No relationship was found between dune length and plant height.

Foredune ridges may be initiated by the formation of multiple shadow dunes along the strandline (Hesp 1983). As the individual shadow dunes grow larger, they coalesce to form a ridge with an undulating crestline. The subsequent pattern of sedimentation depends partly on the growth characteristics of the pioneer vegetation, partly on wind energy and sediment supply conditions, and partly on the extent of modification by wave attack during winter storms. Longshore variations in foredune morphology are often closely related to changes in wave conditions and beach morphology (Short & Hesp 1982). The formation of well developed hummock dune terrain is favoured by a discontinuous plant cover and high winds, which enlarge the cols between adjacent

Figure 6.43 Side view of an elongate parabolic dune which is crossing a vegetated sand plain from right to left, Cape Flattery, North Queensland.

hummocks into deep troughs. Trampling by pedestrians often enhances local wind scouring and leads to the development of broken dune topography and blowouts.

6.4.2 Parabolic and elongate parabolic dunes

Simple *parabolic dunes* are U- or V-shaped in plan with two trailing arms which point upwind (Fig. 6.42). Many have a large sand mound with a steep lee slip face at the downwind end, although some terminate in a low sand ridge or sand lobe without a high slip face. Very large dunes sometimes have multiple crests and slip faces. In all cases, the outside slopes of the trailing arms are partly or wholly vegetated (Fig. 6.43).

A distinction is commonly made between parabolic and *elongate parabolic* dunes. Price (1950, p. 470) defined a parabolic dune as 'an open, bow-shaped structure which has not migrated', and an elongate parabolic dune as 'a larger, clearly developed U-shaped dune developed from a spot blowout which has migrated'. There is clearly an evolutionary transition of parabolic dunes into elongate parabolic dunes. Pye (1980a, 1982a) used a length/width ratio of 3 to differentiate the two types in North Queensland coastal dunefields. Some of the elongate parabolic dunes in this area, which are more than 15 km long and 1.5 km wide, qualify for description as megadunes (Figs 6.44 and 6.45).

Parabolic and elongate parabolic dunes are common in many coastal dunefields around the world, including those of Europe (Landsberg 1956), Oregon (Cooper 1958, 1967), and Australia (Jennings 1957, Pye 1982a,c, 1983b). Inland occurrences of parabolic dunes include the Midwestern United States (Melton 1940, Ahlbrandt 1974), New Mexico (McKee 1966), Saudi Arabia (Anton & Vincent 1986), northwest India (Verstappen 1968, 1970), and northern Canada (David 1977, 1981, Raup & Argus 1982).

The morphology of individual parabolic and elongate parabolic dunes is governed by the strength and directional variability of the wind, the source and amount of sand available, and the nature of the vegetated terrain over which the dunes move. In the coastal dunefields of Cape York Peninsula, North Queensland, for example, unusually large elongate parabolic dunes have a pronounced hairpin shape in plan. This reflects the following factors: (a) the area has a markedly unidirectional wind regime, as illustrated by the Cooktown wind resultant (Fig. 6.46); (b) the shrub vegetation cover surrounding the dunes (Pye & Jackes 1981) resists widening of the dune deflation corridor on a wide front; and (c) the dunes incorporate sand from the underlying sediments as they move downwind. This 'snowball' effect brings about an increase in the height and width of the dune 'nose' as it moves downwind. However, this only continues until the dune moves onto a non-sandy substrate, when fixation of sand in the trailing arms causes a gradual dissipation and narrowing of the nose (Fig. 6.47). Eventually the crest may be lowered sufficiently to allow the wind to break through, forming a low sand lobe which extends beyond the trailing arms.

Parabolic dunes on some central and southern parts of the Queensland coast display a much less pronounced elongate form (Bird 1974, Pye 1983c), reflecting a greater degree of directional variability in the wind regime (Fig. 6.48). These dunes have a lower length/width ratio (typically 3–10) than the North Queensland dunes, but are still characterized by high trailing arms, nose, and terminal slip-face.

Anton & Vincent (1986) described parabolic dunes with a very different morphology and grain size composition in the Jafurah Desert of Saudi Arabia. These parabolic dunes

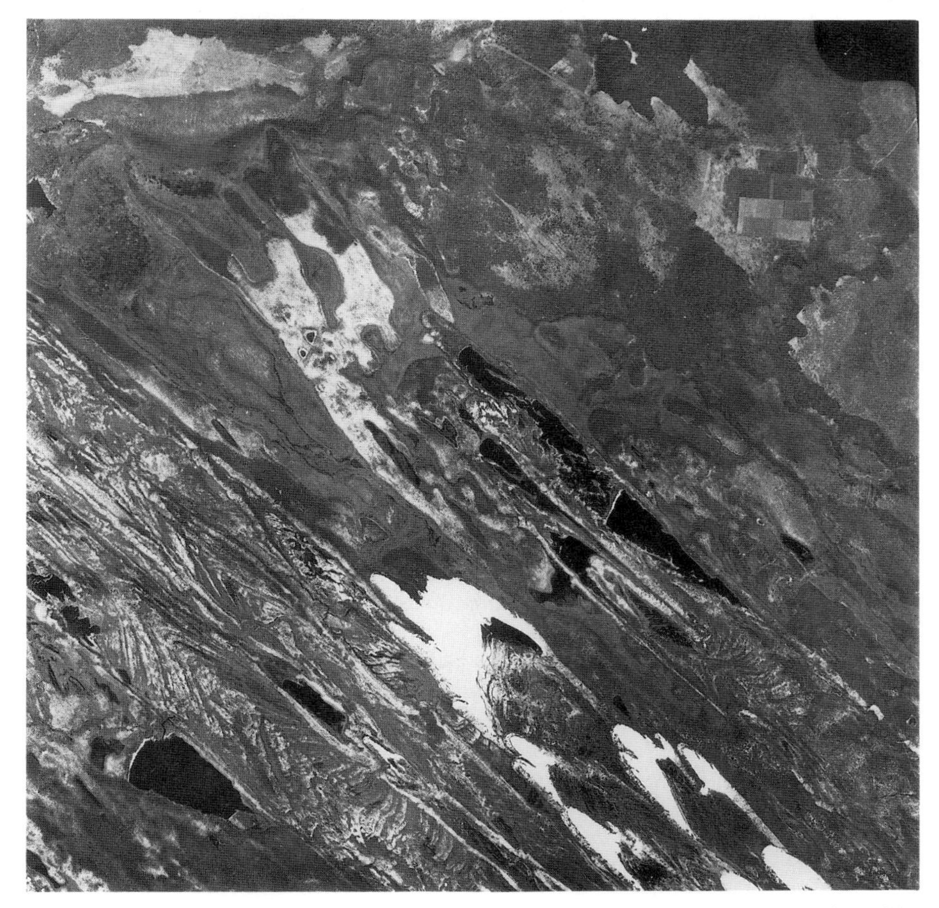

Figure 6.44 Air photograph showing complex active parabolic megadunes crossing older degraded dune terrain, north of Cape Bedford, North Queensland. (Crown Copyright, reproduced by permission of Department of National Mapping, Queanbeyan, Australia.)

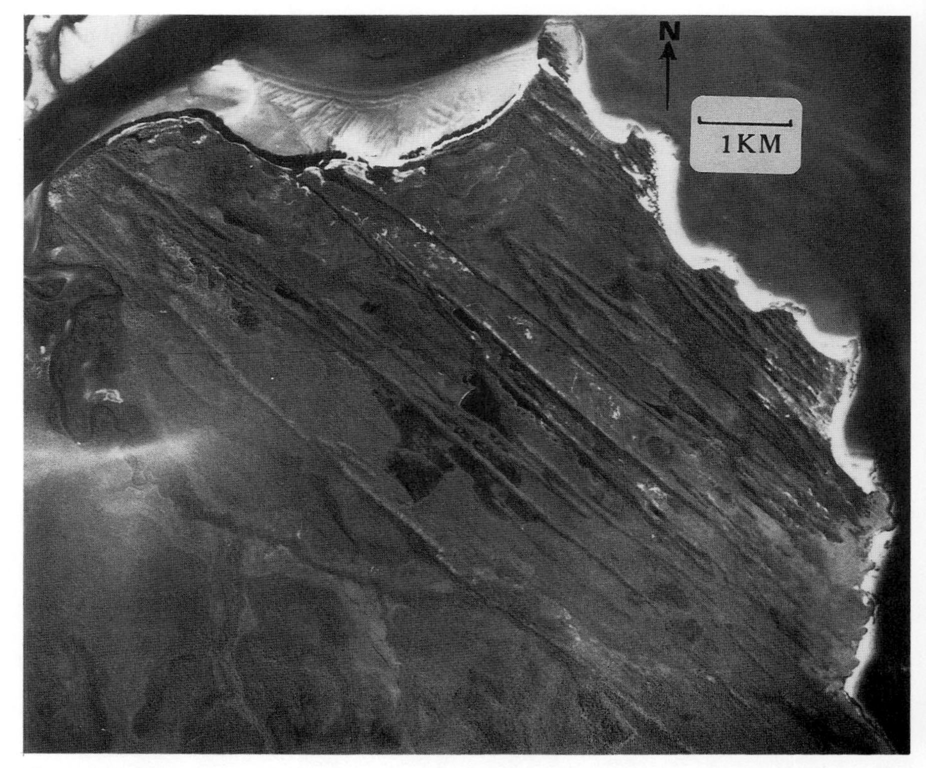

Figure 6.45 Elongate parabolic dunes now stabilized by shrub vegetation, Tern Cliffs, Cape York Peninsula, North Queensland. The noses of several dunes have been truncated by marine erosion (top left of photograph), with the result that the arms appear as linear dunes. (Crown Copyright, reproduced by permission of Department of National Mapping, Queanbeyan, Australia).

are a few hundred metres long and generally rise only a few metres above the level of the deflation basins. The sands comprising the dunes are coarse and markedly bimodal, indicating that the dunes developed by local reworking of poorly sorted sand sheet deposits. The fine to medium sand fraction, which is most readily transported in saltation, is depleted in these dune sediments, apparently owing to winnowing. The broad, low morphology of the parabolic dunes is partly attributable to the coarse nature of the sand. In several areas the parabolic dunes show a downwind transition into transverse and/or barchanoid types.

Parabolic dunes sometimes display marked differences in morphology and orientation within a small geographical area, reflecting local differences in wind climate (Aufrère 1931, David 1977, 1981). The long axes of most coastal parabolic dunes lie almost parallel with the onshore wind resultant (Jennings 1957), but the wind conditions often change only a short distance inland from the beach owing to topographic influences. In areas where parabolic dunes are significantly influenced by cross-winds blowing oblique to the dominant wind direction, they may develop left- or right-handed asymmetry. If strong winds blow from two or more discrete directions at different times of the year, *hemicyclic* or *digitate* forms may develop (Filion & Morisset 1983). These and other plan form variants are shown in Figure 6.49.

Compound parabolic dunes represent coalesced elements of individual parabolic dunes, most frequently arranged *en echelon*. In the Thar Desert of India and Pakistan,

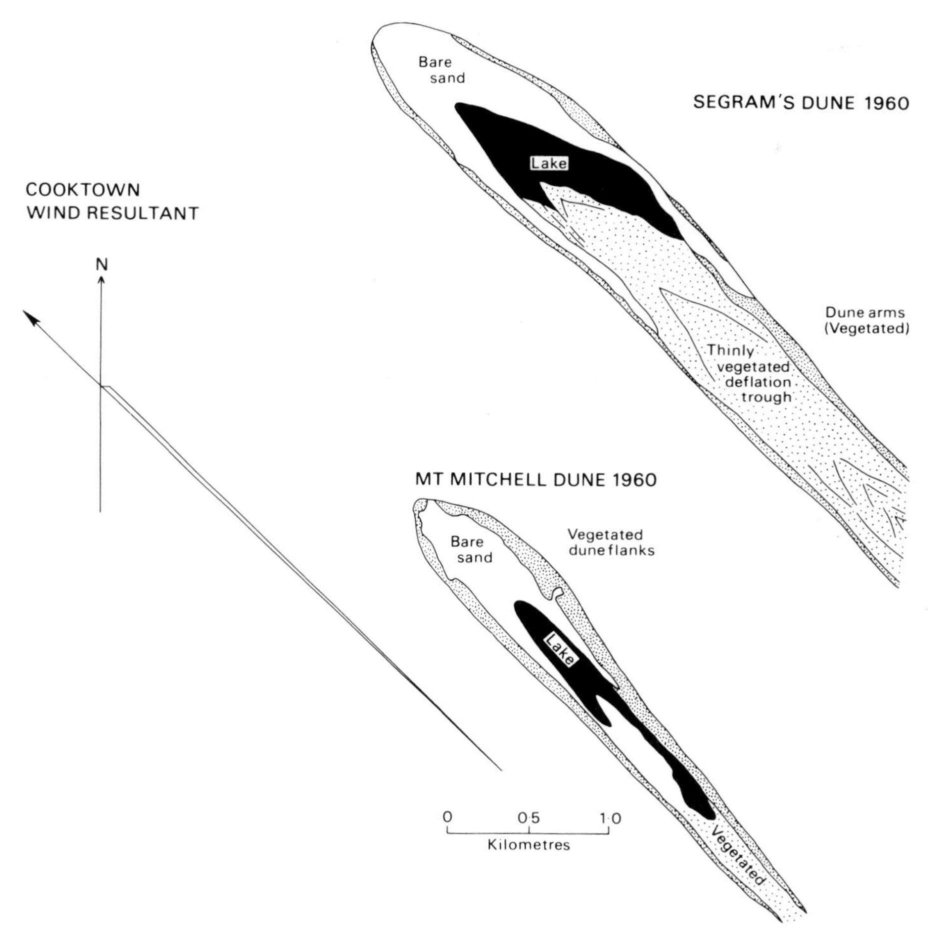

Figure 6.46 Diagram showing the close relationship between the orientation of elongate parabolic dunes at Cape Flattery and the Cooktown wind resultant (winds $>20\,\mathrm{km\,h^{-1}}$). The effective southeast trade winds in this area are highly unidirectional. (After Pye 1982a).

compound parabolic dunes which have a rake-like form cover an area of about 100 000 $\mathrm{km^2}$ (Verstappen 1968). The individual rake-like forms have an average of seven arms, a mean length of 2.6 km, and a mean width of 2.4 km (Breed & Grow 1979).

Complex parabolic megadunes do not uncommonly have secondary barchan or transverse ridge forms superimposed on them, typically in the intra-dune deflation corridor and on the windward sand ramp leading to the dune crest (Pye 1982a).

Parabolic dunes almost always develop from blowouts in a vegetated sand surface. Any local disturbance to the vegetation caused by fire, overgrazing, trampling, disease, or soil changes, including intrusions of saline groundwater into the root zone, can initiate a blowout. Once the surface crust or root mat is breached, enlargement of a

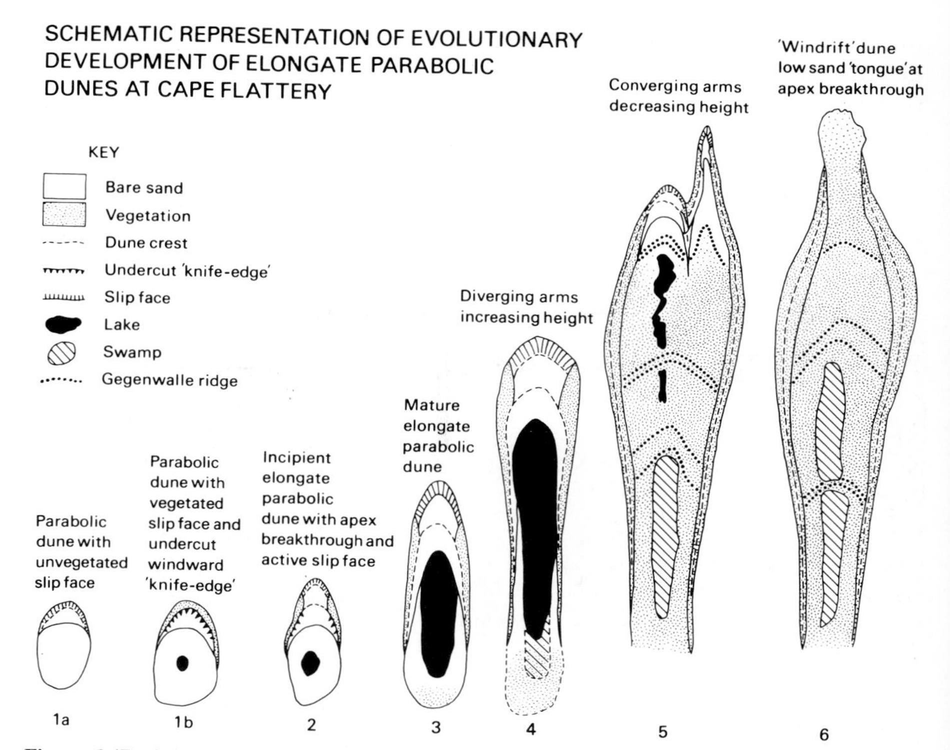

Figure 6.47 Schematic model showing the stages in the growth and eventual dissipation of elongate parabolic dunes in the Cape Flattery area. (After Pye 1982a).

hollow is promoted by turbulent eddying. Sand eroded from the hollow is trapped by vegetation on the downwind side of the blowout. Cooper (1958) made a distinction between shallow, short-lived blowouts, which he termed *saucer blowouts*, and much longer-lived, deeper blowouts, termed *trough blowouts*, which give rise to major transgressive dunes. A trough blowout deepens until the limit of capillary rise is reached and the sand becomes too wet to move. At this stage deflation becomes concentrated on the downwind margin of the blowout and on the windward side of the resulting parabolic dune. Airflow up the windward slope is accelerated towards the crest (see Section 6.3.3.2). Field measurements have shown that flowlines become compressed towards the dune summits (Landsberg & Riley 1943, Olson 1958a, Pye 1980a, 1985b) (Fig. 6.50). Flow along U- and V-shaped windward troughs is especially enhanced because the flow is compressed both laterally and vertically towards the crest.

As the flow passes over the crest it separates and diverges, depositing sand on the lee slope. Most parabolic dunes have a well developed slip face. Elongate parabolic dunes typically have slip faces which are sharply arcuate in plan, but in the case of blunt forms the slip face may be almost straight.

Cooper (1958) reported a maximum rate of forward movement of 2.84 m yr^{-1} for Oregon parabolic dunes based on ground surveys and a maximum of 2.04 m yr^{-1} based on tree ring dating. In North Queensland, a maximum rate of 5.6 m yr^{-1} was reported by Pye (1982a) based on air photograph evidence, while radiocarbon dating of buried wood from one dune suggested a rate of forward movement of about 6.4 m yr^{-1} (Pye & Switsur 1981). These rates of movement are considerably lower than those reported for coastal barchans [e.g. 18 m yr^{-1} in Baja California (Inman *et al.* 1966)].

6.4.3 Precipitation ridges

Cooper (1958) used the term *precipitation ridge* to describe transverse transgressive dunes with laterally extensive slip faces which occur on the landward margins of many of the Oregon coastal dunefields (Fig. 6.51). Similar features which are orientated more or less parallel to the coast in parts of Brazil were referred to as *retention ridges* by McKee & Bigarella (1979). Some of the forms in both Oregon and Brazil are transitional between transverse and parabolic dunes. In southeastern Australia the term *long-*

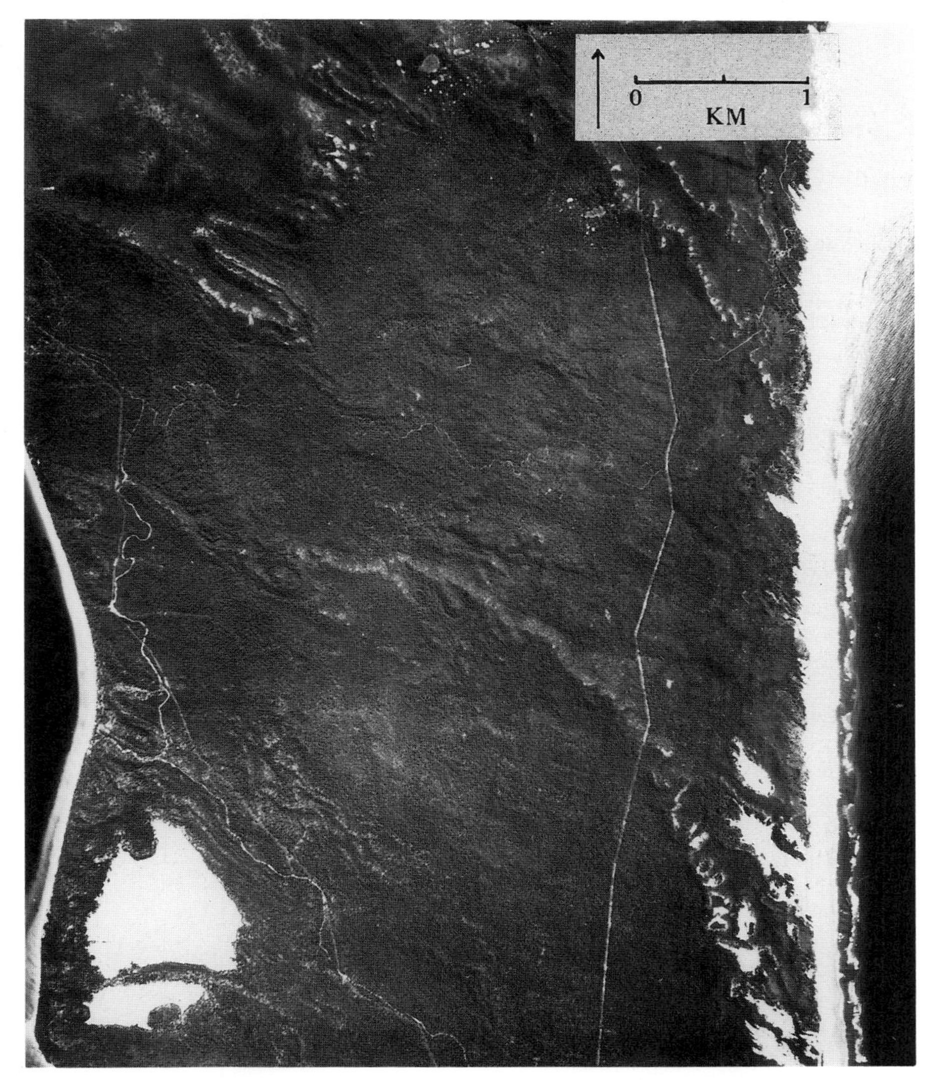

Figure 6.48 Air photograph showing parabolic dunes on North Stradbroke Island, Southern Queensland. Effective sand transporting winds are less unidirectional in this area than on the east coast of Cape York Peninsula, resulting in broader, digitate, parabolic dune forms. (Crown Copyright, reproduced by permission of Department of National Mapping, Queanbeyan, Australia).

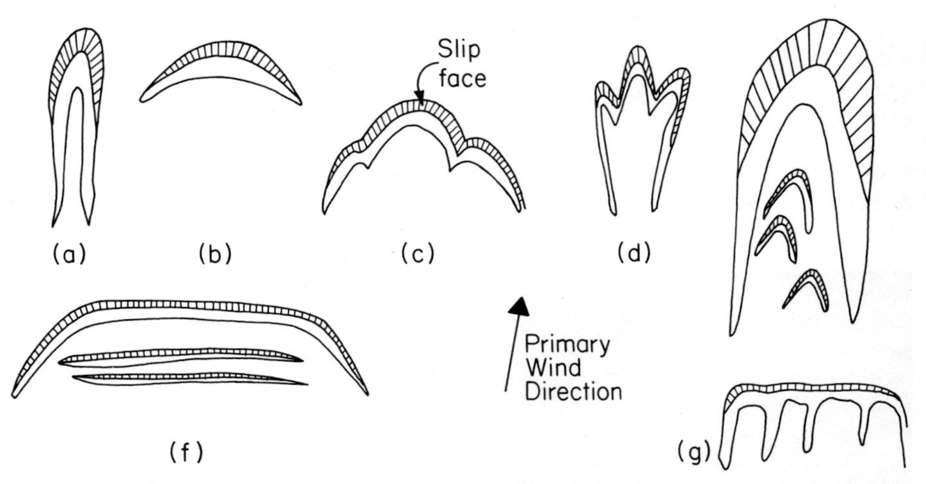

Figure 6.49 Diagram showing seven form variants of parabolic dune: (a) hairpin; (b) lunate; (c) hemicyclic; (d) digitate; (e) nested; (f) long-walled transgressive ridge with secondary transverse dunes; (g) rake-like en-echelon dunes. (a) and (b) are simple forms, (c), (d), (e) and (g) are complex forms, and (f) is a compound form. (Modified after Pye 1982a, David 1981, Verstappen 1970, Filion and Morisset 1983).

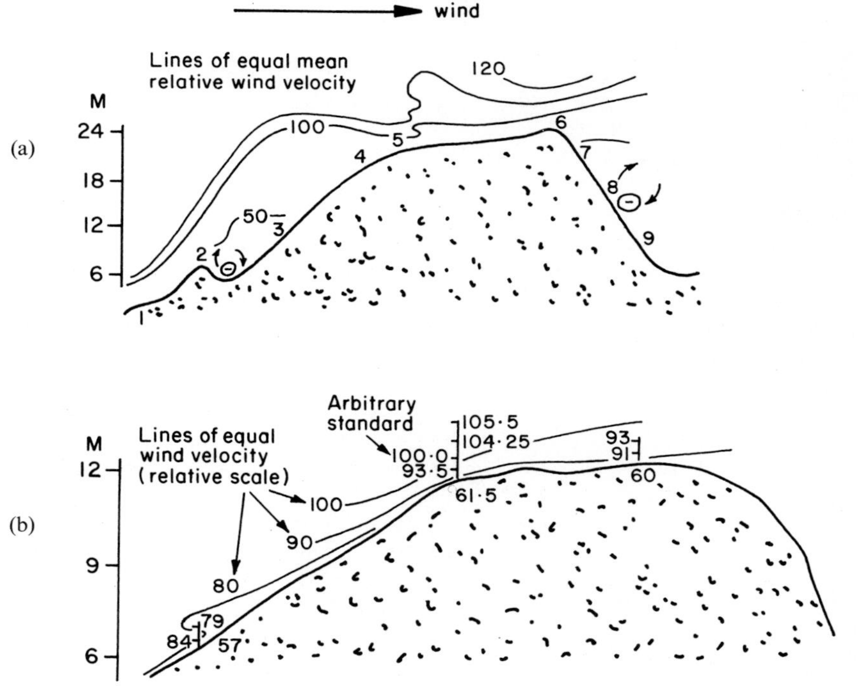

Figure 6.50 Changes in wind velocity over blowout troughs and associated dunes: (a) Stevensville blowout and (b) Marquette Park blowout, Lake Michigan. (After Landsberg & Riley 1943, Olson 1958a).

Figure 6.51 The landward margin of a precipitation ridge on the coast of Oregon.

walled transgressive ridge was applied to related forms by Thom *et. al.* (1981). All of these features form where the sand advances slowly on a broad front, usually due to a combination of factors which include destruction or absence of sand-binding vegetation upwind of the dune ridge, the existence of a belt of thick forest or other movement-resisting vegetation on the leeward side, and a bi- or multi-modal wind regime. Long-walled transgressive ridges and elongate parabolic dunes can be regarded as end members of a family of parabolic dunes.

6.4.4 Lunette dunes

The term *lunette* was first used by Hills (1940) to describe bow-shaped dunes composed of sand, silt, and clay which occur on the downwind margins of ephemeral lakes in semi-arid Australia. Like parabolic dunes, the arcuate plan form of lunettes points downwind, and sedimentation on the surface of the dune is usually enhanced by the presence of vegetation. However, unlike parabolic dunes, lunettes are rarely transgressive.

Some lunettes are composed almost entirely of sand (Fig. 6.44), but many contain a high proportion of silt and clay, which are transported in pellet form from the adjacent pans during periods of low water level (Stephens & Crocker 1946, Lancaster 1978, Goudie & Thomas 1986) (Section 3.8.2.2).

Campbell (1968) suggested that the arcuate shape of most lunettes is determined primarily by wave processes acting on the downwind end of the lake basin when the water level is high. Sediment is then deflated from the exposed beach as the water level falls.

6.4.5 Vegetated linear dunes

Some authors have not made a clear distinction between *vegetated linear dunes* and seif dunes, and the former have sometimes been incorrectly described as seifs (e.g. Verstappen 1968, 1972, Wilson 1973, Langford-Smith 1982, Walker 1986). Other authors have referred to both vegetated linear dunes and seifs as longitudinal dunes (e.g. Folk 1971a,

Mabbutt & Wooding 1983) without demonstrating parallelism of the dune long axes with either the resultant or the dominant wind direction. In Australia, vegetated linear dunes have been widely referred to as *sand ridges* (Madigan 1936, 1946, Buckley 1981, Twidale 1972a,b, 1981) or as *parallel ridges* (Mabbutt 1968). Vegetated linear dunes range in height from several metres to a few tens of metres and typically have a rounded cross-sectional profile (Fig. 6.52). The vegetation cover is thickest on the plinth and lower slopes and is usually sparse or absent on the crest. Some linear dunes appear to be wholly stable, possibly indicating formation under more arid or more windy conditions in the past (Flint & Bond 1968, Lancaster 1981c, Ash & Wasson 1983). Vegetated linear dune ridges may run almost parallel without a break for scores of kilometres. In the Simpson desert of Australia some individual linear dunes exceed 320 km in length (King 1960). In most cases they are asymmetric in cross section, though symmetrical profiles are also found (Mabbutt *et al.* 1969). The symmetry can vary from time to time in accordance with the wind conditions (Twidale 1980).

A common feature of vegetated linear dunes is the tendency to form branching networks in which adjacent ridges converge, forming a Y-junction, before continuing as a single ridge (Fig. 6.53). Y-junctions are of six main types (Fig. 6.54): (a) symmetrical junctions which open upwind, (b) left-handed, (c) right-handed asymmetric junctions which form when only one ridge curves to join the other which continues in a straight line, (d) reversed symmetrical junctions which open downward, (e) reversed left-handed and (f) reversed right-handed junctions. The last two types of junction have been attributed to deflection of the end of one ridge during elongation under the influence of a cross-wind (Madigan 1946, Mabbutt & Sullivan 1968, Thomas 1986a). However, this cannot be a general explanation, since it does not account for the formation of symmetrical Y-junctions which open either upwind or downwind. In the Kalahari Desert, 15.6% of all Y-junctions were found by Thomas (1986a) to be of the reverse type (i.e. the ridges diverge downwind).

Goudie (1969) found that some linear dune networks in the Kalahari are dendritic

Figure 6.52 A vegetated linear dune, northern Negev.

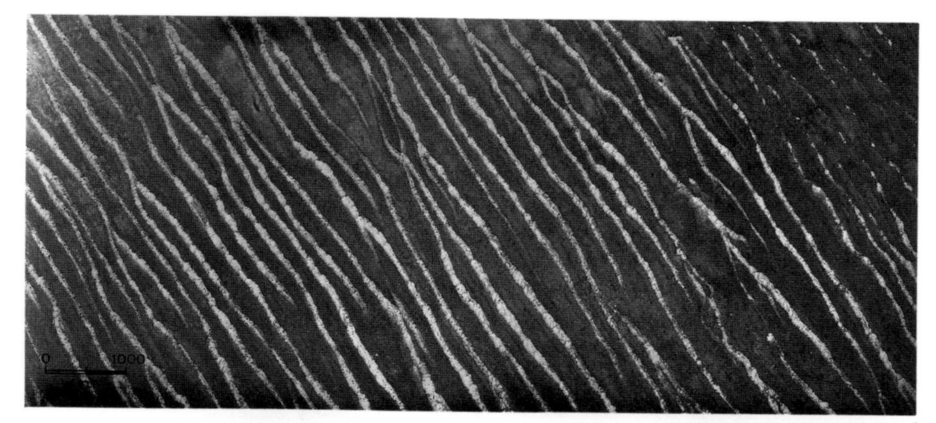

Figure 6.53 Vertical air photograph showing branching linear dune network in the Kalahari desert. These dunes are largely vegetated with active sand movement restricted to the crestal areas. (Reproduced by permission of the Surveys and Mapping Department of South Africa).

and appear to obey Horton's (1945) law of stream numbers. Dune pattern statistics were also used by Mubbutt & Wooding (1983) to analyse the morphometry of dunes in the Simpson Desert of Australia. They concluded that the occurrence of Y-junctions allows the dune pattern to maintain an equilibrium dune spacing. The ratio of continuing ridges to entering ridges (ridges leaving and entering, but not beginning or ending) in a defined area (C/E) was considered by these authors to provide a rough measure of dune spacing equilibrium, with the C/E ratio approximating a value of 1 as this condition is approached.

Destruction of the vegetation cover can turn vegetated linear dunes into seifs or *braided linear dunes* (Fig. 6.55), the latter being linear dunes on which small secondary transverse dunes are superimposed. This has occurred in parts of the Negev and Sinai (Tsoar & Møller 1986), Australia (Madigan 1936, Twidale 1972b, Mabbutt & Wooding 1983) and India (Kar 1987).

Several theories have been proposed to explain the origin of vegetated linear dunes. An early and recently much repeated hypothesis suggested that these dunes are stationary residual features, analogous to yardangs, formed by wind erosion of sand in the interdune depressions (*gassi*) (Frere 1870, Blanford 1876, Medlicott & Blanford

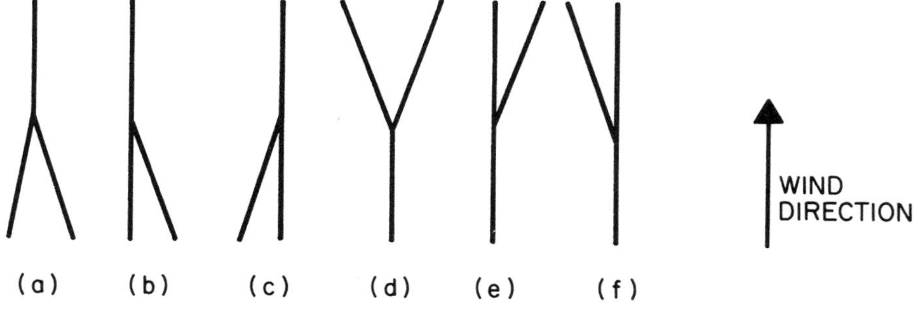

Figure 6.54 Types of Y-junction: (a) normal symmetrical junction (opening upwind); (b) normal left-handed junction; (c) normal right-handed junction; (d) reversed symmetrical junction (opens downwind); (e) reversed left-handed junction; (f) reversed right-handed junction. Types a–c are the most common.

Figure 6.55 Aerial view of braided linear and seif dunes, Sinai Desert. These dunes have formed as a result of destruction of the vegetation on a vegetated linear dune. They are regarded as complex dunes because secondary transverse dunelets are superimposed on the basic linear dune forms.

1879, p. 438, Aufrère 1930, Enquist 1932, King 1960, Folk 1971a, Mainguet 1984a,b). Other authors have suggested that some vegetated linear dunes evolve from parabolic dunes when the wind breaches the nose of a parabolic dune (Hack 1941, Verstappen 1968, 1970). Although there is good evidence that some vegetated linear dunes, such as those in parts of the Rajasthan Desert, have evolved from parabolic dunes (Verstappen 1970, Wasson *et al.* 1983), the majority of linear dunes cannot be explained in this way (Lancaster 1982b, Mabbutt & Wooding 1983, Kar 1987). Many linear dunes have an internal structure (Breed & Breed 1979) which demonstrates beyond doubt that this dune type is a primary depositional aeolian bedform.

In many areas vegetated linear dunes are reported to be aligned approximately parallel to the dominant wind direction (Chudeau 1920, Enquist 1932, Madigan 1936, 1946, Melton 1940, Capot-Rey 1945, Smith 1963, Clarke & Priestley 1970, Folk 1971a,

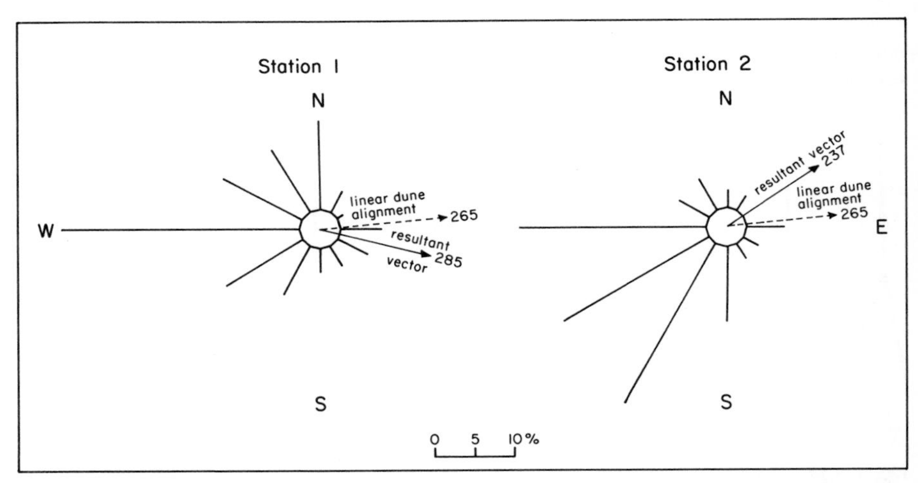

Figure 6.56 Weighted sand-moving wind roses for two meteorological stations close to an area of vegetated linear dunes in the northern Negev. Solid arrow indicates the resultant wind direction and the dashed arrow the average alignment of the vegetated linear dunes. (After Tsoar & Møller 1986).

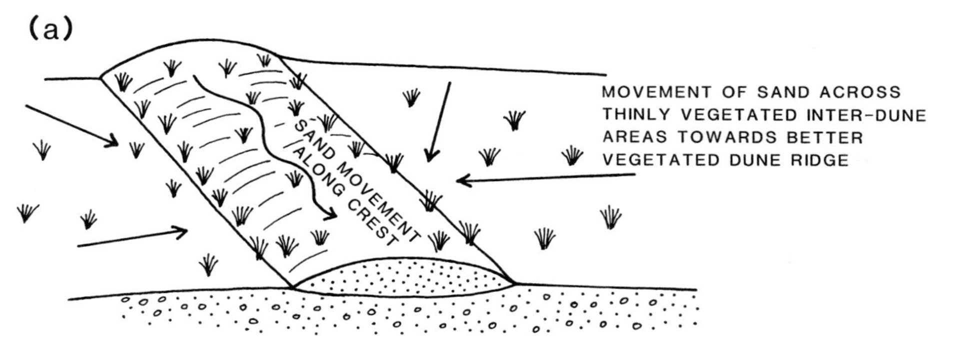

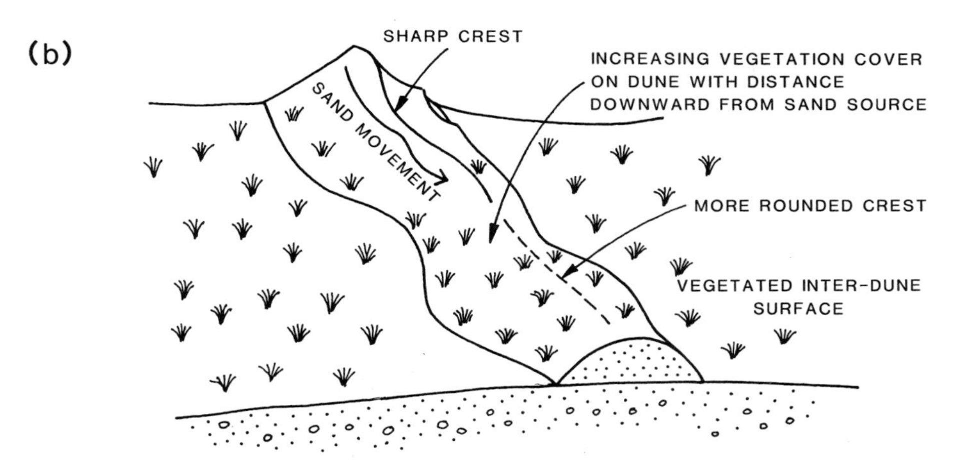

Figure 6.57 Two models for the development of vegetated linear dunes. (a) Slow movement of sand mainly along the crest of a partially vegetated linear dune, with lateral addition of sand to the dune ridge from less well vegetated interdune areas (modified after Tsoar & Møller 1986). (b) Extension of an essentially unvegetated linear dune ridge across a partially vegetated sandplain, fed by sand supplied along the ridge from upwind. Distal parts of the ridge may become vegetated due to low rates of sand supply or increase in rainfall downwind. (Modified after Wopfner & Twidale 1988).

Higgins *et al.* 1974, Breed & Breed 1979, Fryberger & Dean 1979, Lancaster 1981a, 1982b, Mainguet 1984b, Tsoar & Møller 1986, Kar 1987) (Fig. 6.56). Secondary side winds usually exert a modifying influence on the crest and account for either the symmetry or asymmetry of the dune. Other reports indicate a near-parallelism with the annual wind resultant of two or more wind directions (Striem 1954, Twidale 1972a,b, Breed & Breed 1979, Fryberger & Dean 1979).

Most of the Australian vegetated linear dunes display a small deviation from the mean wind direction (Brookfield 1970) (Fig. 2.11). Whereas parallelism to the resultant wind direction is a frequent characteristic of seif dunes, a majority of vegetated linear dunes distinguish themselves by extension parallel to the dominant wind direction. This is related to the fact that, because the presence of a partial vegetation cover raises the threshold for sand entrainment, only strong winds are able to accomplish sand transport and modify the form (Ash & Wasson 1983).

The parallelism of many vegetated linear dunes to the dominant wind direction lent support to the idea that they owe their origin to helicoidal flow in the atmosphere (e.g. Folk 1971b) (see Section 6.3.5).

Some vegetated linear dunes are clearly initiated in the lee of topographic obstacles which generate paired longitudinal vortices as the flow diverges around the obstacle (Tseo 1986, p. 102) (Section 6.3.2.1). Examples are found in the southern and eastern Simpson Desert (Twidale 1972a, 1981), northeastern Arizona (Melton 1940) and the Thar Desert (Wasson *et al.* 1983, Kar 1987).

Vegetation also acts as an obstruction to the wind and is known to encourage the formation of small linear shadow dunes (Hesp 1981) (Section 6.4.1). Tsoar & Møller (1986) have argued that vegetation is a normal component of desert landscapes which receive as little as 50 mm of annual rainfall. Since vegetation in deserts thrives more on sand than other types of desert surface owing to its moisture retention and drainage properties (see Ch. 9), it may be expected that vegetation will cluster along the flanks of the shadow dune and encourage its further development by a process of self-propagation (Tsoar & Møller 1986). Further sand may be added laterally to the growing dune by cross-winds if the inter-dune areas are unvegetated (the *shepherding effect*, Fig. 6.57a). For further extension of the dune to take place, however, the vegetation cover must not be sufficiently dense as to prevent sand movement over the dune surface.

Although some vegetated linear dunes may develop by extension of partially vegetated shadow dunes in the manner described above, evidence from the Simpson Desert provided by Wopfner & Twidale (1988) suggests that others form by downwind extension of essentially bare sand ridges, followed by partial or complete stabilization by vegetation (Fig. 6.57b). These authors maintain that linear dunes in the southern Simpson Desert are still actively extending under the influence of southwesterly and southeasterly winds. Near the upwind margin of the dunefield, linear dune ridges have evolved from barchans, either by downwind elongation of one horn, or by development of a lee projection from the slip face.

Along the upwind extremity of the Simpson Desert, accumulations of wind-blown sand which attain a height of 50 m form a 'source-bordering rampart' along the margins of playas and alluvial plains. Sediment is supplied to this area by ephemeral streams flowing into the Lake Eyre Basin (Wopfner & Twidale 1988).

The active surfaces of the ramparts exhibit barchanoid and transverse forms, often with deep blowout depressions in between. There is a transitional area between the ramparts and the linear dunes which consists of numerous lower linear ridges that converge downwind into a series of active seifs. Downwind of this point the amplitude and wavelength of the dunes increase rapidly while the height of the ridges increases. The mobility of the sand decreases with increasing distance from the source, with the result that there is more vegetation cover on the dunes downwind. The older, more weathered character of the sand downwind is also indicated by its well developed red colour, while the recent sand upwind is white. Most of the dunes terminate abruptly on the gibber plain of the Stony Desert. The only dunes which continue further are those which receive additional supplies of sand from the shores of ephemeral lakes and playas fed by runoff from the Cordillo Dome.

The model of linear dune development proposed by Wopfner & Twidale (1988) implies long-distance transport of sand from lacustrine or alluvial source areas, the sand being transported along the dunes by deflected crest-parallel flow similar to that described on seifs by Tsoar (1978). The model is thus contrary to suggestions that the sands comprising the dunes are locally derived by aeolian reworking of sediments in the interdune depressions (King 1960, Folk 1971a). The dunes form by advancement of largely bare sand ridges over a vegetated sandplain, and not by lateral transfer of sand from a bare interdune sandplain to a vegetated dune ridge (Fig. 6.57b). Vegetation often extends up the lower slopes of the dunes and may partially cover the crest if the rate of along-dune sand transport is sufficiently low. This is most likely to be the case towards the downwind end of dunes furthest from the sand source. A reduction in the sand transport rate, caused by a reduction in either sand supply or wind velocity, is also likely to result in increased vegetation cover. Conversely, overgrazing or fire damage to a vegetated dune may cause reactivation of the crest, involving the development of superimposed seif, barchan, or parabolic forms, as described by Tsoar & Møller (1986) and Kar (1987). Since new phases of aeolian activity at any point on the dune may be initiated by enhanced sand supply leading to increased along-dune transport, such episodes of rapid dune extension need not necessarily be equated with an increase in aridity or windiness. In the Simpson Desert, present-day mean annual rainfall increases from < 100 mm north of Lake Eyre to approximately 150 mm along the northern limit of the dunefields (Purdie 1984). Correspondingly, there is an increase in vegetation cover and reduction in sand transport rate in the downwind direction.

In summary, there appear to be at least two modes of vegetated linear dune development: (a) by slow downwind movement of sand along the crest of a dune ridge whose flanks are largely vegetated, in some instances supplemented by lateral transfer of sand to the ridge from the neighbouring inter-dune areas (Tsoar & Møller 1986), and (b) by more rapid movement of sand along an essentially bare sand ridge, followed by vegetation colonization of the ridge when the sand transport rate is reduced due to an increase in rainfall or reduction in wind energy downwind. In the former case, it is not entirely clear how the ridge was first initiated. As pointed out by Tsoar & Møller (1986) and Thomas & Tsoar (1990), partially vegetated linear ridges displaying crestal sand movement are a normal active type of dune in desert environments and should not always be regarded as palaeoforms, although climatic changes have probably played an important role in their development.

6.5 Sand sheets

6.5.1 Warm climate sand sheets

Sand sheets may be defined as 'areas of predominantly aeolian sand where dunes with slip faces are generally absent' (Kocurek & Nielson 1986, p. 795). *Sand streaks* or *sand stringers* are a sub-type of sand sheet which are markedly elongate in plan (Breed & Grow 1979, p. 281). The surfaces of sand sheets may be rippled or unrippled, flat, regularly undulatory, or irregular. Some show the development of low, usually transverse ridges without slip faces which are known as *zibar* (Section 6.5.2).

Sand sheets can be relatively small features, with an area of only a few square kilometres, or major regional landscape features such as the Selima Sand Sheet of southern Egypt and northern Sudan, which covers an area of almost 100 000 km^2 (Breed *et al.* 1987). Globally, sand sheets cover about 1 520 000 km^2, equivalent to about 38% of the total aeolian depositional surface in the major warm desert sand seas (Breed *et al.* 1979, Fryberger & Goudie 1981). In most cases they occur peripherally to the main dune areas, but occasionally are surrounded by dunes (Warren 1972, Fryberger *et al.* 1979, Lancaster 1982a, Nielson & Kocurek 1986, Porter 1986).

The thickness of sand sheet deposits ranges from a few centimetres to several tens of metres, depending on their environment of formation. In the Selima Sand Sheet, for example, aeolian sands which overlie old alluvial surfaces are in many places only a few tens of centimetres thick (McCauley *et al.* 1982, Haynes 1982, Breed *et al.* 1987). By contrast, sand sheet deposits several metres thick have been described from the Great Sand Dunes of Colorado (Fryberger *et al.* 1979, Andrews 1981) and elsewhere.

The grain size of sand sheet deposits ranges from fine sand, sometimes with a significant silt component, to poorly sorted coarse sand. Many, but not all, sand sheet deposits have a bimodal grain size distribution (Folk 1968, Warren 1971, Binda & Hildred 1973, Tsoar 1978, p. 23, Fryberger *et al.* 1979). With few exceptions, sand sheet deposits can be classified as poorly sorted.

Bagnold (1941, pp. 157 & 243) and Wilson (1973) distinguished between very coarse bimodal sand sheets without any dune forms and with or without ripples, and coarse bimodal sand sheets which develop ripples and dune bedforms without slip faces. In the former case the coarse size mode is > 0.65 mm and in the latter it lies in the range 0.3–0.65 mm. In both cases the fine size mode is in the range of fine sand. Where low bedforms are developed, the sand on the stoss slope is generally coarser than that on the lee slope. In the Algodones dunes of California, for example, the mean grain size of the interdune and stoss slope deposits ranges from 0.2 to 1.0 mm, while that of the lee slope deposits ranges from 0.1 to 0.3 mm (Nielson & Kocurek 1987). These slight differences in grain size give rise to slight colour differences which can be detected on air photographs. Slip face formation on sand sheets is exceptional and takes place only rarely when the relief is greater than usual, leading to flow separation (Lancaster 1982a, Tsoar & Yaalon 1983, Nielson & Kocurek 1986).

The origin of bimodal grain size distributions in sand sheets has been attributed by most authors to selective winnowing of medium and fine sand grains (Binda & Hildred 1973). Warren (1972) suggested that bimodal sediments are composed of lag grains which are too coarse to saltate and very fine particles which are protected in the interparticle voids between the coarse grains. However, bimodal mixtures with one mode in the range of easily saltated particles (0.125–0.25 mm) are also commonly found (Folk 1968, Warren 1972, Breed *et al.* 1987).

As discussed below, the presence of coarse grains appears to be an important factor governing the formation of some sand sheets. Since coarse grains are rare in most well sorted beach deposits which provide the sand source for coastal dunes, sand sheets are poorly developed in coastal dunefields. They do occur, however, where aeolian action has reworked residual or alluvial sands near the coast (Pye 1980a, 1982b).

Sand sheets develop in aeolian environments where conditions do not favour the development of dunes with slip faces. Kocurek & Nielson (1986) identified five factors which they suggested may encourage the formation of small extra-dune and interdune warm climate sand sheets: (a) an evenly distributed vegetation cover, especially composed of grass species, which may encourage a really uniform accretion of low-angle sand laminae; (b) the presence of a surface layer of coarse sand, possibly representing a surface lag deposit formed by winnowing, may prevent sand being mobilized into dunes [cf. the 'killing' of a dune envisaged by Bagnold (1941, p. 180)]; (c) a high groundwater table may limit sand entrainment by keeping the surface sand moist; (d) periodic or seasonal flooding may also prevent dune development by washing away incipient dunes and keeping sand wet for long periods; (e) the development of surface crusts and algal mats may limit sand transport for dune formation.

Fryberger *et al.* (1979) described low-angle sand sheet deposits on the margins of the Great Sand Dunes of Colorado, which are transitional between dune and non-aeolian facies. They suggested that deposition of sand in sheet form is favoured by gentle deceleration of the wind due to sheltering by sand hills, and flow expansion as air passes over them. This area experiences a complex or bimodal wind regime, and reversals in wind direction were suggested by Fryberger *et al.* to encourage uniform spreading of the sheet sand following initial deposition. Vegetation cover and a covering of coarse grains on the surface of the sand sheet were also considered to play a contributory role.

Of the factors discussed above, only the presence of a coarse surface sand layer appears to be important in the formation of large sand sheets such as the Selima Sand Sheet (Breed *et al.* 1987). This flat, almost featureless sand plain consists of laterally extensive, horizontally laminated deposits of sand, silt, granules, and pebbles. Coarse sand and granules typically form an armoured surface layer. Although dune trains periodically migrate across the area, they leave the surface of the sand sheet almost undisturbed. Excavation of shallow pits showed that the upper Selima Sand Sheet is composed of pairs of concordant laminae. The lower lamina consists of a mixture of coarse silt and very fine to medium sand, while the upper layer, typically one grain thick, consists of coarse sand, granules, and small pebbles. No evidence of climbing ripple foresets was detected. Bulk samples of these sediments were found to have a very fine sand mode (0.125 mm) and a very coarse sand mode (1.5 mm) (Breed *et al.* 1987).

6.5.2 Zibar

Zibar are long-wavelength, low-amplitude migrating bedforms without slip faces whose surfaces are usually covered by ripples or megaripples (Holm 1960, Warren 1972, Wilson 1973, Tsoar 1978, Kocurek & Nielson 1986) (Fig. 6.27). The term zibar is derived from the Arabic *zibara*, which means a hard sandy surface that permits the passage of vehicles (Thomas 1932, p. 376). On air photographs zibar appear as chevron-shaped (Maxwell & Haynes 1989), transverse (Nielson & Kocurek 1986), or linear features (Fig. 6.27). On partially vegetated sand sheets zibar may also have a parabolic form (Anton & Vincent 1986).

The low, flat form typical of zibar is governed by the presence of a coarse sand mode in the sediments which comprise them. Bagnold (1941, p. 164) maintained that the profile of a dune depends on the friction Reynolds number (Eqn 2.23), such that dunes composed of fine sand have steeper slopes than dunes composed of coarse sand. Similarly, Nielson & Kocurek (1986) regard the low-angle stoss slope of a zibar as representing the maximum inclination for the transport of grains in a given wind regime. According to Cooke & Warren (1973, p. 309), a layer of coarse grains effectively seals off the zibar surface, and since only strong winds can mobilize coarse grains, bedforms with longer wavelengths are formed because faster winds have turbulent eddies with longer wavelengths. The steep vertical velocity gradients of such winds effectively limit the upward growth of the dune.

6.5.3 Cold climate sand sheets

Relatively small active sand sheets are found in cold regions at the present day (Niessen *et al.* 1984, Good & Bryant 1985, Pissart *et al.* 1977, Ashley 1985, McKenna-Neumann & Gilbert 1986, Dijkmans 1990), and Pleistocene periglacial sand sheet deposits cover extensive areas in northern Europe, North America, and the USSR (Koster 1988) (Fig. 6.58). In Europe they form a belt which extends from northern France to the Baltic states of the USSR (Maarleveld 1960, Nowaczyk 1976, Kolstrup & Jorgensen 1982, Koster 1982, Kolstrup 1983, Ruegg 1983, Schwan 1986, 1987, 1988). These deposits are widely referred to as *coversands* because they form a surficial blanket ranging in thickness from about 50 cm to several metres (Catt 1977, Buckland 1982, Koster 1982). The surface morphology is partly dependent on that of the underlying surface. In many places low ridges on the sand sheet surface reflect the presence of push moraine deposits beneath the sand.

Syn-depositional dune forms are not extensive in northwest Europe, although they are more common in Poland and neighbouring areas (Hogbom 1923). In many parts of northwest Europe, however, coversands have been reworked into low dunes or sand drifts during the Holocene (Mathews 1970, Peeters 1983, Castel *et al.* 1989) (Fig. 6.59).

In The Netherlands, sand sheet deposits accumulated in at least six glacial periods of the Pleistocene. They locally form stacked sequences up to 40 m thick (Ruegg 1983). Extensive sand sheet deposition which took place in the later part of the last glacial

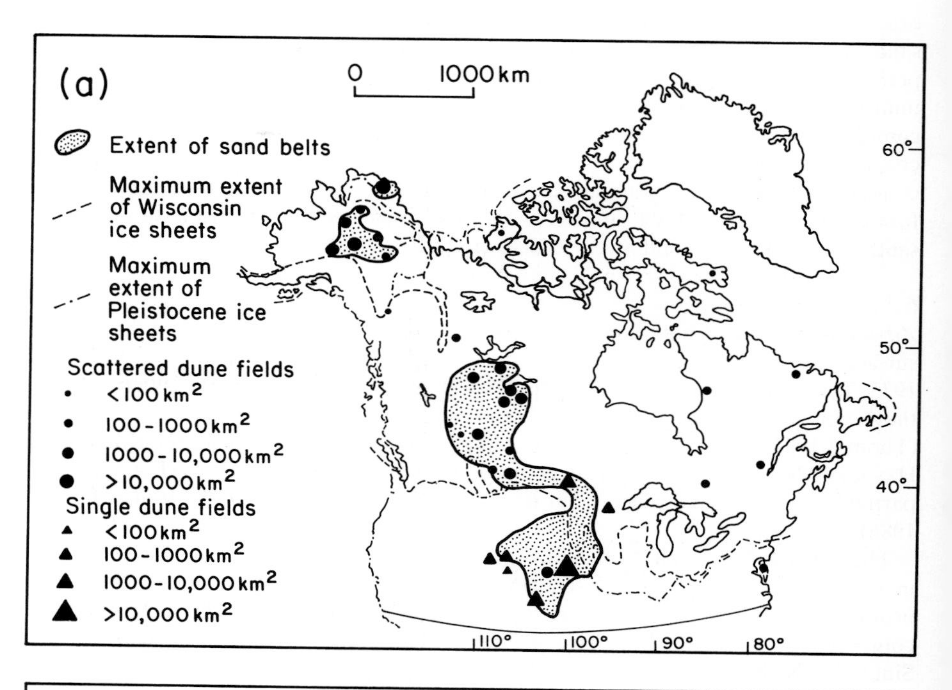

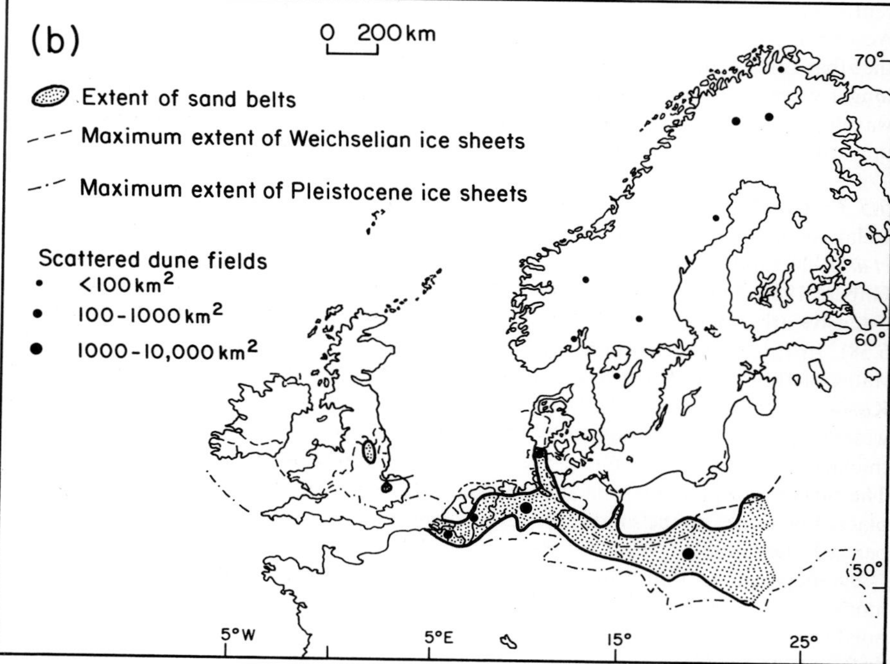

Figure 6.58 Distribution of cold climate sand sheets and dunefields in (a) North America and (b) Europe. The age and timing of aeolian activity vary in different areas. (Modified after Koster 1988).

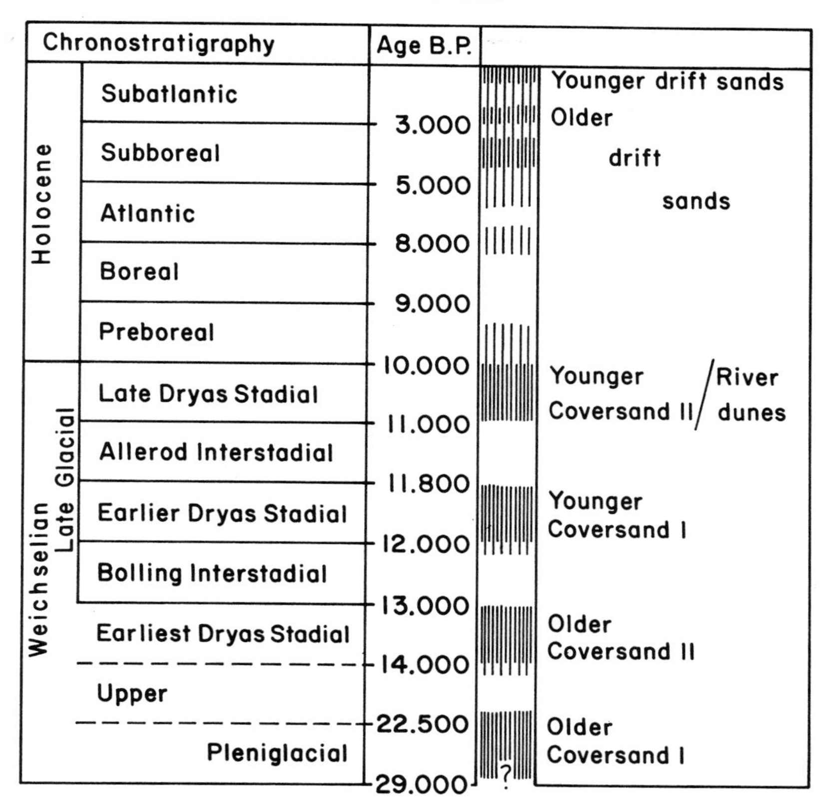

Figure 6.59 Timing of aeolian events in the European Lowlands during the late Pleistocene and Holocene. Heavy shading represents strong aeolian sand deposition; light shading represents weak deposition. (Modified after Koster 1988).

period covered at least 30 000 km^2 in The Netherlands alone.

Two major types of contrasting aeolian sand sheet sub-facies were recognized by Ruegg (1983) and Schwan (1988): evenly laminated sand deposits without silty laminae, and sands with alternating silty laminae which are transitional between the sand-only facies and loess deposits which occur beyond the downwind margin of the sand sheets (Fig. 3.16). The two sand sheet sub-facies were interpreted by Ruegg (1983) as indicating aeolian deposition on dry only and alternating wet and dry surfaces, respectively. A similar interpretation was made by Schwan (1988).

Fluvioglacial outwash channels and braided sandur (outwash plain) deposits provided the source of these aeolian sand sheet sediments. However, mineralogical evidence suggests that some of the material was derived from distal sources on the exposed floor of the North Sea basin (Schwan 1988).

Hobbs (1943) proposed that aeolian transport was accomplished mainly by anticyclonic winds emanating from a high-pressure cell centred over the Fenno–Scandinavian ice sheet, but grain size trends and morphological evidence indicates that mid-latitude westerlies must also have played a role in transporting sediment from northwesterly sources (Ruegg 1983).

The sand sheet deposits without silt laminae typically show alternating sequences of horizontally laminated sands and adhesion ripple sands, indicating some sand deposition on wet surfaces. Occurrences of inclined bedding are usually related to small, isolated dome dunes, sub-surface irregularities, or scoured depressions formed either by wind deflation or water erosion (Schwan 1988). The scoured surfaces are frequently lined by pebble stringers and ventifacts. Cryoturbation features, including well developed ice-wedge casts, are also common in the sequences.

The reasons for the extensive development of sand sheets, rather than dunes, in northwest Europe during the Pleistocene has not been fully explained. Bagnold (1941, p. 151) noted that ripple wavelengths flatten out and eventually disappear when the wind velocity rises above a certain strength, producing a plane bed analogous to the subaqueous plane bed of the upper flow regime (Simons *et al.* 1965). Hunter (1977a) suggested that 'planebed' laminae may be formed by strong winds in excess of 18 m s^{-1}. As pointed out in Section 2.3.1, high latitude areas have relatively high wind energy, and wind drag is relatively more effective owing to the higher density of air at low temperatures. However, strong winds in other areas do not prevent the formation of dunes; indeed, high, widely spaced dunes are typical of strong wind environments (Wilson 1972a, Lancaster 1982b). Further research is required to clarify this issue.

Schwan (1988) suggested that sand sheet deposition was favoured by a rarity of topographic barriers, sparseness of vegetation cover, and a high ratio between wind energy and sand availability during transport and deposition. A combination of these conditions may have given rise to a situation where transport of sand predominated over deposition. The contemporaneous development of dune morphology on a larger scale in northeast Europe may, according to Schwan, have been due to a greater degree of climatic aridity and deceleration of the regional airflow in the vicinity of the Oder–Niesse floodplain, leading to more rapid sand aggradation in this area.

There is little evidence to suggest that the formation of dunes in northwest Europe was suppressed by the presence of a coarse-grained sand component. Neither of the major aeolian sand facies recognized by Ruegg (1983) and Schwan (1988) contain more than 2% of grains larger than 0.5 mm. The facies sub-type which is composed only of sand is typically unimodal, with a mode in the 0.18–0.25 mm size range, while second facies sub-type, which consists of alternating sand and silt laminae, is characterized by bimodal sediments with one mode in the 0.15–0.18 mm (medium sand) range and a finer mode in the 0.025–0.035 mm (coarse silt) range (Ruegg 1983, p. 477). The absence of a coarse sand mode probably explains the non-development of zibar in these deposits.

6.6 Summary of factors determining the morphology of aeolian sand accumulations

Despite more than a century of research, considerable uncertainty still surrounds the relative importance of factors which determine the morphology of aeolian sand accumulations. In part this arises because there have been too few field studies of aeolian processes and sediment transport, and reliable wind data from dune areas are scarce. Although some steps have been taken to rectify these deficiencies in recent years, models of dune development still rely heavily on deductive interpretations based on dune form and, to a lesser extent, on internal structures and grain size distributions.

Laboratory experimental work on dune dynamics is handicapped by scaling problems and by difficulties in reproducing process variability relevant to natural systems. Field experimental and monitoring work has also been restricted by limited resources and the logistic difficulties encountered when operating sophisticated equipment in remote areas.

At a broad level, the form and scale of aeolian sand accumulations is governed by at least six factors: (a) sand availability, (b) grain size distribution, (c) wind energy, velocity distribution, and directional variability, (d) vegetation cover, (e) the presence or absence of topographic obstacles, and (f) sequential climatic changes which may bring about fluctuations in any of the first four factors and lead to the modification of existing dune forms. Perhaps the greatest uncertainty concerns the role of secondary atmospheric circulations at different scales. It has long been hypothesized that longitudinal and transverse vortices render a plane sand bed unstable, except where the surface sand is too coarse to move (Folk 1971a, Wilson 1972a,b). Some support for this hypothesis is provided by the widely observed regularities in dune spacing and relationships between dune height and spacing (Lancaster 1981a). However, the existence in some areas of sand sheets composed of well sorted medium to fine sand throws doubt on the universal applicability of this mechanism as an explanation for dune initiation. Field observations show that many small dunes are initiated by chance topographic irregularities or changes in surface roughness, which give rise to spatial variations in the sand transport rate. Once initiated, the growth of sand mounds is encouraged by positive feedback until a condition of dynamic equilibrium is attained with long-term average wind and sediment transport conditions. Groups of dunes gradually evolve to produce regular patterns, but whether these patterns also represent some form of dynamic equilibrium, or are simply a statistical phenomenon, remains to be proved. Much work is also required to clarify the relationship between dune morphology and average or dominant flow conditions.

Difficulties in assessing the factors which determine dune morphology arise because several of the factors are interdependent. For example, the stronger or more frequent the wind, the less vegetation is found on dunes. This is especially true for transverse and barchan dunes (e.g. Illenberger & Rust 1988). On the other hand, vegetation cover raises the threshold velocity, with the result that only strong winds can move sand on partially vegetated dunes. For this reason, assessments of wind directional variability and sand transport capacity, as performed by Brooks & Carruthers (1953) and Fryberger & Dean (1979), have little meaning unless the actual threshold value for particular groups of vegetated dunes can be specified. Further, in humid climates thresholds may differ significantly between different times of the year owing to variations in rainfall, temperature, and humidity. The morphology of vegetated dunes is usually influenced only by the strongest winds, which are often almost unidirectional. Classification of such dunes as transverse, longitudinal, or oblique is therefore meaningful only if related to the effective transporting winds, rather than to a hypothetical resultant calculated using an assumed threshold value which is only appropriate for bare, dry sand.

The balance of evidence now available indicates that large barchans, transverse dunes, and dome dunes form in almost unidirectional wind regimes. Where there is a significant secondary cross-wind, these forms may display asymmetry, reversing behaviour, superimposed secondary bedforms with a different orientation, or some degree of sideways movement. In markedly acute bidirectional wind regimes, linear dunes of the seif type are most likely to develop, whereas in obtuse bimodal and multimodal wind regimes, reversing transverse dunes and star dunes are the most likely forms. In humid areas where vegetation is present, formation of elongate parabolic dunes is favoured by a unidirectional effective wind regime, whereas in areas of bimodal and complex wind regimes the parabolic dunes will typically be broader and may also display asymmetric or digitate forms.

Contrary to the suggestion by Wilson (1972a,b), grain size appears not to exercise a general control on the morphology and size of dunes, with the exception of zibar formation (Wasson & Hyde 1983b, Thomas 1988a).

Sand availability exercises some control on dune morphology, but is not directly related to the size or spacing of individual dune forms in an area. For example, isolated barchan dunes tend to form where sand availability is restricted, whereas barchanoid ridges or transverse ridges form where more sand is available. However, even where sand is very scarce, as indicated by the equivalent sand thickness (EST) calculated according to the method of Wasson & Hyde (1983a), individual barchans may assume megadune proportions (e.g. Simons 1956). In many sand seas the largest, most widely spaced dunes are found furthest away from the sand source, as described in the Simpson Desert by Wopfner & Twidale (1988).

7
Internal sedimentary structures of aeolian sand deposits

7.1 Introduction

The sedimentary structures found in aeolian sand deposits fall into two broad groups: *primary structures* and *secondary structures*. Primary structures reflect the processes responsible for transport and initial deposition of the sand, whereas secondary structures form syn- or post-depositionally due to disturbance of the primary depositional fabric. Hunter (1977a) recognized three groups of processes which are responsible for the formation of primary aeolian sedimentary structures (Table 7.1): (a) *grainflow deposition*, (b) *grainfall deposition*, and (c) *tractional deposition*. The structures produced by the first two processes were referred to by Hunter as *sandflow cross-stratification* and *grainfall lamination*. However, following Kocurek and Dott (1981), the term *grainflow cross-stratification* is used here in preference to sandflow cross-stratification. Hunter (1977a) suggested that tractional deposition may produce five possible stratification types, which he named *subcritically climbing translatent stratification, supercritically climbing stratification, ripple foreset cross-lamination, rippleform lamination*, and *planebed lamination*. Planebed lamination, which forms at wind velocities too high for ripple formation (Hunter 1977a, 1980), is found relatively rarely in aeolian dune deposits. Horizontal or low-angle planar laminated sands are common in interdune areas and sand sheets, but not all are true planebed deposits (see Section 6.4.4).

Secondary sedimentary structures form in a wide variety of ways including slumping, flowage of wet sand, as a result of tectonic disturbance, bioturbation, cryogenic processes, and erosional episodes involving wind or water (McKee *et al.* 1971, McKee & Bigarella 1972, Ahlbrandt *et al.* 1978, Horowitz 1982, Pye 1983f).

Relatively little attention has been given to the internal structures of modern dunes, although the internal geometry of ancient aeolian sand bodies has been extensively studied. This largely reflects the difficulty of obtaining suitable sections in dry, unlithified dune sands. Most work on modern dunes has been based on the excavation of shallow pits and trenches following drenching of the sand with large quantities of water (e.g. McKee & Tibbitts 1964). Since it is rarely practical to excavate complete sections through large active dunes even using a bulldozer, most of the available information about the internal structures of modern dunes relates to small dunes < 10 m high (e.g. McKee 1966).

The internal structures of dunes are of interest for four main reasons. First, they provide clues to the growth and dynamics of modern dunes, and allow comparisons with resultant sand drift directions calculated using present-day wind data. Second, internal structures have been extensively used, with varying degrees of success, to distinguish between aeolian and subaqueous sands and to identify the bedform types represented by ancient aeolian deposits (e.g. Kocurek & Dott 1981, Glennie 1972, 1983a, Rubin & Hunter 1983, 1985, Steele 1983, 1985). Third, internal structures in fossil dunes provide a means of reconstructing palaeowind directions and testing global circulation models for

Table 7.1 Characteristics of basic types of aeolian stratification. (After Hunter 1977a).

Depositional process	Character of depositional surface	Type of stratification	Dip angle	Thickness of strata, sharpness of contacts	Segregation of grain types, size grading	Packing	Form of strata
Tractional deposition	Rippled	Subcritically climbing translatent stratification	Stratification: low (typically 0–20°, maximum ~30°) Depositional surface: similarly low	thin (typically 1–10 mm, maximum ~5 cm) Sharp, erosional	Distinct Inverse	Close	Tabular planar
		Supercritically climbing translatent stratification	Stratification: variable (0–90°) Depositional surface: intermed.(10–25°)	Intermediate (typically 5–15 mm) Gradational	Distinct Inverse except in contact zones	Close	Tabular commonly curved
		Ripple-foreset cross-lamination	Relative to translatent stratification intermed. (5–20°)	Individual laminae: Thin (typically 1–3 mm) Sharp or gradational, nonerosional. Sets of laminae: Intermediate (typically 1–10 cm) Sharp or graditional, nonerosional	Individual laminae and sets of laminae: Indistinct Normal and inverse, neither greatly predominating	Close	Tabular, concave-up or sigmoidal
		Rippleform lamination	Generalized: intermediate (typically 10–25°)			Close	Very tabular, wavy
	Smooth	Planebed lamination	Low (typically 0–15° max?)			Close	Very tabular, planar
Largely grainfall deposition	Smooth	Grainfall lamination	Intermediate (typically 20–30° min. 0° max. ~40°)			Intermediate	Very tabular follows pre-existent topography
Grainflow deposition	Marked by avalanches	Sandflow cross-stratification	High (angle of respose) (typically 28–34°)	Thick (typically 2–5 cm) Sharp, erosional or nonerosional	Distinct to indistinct Inverse except near toe	Open	Cone-shaped tongue-shaped, or roughly tabular

earlier periods in Earth history (e.g. Bigarella & Salamuni 1961, Parrish & Peterson 1988). Finally, sedimentary structures and associated textural variations exert an important influence on the porosity and permeability characteristics of hydrocarbon reservoirs and aquifers (Lupe & Ahlbrandt 1979, Ahlbrandt & Fryberger 1981, Weber 1987, Chandler *et al.* 1989).

7.2 Internal structures of sand dunes

7.2.1 Primary structural features common to most dune types

All dunes with slip faces display well developed *foreset laminae* which dip downwind at a maximum angle of 32–34°. Near the base of the slip face the foreset laminae often flatten out, giving rise to a concave-upwards profile. Individual foreset laminae, which are usually 2–5 cm thick, are grouped together into *crossbed sets* which are defined by *bounding surfaces*. The latter may be *planar* or *curved*, while the cross-bed sets may be either *tabular*, *wedge-shaped*, or *trough-shaped* in cross-section (Fig. 7.1). The orientation and angle of dip of the bounding surfaces, and of the cross-sets which they define, vary with dune type, position on the dune, the complexity of the wind regime, and the plane in which the section is viewed. Tabular planar cross-bedded sets tend to be more common in the lower parts of dunes while thinner, wedge-planar sets become more common towards the top. This reflects the fact that the upper levels of a dune are affected more by cutting and filling events in response to changing wind conditions and by the passage of small superimposed dune bedforms.

Steeply dipping (32–34°) foresets are produced mainly by grainflow deposition on the slip face. They correspond to the encroachment deposits recognized by Bagnold (1938b). Grainflow deposits formed in dry sand consist of a series of tongue-shaped bodies, typically 2–10 cm thick and 5–30 cm wide (Fig. 4.25). They often truncate underlying foreset laminae at low angles. In the upper part of the flow the grain size of individual sand tongues coarsens upwards since larger grains are moved towards the top of the flow owing to dispersive stress (Bagnold 1954a, Sallenger 1979). The grain size may also increase in the downflow direction since the coarser surface grains roll faster and further than the finer grains beneath (Kocurek & Dott 1981, Pye, 1982b).

Especially on small dunes, and on the lower slip faces of large dunes, grainflow

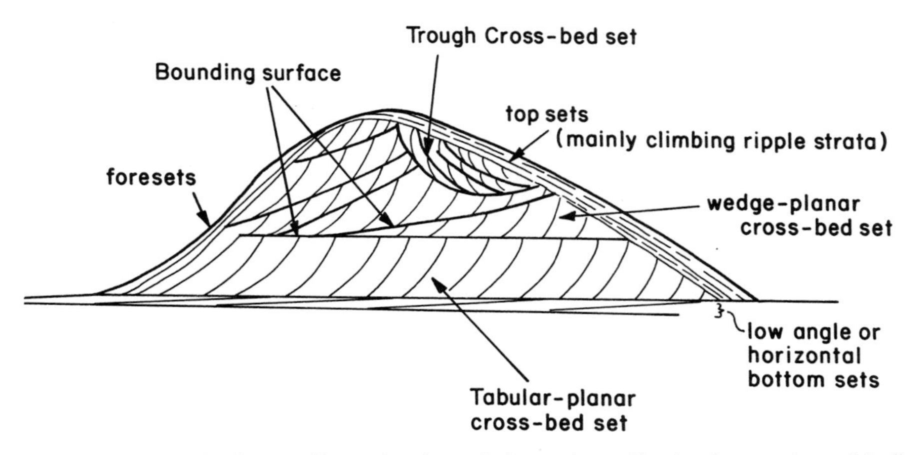

Figure 7.1 Schematic diagram illustrating the main internal stratification features in an 'ideal' simple barchan dune.

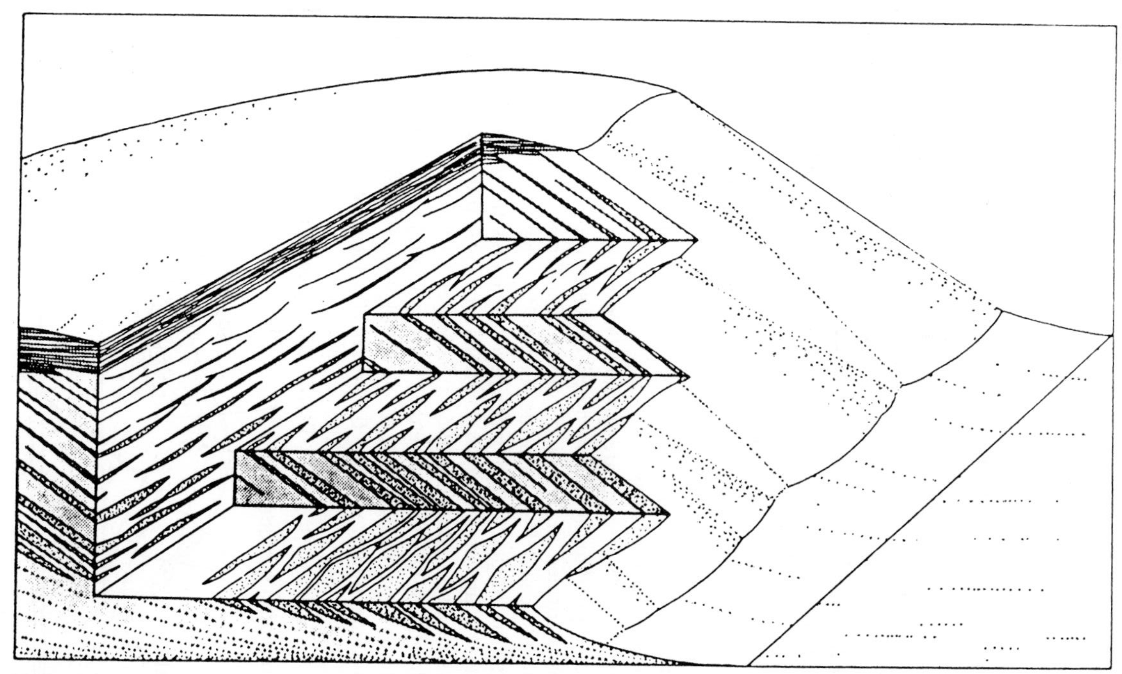

Figure 7.2 Schematic diagram showing the distribution of stratification types in a small dune. Grainflow cross-strata heavily stippled, grainfall laminae unshaded (foreset cross-strata) or lightly stippled (bottomset strata). Subcritically climbing translatent strata (topset strata) thin lined. (After Hunter 1977a).

laminae frequently alternate with inclined grainfall laminae (Fig. 7.2). The latter are produced by the accumulation of saltating grains which pass over the crest and then settle out in the leeward zone of reduced wind velocity. These deposits lie concordantly on the existing dune lee-slope topography. As defined by Hunter (1977a), grainfall lamination includes deposits formed by a complex of processes including pure grainfall and processes which are transitional between pure traction and pure grainfall, i.e. grainfall includes any mode of deposition excluding grainflow and climbing ripple migration. Internally, grainfall laminae display relatively little vertical size grading, although they sometimes become finer in the downslope direction because smaller grains are carried further from the crest than larger grains before settling out. The packing density of grainfall laminae is intermediate between the loosely packed grainflow cross-strata and the more tightly packed climbing ripple laminae.

Grainfall cross-strata can only be preserved if the angle of the lee slope is insufficient to cause complete reworking by avalanching, and if the lower lee slope deposits are not reworked by ripples migrating across the dune slope. For example, on simple barchan dunes favourable conditions for the preservation of grainfall deposits may be found on the lower part and lateral margins of the slip face (Hunter 1977a) (Fig. 7.3).

If very strong winds are experienced at the dune crest, some of the finer grains may be blown beyond the downwind limit of the slip face and accumulate as a low-angle sand platform several metres wide. As the dune advances, these deposits are buried, forming fine-grained, horizontal to gently leeward-dipping *bottomsets*. (Figure 7.1).

Climbing ripple strata are formed by the migration of ripples under conditions of net deposition. A continuous lamina is generated by the migration of each ripple (Rim 1951, Yaalon 1967, Hunter 1977a,b, Rubin & Hunter 1982). The angle of ripple climb can vary greatly depending on the rate of ripple migration and the rate of net sedimentation. If the angle of ripple climb (A) is less than the angle between the ripple stoss slope and the general depositional surface (B), the ripples are referred to as *subcritical climbing ripples* (Hunter 1977a). If $A = B$ the angle of climb is referred to as *critical* and if $A > B$ it is described as *supercritical* (Fig. 7.4). This classification is analogous to that proposed by Allen (1973) for subaqueous climbing ripples.

Climbing ripple structure may be composed of wavy layering parallel to successive rippled depositional surfaces or of even layering parallel to the vector of ripple climb. The two types are referred to as *rippleform laminae* and *translatent strata*, respectively (Hunter 1977a) (Fig. 7.4). Subcritical climbing translatent strata are the most commonly found type in aeolian sediments. Cross-laminated ripple foresets are rarely visible in such strata unless they are abnormally thick. Where visible, the dip of the truncated

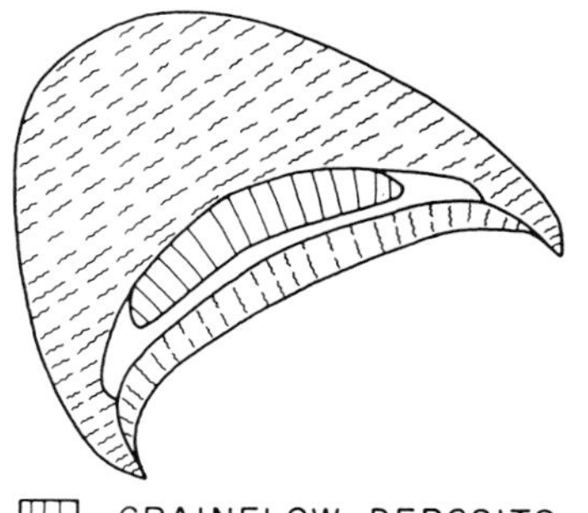

Figure 7.3 Schematic diagram showing the distribution of stratification types on the surface of a simple barchan dune. (After Hunter 1977a, Kocurek & Dott 1981).

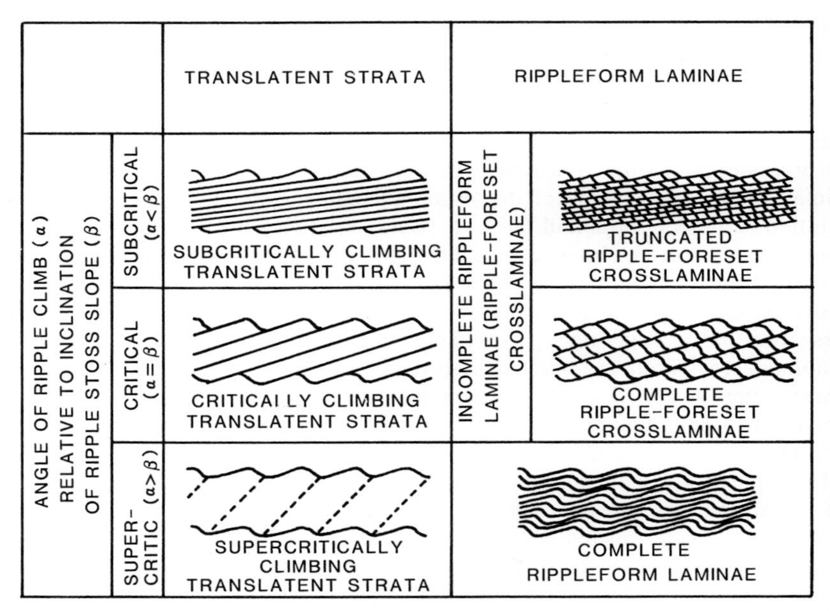

Figure 7.4 Types of structures produced by aeolian climbing ripples at various angles of climb. (After Hunter 1977a).

foreset laminae is considerably less than the angle of repose, in accordance with the low steepness of the lee slopes of wind ripples.

Subcritical translatent strata are characterized by thin, sharply defined, inversely graded laminae with few visible foresets, reflecting the large-wavelength, low-amplitude nature typical of wind ripples, and the concentration of coarser grains at their crests. At supercritical angles of climb, the contacts of climbing translatent strata are gradational rather than sharp and erosional. The grain size grading in the thin gradational contact zones is normal rather than inverse. Since stratification parallel to the depositional surface generally becomes visible in aeolian sands when the angle of ripple climb approximates critical, supercritically climbing translatent stratification is usually accompanied by rippleform lamination (Hunter 1977a).

Translatent climbing ripple deposits are characterized by a relatively dense packing arrangement. They are broadly equivalent to the accretion deposits recognized by Bagnold (1938b). On migrating dunes they often form relatively thin topsets on the windward slope and crestal areas. Thin translatent strata are also commonly found interbedded with grainfall strata on the lower part of the lee slope below the slip face (Fig. 7.3).

The angle of ripple climb on dunes can vary significantly in response to daily or longer term wind fluctuations, resulting in cyclic variations in the texture and composition of the deposited sands (Hunter & Richmond 1988).

During aeolian ripple migration, fine sand and silt particles tend to become concentrated in the ripple troughs, thereby forming a very fine grained layer at the base of each climbing translatent ripple stratum (Fryberger & Schenk 1988). These fine deposits often become preferentially cemented during early diagenesis (e.g. White & Curran 1988), giving rise to a distinctive *pin-stripe lamination* when the deposits are exposed in outcrop. Pin-stripe lamination may also form in grainflow deposits owing to the concentration of fine grains near the basal shear plane (Fryberger & Schenk 1988).

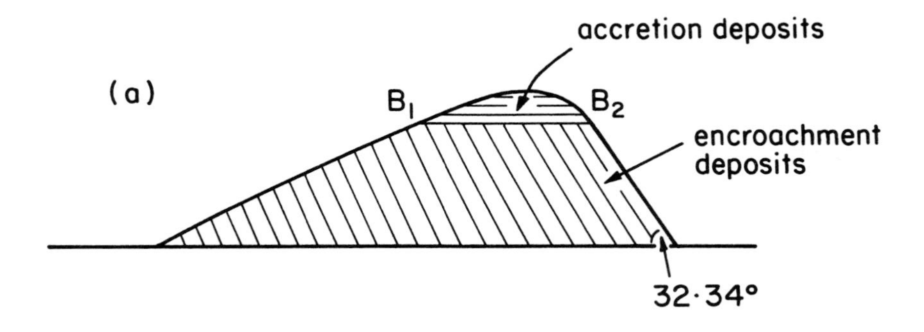

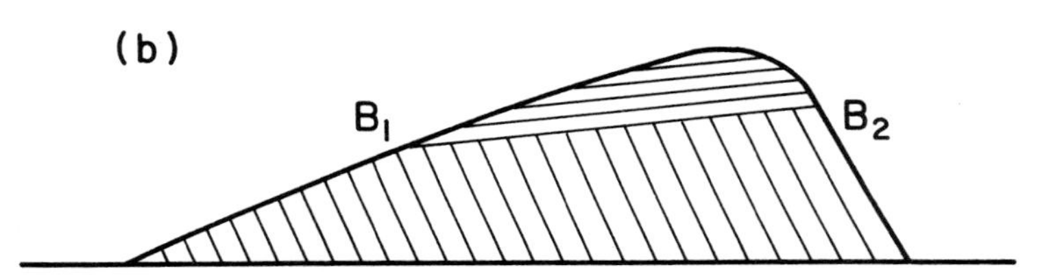

Figure 7.5 Idealized longitudinal section showing the internal structure of a barchan dune which (a) has maintained a constant size during migration and (b) increased in height during migration. The changing position of the brink is marked by the line B_1–B_2. (Modified after Bagnold 1941, p. 241).

7.2.2 Internal structure of barchans

Bagnold (1941, p. 241) presented a simplified model of the internal structure of a barchan dune with a separate brink and crest, and with a slip face extending to the base of the lee slope (Fig. 7.5). All of the sand above the present and past level of the brinkline (B_1–B_2 in Fig. 7.5) should represent tractional deposits, whereas below this level the sand should consist entirely of grainflow laminae.

Field observations have shown that the structure of many barchans is more complex than suggested by this basic model, depending on whether a dune has a coincident brink and crest, depending on whether or not the slip face extends to the base of the lee slope, the degree of seasonal and longer term wind variability, and changes in the size and shape of the dune over time (McKee 1957, 1966, Hunter 1977a). If the slip face does not extend to the base of the slip face, the lower part of the dune will consist of interfingered grainfall laminae and climbing ripple strata. There are also significant lateral variations which reflect the varying importance of different sand transport processes on different parts of the dune. On the barchan horns, for example, grainflow sedimentation is relatively unimportant, resulting in a dominance of low-angle translatent strata and grainfall laminae. Many barchans undergo major changes in size and shape as they migrate (Hastenrath 1987, Haynes 1989), adding to the complexity of the internal structure.

McKee (1966) excavated trenches both parallel and transverse to the dominant wind direction on part of a barchanoid ridge at White Sands National Monument, New Mexico. Trenching parallel to the dominant wind direction revealed a sequence of nearly flat-lying tabular planar sets of cross strata, each 0.9–1.2 m thick, containing foresets

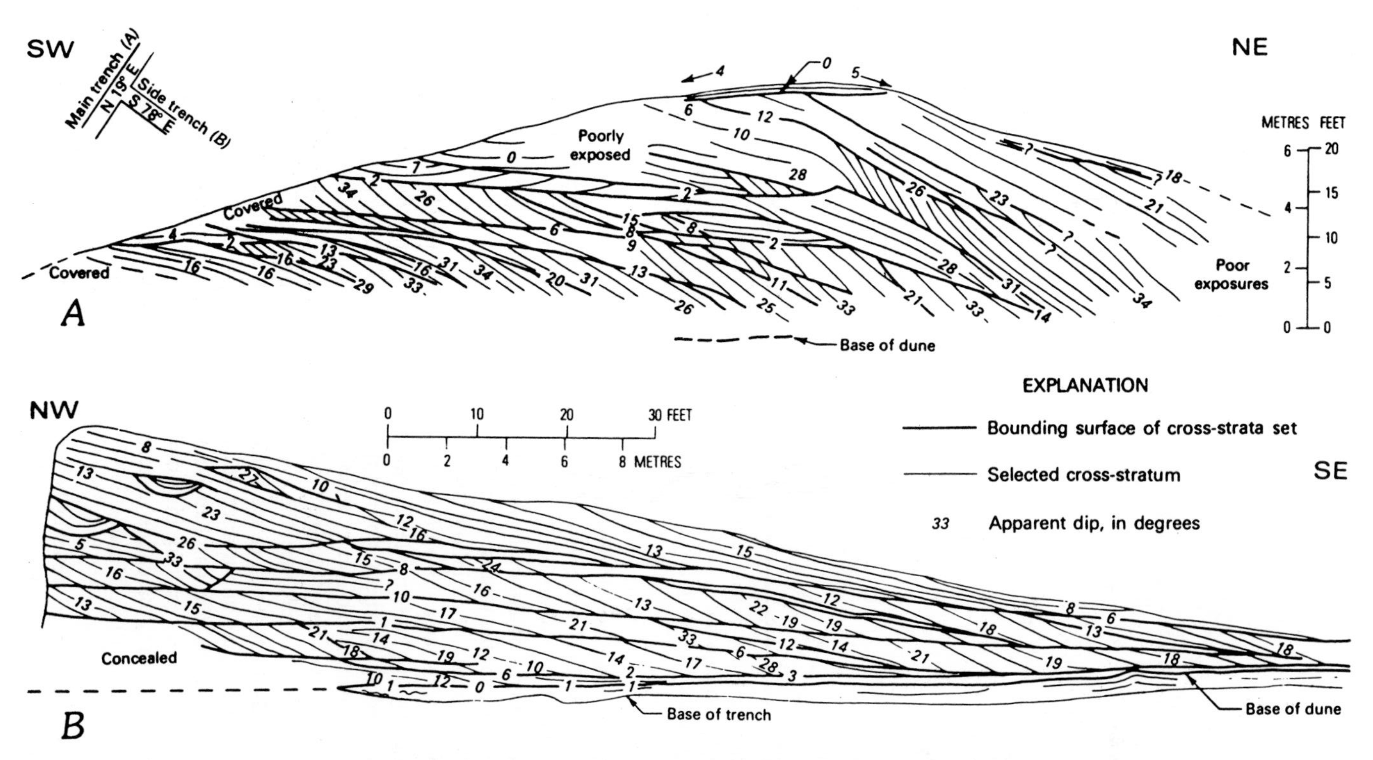

Figure 7.6 Internal structure revealed by a longitudinal cross-section through a barchanoid ridge dune at White Sands, New Mexico. Section A parallel to dominant wind direction, section B normal to dominant wind direction. (After McKee 1966).

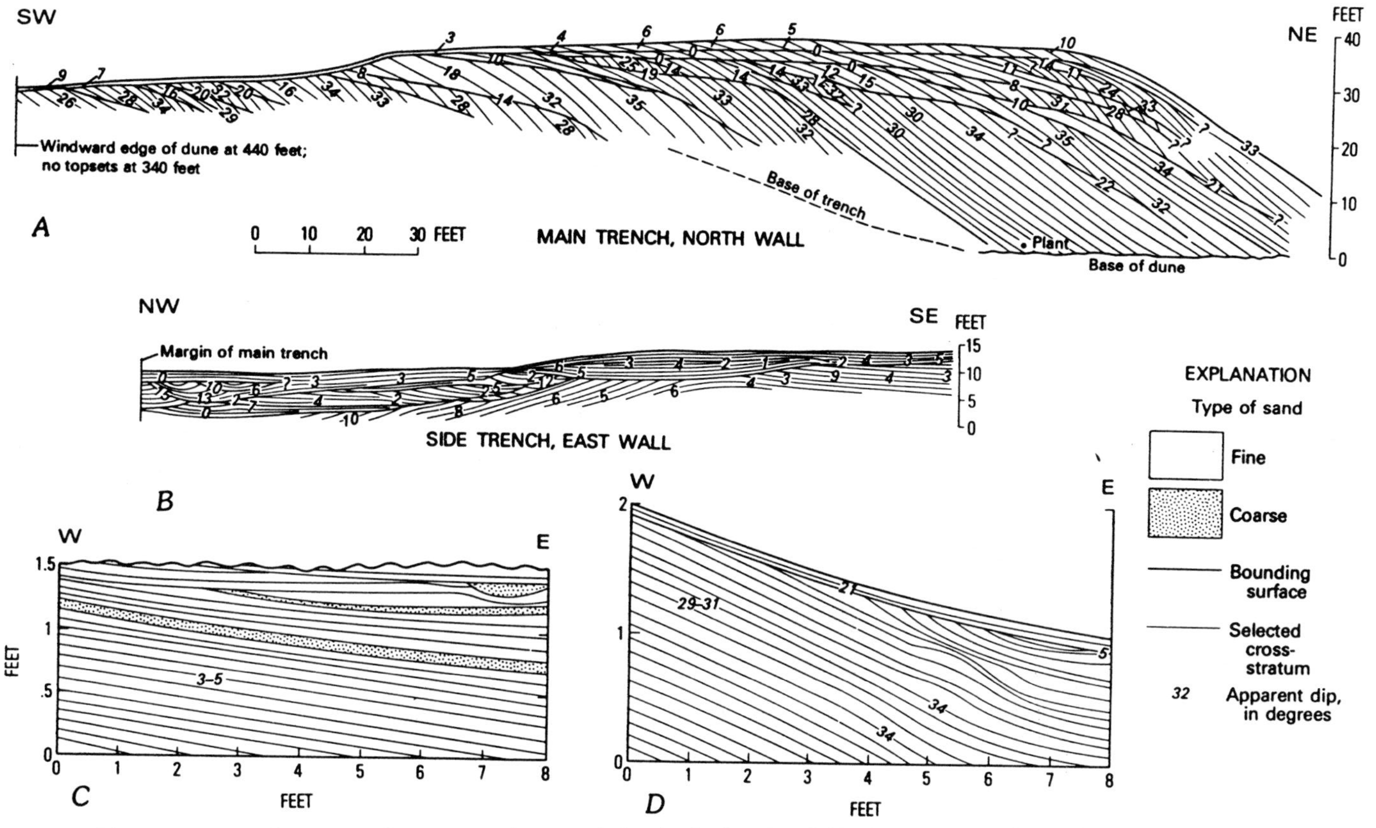

Figure 7.7 Internal structure of a transverse dune seen in longitudinal section, White Sands, New Mexico. (After McKee 1966).

that dipped downwind at 26–34° (Fig. 7.6). Towards the downwind end of the trench the bounding surfaces of each set changed from nearly horizontal to steeply dipping. The trench cut transverse to the dominant wind direction revealed cross-bed sets with nearly horizontal bounding surfaces near the middle of the dune, while on the flanks both the cross-sets and the bounding surfaces were observed to dip outward at low angles as a result of curvature of the horns. Individual cross-strata sets tapered towards the dune margins, and foresets showed apparent dips of 12–23°.

7.2.3 *Internal structure of transverse dunes*

Simple transverse dunes possess a less complex three-dimensional geometry than barchanoid ridges and consequently their internal structures show a simpler pattern (Fig. 7.7). Unless they experience a reversing wind regime, the cross-bed sets of transverse dunes all dip in the same general direction with a relatively narrow range of apparent dip angles. McKee (1966) found that the most diagnostic feature of transverse dunes at White Sands National Monument was the great lateral extent of nearly horizontal parallel laminae formed by the apparent dip of strata as seen in cross-sections cut transverse to the dominant wind direction. A transverse dune at Killpecker dunefield, Wyoming, trenched by Ahlbrandt (1973), showed similar features.

7.2.4 *Internal structure of seif dunes*

Bagnold (1941, p. 242) reasoned that in the case of seif dunes, where the slip face shifts alternately from one side of the dune to the other in response to periodic wind changes,

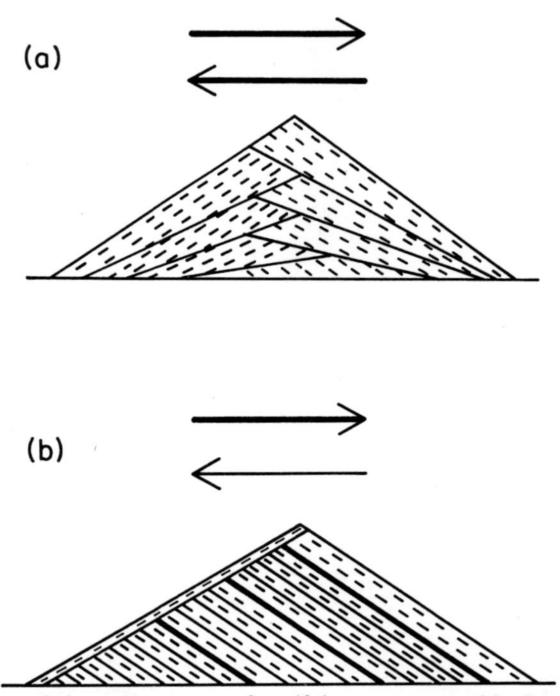

Figure 7.8 (a) Schematic internal structure of a seif dune, as proposed by Bagnold (1941). The central part is composed of bidirectional high-angle cross-beds (encroachment deposits) and the flanks of low-angle cross-strata (accretion deposits). (b) Suggested internal structure of a longitudinal dune in an asymmetric bidirectional wind regime, according to Rubin & Hunter (1985).

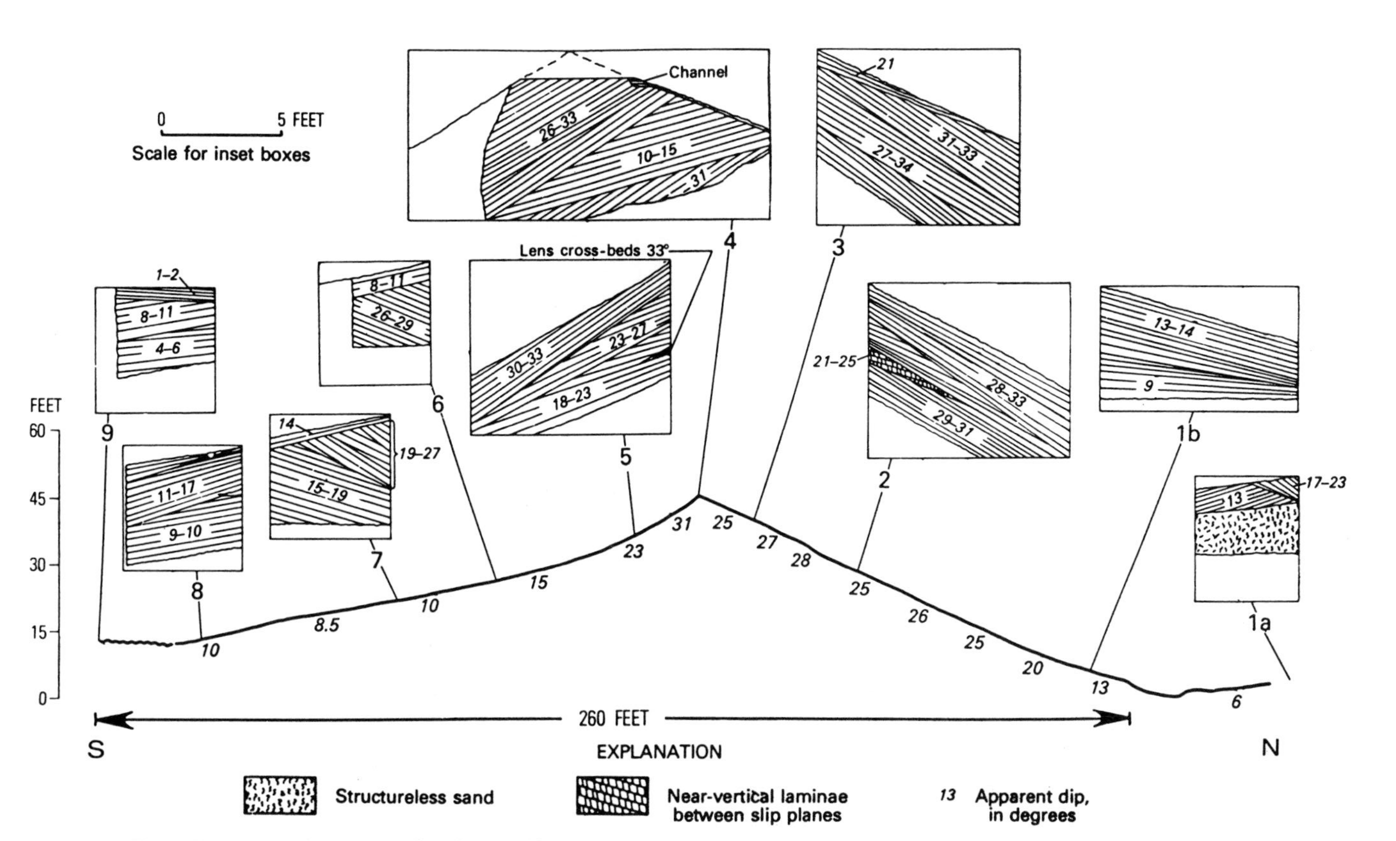

Figure 7.9 Internal structure of a Libyan seif dune, revealed by excavation of trial pits. (After McKee & Tibbitts 1964).

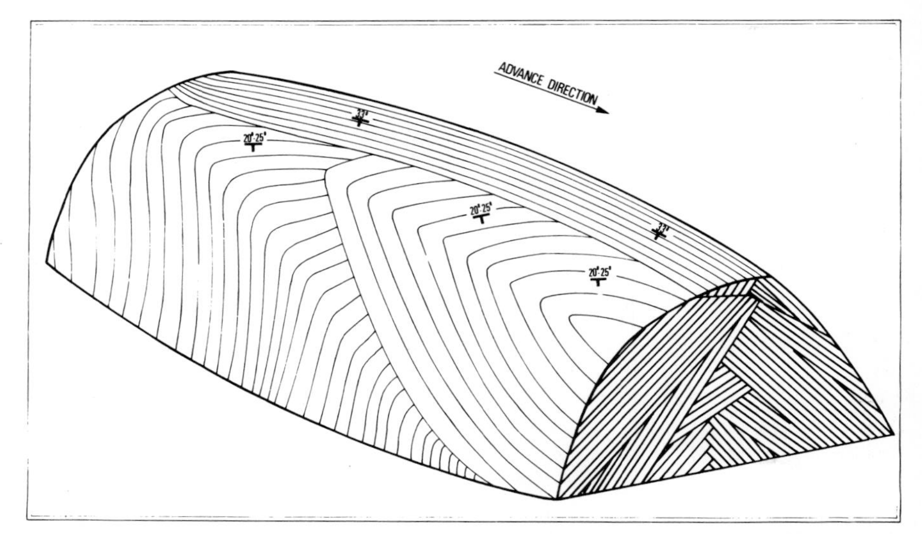

Figure 7.10 Model of the three-dimensional internal structure of a seif dune developed in a bimodal wind regime, based on field observations in northern Sinai. (After Tsoar 1982).

grainflow strata should show a bimodal dip distribution (Fig. 7.8a). Since the slip face rarely extends to the dune toe, the flanks of the dune should consist of grainfall laminae and/or translatent ripple laminae which interfinger in mid-slope with grainflow laminae.

A bimodal distribution of foreset dip directions was found in the Libyan seif dune trenched by McKee & Tibbits (1964) (Fig. 7.9). They argued that because the dip of the grainflow deposits on a seif dune lies at right-angles to the dune crest, regardless of the prevailing wind directions, the deposits should form two groups of high-angle cross-laminae dipping in nearly opposite directions. This is in contrast to the more unidirectional ($< 120°$) spread of dips of grainflow laminae observed in barchan dunes. A similar conclusion was reached by Glennie (1970). However, Rubin & Hunter (1985) pointed out that preservation of symmetrical bimodal dip structures as shown in Figure 7.8a requires that winds of equal magnitude blow from each side of the dune, such that there is no net lateral movement of the dune. Any slight asymmetry in the wind regime would produce a corresponding asymmetry in the dune form and produce internal stratification dipping in the direction of the asymmetry (Fig. 7.8b). This, they suggested, may partly explain the apparent rarity of longitudinal dune deposits preserved in the geological record.

Tsoar's (1982) observations on a sinuous seif dune in northern Sinai showed that the internal geometry is relatively complex when viewed in three dimensions, reflecting seasonal and spatial variations in the pattern of erosion and deposition along the dune. Since net sand deposition takes place principally where the wind crosses the meandering crest almost perpendicularly, resulting in a reduction in wind velocity on the lee side (see Section 6.3.5.1), the angle of maximum dip is actually oblique to the main longitudinal axis of the dune. Viewed in section transverse to the longitudinal axis, the apparent dips are bimodal (Fig. 7.10). On the seif dune studied by Tsoar, slip faces formed only on the upper part of the lee slope, and most of the preserved windward deposits represent climbing ripple strata and grainfall laminae (Tsoar 1982). At the beginning of each season, deposition starts on a deflational surface which slopes mainly at an angle of 10–20°. The deposits which accumulate first consist predominantly of translatent cross-strata and grainfall laminae. As progressively more sand is deposited, the upper part of the leeward slope steepens, and at the end of the season grainflow cross-strata predominate, especially on those parts of the dune where the crest line is perpendicular to the prevailing seasonal wind. Owing to the seasonal wind reversal, each episode of accretion is separated by an erosional unconformity which takes the form of a curved bounding surface (Fig. 7.11).

7.2.5 Internal structure of unvegetated dome dunes

McKee (1966) observed that the upwind part of a dome dune at White Sands National Monument consisted largely of tabular planar cross-strata with foresets dipping downwind at high angles (Fig. 7.12). In the downwind part of the dune, foresets dipped less steeply (14–27°). A trench dug transverse to the dominant wind direction revealed stratification with a very low dip towards the dune margins. A series of sand-filled scours was observed parallel to the dominant wind direction, while the rounded dune top was found to be capped by a layer of horizontal or nearly horizontal laminae. Broadly similar structures were observed by Ahlbrandt (1973) in a dome dune at Killpecker dunefield, Wyoming.

7.2.6 Internal structure of reversing dunes and star dunes

Merk (1960) described reversing dunes in the Great Sand Dunes of Colorado which showed two distinct foreset dip directions. The basal parts of the dunes indicated deposition by prevailing southwesterly winds, while the upper parts showed evidence of reforming under the influence of easterly storm winds. Preserved sets of low-angle laminae, which probably formed as windward slope deposits, were also observed to have been buried within the lower parts of the dunes during periods of wind reversal.

McKee & Bigarella (1972) also observed steep grainflow foresets, dipping in two almost opposite directions, in a large reversing dune at Lagoa dunefield on the southern coast of Brazil.

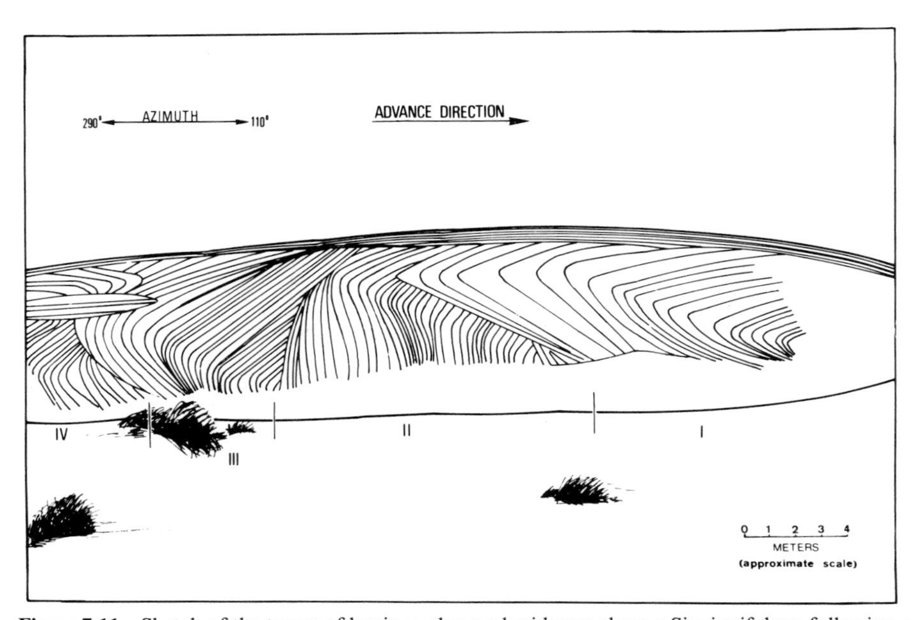

Figure 7.11 Sketch of the traces of laminae observed mid-way along a Sinai seif dune following a period of heavy rain. The Roman numerals indicate four annual cycles of seasonal deposition. (After Tsoar 1982).

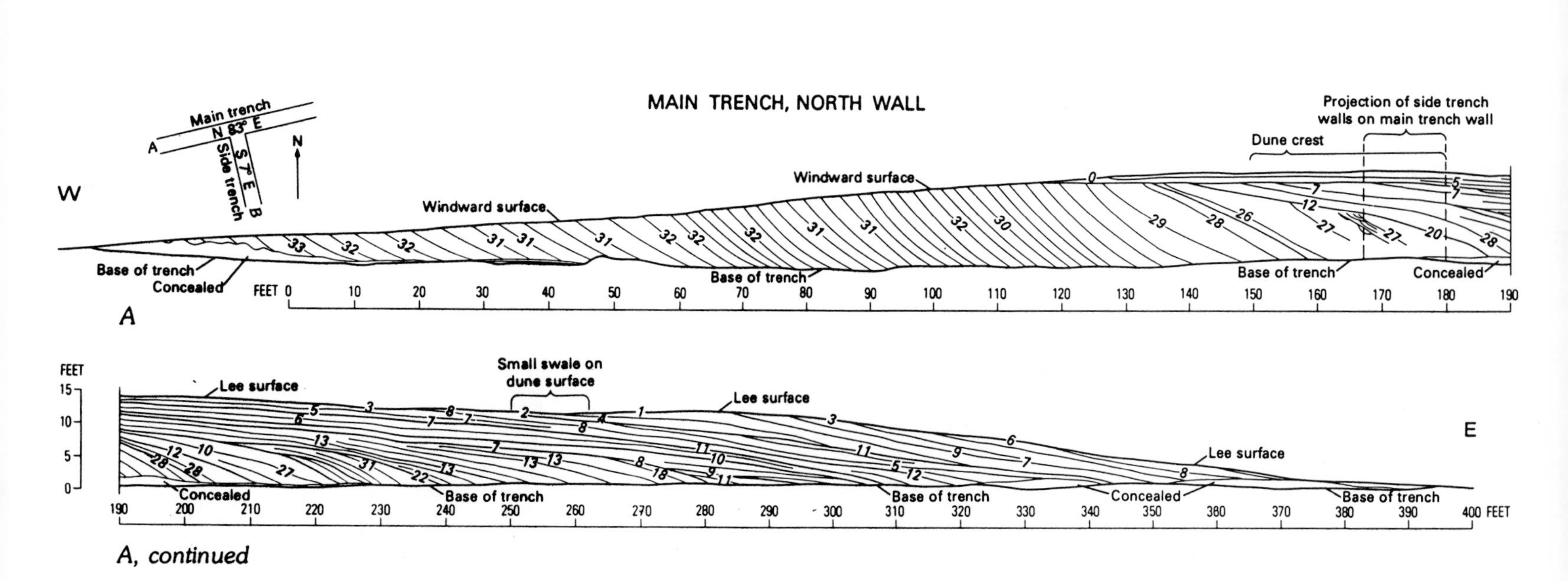
MAIN TRENCH, NORTH WALL
Main trench
N 83° E
Side trench
S 7° E
A
B
N
W
E
Windward surface
Windward surface
Dune crest
Projection of side trench walls on main trench wall
Base of trench
Concealed
Base of trench
Base of trench
Concealed
FEET 0
10
20
30
40
50
60
70
80
90
100
110
120
130
140
150
160
170
180
190
A
Lee surface
Small swale on dune surface
Lee surface
Lee surface
Concealed
Base of trench
Base of trench
Concealed
Base of trench
FEET
15
10
5
0
190
200
210
220
230
240
250
260
270
280
290
300
310
320
330
340
350
360
370
380
390
400 FEET

A, continued

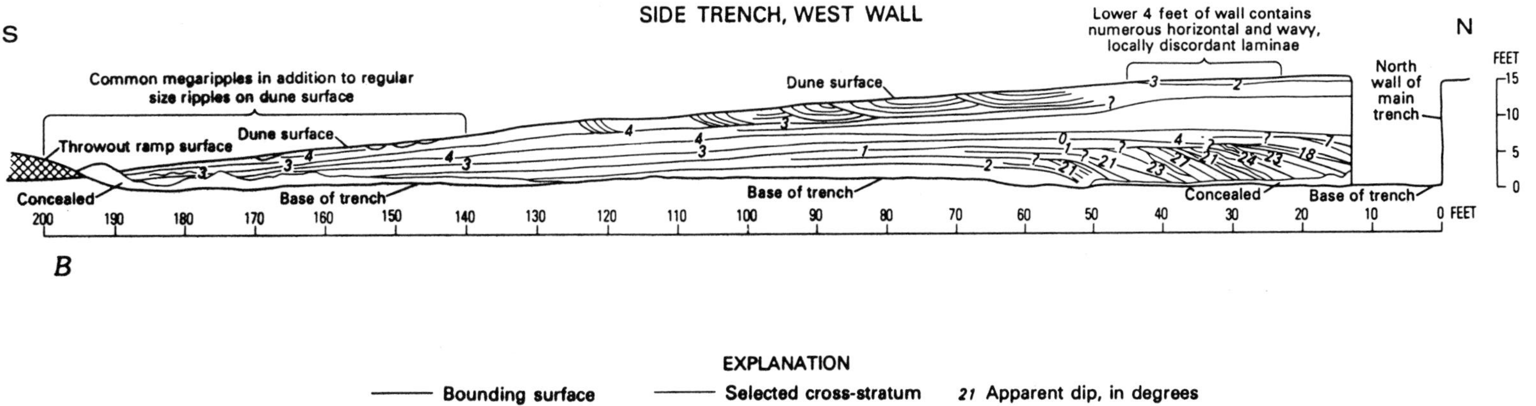

Figure 7.12 Internal structure of a dome dune at White Sands, New Mexico. (After McKee 1966).

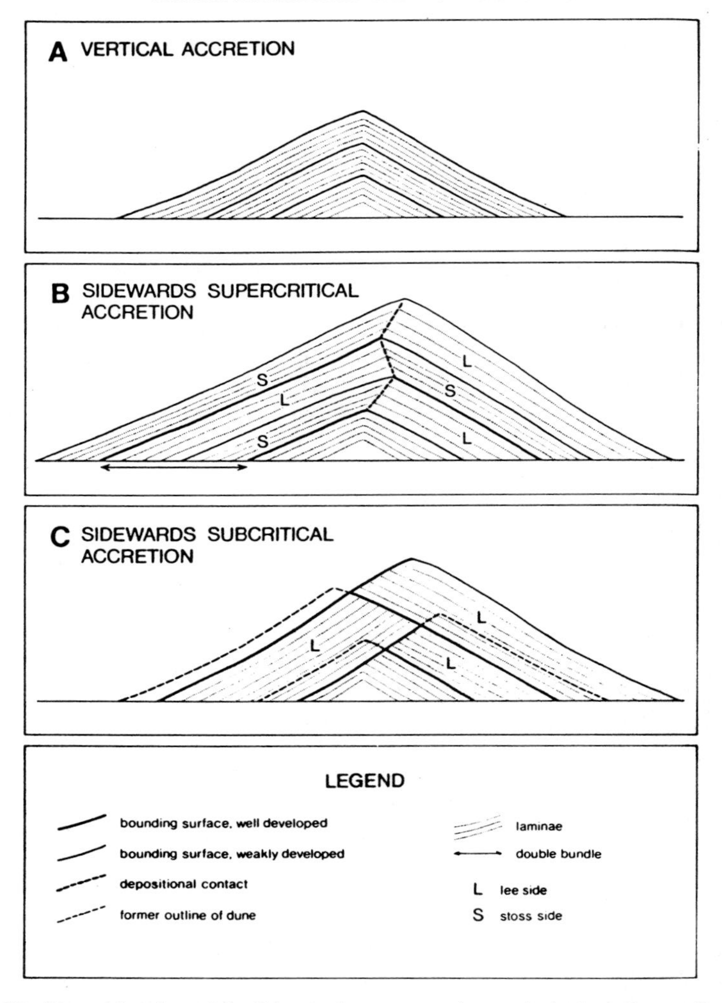

Figure 7.13 Depositional model of internal structures in sand shadow dunes. See text for explanation. (After Clemmensen 1986).

Star dunes, which form in complex wind regimes, typically display cross-bed sets which dip in several directions. Structures in the crestal areas of star dunes in the Namib Sand Sea were found by McKee (1982) to display clear evidence of seasonal wind reversal. Very complex structures were observed on the arms of the star dunes.

Nielson & Kocurek (1987) found that the lower parts of star dunes in the Dumont dunefield, California, are dominated by low-angle climbing translatent strata, with grainflow deposits restricted to a relatively small area near the crest. By virtue of their topographic position, the latter deposits probably have a low preservation potential and internal structures diagnostic of star dunes may not be recognizable in the fossil record (Kocurek 1986, Nielson & Kocurek 1987).

Clemmensen (1987) identified two types of cross-bedding in complex star dune deposits of the Permo-Triassic Hopeman Sandstone, Scotland. The first type consisted of large-scale or giant-scale, mainly trough-formed high- to medium-angle cross-sets, which he interpreted as slip-face deposits formed by actively migrating star dunes. The second type consisted of bimodally dipping more wedge-shaped large-scale or giant-scale sets, which apparently formed on the opposite flanks of relatively stationary star arm segments.

7.2.7 Internal structures of shadow dunes

Clemmensen (1986) demonstrated that shadow dunes possess a characteristic internal structure with two groups of foreset laminae dipping in opposite directions away from the median ridge. If the wind blows with equal strength across both flanks of the dune, deposition occurs on both dune flanks and the laminae are deposited continuously across the ridge, forming a tepee-like structure (Fig. 7.13a). If the strength of the wind from each side of the ridge fluctuates over time, the crest of the dune migrates from side to side and a chevron-like pattern of cross-bedding results (Fig. 7.13b,c). In the case of supercritical lateral accretion, both stoss-side and lee-side deposits are preserved (Fig. 7.13b). When lateral accretion is subcritical, only lee-side laminae are preserved and a simpler chevron-like pattern is formed (Fig. 7.13c).

7.2.8 Internal structures of vegetated coastal dunes

Several authors have observed that the azimuths of high-angle foresets in coastal foredunes are commonly bimodal, suggesting that some such dunes originated as shadow dunes (McBride & Hayes 1962, Bigarella *et al.* 1969b, Yaalon & Laronne 1971, Goldsmith 1973, 1985, Hunter 1977a, Shideler & Smith 1984). However, many other coastal dunes are characterized by low-angle dips which show a unimodal azimuth that correlated directly with the prevailing wind (Land 1964, Bigarella *et al.* 1969b, Yaalon & Laronne 1971, Goldsmith 1973, 1985). The dominance of low-angle cross-beds has been considered to reflect sand accumulation on a well vegetated surface (Goldsmith 1973, Yaalon 1975, Oertel & Larsen 1976).

Vegetated dome dunes on the coast of Brazil were reported by Bigarella (1972, p. 25) to be characterized by low-angle strata with dip angles distributed completely around the compass (Fig. 7.14). Many hummock dunes in coastal environments also show a similar wide range of dip directions. Erosional trough and fill structures are also a common feature of such dunes.

Hesp (1988) compared the internal structures of established foredune ridges in different stages of development on the coast of New South Wales, Australia. Foredunes which had not been affected by wave erosion were found to be characterized by large-scale, low-angle ($<15°$) cross-beds which are almost continuous across the dune. They dip in two main directions, seawards and landwards, if the ridge is laterally extensive. Where the ridge consists of a series of overlapping mounds, a wider range of dip directions was encountered. In the case of foredunes which have been affected by wave erosion or blowout development, more complex structures are developed, including buried wave-cut scarps, slump deposits, infilled troughs, and localized development of high-angle cross-beds. Buried palaeosols may also be present.

7.2.9 Internal structure of parabolic dunes

A parabolic dune at White Sands, New Mexico, trenched by McKee (1966), showed a dominance of high-angle foresets in the lower part of the dune, similar to barchanoid and other dune types in the area. Thin, low-angle to horizontal cross-sets were found in the upper part of the dune, expecially on the windward side (Fig. 7.15). A few sets of symmetrically filled trough cross-strata, 7–9 m wide and about 1 m deep, were also present near the top. Some of the foresets were found to be convex; this may reflect oversteepening of the foresets by crosswinds blowing across the base of the dune nose (McKee & Bigarella 1979, p. 95). Other features apparently unique to this dune type

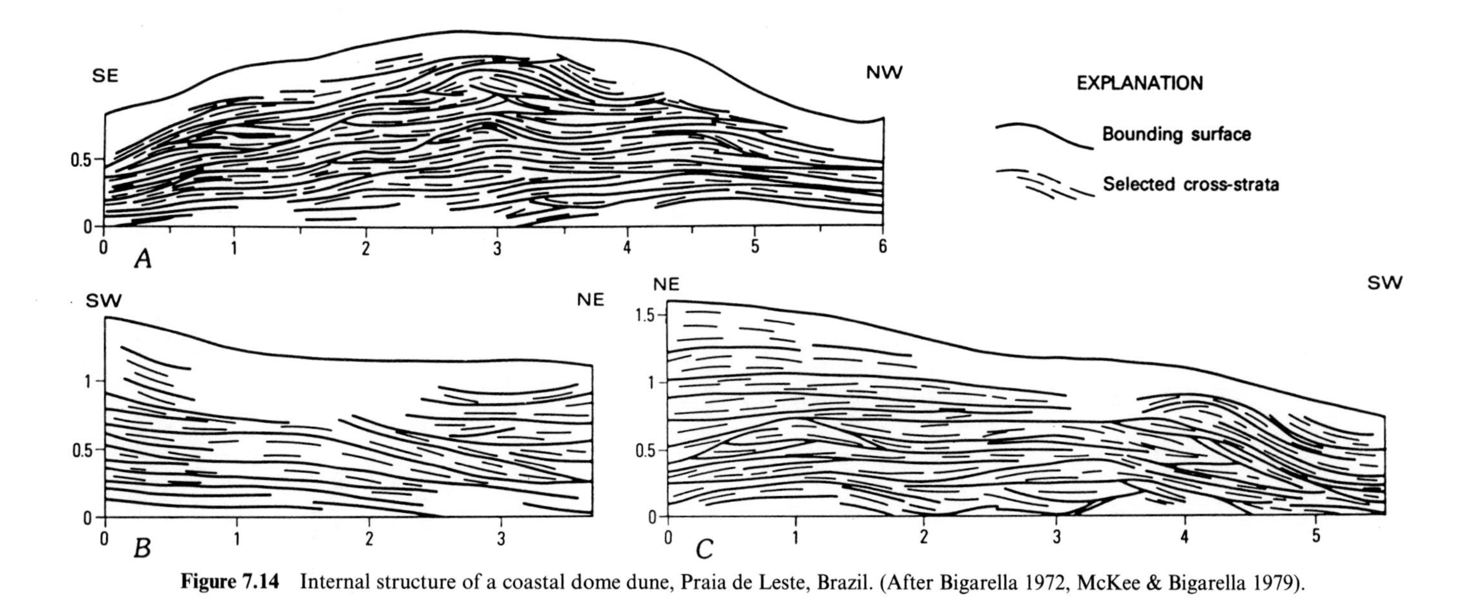

Figure 7.14 Internal structure of a coastal dome dune, Praia de Leste, Brazil. (After Bigarella 1972, McKee & Bigarella 1979).

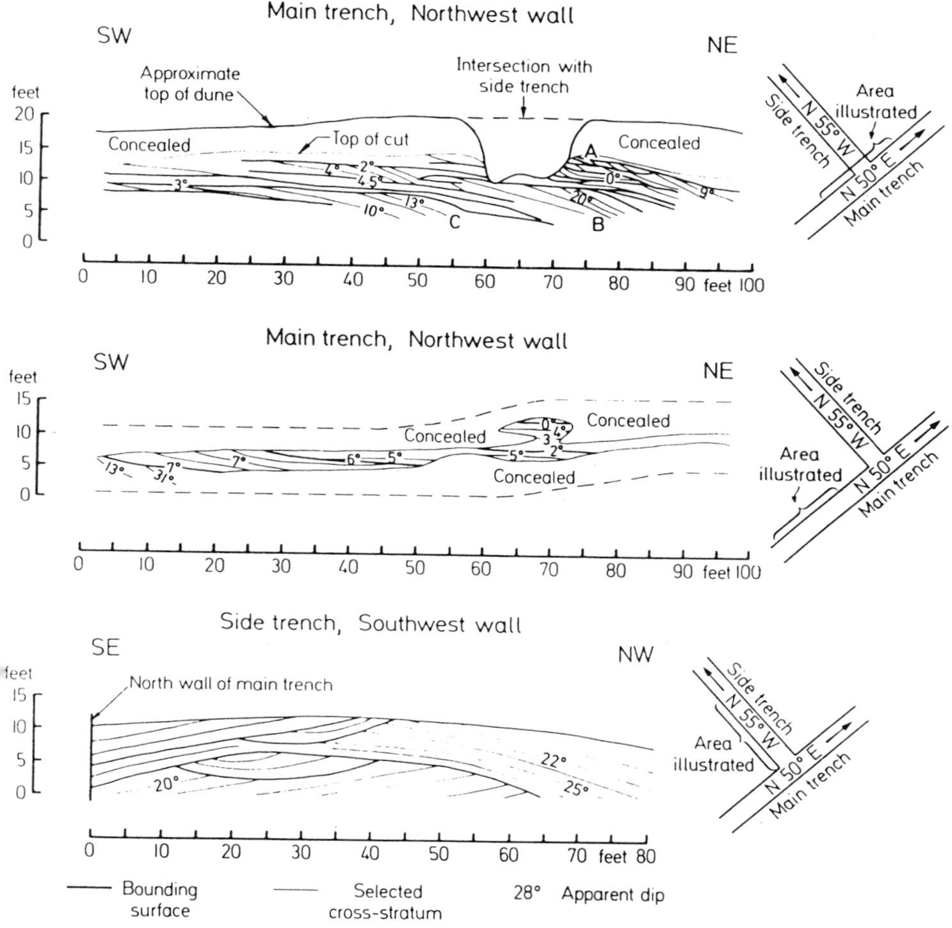

Figure 7.15 Variously oriented sections through a parabolic dune at White Sands, New Mexico. (After McKee 1966).

were disturbances to bedding surfaces caused by root growth, and an unusually wide spread (>200°) of cross-bed dip directions (McKee 1966).

Detailed studies of a parabolic dune at Lagoa dunefield, southeastern Brazil, also showed that cross-strata are large scale in the basal parts of the dune and become thinner and flatter towards the top (Bigarella 1975a). In the nose the strata dip at high angles (29–34°) in a broad arc, while in the arms they were found to be bidirectional normal to the dune axis (Fig. 7.16). Several erosion surfaces, buried beneath high-angle cross-strata, were found in the crestal area.

Studies of parabolic dunes at Cape Flattery, North Queensland, by Pye (1980a) indicated a similar pattern. Accumulation of grainflow and grainfall deposits was found to occur only on the outer slopes of the arms near the nose, whereas further upwind these slopes were well vegetated with a well developed soil A horizon. The inner slopes of the arms all along the dune were deflational or covered with a thin veneer of sand in transport parallel to the dune axis. Buried soil A1 horizons were exposed along the inner slopes of the arms, especially close to the dune nose where the blowout troughs were being widened (Fig. 7.17).

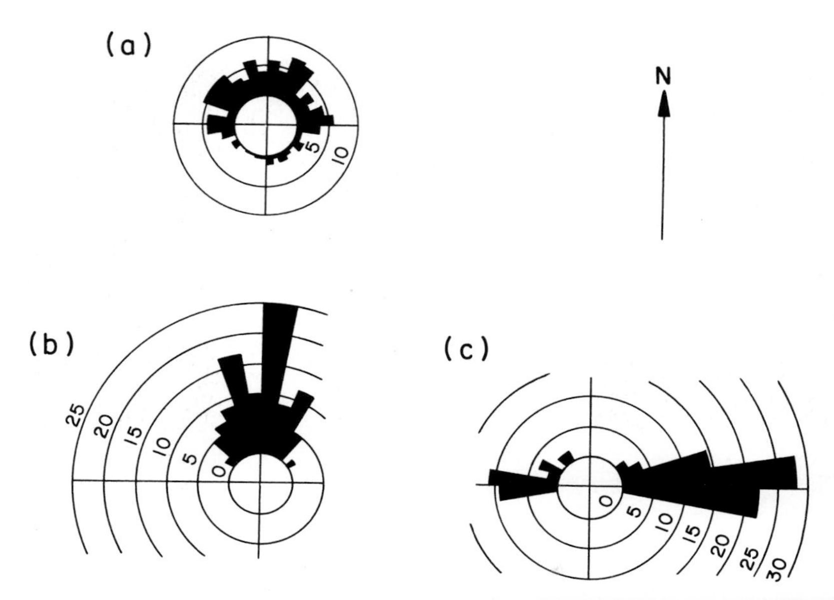

Figure 7.16 Rose diagram showing cross-strata dip directions for (a) the entire dune, (b) the central part of the nose, and (c) one of the arms of parabolic dune at Lagoa dunefield, Brazil. (Modified after Bigarella 1975a).

Figure 7.17 Organic-rich soil Al horizon exposed on the inward-facing slope of a parabolic dune trough, Cape Flattery, North Queensland.

7.2.1 *Nature and origin of bounding surfaces*

Bounding surfaces are erosional discontinuities which separate sets or cosets of cross-strata. They have long been recognized as an important feature in aeolian deposits (e.g. Shotton 1937). Stokes (1968) suggested that large-scale bounding surfaces, which he termed *multiple truncation bedding planes*, are essentially deflation surfaces whose level is controlled by the water table. This interpretation was questioned by McKee & Moiola (1975), who proposed that major bounding surfaces represent the floors of migrating interdune areas which have truncated the upper surface of a series of dune cross-strata. Brookfield (1977) subsequently recognized three orders of bounding surfaces in aeolian deposits (Fig. 7.18). First-order bounding surfaces are flat-lying or convex-up bedding planes which cut across cross-bedding and other dune structures. Second-order surfaces are low to moderately dipping flat or convex-up surfaces which bound sets of cross-strata. They mainly, although not invariably, dip downwind and may be truncated by first-order bounding surfaces. Third-order bounding surfaces are relatively small-scale features which separate thin groups of laminae within cross-laminated sets.

Brookfield (1977) rejected the earlier explanations for the origin of bounding surfaces and suggested instead that they are related to the migration of climbing bedforms of differing hierarchical order. First-order bounding surfaces were attributed to the migration of complex or compound megadunes (draas) and second-order surfaces to the migration of dunes across the draa surfaces. Third-order bounding surfaces were suggested to be due either to short-term changes in wind distribution and velocity or to local airflow modifications induced by the dune bedforms themselves.

Although subsequent evaluations have not ruled a deflational origin for at least some first-order bounding surfaces (Loope 1984, 1985a, Kocurek 1984, Rubin & Hunter 1984, Fryberger *et al.* 1988), many workers hold the view that a majority are formed by climbing bedform migration (Rubin & Hunter 1982, Kocurek 1988b). Very extensive bounding surfaces, of regional or sub-regional extent, termed *regional* or *super-bounding surfaces* (Talbot 1985), probably reflect large-scale changes in erg surface processes brought about by changes in climate, basin tectonics, or sea level (Blakey & Middleton 1983, Loope 1985, Chan & Kocurek 1988, Kocurek 1988b). Large, concave-up erosional surfaces which are common in Pennsylvanian to Jurassic age aeolian sandstones of the Colorado Plateau, were termed *superscoops* by Blakey (1988).

According to Kocurek (1988b), climbing simple dunes produces a single set of cross-strata between two first-order bounding surfaces, each of which marks the floor of an interdune area upon which interdune deposits may accumulate (Kocurek 1981b). Localized scouring or reworking of the dune face during migration may produce

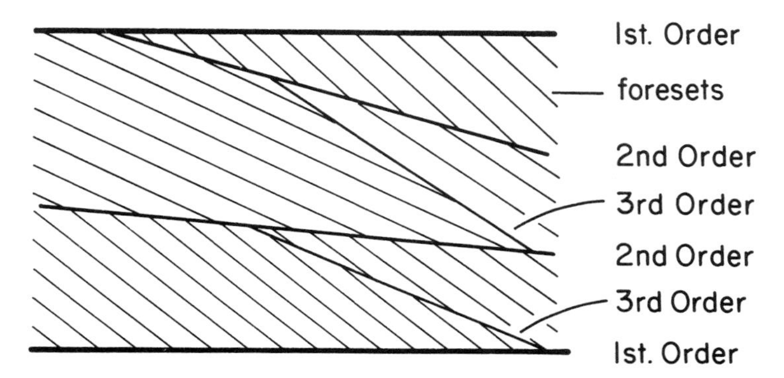

Figure 7.18 Schematic diagram showing three orders of bounding surface.

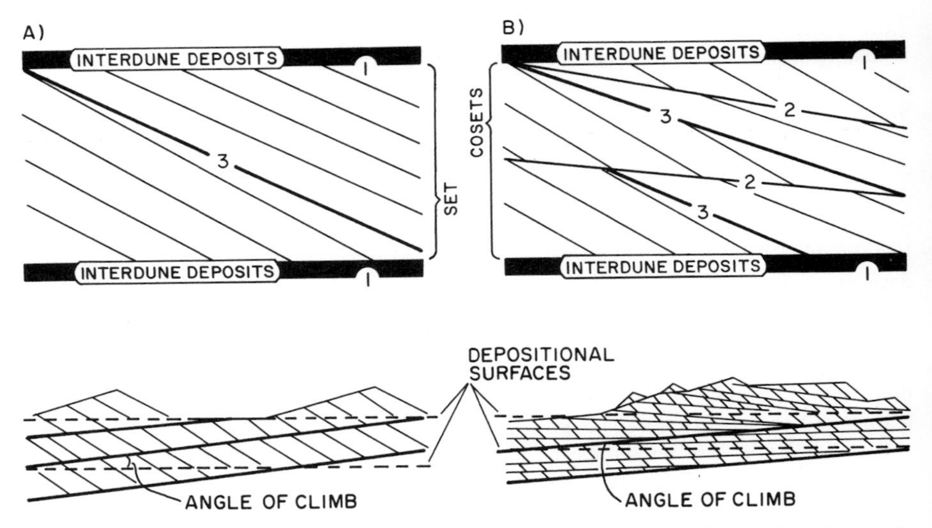

Figure 7.19 Model for (A) the formation of first- and third-order bounding surfaces during the migration of simple dunes and interdune areas and (B) the formation of first-, second-, and third-order bounding surfaces during the migration of draas and interdune areas. Angles of climb are measured with respect to the depositional surfaces. (After Kocurek 1988b).

third-order reactivation surfaces. Two orders of bounding surface are, therefore, ideally represented in deposits formed by climbing simple dunes (Fig. 7.19). In the case of draas, three orders of bounding surface may be present, since the surface of the megadune is covered by superimposed dune bedforms which migrate faster than the megadune itself (Brookfield 1977, Rubin & Hunter 1983, Steele 1983, Mader & Yardley 1985). However, not every point on a draa may be covered by superimposed dunes, so that parts of a draa may generate simple cross-strata with only two orders of bounding surface (Havholm & Kocurek 1988), where as others generate compound cross-bedding (Rubin & Hunter 1983).

7.3 Secondary sedimentary structures in dunes

In addition to sedimentary structures formed by primary aeolian deposition and bedform migration, dune deposits commonly display a wide range of syn- and post-depositional deformation structures. The primary causes of deformation are slumping of weakly coherent sand blocks, flowage of saturated sand, pressure loading due to sediment overburden, scour and fill by wind or water, root growth, burrowing by animals, seasonal freezing and thawing, and seismic shocks. All of these processes result in the formation of contorted bedding.

A classification and suggested terminology for deformation structures in aeolian sands was proposed by McKee *et al.* (1971), based on field observations at White Sands National Monument and laboratory experiments. The principal structures recognized by these authors are *rotated structures*, *warps* or *gentle folds*, *drag folds* and *flame structures*, *high-angle asymmetric folds*, *overturned folds*, *overthrusts*, *break-apart structures*, *breccias*, and *fade-out laminae* (Fig. 7.20).

Grain flows in non-cohesive dry sand tend to produce a shallow, spoon-shaped depression on the upper slope where the flow originates and a flat to slightly convex tongue-shaped mound downslope where the flow comes to rest. Associated tensional features near the top of the flow include stretched laminae, warps or gentle folds, and

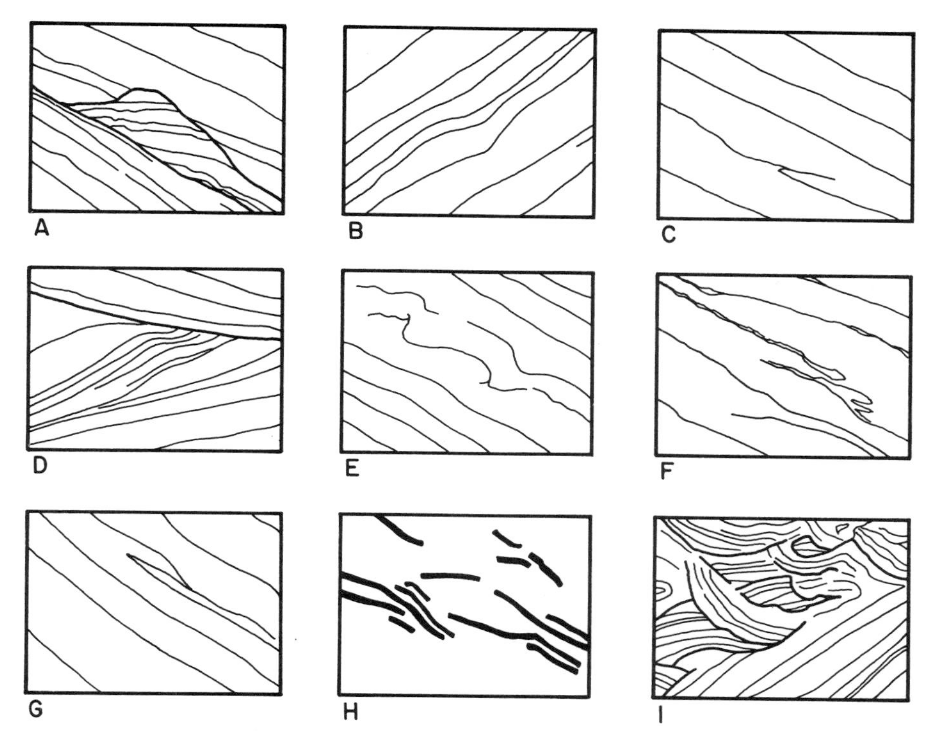

Figure 7.20 Principal types of deformational structures in avalanche deposits of dunes: A, rotated structures; B, warps or gentle folds; C, flame structure; D, drag fold; E, high-angle asymmetric folds; F, overturned folds; G, overthrust; H, break-aparts; I, breccias. Each block represents an area of ca 15 × 10 cm. (After McKee & Bigarella 1979).

Figure 7.21 Slump structures in slightly damp coastal dune sand, Oregon.

drag folds and flames (McKee *et al.* 1971). Features associated with the lower part of the flow include drag folds and flames, high-angle asymmetric folds, and overturned folds. Rotated blocks and plates, consisting of weakly coherent sand, may be incorporated within the flow. Near the top of the flow inverse grading stratification is usually well developed, but this becomes less distinct downflow. Laminae which lose their distinctiveness in this way are referred to as fadeout laminae.

Slump-type mass movements occur when the sand is weakly cohesive, due either to the presence of surface moisture films or thin salt crusts (Fig. 7.21). Many of the grains in the upper part of the slump broadly retain their relative position, although the laminae may be contorted and small step-faults may be formed. The basal shear plane may be planar or rotational. Within the slump, break-apart structures, overthrusts, and high-angle asymmetric folds may be present.

If the infiltration capacity of the sand is exceeded during heavy rainfall, or if the dune surface is covered by runoff from higher ground, the surface sand layers may become saturated and move as a liquified flow. Structures associated with saturated sand flows include drag and flame structures, fadeout laminae, and recumbent folds (McKee *et al.* 1971). Evidence of surface runoff from higher ground includes scour features and beds of pebbles or silts (Bigarella 1975b, Pye 1983f). In weathered, slightly cohesive sands, the introduced pebbles may protect the underlying sand from raindrop erosion, giving rise to sand pedestals (Fig. 7.22) (see also Gees & Lyall 1969).

Bigarella (1975b) introduced the term *dissipation structure* to describe deposits in the Lagoa coastal dunefield, Brazil, in which the primary aeolian depositional structures have been substantially obscured owing to reworking by running water and infiltration of

Figure 7.22 Pedestals of slightly coherent sand capped by small pebbles, Cape Flattery, North Queensland. The pebbles have protected the underlying sand from raindrop erosion.

Figure 7.23 Wavy laminations (enhanced by iron oxyhydroxide precipitation) formed by deformation of saturated dune sand deposits owing to overburden pressure, Cape Bedford, North Queensland.

allochthonous fine-grained material. Ahlbrandt & Fryberger (1980) adopted this term more specifically to describe concentrations of infiltrated fines which are superimposed on the primary depositional lamination in parts of the Nebraska Sandhills. As pointed out by Pye (1983f), these latter features are probably more appropriately described as *infiltration structures*.

Saturated dune sands below a seasonally high water table often develop wavy laminations due to compressional loading (Fig. 7.23). Laboratory experiments by Rettger (1935) and McKee *et al.* (1971) showed that saturated sands and silts develop such structures when subjected to compressional stress.

Generally, cohesive slump features and saturated compressional deformation structures are more common in coastal dunes than in desert dune sands owing to the higher rainfall and groundwater levels to which they are exposed (McKee & Bigarella 1972).

Larger scale deformational structures, involving displacements of several metres or tens of metres, have been recorded in many ancient aeolian sandstones (Doe & Dott 1980, Horowitz 1982), but there is no general agreement about their origin. Suggested explanations include gravity slumping of over-steepened dunes, collapse of storm-wetted foresets, and earthquake-induced liquefaction.

Root growth structures are found both in coastal and desert dunes, but generally are more common in better vegetated coastal dune deposits (Ahlbrandt *et al.*, 1978). Calcified root moulds are common in vegetated dune sands which contain more than about 8% calcium carbonate.

Bioturbation of aeolian deposits may also be accomplished by a wide range of insects and animals ranging from ants and spiders to rabbits and gophers (Ahlbrandt *et al.* 1978). Deformed laminations and erosion residuals formed by human and animal foot impressions have been described by Lewis & Titheridge (1978) and Ehlers (1988, p. 158).

7.4 Sedimentary structures of interdune areas and sand sheets

7.4.1 Interdune areas

Interdune areas can be divided into two broad categories: those which are dominated by deflation, and those which are dominated by deposition (Ahlbrandt & Fryberger 1981). Deflationary interdunes may be either essentially devoid of sandy sediment, exposing bedrock, clay, or other non-aeolian sediments, or they may expose truncated aeolian sands in the process of being removed by the wind. Sometimes a thin veneer of windblown sand in temporary storage overlies the non-aeolian sediments or truncated dune deposits. Such surfaces may also be partly covered by residual coarse grain lag deposits, shell pockets or shell pavements (Carter 1976) (Fig. 7.24), fulgurite fragments (Pye 1982e), ventifacts, and crop stones dropped by birds. Small migrating dunes, ripples, or fixed shadow dunes formed in the lee of isolated vegetation clumps may also be present (Lancaster & Teller 1988).

A further feature of some interdune areas is the presence of broad, low-amplitude transverse or chevron-shaped ridges composed of medium to fine sand (Fig. 6.44). Such features, which were termed *gegenwalle* ridges by Paul (1944), have several different origins, but many appear to reflect the control exerted by a seasonally fluctuating groundwater table on the level of deflation (Pye 1982a).

Depositional interdunes may be classified according to whether the surface is predominantly dry, damp, or wet (Kocurek 1981b). The main features found in each type are summarized in Table 7.2. Dry depositional interdune areas are normally dominated by deposits formed by migrating ripples and small dunes. Dry interdune sediments generally are less well sorted and contain more fines than adjacent dune sands (Ahlbrandt 1979). The deposits are commonly discontinuously laminated or structureless due to secondary processes (McKee & Bigarella 1979).

Figure 7.24 Shell pocket formed on a deflational coastal interdune surface, Burdekin Delta, North Queensland. Note also the small pyramidal shadow dunes and low-amplitude ripples formed in coarse sand lag deposits.

Table 7.2 Summary of sedimentary structures and other features characteristic of interdune deposits and the range of depositional conditions under which they form. (Modified after Kocurek 1981a).

Structure/feature	Dry interdune	Damp interdune	Wet interdune
wind ripples	—		
dune cross-strata	—	—	
lag grain surfaces	—	—	
deflation scours	—	—	
bioturbation structures	—	—	—
plant root structures	—	—	—
sand shadows and streaks	—	—	
adhesion laminae		—	
microtopography		—	
rain impact ripples		—	
brecciated laminae		—	—
adhesion ripples		—	—
adhesion warts		—	
evaporite structures		—	—
algal structures		—	—
fenestral porosity		—	—
contorted structures		—	—
rill marks		—	—
wavy laminae		—	—
wrinkle marks			—
channels			—
small deltas			—
water ripples			—
subaqueous cross-strata			—

Normal wind ripple deposits in depositional interdune areas are of two main types: (a) thin (1–50 mm), more or less continuous, commonly inversely graded laminae with few preserved ripple foresets which represent subcritically climbing translatent strata, and (b) discontinuous, undulating ripple form deposits, formed by ripple migration under conditions of restricted sand supply from upwind (Korcurek 1981b). Clemmensen (1989) reported coarse-grained ripple strata up to 9 m thick in inter-draa deposits of the Lower Permian Yellow Sands, northeast England.

On damp interdune surfaces, adhesion structures commonly develop (Hummel & Kocurek 1984, Kocurek & Fielder 1982) (see Section 6.2.4). A partial cover of vegetation may also be present (Fig. 7.25), resulting in the formation of plant root structures and related bioturbation structures. The surface of damp interdune areas is often marked by the presence of microtopography, reflecting the irregular distribution of deflation and deposition. Since the degree of surface moisture present varies over time, most damp interdune deposits experience repeated cycles of local scouring and deposition by ripples or small dunes (Simpson & Loope 1985).

Wet interdune areas are frequently occupied by temporary lakes (Fig. 7.26). Consequently, water ripple structures and beach deposits are well represented. Both wave ripples and current ripple structures may be present. Around the margins of such depressions fluvial scour channels and small-scale deltaic depositional lobes consisting

Figure 7.25 Seasonally damp to wet interdune area, Cape Flattery, North Queensland. Note the water-filled channel, the presence of reeds, and the generally irregular surface microtopography.

Figure 7.26 Intra-dune lake occupying deflation trough, Cape Flattery, North Queensland.

Figure 7.27 Section exposed by wave erosion through interdune deposits, Formby Point, Merseyside, UK. Horizontally bedded organic-rich sands over an erosional hummocky surface formed by deflation under dry conditions. A second hummocky deflation surface is overlain by a 10 cm thick dune slack peat, representing slow accretion under wet conditions. Mottling and weak iron oxyhydroxide cementation around decayed root channels extends up to 1 m below the surface of the buried dune slack peat. The top 20 cm of the exposure consists of very recent windblown sand deposited during transgression of a foredune ridge over the dune slack deposits.

of reworked aeolian sand may also occur (Talbot & Williams 1978, McKee & Bigarella 1979).

In humid regions, organic-rich soil horizons (dune slack peats) commonly develop when the water level in the interdune area drops below the sand surface (Fig. 7.27). During high water stands, freshwater peats and organic-rich lake deposits accumulate (Ahlbrandt & Fryberger 1981).

In arid regions the waters in the interdune depression may become highly saline owing to seasonal evaporation, resulting in the formation of evaporite crusts and thrust polygon structures. Algal mats, associated with vesicular sand layers formed by release of gas from the algae, may occur with the evaporites (Fryberger *et al.* 1988).

The hydrological regime of arid interdune areas undergoes rapid changes following a rainfall event. Surface runoff and infiltration through the dunes is responsible for washing large amounts of fines into the interdune depressions, where a temporary lake may form. The fines settle out over a period of a few days or weeks, eventually forming a planar silt/clay crust or low-angle clay drape over the inter-dune surface and footslopes of the dunes (Fig. 7.28) (Langford 1989, Langford & Chan 1989). Some suspended fine material may also infiltrate laterally into the basal dune deposits (Pye & Tsoar 1987). Polygonal crack patterns and curls are characteristic features of the interdune clay crusts.

Figure 7.28 Accumulation of water-transported silt and clay in an interdune depression, Stovepipe Wells, Death Valley, California.

7.4.2 Extra-dune sand sheets

Only limited information is available concerning the internal structures in warm climate sand sheets. Breed *et al.* (1987) reported that the Selima Sand Sheet of the eastern Sahara is composed almost entirely of horizontal, near-parallel laminae. No evidence was found of climbing ripple foresets or truncated dune cross-strata, even though giant surface ripples were noted in some areas and the area is crossed periodically by migrating barchan trains. The surface of the sand sheet clearly represents a condition of slow net sand deposition, although it is uncertain whether accretion is active at the present day and, if so, whether it is continuous or episodic. The horizontal laminae display inverse size grading (Breed *et al.* 1987), suggestive of deposition by migrating large-wavelength ripples. The surface of the Selima Sand Sheet is essentially free of vegetation and no roots or bioturbation structures were observed.

By contrast Kocurek & Nielson (1986) reported that both climbing ripple structures and bioturbation structures are well represented in the dune-marginal sand sheets of the

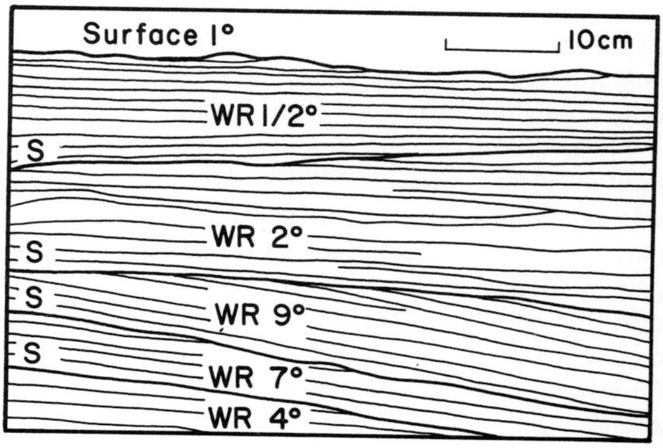

Figure 7.29 Internal structure of a warm climate extradune sand sheet, Algodones, California, revealed by trenching. Note units of wind ripple laminae (WR) with low-angle apparent dips. S denotes truncating surface separating units of diversely oriented wind ripple laminae. After Kocurek & Nielson (1986).

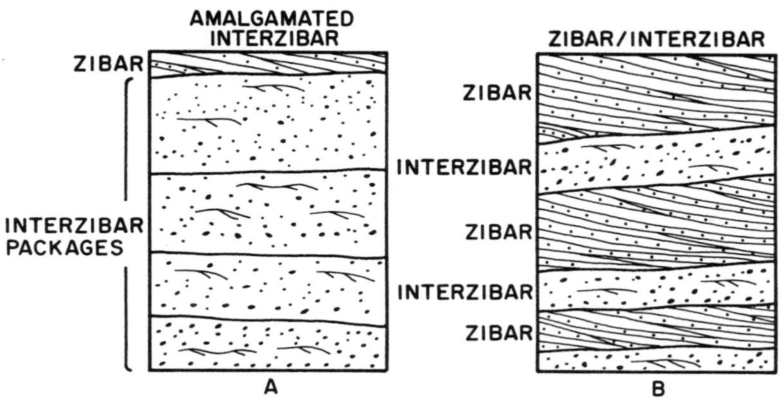

Figure 7.30 Idealized depositional sequences of (A) amalgamated interzibar deposits and (B) zibar/interzibar deposits. The tangential bottom contact of zibar lee-face wind ripple laminae and their sharp truncation by coarser interzibar wind ripple laminae are distinctive characteristics of the zibar/interzibar sequence. Zibar wind ripple laminae are distinctive characteristics of the zibar/interzibar sequence. Zibar wind ripple laminae dip less than 15° and may be slightly concave. Interzibar wind ripple laminae typically dip less than 5°. The amalgamated interzibar sequence lacks distinctive structural or textural characteristics and thus cannot be distinguished from the deposits of coarse-grained sand sheets devoid of zibars. (After Nielson & Kocurek 1986).

Algodones dunefield, California (Fig. 7.29). In this area, the surfaces of the sand sheets are covered by zibar which have an amplitude of 2–3 m and an average spacing of 60 m. Since zibar lack slip faces, the internal structures represent climbing ripple structures whose foreset laminae dip at angles of < 15° (Nielson & Kocurek 1986) (Fig. 7.30). The widespread occurrence of root mottling and bioturbation structures reflects the existence of a partial vegetation cover.

In summary, warm climate aeolian sand sheets are dominated by low-angle or horizontally laminated deposits which are laterally more extensive than interdune deposits. They sometimes contain bioturbation structures but lack the lateral systematic variations typical of interdune deposits (Kocurek 1986).

7.5 Niveo-aeolian deposits and cryogenic structures in cold-climate dunes

Niveo-aeolian deposits are mixed deposits of windblown snow and sand, silt, or other detritus which form in climates that are cold at least seasonally (Pissart *et al.* 1977, Cailleux 1978, McKenna-Neuman & Gilbert 1986, Koster & Dijkmans 1988, Koster 1988, Dijkmans 1990). They are common at high latitudes but also occur in mid-latitude, high-altitude cold regions (Steidtmann 1973, 1982, Ahlbrandt & Andrews 1978, Ballantyne & Whittington 1987). Most are annual, with all the snow dissipating in summer, but in exceptionally cold environments, such as Antarctica, perennial niveo-aeolian deposits also occur (Cailleux 1967, Calkin & Rutford 1974).

Niveo-aeolian deposits normally form irregular patches or blankets rather than well formed dunes. However, interbedded mixtures of aeolian dune and niveo-aeolian deposits occur in some areas (Koster & Dijkmans 1988).

Many niveo-aeolian deposits consist of alternating layers of relatively pure sand and snow, each layer typically being 0.2–60 cm thick. Others contain mixed layers of sand

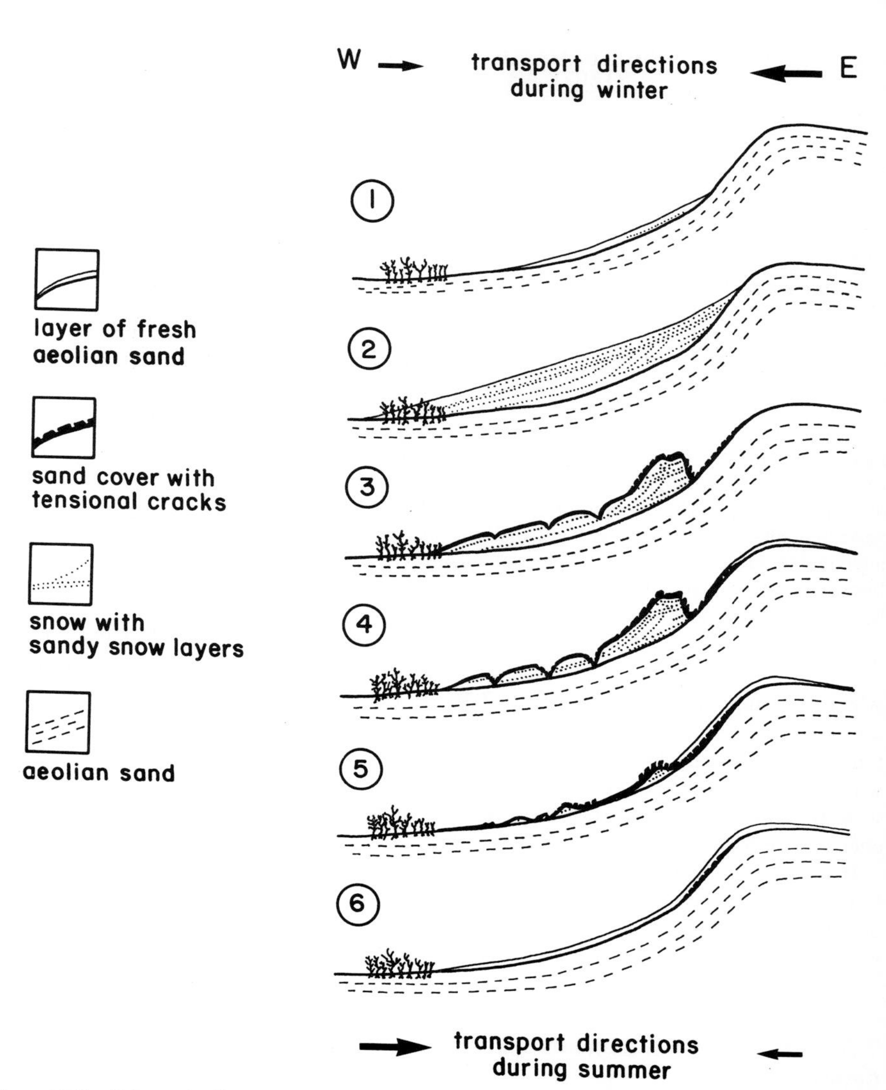

Figure 7.31 Schematic diagram showing the formation of niveo-aeolian deposits in winter (1 and 2) and denivation forms in springs and summer (3–6) on the lee side of a 5–10 m high dune in the northwestern part of the Great Kobuk Sand Dunes, Alaska. Large and small arrows indicate prevailing and secondary wind directions, respectively. (After Koster & Dijkmans 1988).

and snow. In summer, the surface of the deposits is always formed by a layer of sand 20–30 cm thick, any snow originally present having melted or sublimated.

The surfaces of fresh niveo-aeolian deposits are frequently covered by mixed sand–snow ripples. In summer, *denivation forms* are produced by melting or sublimation of the snow. Features which are produced include sand pellets and sand rolls, surface cones and hillocks with a surface of cracked sand, and sink holes surrounded by tensional cracks (Cailleux 1978, Koster & Dijkmans 1988).

Figure 7.31 shows a schematic representation of the development of niveo-aeolian and denivation features, based on field observations in the Great Kokuk Sand Dunes,

Alaska (Koster & Dijkmans 1988). Each winter a niveo-aeolian bed is formed on the lee side of a dune (Fig. 7.31, 1 and 2). During the spring and early summer the snow melts and a number of denivation features develop sequentially. Tensional cracks and slump features are prominent (Fig. 7.31, 3–5). After complete melting of the snow (Fig. 7.31, 6), the sand cover dries out and is largely reworked by the wind. Since the advance rate of the dunes is low, the preservation potential of the denivation features is low.

Observations in the Sondre Strømfjord area, western Greenland, indicated that deposition of niveo-aeolian beds occurs mainly in the early winter, before a continuous snow cover develops (Dijkmans 1990). On low-angle inclined surfaces niveo-aeolian deposits are usually reworked after the sand dries in the following summer. However, dune slip face denivation structures are sometimes preserved as deformed cross-strata. These deformation structures are similar to those found in wet sand in warm climates (McKee *et al.*, 1971), but the alternation of packets of deformed and undeformed laminae may be diagnostic of cold climate dune deposits (Dijkmans 1990). Denivation deposits also commonly have a relatively porous structure due to entrapment of air following melting and sublimation of the snow (Dijkmans & Mucher 1989).

Any aeolian deposit formed in cold climates, including dune sands, sand sheets, and loess, may experience syn- and post-depositional cryogenic deformation due to the development of seasonal or perennial permafrost (e.g. Ahlbrandt & Andrews 1978, Ahlbrandt & Fryberger 1980, Ruegg 1983, Schwan 1988). Features which may be produced include frost cracks, ice-wedge casts, sand wedges, involutions, and infiltration structures governed by the position of the frozen layer.

Periglacial aeolian deposits have been identified in the Permo-Carboniferous of South Australia (Williams *et al.* 1987) and in the Late Proterozoic of Western Mali (Deynoux *et al.* 1989).

8
Post-depositional modification of dune sands

8.1 Introduction

The primary textural and mineralogical characteristics of aeolian dune sands are commonly modified significantly by post-depositional changes. These include physical reworking by surface processes, bioturbation, compaction, weathering, pedogenesis, and cementation (Table 8.1). These changes exert a significant influence on the residual permeability and porosity characteristics of the sands, on their engineering behaviour, and on their potential as economic resources (Pye 1983f).

The nature and relative importance of different post-depositional changes are governed principally by the accumulation rate and initial mineralogical composition of the sands, by the climatic conditions at the deposition site, and by hydrological and geochemical changes experienced during burial and uplift. In the following discussion attention is focused mainly on mineralogical and textural changes which result from sub-aerial weathering and early diagenesis (i.e. those which take place at depths of up to a few tens of metres below the surface). Examples of later diagenetic changes during deep burial and uplift of aeolian deposits were described by Waugh (1970), Glennie *et al.* (1978) and Pye & Krinsley (1986).

Table 8.1 Principal early post-depositional changes affecting aeolian dune sands.

addition of allochthonous components (airborne dust, phytoliths, salts)
crust formation by raindrop impact and growth of algal/fungal mats
erosion by surface wash, soil creep, slumping and flowage of saturated sand
bioturbation
compaction
gullying
physical and chemical weathering
vertical translocation of fine particles and dissolved ions
lateral movement of particles and ions due to subsurface water movement
formation of soil profiles and catenas
vadose zone cementation
formation of rhizoconcretions
cementation in the zone of capillary rise
phreatic zone cementation
phreatic zone intrastratal dissolution/replacement of detrital minerals

8.2 Denudation by rain splash, surface wash, soil creep, and gullying

Water erosion is a process of major importance which modifies the surface of stabilized dunes, particularly in humid regions (Rutin 1983, Bridge & Ross 1983, Thompson 1983, Thompson & Bowman 1984, Jungerius & van der Meulen 1988). Evidence of its effectiveness is provided by sand and organic debris trapped on the upslope side of obstacles, such as logs and clumps of vegetation, by emergent tree roots, pebble-capped sand pedestals, gullies, and associated fan deposits. Measurements on dune slopes at Cooloola, southeast Queensland, showed that downslope sand transport is accomplished both by rain splash and by surface wash at times when the infiltration capacity of the sand is exceeded (Bridge & Ross 1983). Rates of sand movement were found to be determined by the degree of water repellence of the surface sand, the nature and amount of vegetation cover, and the amount and intensity of rainfall. Slope angle was found to have a relatively minor effect on transport by surface wash and rain splash.

The water repellence of surface dune sands depends on the proportion and distribution of fine particles in the surface sand layers, the absence of an abundant surface microflora, and the thickness of any organic litter layer. Like other loamy soils, bare dune sands which contain fines are particularly prone to crusting. Three processes are involved (Le Bissonnais *et al.* 1989): breakdown of fine aggregates, mechanical compaction by impacting raindrops, and formation of clay bridges which cement larger particles following repeated wetting and drying cycles.

Surface sands with or without fines may develop algal or fungal crusts (Bond 1964, Bond & Harris, 1964, van den Ancker *et al.* 1985). Pluis & de Winder (1989) observed that, in coastal dune blowouts of The Netherlands, development of algal mats was initiated during wet periods of the year by colonization of cyanobacteria (mainly *Oscillatoria* and *Microcoleus*), and was reinforced by subsequent colonization by the green alga *Klebsormidium flaccidum*. The effect of algal and fungal crusts is to increase the resistance of the surface to wind erosion and to increase the chances of colonization by higher plants. However, such crusts commonly increase the surface water repellence and the likelihood of erosion by surface wash (Talbot & Williams 1978, Rutin 1983, Jungerius & de Jong 1989).

The overall effect of rain splash and surface wash is to move grains downslope under the influence of gravity, thereby rounding dune crests and forming thick colluvial mantles on the lower dune slopes and in the interdune swales. Where dunes are affected by bioturbation, or are subject to frost action, soil creep may also contribute significantly to the downslope movement of grains into the swales.

Progressive physical degradation of dunes with increasing age is clearly seen in the coastal dunefields of eastern Australia. At Cooloola in southeast Queensland, Thompson (1983) noted that the three youngest dune systems, which are of Holocene age, show evidence of modification by raindrop impact, surface wash, and gully erosion, but the basic forms of the parabolic dunes are still readily recognizable. Parabolic dunes of late Pleistocene age are much more degraded, with widespread dissection of the original dune forms by headward gully erosion and formation of well developed stream networks. The oldest Pleistocene dunes in the area are characterized by low whaleback forms which show little evidence of the original aeolian morphology.

8.3 Near-surface compaction

Most dune sands have initial porosities ranging from 30 to 50%, depending on the size distribution, sorting, shape, and packing arrangement of the grains. Following deposition, readjustment of the grains into a tighter packing arrangement may be induced by earthquake shocks, overburden pressure, moisture saturation, and selective weathering of unstable minerals leading to secondary porosity formation. Overburden pressures in the near-surface environment are relatively small compared with those experienced during deep burial, but may nevertheless be sufficient to cause some grain rotation, sliding, and plastic deformation of softer constituents. Highly angular or very brittle grains may also experience breakage at points of inter-grain contact. The silt and clay-size particles produced then migrate into the pore spaces between larger grains.

In practice, the effects of compaction are difficult to separate from settlement due to volume loss induced by chemical leaching and from lowering due to erosion by surface processes.

8.4 Addition of allochthonous components

Fine allochthonous sediment components may be introduced into a dune sand body by deposition of airborne dust, by surface wash from higher ground, by laterally migrating ground water, and by release of opal phytoliths from dune vegetation. In the case of unvegetated dunes, fines settle only temporarily during periods of sand stability and are periodically injected back into the atmosphere to be carried away in suspension. The average content of fines in active dune sands therefore rarely exceeds 1%, although thin clay crusts and drapes may form in interdune hollows where rain water accumulates. In the case of stabilized dunes, deposited fines are able to accumulate and form distinct silty horizons near the sand surface.

The importance of airborne dust additions in contributing to the post-depositional modification of aeolian sands has been demonstrated by several studies (Lutz 1941, Olson 1958c, Yaalon & Ganor 1973, Sidhu 1977, Walker 1979, Danin & Yaalon 1982). Tsoar & Møller (1986) found that a stabilized linear dune in the northern Negev contained 10% silt and clay at the surface and 5–7% fines at a depth of 40 cm (Fig. 8.1).

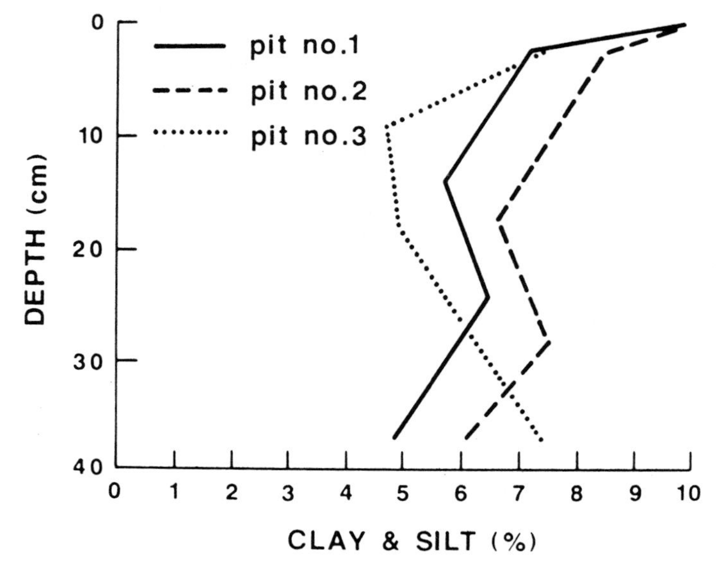

Figure 8.1 Variations in silt and clay content with depth in a northern Negev linear dune. (After Tsoar & Møller 1986).

Breed & Breed (1979) also reported that vegetated linear dunes in Rice Valley, California, contained 4.6–6% fines, and in central Australia up to 5% fines. Dunes which contain > 30% fines were reported from northwest India and Pakistan by Goudie *et al.* (1973).

The depth of dust infiltration depends partly on the amount, frequency, and intensity of rainfall, on the pore size distribution of the dune sand, and on the size of the deposited dust. Leaching experiments have shown that fine silt and clay is more easily transported through sand columns than medium and coarse silt particles, and that the translocation of silt becomes more difficult with decreasing sand size (Wright & Foss 1968). Naturally deposited dust often displays mineralogical variation with grain size, such that the larger silt particles composed mainly of quartz, feldspar and calcite are retained closer to the sand surface than the finer dust fractions which are dominated by clay minerals and micas (Pye & Tsoar 1987). The mineralogical composition of airborne dust varies considerably between different regions depending on the nature of exposed source rocks and sediments (Pye 1987).

Relatively few field data are available concerning the infiltration of moisture and fine particles in natural dune sands. Dincer *et al.* (1974) found that 1 mm of rain penetrated to a depth of 7 mm in well graded dune sand with a mean size of 150 μm and to a depth of 20 mm in sand with a mean diameter of 300 μm. Observations in parts of the northern Negev which receive approximately 100 mm of annual rainfall suggest that the maximum wetting depth in dune sand does not normally exceed 100 cm (Tsoar & Zohar 1985). By contrast, in humid tropical and subtropical dunefields, such as those in eastern Australia which receive > 1800 mm of annual rainfall, the wetting depth may exceed 20 m (Pye 1980a, Thompson 1983).

Thin horizontal bands of fine material, termed *textural subsoil lamellae* (Dijkerman *et al.* 1967), are commonly observed in soil profiles developed on dune sand and other sandy parent materials. Laboratory experiments suggest that a majority of such bands are formed by translocation of fines (Robinson & Rich 1960, Dijkerman *et al.* 1967, Bond 1976). Deposition of the fines appears to be enhanced when the downward-moving fines encounter a finer textured sand layer, but the possibility of other, possibly electrochemical, controls on deposition has not been ruled out.

8.5 Weathering and pedogenesis of siliceous dune sands

8.5.1 Leaching of soluble salts and carbonates

Typical freshly deposited siliciclastic dune sands are composed predominantly of quartz with smaller amounts of feldspars, heavy minerals, biogenic carbonate fragments, and soluble salts. In humid regions leaching of salts occurs very quickly, taking only a few months or years. Leaching rates of shell carbonate are also high on dunes where organic acids are produced from decaying vegetation. On English coastal dunes, for example, Salisbury (1922, 1925) found that almost complete decalcification of the surface sands occurred within about 300 years. In the Lake Michigan sand dunes, Olson (1958c) found that about 75% of the original carbonate content of 1.4% $CaCO_3$ in the surface 10 cm of sand was lost in 300 years. He concluded that it takes about 1000 years to leach carbonate fully out of the uppermost 2 m of sand. The rate of decalcification is controlled partly by the rate of shell dissolution, which in turn is governed by its particle size, the amount of rainfall, soil pH, and the rate of surface carbonate replenishment from allochthonous sources.

In arid regions, substantial quantities of airborne salts may also accumulate in the near-surface layers of stable and semi-stable dune sands. The salts can be introduced both as dissolved species in rain or fog and as solid particles (e.g. Schroeder 1985). Once deposited, the salts show differing susceptibilities to leaching, with $Cl^- > SO_4^{2-} > HCO_3^-$ (Yaalon 1964). In areas which receive less than 100 mm of rainfall the salts are retained within the upper 1 m of sand. The high salinity and alkalinity which they induce enhance silica dissolution and reprecipitation, leading to the formation of scaly, impure alumino-silicate grain coatings on framework sand grains (Pye & Tsoar 1987).

8.5.2 Chemical weathering of silicates and oxides

Minerals vary in their degree of thermodynamic stability under earth surface conditions and hence the rates at which they may be expected to break down to form more stable products. Silicates with relatively few Si—O bonds, such as pyroxenes and amphiboles, break down much more quickly than minerals with relatively large numbers of Si—O bonds, such as quartz (Loughnan 1969, Carroll 1970). The actual rates of mineral decomposition reactions are controlled by a range of environmental factors including interstitial porewater chemistry, particle size, temperature, and the rate at which weathering products are removed from the system (McClelland 1950, Berner 1978).

Under very arid conditions, rates of near-surface weathering are slow, since both hydrolysis reactions and flushing of weathering products from grain surfaces are limited by moisture availability. With increasing moisture availability, hydrolysis becomes more rapid and leaching more effective. Under very humid conditions, and especially where organic acids are abundant, feldspars and ferromagnesian heavy minerals are destroyed very rapidly, and grains of quartz, kaolinite, and heavy minerals such as zircon and ilmenite may be attacked.

8.5.2 Heavy minerals

Experiments by Williams & Yaalon (1977) 'using Soxhlet columns' demonstrated that leaching of dune sand by hot and cold water under free-draining conditions is capable of causing significant alterations of some heavy minerals (mainly hornblende) within a period of 3 months. Leaching resulted in notable changes in surface texture and loss of Na, Ca, Mg, K, and Al ions in solution. Fe released by leaching was observed to precipitate within the sediment column as a thin oxide coating on the quartz grains.

Walker (1979) observed pitting on the surfaces of augite and hornblende grains from Libyan dune sands which he attributed to *in situ* weathering. The grains showed a progressive increase in the degree of alteration of both augite and hornblende with distance from the sand source (i.e. with inferred dune age).

In aeolian sands of semi-arid southeast India, feldspars and almandine garnet have been destroyed by weathering to form authigenic haematite, kaolinite, and illite (Gardner 1981, 1983a). Pyroxenes and amphiboles were initially present in only low concentrations in these sands. Under the more humid and intense acid leaching conditions of tropical North Queensland, chemical decomposition of ilmenite, zircon, and rutile has occurred in the A horizons of giant podsol soil profiles, resulting in a relative enrichment of tourmaline, andalusite, staurolite, and anatase which are more stable under these conditions (Pye 1983g).

8.5.2.2 Feldspars Although feldspars are common constituents of aeolian sands, no detailed studies of feldspar alteration in dunes have been published. However, evidence from other weathering studies suggests that potassium feldspars, such as orthoclase and microcline, should be more stable in dune weathering environments than plagioclase feldspars (Wollast 1967, James *et al.* 1981). Preliminary observations on southern

Queensland coastal dune sands suggested that orthoclase is the dominant feldspar present, and that its abundance progressively declines in older, more deeply weathered sand units (Thompson & Bowman 1984). Feldspar is a rare constituent in many of the North Queensland dune formations which have experienced long periods of podsolization and several episodes of reworking during the Quaternary (Pye 1983g).

8.5.2.3 Quartz A number of studies have shown that, although quartz is a mineral which is relatively resistant to weathering, some varieties are prone to break down during post-depositional weathering. Little *et al.* (1978) recognized four different quartz grain types in the coastal dunes of Fraser Island, southern Queensland, which they described as *clear and unetched, milky, saccharoidal*, and *microgranular*. The saccharoidal and microgranular types were found to be highly weathered with deeply etched surfaces. Many could be completely disintegrated by gentle pressure. Some etching was also observed on the surfaces of the milky grains. Little *et al.* (1978) also noted that the proportion of highly weathered grains increased with dune age, and that within any given generation of dunes the proportion of weathered grains is highest in the B and C horizons. They were uncertain, however, whether this reflected the fact that most of the weathered grains in the A horizons had been destroyed or whether the more moist conditions in the B/C horizons preferentially favoured the formation of weathered grains. Observations by Pye (1983a) in the similarly podsolized dunes of North Queensland indicated that granular disintegration is most intense in the A1 and A2 horizons. Many of the weathered grains in these horizons disintegrate completely into silt particles which are translocated down the profile and are deposited at the top of the B horizon. A similar conclusion was reached by Thompson & Bowman (1984) based on observations at Cooloola in southern Queensland.

The evidence currently available suggests that polycrystalline and strained monocrystalline quartz grains are more susceptible to post-depositional disintegration than unstrained monocrystalline quartz grains (Pye 1983a). However, virtually all quartz grains contain microfractures (Moss *et al.* 1973, Moss & Green 1975) and lines of crystallographic weakness which can be exploited, given sufficient time and a sufficiently aggressive weathering environment. SEM examination of weathered grains by Pye (1983a) showed that grain disintegration takes place by dissolution of silica along incipient fractures and crystallographic flaws such as chains of fluid inclusions (Figs 8.2 & 8.3). Although the solubility of quartz in pure water is low at pH < 9 (Morey *et al.* 1962, Siever 1962, Iler 1979), it is increased significantly by the presence of organic acids (van der Waals 1967, Crook 1968). At present, however, the relationship between silica dissolution and crack propagation in natural quartz grains is not fully understood.

8.5.3 Physical weathering processes

Laboratory experimental studies have shown that crystallization, hydration, and thermal expansion of salt crystals can cause mechanical disintegration of sand-size particles under simulated warm desert conditions (Goudie *et al.* 1979, Pye & Sperling 1983). Feldspar and mica grains are more susceptible to salt damage than quartz grains, apparently because of their better developed cleavage, and first-cycle quartz grains containing many inherent flaws are more susceptible to breakdown than mature quartz grains which have experienced several sedimentary cycles. Wetting and drying alone were found by Pye & Sperling (1983) to be much less effective in accomplishing breakdown than combined wetting and drying in the presence of salts. Sodium sulphate, magnesium sulphate, sodium carbonate, and calcium chloride are the most destructive salts (Goudie *et al.* 1970, Goudie 1974, 1985). Halite and gypsum, which are the two

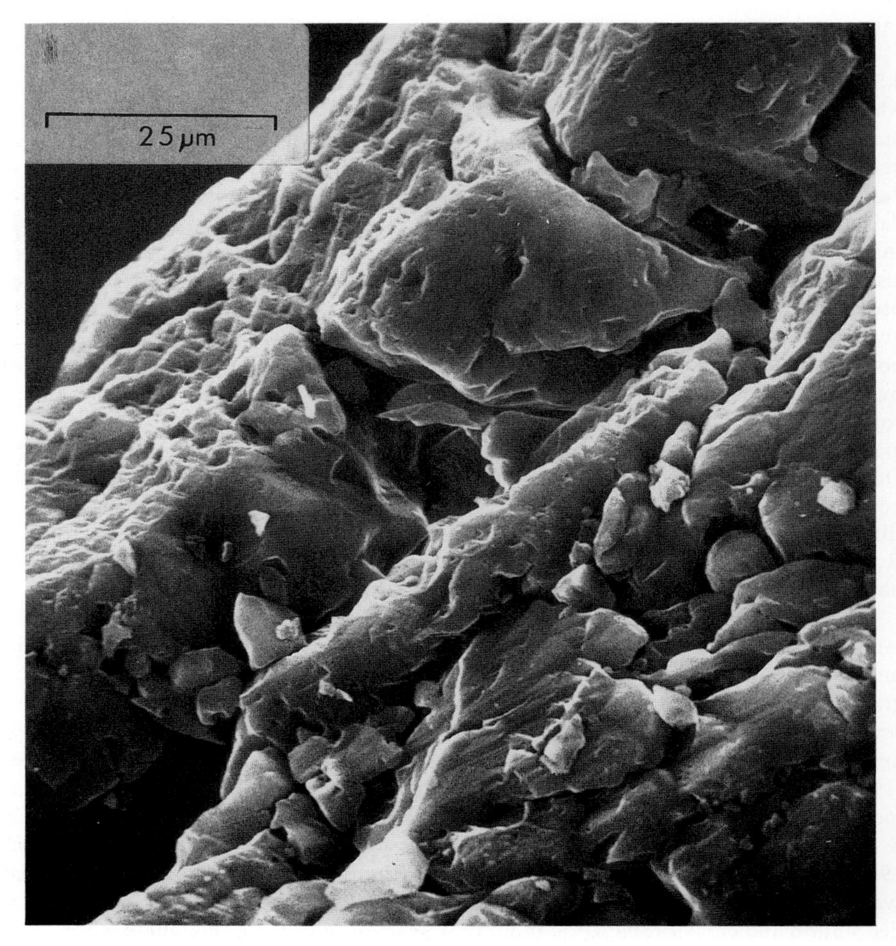

Figure 8.2 Scanning electron micrograph showing the surface texture of a weathered quartz grain from a podsolized dune, North Queensland. Dissolution of silica along lines of crystallographic weakness has broken the grain surface up into a series of residual silt-size particles.

most common salts founds in nature, are relatively less destructive but still effective. Salt weathering is potentially a significant process affecting desert dune sands in the zone of groundwater capillary rise on playa margins, although its precise effects on grain size distributions remain to be documented by field evidence.

Laboratory experiments have also shown that frost action can induce fracture of first-cycle plutonic sand grains (Moss *et al.* 1981), and there is considerable field evidence for the breakdown of sand grains under cryogenic conditions (Zeuner 1949, St. Arnaud & Whiteside 1963, Konischev 1982). However, the detailed effects of frost action on natural dune sands remain to be documented.

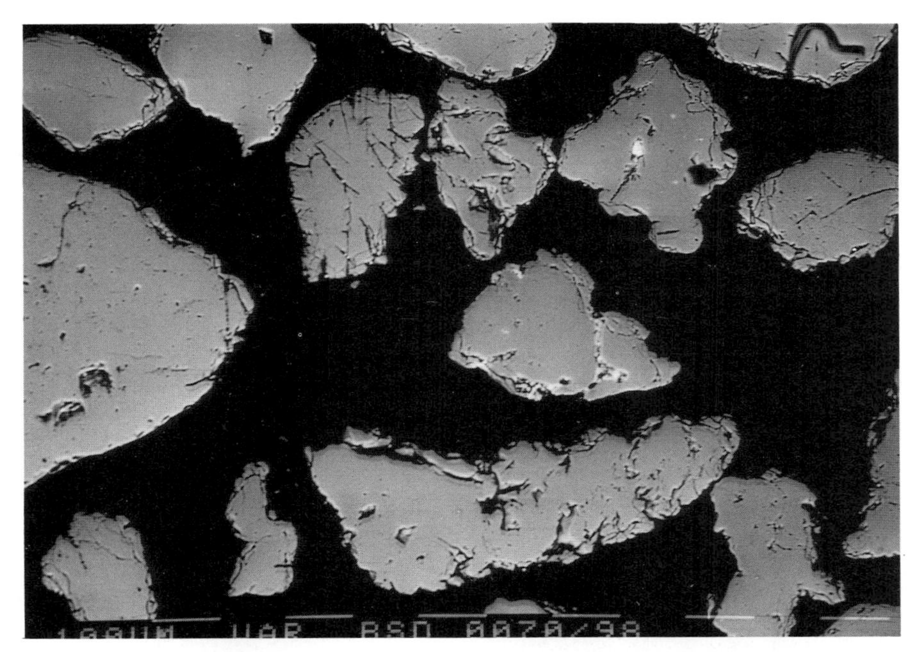

Figure 8.3 Backscattered scanning electron micrograph of a polished section showing a number of quartz grains from the A horzion of a podsolized dune, North Queensland. Solution of silica along cracks and crystallographic flaws is clearly seen. Scale bars = 100 μm.

8.5.4 Chemical weathering and reddening of siliciclastic dune sands

Red and orange dune sands occur widely in both coastal and continental settings. The origin of the red coloration has been widely discussed in terms of its origin, palaeoenvironmental significance, and use as a means of dating sand dunes (Norris 1969, Folk 1976b, Walker 1979, Gardner & Pye 1981). Some red sands have clearly inherited their colour from red parent sediments or rocks (Anton & Ince 1986), but the question of whether others become redder during downwind transport remains controversial (Wopfner & Twidale 1967, Norris 1969, Walker 1979). During the movement of very large active dunes, a substantial part of the sand body may remain at rest for several tens or even hundreds of years. During this time iron oxides may form by chemical alteration of detrital iron-bearing minerals or by alteration of infiltrated airborne dust (Walker 1976, 1979). However, Gardner & Pye (1981) have pointed out that the degree of redness which can be attained during transport is limited by abrasion. Anton & Ince (1986) and Wasson (1983a) have also concluded that the concept of progressive reddening with time is not fully supported by field evidence. Some reddening is clearly possible under arid conditions, but it occurs much more rapidly in destabilized sands under semi-arid and humid conditions. Many of the currently active red sand areas were formerly stabilized during more humid periods of the Quaternary, when they experienced pedogenetic rubefaction (Gardner & Pye 1981).

Pye (1983h) recognized four broad groups of pedogenetic red beds: (a) red latosols, which include all leached red soils of the tropics except those with a bleached A2 horizon; (b) red podsols, which occur in humid regions and are characterized by bleached A horizons overlying a reddened B horizon; (c) red desert soils, which generally show weak horizon differentiation; and (d) red Mediterranean soils which typically have a reddish brown A horizon and reddened clay-rich B horizon with calcareous nodules in the lower part. Variants of all four types are developed on stabilized dunes in different parts of the world.

The reddish pigment in modern dune sands is often a mixture of poorly crystallized iron (III) oxides and oxyhydroxides. Well crystallized haematite is rare unless the pigment is derived from older rocks. More commonly, the pigment is a mixture of poorly crystallized haematite and goethite. Maghemite, lepidocrocite, and ferrihydrite are also sometimes present (Pye 1983g, h).

Figure 8.4 illustrates the typical surface texture of a reddened sand grain from a podsolized dune at Cape Flattery, North Queensland. The surface is covered by a large number of small kaolinite flakes and granular aggregates of iron oxide/oxyhydroxide. Si, Al, and Fe are the only important elements present in the surface coating (Fig. 8.5). By contrast, Fig. 8.6 shows a much smoother type of surface texture on a reddened quartz

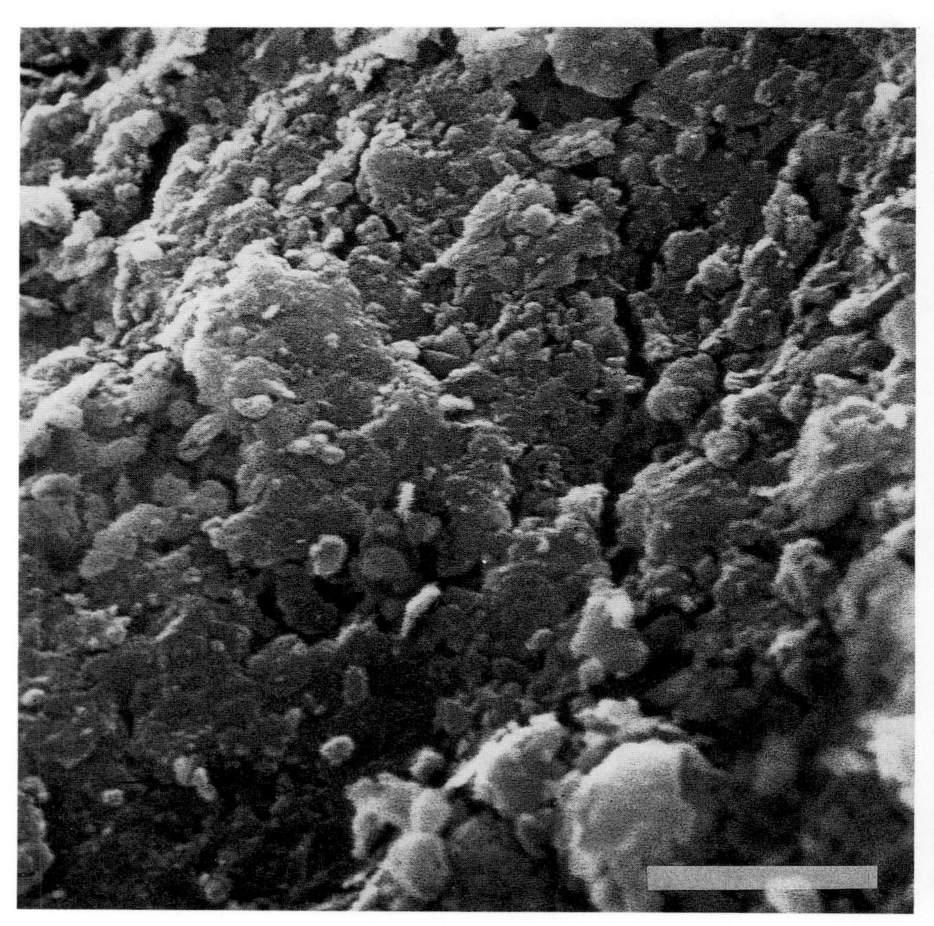

Figure 8.4 Scanning electron micrograph showing the surface texture of a reddened grain from the B horizon of a podsolized dune in North Queensland. The coating, which is more than 50 μm thick is composed mainly of flakes of kaolinite and iron oxide/oxyhydroxide. Scale bar = 10 μm.

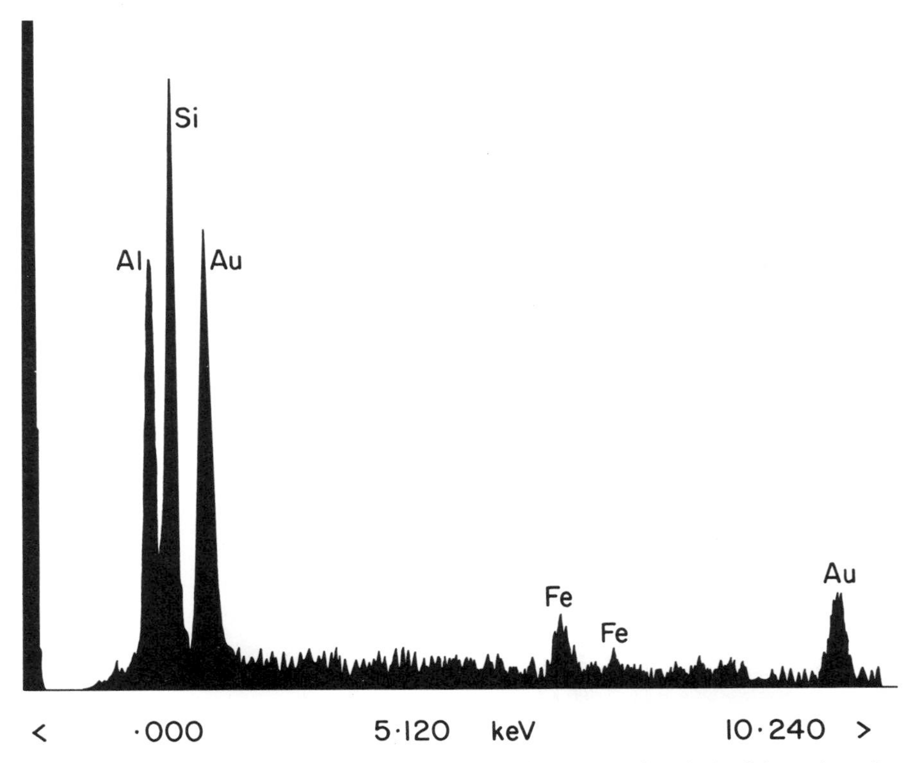

Figure 8.5 Energy-dispersive X-ray spectrum obtained from an areal analysis of the grain surface shown in Figure 8.4 (gold-coated specimen).

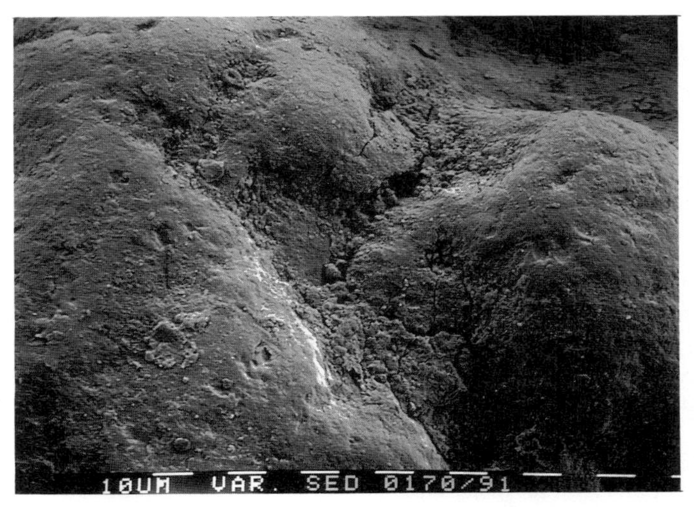

Figure 8.6 Scanning electron micrograph showing the surface texture of a reddened quartz dune sand grain from northern Sinai. The surface is coated with amorphous aluminosilicate material, calcium carbonate and iron oxide/oxyhydroxide. Scale bars = 10 μm.

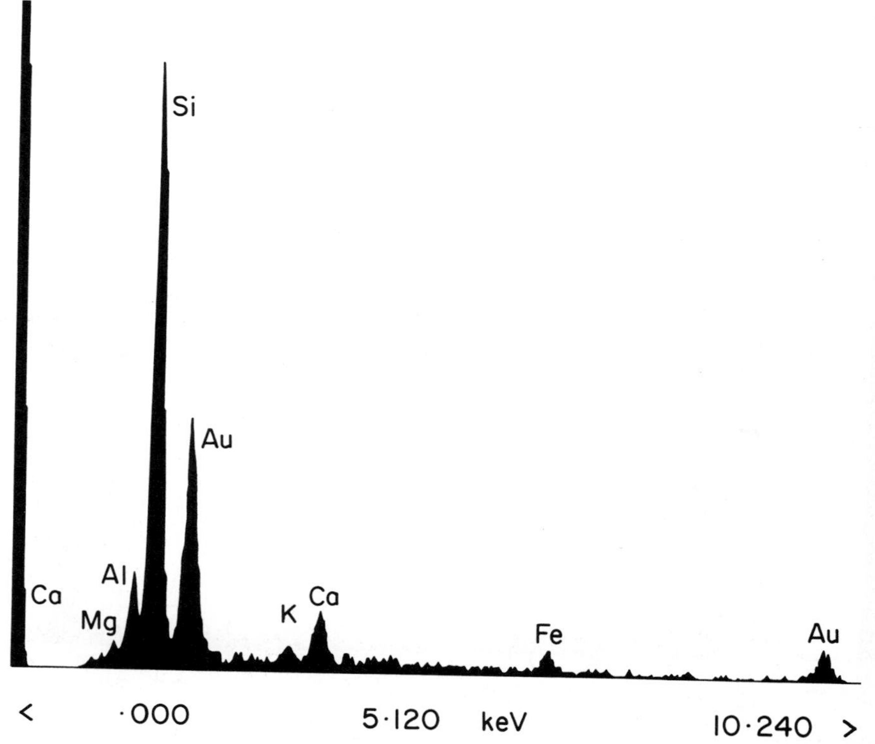

Figure 8.7 Energy-dispersive X-ray spectrum obtained from an areal analysis of the grain surface shown in Figure 8.6 (gold-coated specimen).

grain from a stabilized dune in northern Sinai. The coating on this grain consists of X-ray amorphous aluminosilicate material mixed with traces of illite–smectite clay, calcium carbonate, and iron oxide/oxyhydroxide. This is reflected by the presence of Mg, Al, Ca, K, and Fe (Fig. 8.7).

Some red podsolic weathering profiles contain indurated petroferric layers. In simple profiles there is a single layer near the top of the B horizon, but complex profiles may contain multiple petroferric layers at different levels. In North Queensland dunes the petroferric layers are sometimes > 15 cm thick and are laterally extensive. They often show internal colour banding which reflects variations in the iron oxide cement mineralogy. The two most common types of cement are dense granular haematite (Fig. 8.8a) and fibrous aluminous goethite (Fig. 8.8b). Many of the cemented grains show evidence of marginal corrosion by the cement (Fig. 8.9) and some show signs of complete disintegration and replacement (Fig. 8.10).

8.5.5 Silica coatings and cementation

Folk (1978) observed that many dune sand grains in the Simpson Desert of Australia have a 'greasy' surface texture due to the presence of a silica-rich surface coating. He suggested that this coating, which he termed a 'turtle-skin silica coat', is formed by dissolution of opal phytoliths by alkaline dew and reprecipitation of amorphous silica on the surfaces of quartz grains. Other authors have also noted that textural features indicative of aeolian abrasion are relatively rare on desert sand grains owing to the rapidity with which the surfaces are chemically altered in the presence of alkaline moisture (Pye & Tsoar 1987).

Dissolution of silica on quartz grain surfaces also occurs in soil profiles where organic acids are abundant (Crook 1968, Cleary & Conolly 1971). In the podsolized dunes of North Queensland, some of this silica was observed to be reprecipitated in the lower part of the A2 horizon, both as grain coatings and as intergranular cement at points of grain contact (Pye 1983a, f). However, no layers were sufficiently indurated to be termed silcretes.

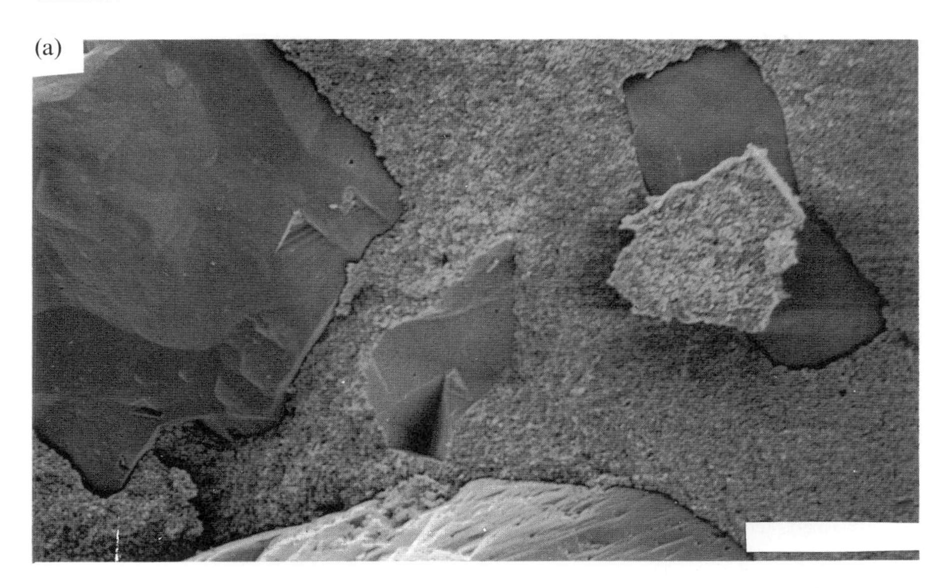

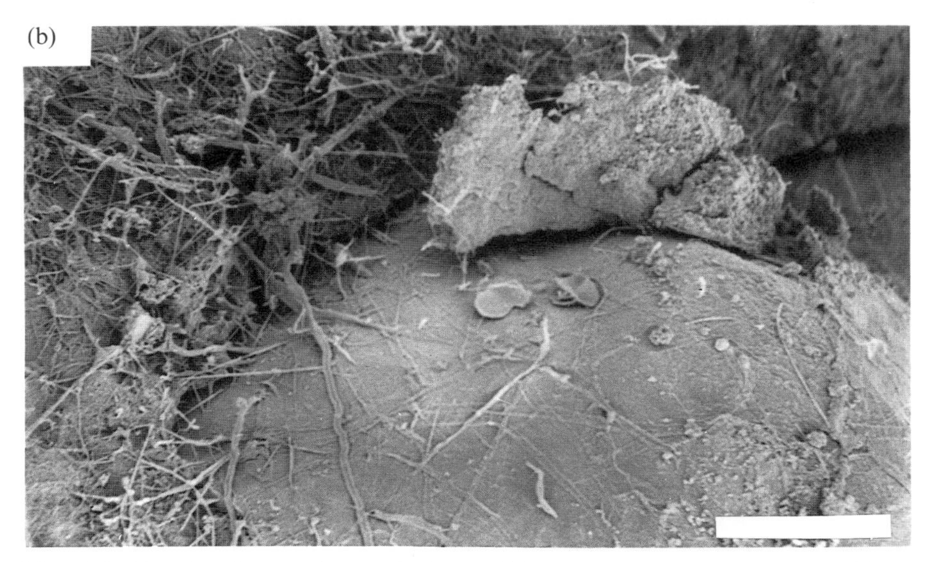

Figure 8.8 Scanning electron micrographs showing (A) granular haematite cement and (B) fibrous aluminous goethite cement in a petroferric layer from a podsolized dune, North Queensland. Scale bars = 10 μm.

Figure 8.9 Scanning electron micrograph showing the etched surface of a quartz grain in a petroferric layer from North Queensland. The etching occurs by dissolution of silica and simultaneous replacement by iron (III) oxide. Scale bar = 10 μm.

Silicified plant root structures and silica rhizoliths in dune sands from southeast India were described by Hendry (1987).

8.5.6 Formation of soil profiles in dune sands

The formation of soil profiles in dune sands, as on other parent materials, is influenced by the mineralogy and texture of the unweathered sand, by the surface topography and drainage conditions, climate, vegetation, and addition of allochthonous material. All of these factors vary over time in response both to extrinsic factors and intrinsic changes brought about by soil development itself. Pedogenesis is characterized by vertical horizon differentiation and by the emergence of spatial toposequences of different soil types, known as *catenas*. Soil horizon differentiation is primarily due to the formation of weathering products in the near-surface zone, addition of biological material, introduction of airborne dust and salts, and the variable translocation and subsequent deposition of these materials in the subsoil. Catena development, in addition, is influenced by lateral variations in drainage conditions and by downslope movement of soil constituents.

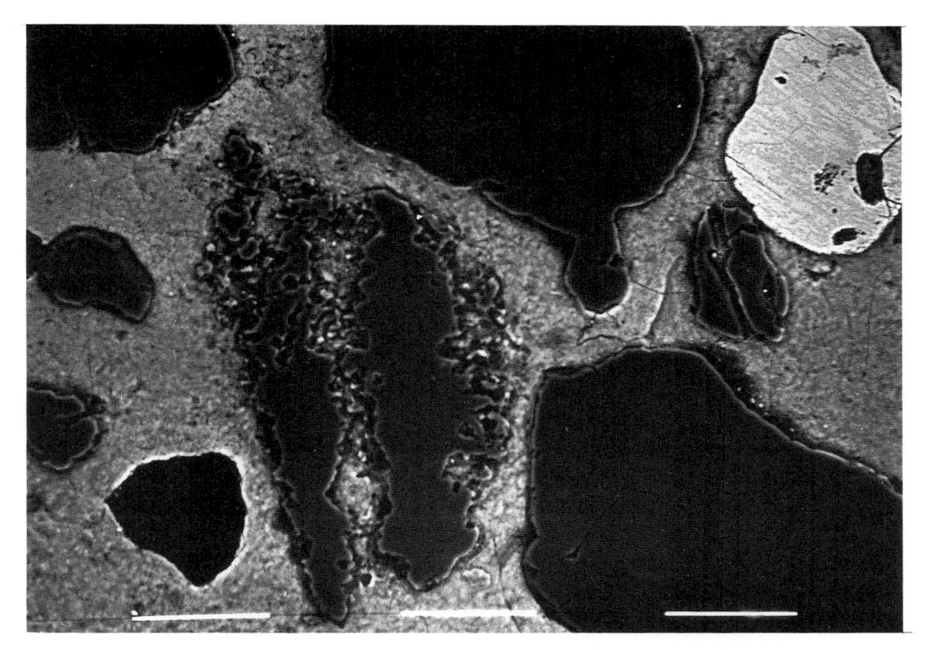

Figure 8.10 Backscattered scanning electron micrograph of a polished section through part of a petroferric layer from North Queensland, showing partial replacement of a detrital quartz grain. The cement is composed of poorly crystallized haematite. Scale bars = 100 μm.

Virtually all fixed dune landscapes are characterized by the development of soil catenas. For example, in the Gezira area of Sudan, Williams (1968) described catenas on late Pleistocene fixed dunes which consist of leached sands on the dune crests and illuvial loams and clays in the swales. The swale soils contain both pedogenetic and relict lacustrine carbonate. Many of the profiles are polygenetic, reflecting repeated phases of downslope sediment movement by sheetflow followed by horizon differentiation. In the Coastal Plain of Israel, catenas have developed on mixed siliclastic–carbonate dune sand parent material which has been enriched with airborne dust (Dan *et al.* 1968/69). Red *hamra* soils are developed on the crests and upper slopes of the dunes, sandy clay loam soils on the middle slopes, pseudogley on the footslopes and hydromorphic grumosols on the swampy toe slopes and inter-dune depression surfaces (Fig. 8.11). Textural differentiation in well drained profiles on the upper and crest slopes is pronounced, indicating a high intensity of leaching and clay mobility.

8.5.7 Podsolization and humate cementation

Podsol soil profiles commonly develop on siliceous dune sands in humid region dune sands. Podsolization is a progressive process in which the depth of the bleached A2 horizon increases with time (Thompson 1981, 1983) (Fig. 8.12). Sesquioxides and clay minerals in the upper part of the B horizon are continually being remobilized and redeposited lower in the profile. Mobilization of metals, particularly Fe and Al, has been widely attributed to chelation by organic acids, while deposition has been considered to be due to changing metal saturation, drying, and changes in potential (De Coninck 1980), followed by further reactions after decomposition of the organic complexes. Other workers have suggested that at least some Al and Fe may be transported independently of organic matter (Anderson *et al.* 1982), but covariance of

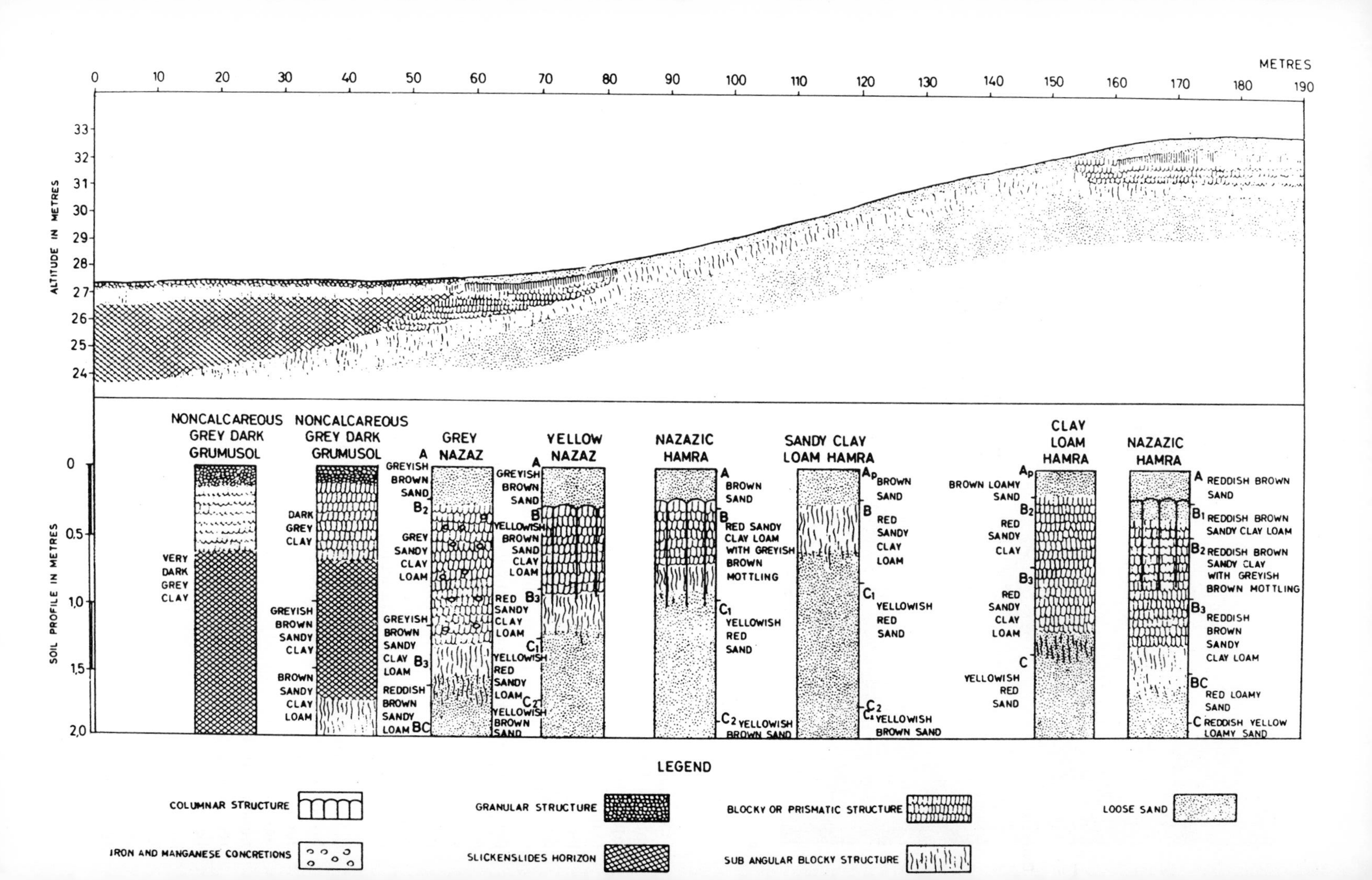
METRES
ALTITUDE IN METRES
SOIL PROFILE IN METRES
NONCALCAREOUS GREY DARK GRUMUSOL
VERY DARK GREY CLAY
NONCALCAREOUS GREY DARK GRUMUSOL
DARK GREY CLAY
GREYISH BROWN SANDY CLAY
BROWN SANDY CLAY LOAM
GREY NAZAZ
A GREYISH BROWN SAND
B2 GREY SANDY CLAY LOAM
GREYISH BROWN SANDY CLAY LOAM
B3 REDDISH BROWN SANDY LOAM
BC
YELLOW NAZAZ
A GREYISH BROWN SAND
B YELLOWISH BROWN SAND CLAY LOAM
B3 RED SANDY CLAY LOAM
C1 YELLOWISH RED SANDY LOAM
C2 YELLOWISH BROWN SAND
NAZAZIC HAMRA
A BROWN SAND
B RED SANDY CLAY LOAM WITH GREYISH BROWN MOTTLING
C1 YELLOWISH RED SAND
C2 YELLOWISH BROWN SAND
SANDY CLAY LOAM HAMRA
Ap BROWN SAND
B RED SANDY CLAY LOAM
C1 YELLOWISH RED SAND
C2 YELLOWISH BROWN SAND
CLAY LOAM HAMRA
Ap BROWN LOAMY SAND
B2 RED SANDY CLAY
B3 RED SANDY CLAY LOAM
C YELLOWISH RED SAND
NAZAZIC HAMRA
A REDDISH BROWN SAND
B1 REDDISH BROWN SANDY CLAY LOAM
B2 REDDISH BROWN SANDY CLAY WITH GREYISH BROWN MOTTLING
B3 REDDISH BROWN SANDY CLAY LOAM
BC RED LOAMY SAND
C REDDISH YELLOW LOAMY SAND
LEGEND
COLUMNAR STRUCTURE
GRANULAR STRUCTURE
BLOCKY OR PRISMATIC STRUCTURE
LOOSE SAND
IRON AND MANGANESE CONCRETIONS
SLICKENSLIDES HORIZON
SUB ANGULAR BLOCKY STRUCTURE

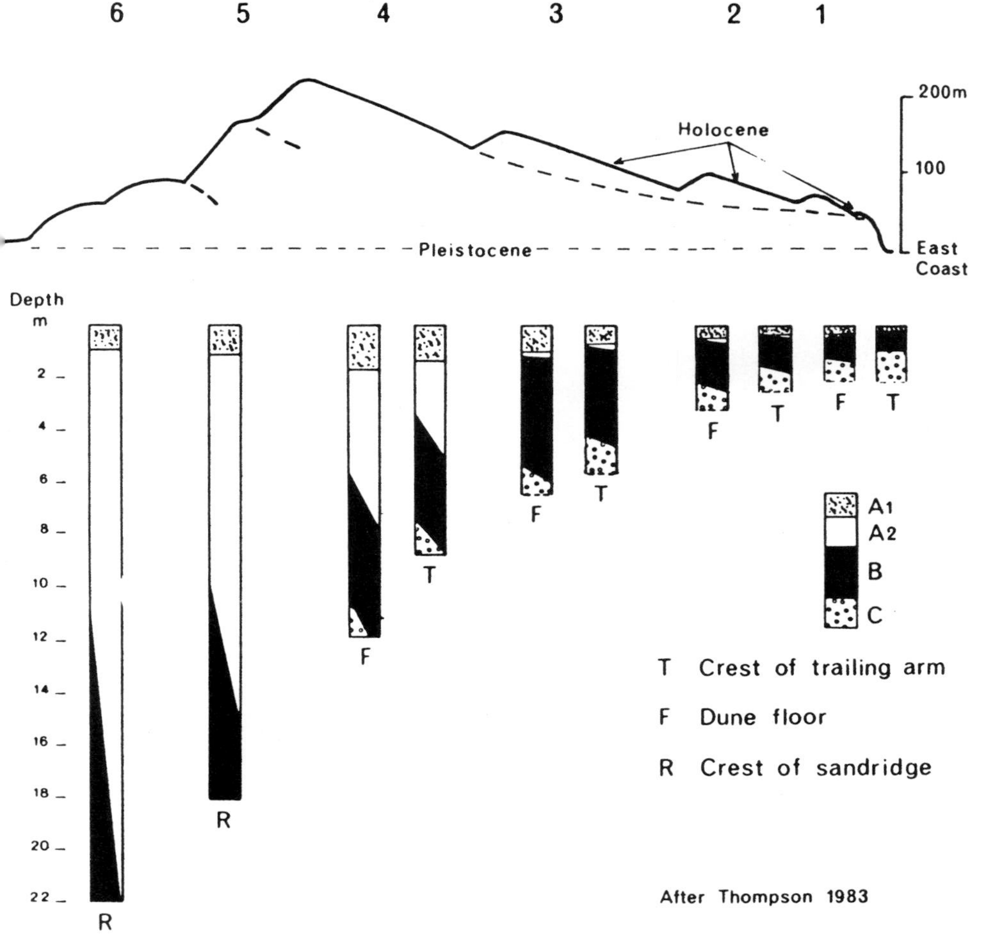

Figure 8.12 Schematic cross-section of the parabolic dune systems at Cooloola, southern Queensland, showing progressive development of podsol profiles with time. Diagonal lines indicate variations in depth to the top of the underlying horizon. (After Thompson 1983).

these elements with organic carbon abundance has been observed in many podsol profiles (e.g. Little 1986).

Andriesse (1969/70) inferred from podsol chronosequences in Sarawak that a preliminary reddening phase is followed by progressive bleaching and finally by the accumulation of dark brown organic colloids (humate) in the subsoil. A similar sequence was documented in dune sands at Cape Flattery in North Queensland by Pye (1983f). Indurated sands cemented mainly by humate and aluminium hydroxides (Fig. 8.13) were termed *humicretes* by Pye (1982d). In weakly cemented humicretes the humate forms black, discontinuous coatings around the quartz grains (Figs 8.14 & 8.15). In better indurated examples virtually all of the intergranular porosity is filled by humate, kaolinite, and gibbsite. An even later stage of podsol evolution, in which the dark brown coloration is lost by oxidation of humate to leave residual grey–white gibbsite-rich

Figure 8.11 Section across the slope of the Netanya dune catena, Coastal Plain of Israel, showing the spatial distribution of soil types (after Dan *et al.* 1968/9).

Figure 8.13 Humicrete exposed by marine erosion of weathered dunes, Rainbow Beach, southern Queensland.

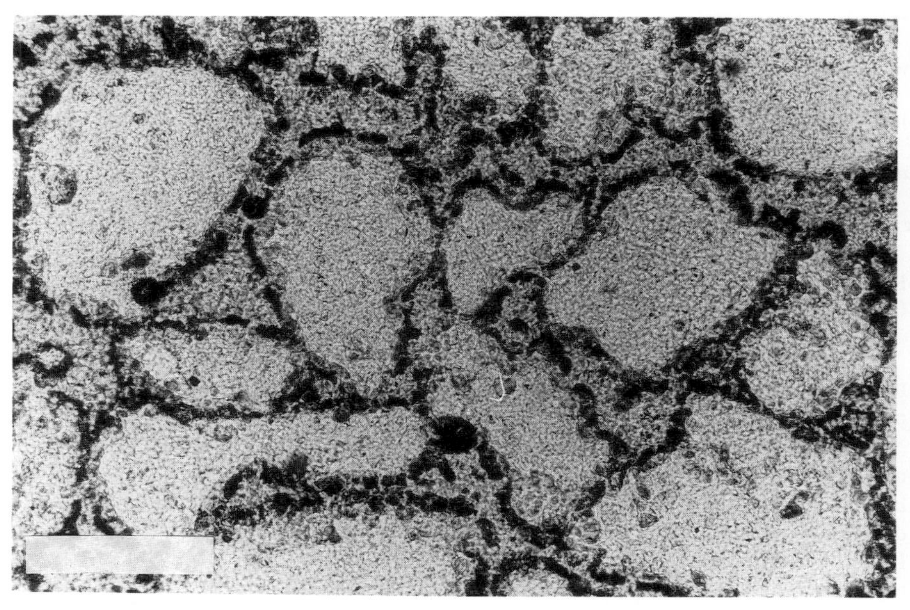

Figure 8.14 Optical micrograph (plane light) showing humate coatings on quartz sand grains in a weakly cemented humicrete from North Queensland. The discontinuous nature of the coatings is due to drying shrinkage. Scale bar = 100 μm.

Figure 8.15 Scanning electron micrograph showing humate coating on the surface of a quartz grain from a groundwater podsol, Cape Flattery, North Queensland. The coating is composed mainly of humic acids with subsidiary kaolinite and gibbsite. Scale bar = 10 μm.

sands, has been recognized in southern Queensland dune sands (Ward *et al.* 1979). A schematic model, showing the stages in progressive podsolization, is shown in Figure 8.16. Representative whole-rock compositional data for sands from different podsolic horizons in North Queensland are shown in Table 8.2. The high purity of the bleached A2 horizons in particular makes these sands an important raw material for glass and ceramics manufacture. Even in the reddened B horizons, levels of HCl-extractable Fe and Al rarely exceed 2% by weight (Fig. 8.17). Humicrete layers in coastal Queensland generally contain up to 7% organic carbon and 2% Al (Ward *et al.* 1979, Pye 1982d).

In favourable circumstances of high rainfall, abundant organic acids, and highly siliceous parent materials, podsolization can be a rapid process. Paton *et al.* (1976) observed that miniature podsol profiles developed in less than 5 years on dune sands replaced after mining disturbance. Radiocarbon dating at Cape Flattery showed that profiles equivalent to stage 3 shown in Figure 8.16 are developed in dune sands which are less than 7500 years old (Pye 1981). Dunes containing profiles equivalent to stages 6 and 7 of this model yielded radiocarbon ages older than 48 000 years. The more degraded

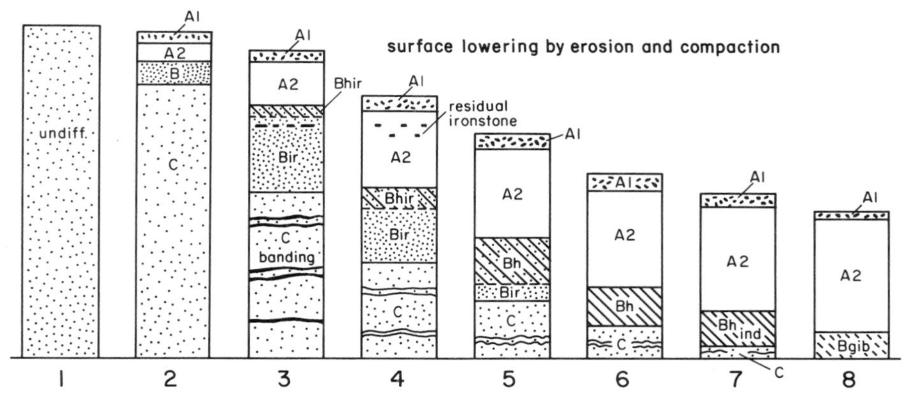

Figure 8.16 Schematic model showing the stages in the progressive podsolization of dune sands under humid tropical conditions. (After Pye 1983g).

Table 8.2 Major and trace element analyses of Cape York Peninsula dune sands*, determined by X-ray fluorescence spectrometry. Major elements in weight-% oxide, trace elements in ppm.

	CB77	CF5A	CF1P	OR51	AY40	CF212	AY18	CF238	AY34	AY7
SiO_2	99.13	99.65	98.73	99.72	93.07	91.31	93.71	94.70	65.91	19.46
Al_2O_3	0.07	0.02	0.09	0	4.14	0.37	2.57	2.78	3.21	0.31
TiO_2	0.15	0.10	0.74	0.03	0.68	0.44	0.44	0.44	0.16	29.27
Fe_2O_3	0.06	0.04	0.38	0	0.36	0.24	1.87	0.95	25.15	24.15
MgO	0.04	0.12	0.01	0	0.03	0.23	0.01	0.02	0	0.29
CaO	0.03	0.04	0.03	0.03	0.02	0.15	0.03	0.02	0.03	0.04
Na_2O	0	0	0	0	0.06	0.24	0	0.02	0	0
K_2O	0	0.01	0.01	0.01	0	0.01	0.02	0.05	0	0
MnO	0.01	0.01	0.01	0.01	0.03	0.01	0.02	0.02	0.01	1.34
P_2O_5	0	0	0.01	0	0.01	0	0.02	0	0.20	0.08
LOI	0.23	0.18	0.16	0.26	1.65	7.17	1.30	1.16	5.35	0.30
Total	99.74	100.18	100.16	100.06	100.05	100.17	100.00	100.16	100.02	87.16
As	0	0	-	0	0	2	22	3	257	111
Ba	0	0	-	12	24	15	19	25	16	41
Ce	0	0	-	0	0	0	0	0	16	1193
Cl	0	0	-	12	100	8	19	7	110	0
Co	0	3	-	0	7	0	6	0	9	37
Cr	5	9	-	9	24	13	60	40	170	490
Cu	3	4	-	14	4	2	4	16	10	78
Ga	0	0	-	0	3	3	2	3	4	0
La	7	0	-	0	6	5	4	10	0	152
Ni	7	6	-	4	5	15	7	8	10	49
Nb	3	2	-	1	8	3	6	6	3	661
Pb	14	9	-	5	15	11	14	14	13	77
Rb	0	0	-	1	0	2	0	2	3	11
Sr	0	0	-	0	4	18	4	5	2	15
Th	0	3	-	2	2	4	0	1	4	123
U	0	2	-	2	0	0	2	0	3	52

V	5	4	-	4	22	5	30	22	216	2267
Y	0	3	-	2	0	0	0	3	0	272
Zn	0	0	-	6	0	0	4	5	5	242
Zr	115	35	-	38	233	3550	195	230	60	115 569

* Key to samples analysed:

CB77 Leached white sand from A2 horizon (depth 2 m) in a stablilized dune, Cape Bedford.
CF5A White quartz sand from 1.8 m depth near the crest of an active dune, Cape Flattery.
CF1P White sand with heavy mineral seams, crest of an active dune, Cape Flattery.
OR51 White quartz sand from the crest of an active dune, Temple Bay.
AY40 Whitish grey kaolinitic sands from the B horizon of a weathered dune, Cape Bedford.
CF212 Black humate-cemented sands from the B horizon of a groundwater podsol, Cape Flattery.
AY18 Orange iron-stained sand from the B horizon (depth 6 m) of a stabilized dune, Cape Bedford.
CF238 Red iron-stained sand from the B horizon (depth 8 m) of a stabilized dune, Cape Flattery.
AY34 Iron oxide-cemented petroferric horizon from the B horizon of a weathered dune, Cape Bedford.
AY7 Localized heavy mineral concentration, Cape Bedford.

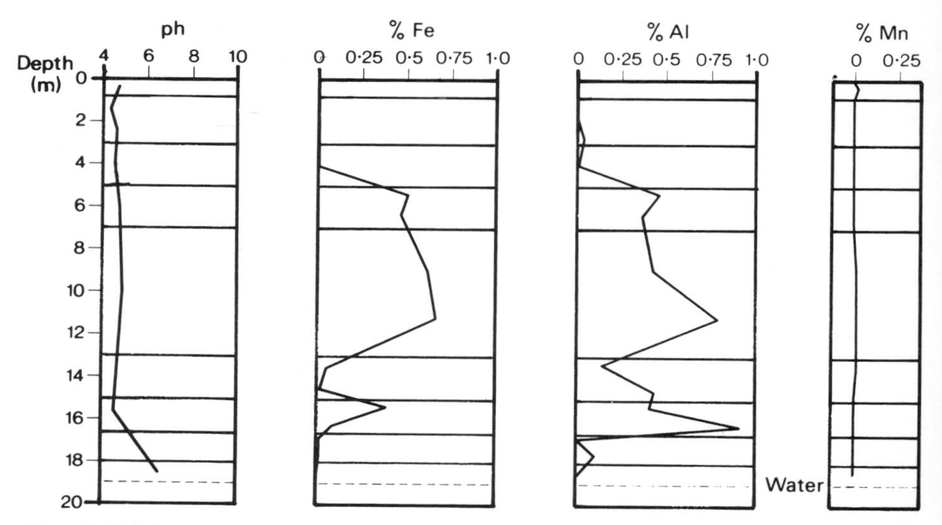

Figure 8.17 (left and right) Variations in pH and the concentration (% dry weight of bulk sand) of extractable Fe, Al, Mn, Ca, Mg, K, and Ti with depth in a podsolized dune, Cape Flattery, North Queensland.

dune units which contain groundwater humicretes in southern Queensland have not been precisely dated but are probably several hundred thousand years old.

Podsolization on these dune sands was favoured by the highly siliceous composition of the parent sands, by the acidophyllous nature of the natural dune vegetation (Pye & Jackes 1981), and by the high rainfall (1800–2000 mm) in the area. Where conditions are less extreme, leaching is less severe and profile differentiation less marked. For example, chronosequences of coastal dune soils which show only moderate leaching have been described from the Manawatu area of North Island, New Zealand (Cowie 1968). The parent sands in this area, which receives 800–1000 mm of annual rainfall, contain several percent feldspar, ferromagnesian minerals, shell, and pumice fragments. The progressive soil changes observed with increasing dune age include increasing depth and organic matter content of the A horizon, increase in silt and clay content of the A and B horizons, progressive leaching of carbonate and bases, and reduction of soil pH. However, leaching conditions have not been sufficiently intense or prolonged to form sandy podsols with bleached A2 horizons.

8.6 Formation of carbonate aeolianites

8.6.1 Definition and occurrence of aeolianites

The term *aeolianite* was originally used by Sayles (1931) to describe 'all consolidated sedimentary rocks which have been deposited by the wind', but most later workers have restricted the term to describe only aeolian sands cemented by early diagenetic calcite. Some English-speaking geologists have used the term *aeolian calcarenite* in preference to aeolianite (e.g. Milnes & Ludbrook 1986). Other terms which have been used include *kurkar* (Israel), *grès dunaire* (French North Africa), *miliolite* (India, Persian Gulf), *dune limestone*, *aeolian limestone* and *dunerock* (Australia).

Aeolianites vary considerably in composition and texture. One end-member consists entirely of siliclastic framework grains cemented by calcite, while at the other extreme both framework grains and cement may be composed entirely of calcium carbonate. A

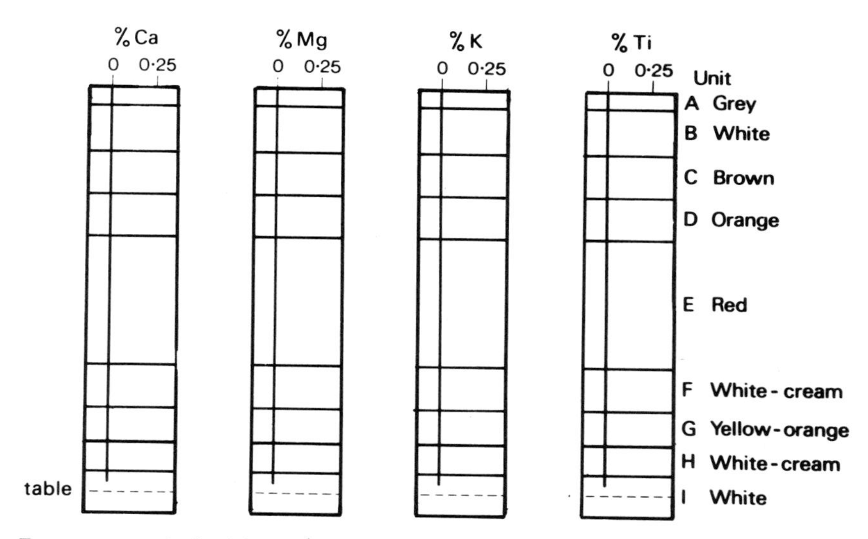

Extracts were obtained by boiling the oven-dried samples in 10% HCl for 20 min; analyses by atomic absorption spectrometry. Data from Pye (1980a, 1983g).

majority of aeolianites contain both carbonate and non-carbonate framework grains. Fairbridge & Johnson (1978) made an arbitrary distinction between *quartzose aeolianite* (< 50% $CaCO_3$) and *carbonate aeolianite* (> 50% $CaCO_3$).

Aeolianites of Quaternary age have attracted scientific attention for more than a century (Branner 1890, Agassiz 1895, Chapman 1900, Evans 1900), and their early diagenesis have been investigated in considerable detail. The most comprehensive studies have been undertaken in the Caribbean (Ball 1967, Land 1970, Ward 1973, 1975, Longman *et al.* 1983, Beier 1987, White & Curran 1988), Western and South Australia (Fairbridge & Teichert 1953, Reeckman & Gill 1981, Semeniuk & Meagher 1981, Semeniuk 1986, Warren 1983, Milnes & Ludbrook 1986), the Mediterranean (Gavish & Friedman 1969, Yaalon 1967, Selim 1974, Amiel 1975, Klappa 1978, 1980, Calvet *et al.* 1980), and southeast Africa (McCarthy 1967, Maud 1968, Coetzee 1975/6a, b). However, aeolianites occur widely on oceanic islands and continental shorelines in arid and semi-arid parts of the world where the supply of siliciclastic sediment is restricted and rates of biogenic carbonate productivity or ooid formation are high [see reviews in Gardner (1983b) and McKee & Ward (1983)]. In a few places, notably the Thar Desert of India (Sperling & Goudie 1975, Goudie & Sperling 1977) and the Wahiba sands of Oman (Allison 1988, Goudie *et al.* 1987), marine biogenic carbonate grains have been blown long distances inland, but most major aeolianite occurrences show a close relationship to present or former marine shorelines. Lacustrine shoreline dunes cemented by calcium carbonate are known (e.g. Jones 1938) and localized carbonate cementation in some predominantly quartzose continental dunes has also been recorded (e.g. Galloway *et al.* 1985, Dijkmans *et al.* 1986).

There is no generally agreed definition of the degree of carbonate cementation required to distinguish aeolianite from unlithified carbonate or polymineralic dune sands. Many young carbonate dunes, such as those of Quintana Roo, Mexico (Ward 1973, 1975), are very weakly cemented except in near-surface sand layers which have been affected by subaerial weathering and pedogenesis. Elsewhere, as in parts of northwest Britain (Roberts *et al.* 1973) and Western Australia (Semeniuk & Meagher 1981), essentially unlithified dune sands are cemented only in the basal parts of dunes which are affected by groundwater, or around plant roots and burrows in the near-surface zone. Such cementation phenomena have generally been described as types of calcrete, but the distinction between dune sand-hosted calcretes and aeolianites is not well defined. Some lithified carbonate dune sand sequences consist of a series of superimposed calcrete horizons, reflecting episodic sand accretion and pedogenesis over a long period. In other instances pervasive carbonate cementation has occurred within the vadose zone of an aeolian dune formed during a single phase of aeolian sand encroachment, with minimal development of calcrete or other pedogenic phenomena. Pedogenesis and calcrete development often destroys the primary depositional textures present in aeolian sands. Consequently, the term aeolianite should be restricted to carbonate-cemented sands in which primary aeolian depositional features, such as cross-bedding and grain size lamination, are readily visible (Fig. 8.18). Cemented sands in which these features are obliterated or heavily overprinted are more appropriately described as calcretes.

8.6.2 Controls on carbonate cementation in aeolianites

The extent, vertical distribution, and composition of carbonate cement reflects the abundance and composition of carbonate grains in the host sediment, the amount of water passing through the sand column, the effect of vegetation on the soil moisture regime, and the episodic nature or otherwise of aeolian sedimentation.

8.6.2.1 Effects of carbonate mineralogy Carbonate grains in aeolianites are mainly composed of aragonite, high-Mg calcite (containing >5 mol-% $MgCO_3$), low-Mg calcite, or mixtures of these minerals. Many ooids, gastropods, corals, and algae are predominantly composed of aragonite, which is the dominant mineral found in many low-latitude littoral deposits. However, on some shores, high- or low-Mg biogenic constituents are dominant (e.g. southern South Australia) (Warren 1983).

All calcium carbonate minerals undergo dissolution when exposed to rain and soil

Figure 8.18 Quaternary coastal aeolianite, Spencer Gulf, South Australia (photograph by V. P. Wright).

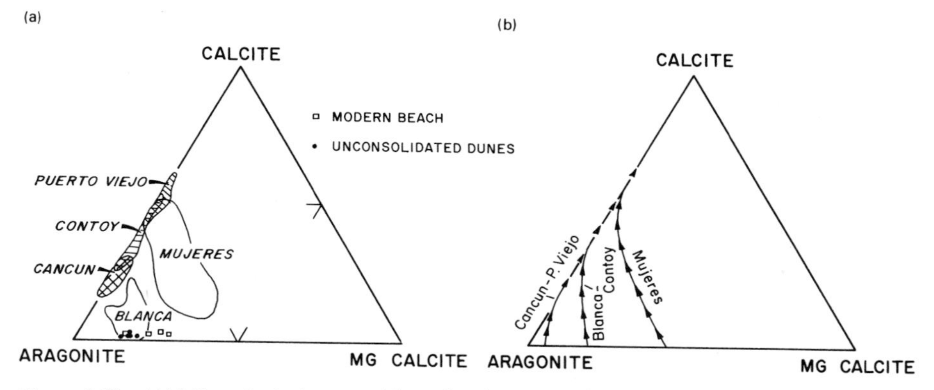

Figure 8.19 (A) Mineralogical composition of carbonate sands, Holocene aeolianites (unshaded), and Pleistocene aeolianites (shaded) at different sites on the northeast coast of Yucatan Peninsula, Mexico. (B) Generalized path of mineralogical changes during progressive aeolianite diagenesis at these locations. (After Ward 1975).

water in the meteoric diagenetic environment, but aragonite and high-Mg calcite are considerably more soluble than low-Mg calcite [see James & Choquette (1984) for a useful review of meteoric diagenesis]. High-Mg calcite which contains more than 12 mol-% $MgCO_3$ is the most soluble phase in pure water (and also in water containing soil-derived CO_2), followed by aragonite and Mg calcite containing less than 12 mol-% $MgCO_3$. The least soluble phase is calcite containing virtually no magnesium.

Rainwater which enters the top of a column of dune sand is initially undersaturated with respect to all carbonate minerals. However, since aragonite and high-Mg calcite are more soluble than low-Mg calcite, the downward-percolating waters become saturated with respect to low-Mg calcite, while still undersaturated with respect to aragonite and high-Mg calcite. Precipitation of low-Mg calcite cement crystals may therefore take place simultaneously with dissolution of aragonite and high-Mg calcite. Low-Mg calcite precipitation keeps the solution undersaturated with respect to aragonite, thereby ensuring its continued dissolution. Dissolution of aragonite and high-Mg calcite allochems may form large voids, either before or after filling of the surrounding pores by calcite cement. In the latter case large mouldic pores are formed which may be filled by later cement. Dissolution of aragonite or Mg calcite and replacement by calcite may also occur almost simultaneously. In this case fine structural detail is often preserved. Formation of mouldic porosity and infilling by later low-Mg calcite spar occurs most commonly as a result of congruent dissolution of aragonite. High-Mg calcite may undergo either congruent dissolution or incongruent dissolution, depending on the concentration of dissolved calcium and magnesium in the pore waters. During incongruent dissolution only Mg is removed from the crystal lattice (Land 1967).

Several empirical studies have shown that the abundance of aragonite and high-Mg calcite in aeolianites declines relative to low-Mg calcite with increasing sediment age (Land *et al.* 1967, Gavish & Friedman 1969, Ward 1975, Calvet *et al.* 1980, Reeckman & Gill 1981) (Fig. 8.19). The relative rates of loss of aragonite and high-Mg calcite are dependent on the initial relative abundance of these minerals and on the availability of meteoric water. Gavish & Friedman (1969) reported a total loss of high-Mg calcite in Israeli aeolianites within 10 000 years, and complete loss of aragonite within 50 000 years. A much longer timescale was envisaged by Reeckman & Gill (1981), who estimated that it has taken 100 000 years for the disappearance of high-Mg calcite and 600 000 years for the disappearance of aragonite in the coastal aeolianites of southern

Figure 8.20 Aeolianite from southern Gaza showing preferential cementation of finer-grained laminae. Scale bar = 10 mm.

Victoria. It needs to be emphasized, however, that few pre-late Pleistocene aeolianites have been accurately dated.

Stable isotope analyses of whole rock samples have shown that $\delta^{13}C$ values become more negative during progressive diagenesis of aeolianites as marine allochems are destroyed and more soil carbonate becomes incorporated in low-Mg calcite cement (Gross 1964, Magaritz *et al.* 1979, Reeckman & Gill 1981, Beier 1987). However, little work has been undertaken to determine time-dependent changes in the stable isotope characteristics of individual mineral constituents of aeolianites.

Calcite cement crystals precipitated in the vadose zone typically show an irregular, patchy distribution in homogeneous sands. Where grain size lamination is pronounced, the finer sand layers often become preferentially cemented (Fig. 8.20). In weakly cemented aeolianites evidence of vadose cementation may be provided by pendulous cement crystals, formed on the underside of grains where droplets of water are retained by gravity and surface tension (Fig. 8.22). Meniscus cements may also form at points of grain contact where moisture is retained by surface tension. In better cemented aeolianites, virtually all of the porosity, both primary and mouldic, may be filled with calcite (Gardner 1981). The distribution and grain size of such pore-occluding calcite is often highly patchy (Figs 8.22–24). Syntaxial overgrowths may also form on detrital echinoderm plates which are composed of low-Mg calcite (McKee & Ward 1983). Since conditions in the vadose zone are predominantly oxidizing, the precipitated cements are almost always non-ferroan.

Yaalon (1967) concluded that an initial calcium carbonate content of 8–10% is required for carbonate cementation to proceed in sands on parts of the coast of Israel which receive 300–600 mm of annual rainfall. This minimum initial carbonate content increases in regions with higher rainfall or less evaporation, and was judged by Yaalon (1982) to be 25% in sub-humid coastal Natal. According to Yaalon (1967), cementation proceeds from the top downwards, as $CaCO_3$ derived from dissolution of skeletal

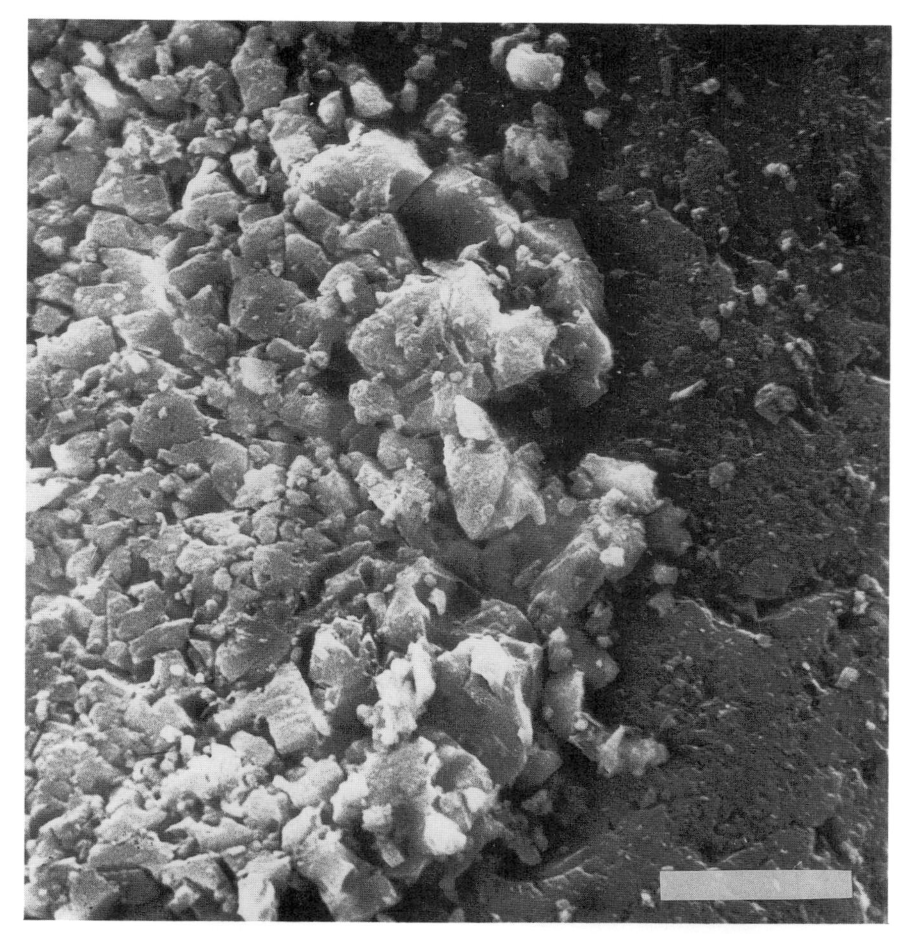

Figure 8.21 Scanning electron micrograph showing low-Mg-calcite cement crystals on the surface of a sand grain in aeolianite from Double Island Point, southern Queensland. Scale bar = 10 μm.

fragments is reprecipitated in the sub-surface. The cemented layers in Israeli aeolianites were thus interpreted by Yaalon as B_{Ca} or BC_{Ca} soil horizons. On many Israeli aeolianites the largely decalcified A horizon sands have been stripped away, allowing calcrete development and karstification to occur on the subaerially-exposed cemented B horizon sands before being followed, in some places, by renewed aeolian sedimentation and further vadose cementation.

8.6.2.2 Effects of rainfall and evaporation Climate has an important effect on aeolianite diagenesis since it controls the availability of meteoric water and therefore the intensity and rate of carbonate mineral alteration. Under hot, arid conditions, water penetration into the dune sand column is limited, and mineral alteration in the vadose zone is extremely slow. Thin near-surface calcrete horizons may develop, but much of the sediment in the vadose zone may remain essentially unaltered. Under semi-arid conditions, where there is a net moisture surplus (excess of rainfall over evaporation) for at least 3 months of the year, mineral alteration will occur more quickly and will extend to a greater depth. Thicker pedogenetic calcrete profiles develop under such conditions,

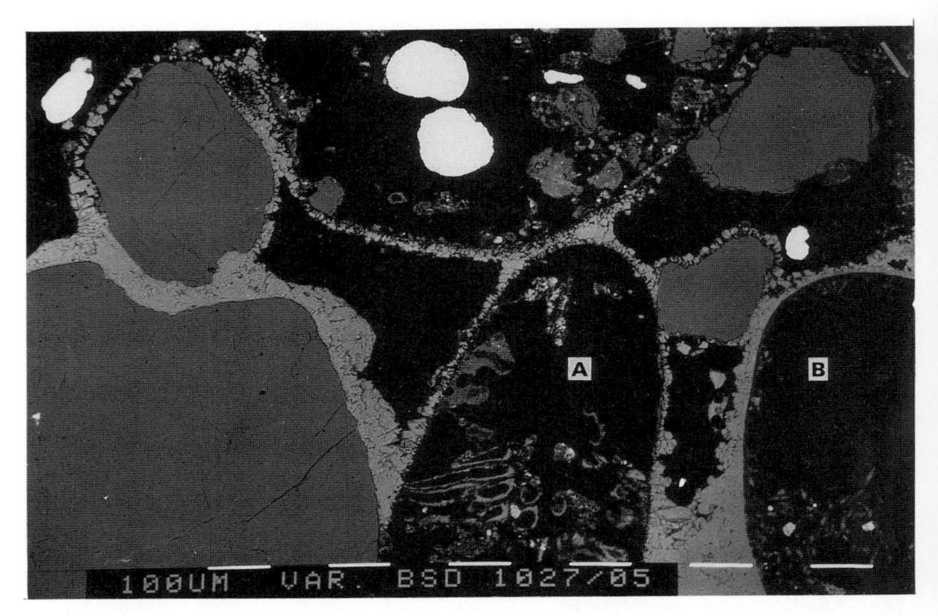

Figure 8.22 Backscattered scanning electron micrograph of a polished section of aeolianite from Double Island Point, southern Queensland, showing meniscus cement and partial pore-filling low-Mg-calcite cement. The detrital allochems in the lower part of the photograph (A and B) have been almost wholly dissolved following formation of the rimming cements and the voids partially filled with translocated silt and fine sand. The white grains are heavy minerals. Scale bars = 100 μm.

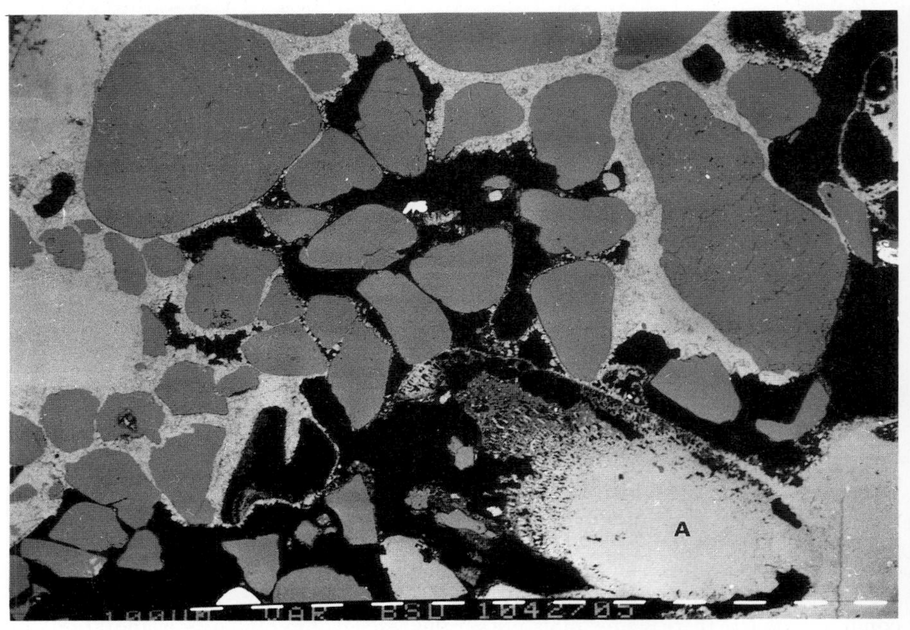

Figure 8.23 Backscattered scanning electron micrograph showing patchy distribution of vadose cement in aeolianite from the Burdekin Delta, North Queensland. Note the selective dissolution of allochem labelled A. Scale bars = 100 μm.

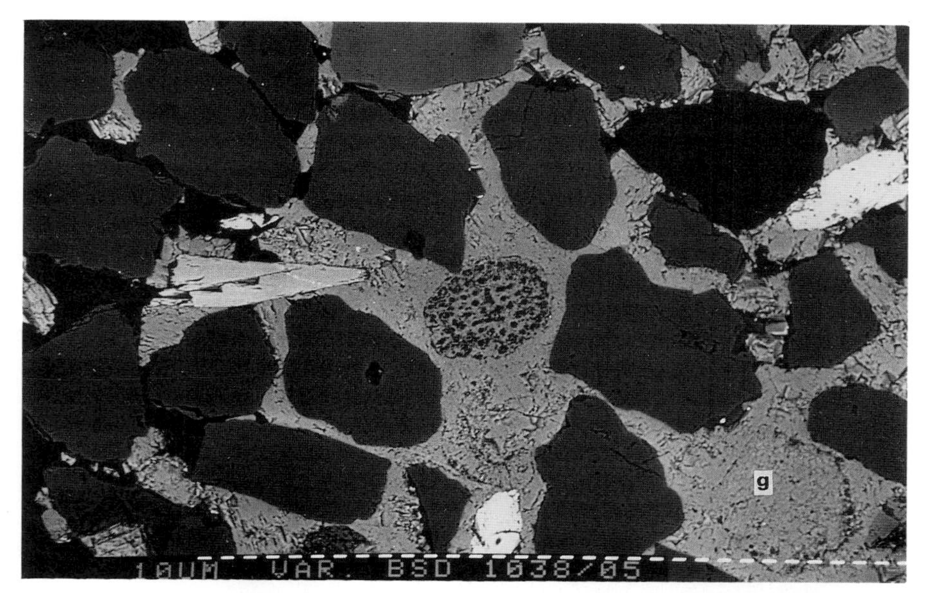

Figure 8.24 Backscattered scanning electron micrograph showing almost total cementation by low-Mg-calcite in aeolianite, Burdekin Delta, North Queensland. Note the 'ghosts' (labelled g) which represent allochems completely replaced by low-Mg-calcite. Scale bars = 10 μm.

and pervasive vadose zone diagenesis is possible. Under humid conditions, where water is able to pass right through the dune sand column to the water table, pervasive vadose diagenesis is the rule. In the case of impure carbonate sands, a partly decalcified residual soil may form at the top of the profile, and karst features may develop at the top of the cemented zone (Day 1928, Coetzee 1975/6b).

8.6.2.3 Effects of vegetation Precipitation of carbonate cement in the vadose zone is enhanced by the removal of moisture from the subsoil through evapotranspiration. During periods of net moisture deficit, water from the subsoil is returned to the atmosphere through plant roots and leaves. Any dissolved ions which are not taken up by plants are then precipitated in the soil as the residual soil moisture becomes supersaturated.

Cementation is often particularly marked around roots and trace fossils which contain organic matter (Fig. 8.25). Several different names have been used to describe cemented root structures, including *rhizoconcretions* (Kindle 1923), *pedotubules* (Brewer 1964), rhizoliths (Klappa 1980, Loope 1985b), *rhizocretions* (Steinen 1974, Esteban 1976), *dikaka* (Glennie & Evamy 1968) and *rhizoids* (James & Choquette 1984). Some rhizoconcretions are several metres long and tens of centimetres in diameter, apparently representing calcification around tree roots. At the other end of the scale small calcified tubules about 15 μm in diameter have been interpreted as calcified root hairs (Ward 1973, Klappa 1979). Many of these microtubules are associated with the development of needle-fibre cement which is thought to precipitate along or within fungal hyphae (Ward 1975, Klappa 1980). Spherical, elliptical or sheet-like bodies composed of small calcite prisms, referred to as *microdium*, which have been observed in some aeolianites, are also thought to result from calcification of mycorrhizal (root-fungus) associations (Klappa 1978). Other calcified biological structures reported in aeolinites include insect puparia (Fairbridge 1950, Milnes & Ludbrook 1986), faecal pellets, and trace fossils (Longman *et al.* 1983, White & Curran 1988).

Figure 8.25 Rhizoconcretions in carbonate dune sands, Spencer Gulf, South Australia (photograph by V. P. Wright).

The mechanisms which form rhizoliths and calcified burrows are not well understood. Calvet *et al.* (1975) maintained that rhizoliths form only around decaying root material, and that soil microorganisms play an important role in bringing about carbonate precipitation around or within the decaying root. Other authors [e.g. Klappa (1980)] have maintained that rhizoliths can also form around living roots.

8.6.3 Calcrete horizons in carbonate dune sands

Several different types of calcrete occur in carbonate dune sands. Pedogenetic calcretes display a wide range of textures and structural features (Read 1974, Klappa 1978, 1980, Warren 1983, Beier 1987). They include relatively thin, continuous surface crusts, discontinuous sub-surface, nodular calcrete, rhizoliths, pisoid layers, laminar calcrete, micritized horizons, brecciated boulder calcrete, and massive hardpan calcrete horizons. Groundwater calcrete profiles show considerable variation, but in dune sands commonly develop a sheet-like morphology.

In the carbonate dunes of southwestern Australia, calcrete occurs both as pedogenetic rhizoconcretions in the vadose zone and as a sheet (up to 0.5 m thick, with a profile consisting of mottled, massive, and laminar structures), in the zone of capillary rise above the water table (Semeniuk & Meagher 1981, Semeniuk & Searle 1985, Semeniuk 1986). Both calcrete types are related to vegetation. Areas of lower dune relief are vegetated by tree species which exploit shallow groundwater and induce massive precipitation of a calcrete sheet in the zone of capillary rise. Areas of higher dune relief are covered by coastal heath scrub, which exploits vadose water and encourages the formation of rhizoconcretions (Fig. 8.26). At the regional scale the distribution of vegetation types is related to climate, with woodland forest species being restricted to humid and sub-humid areas (Semeniuk 1986).

Phreatic groundwater calcretes (i.e. formed by carbonate precipitation below the groundwater table) have not yet been documented in carbonate dune sands.

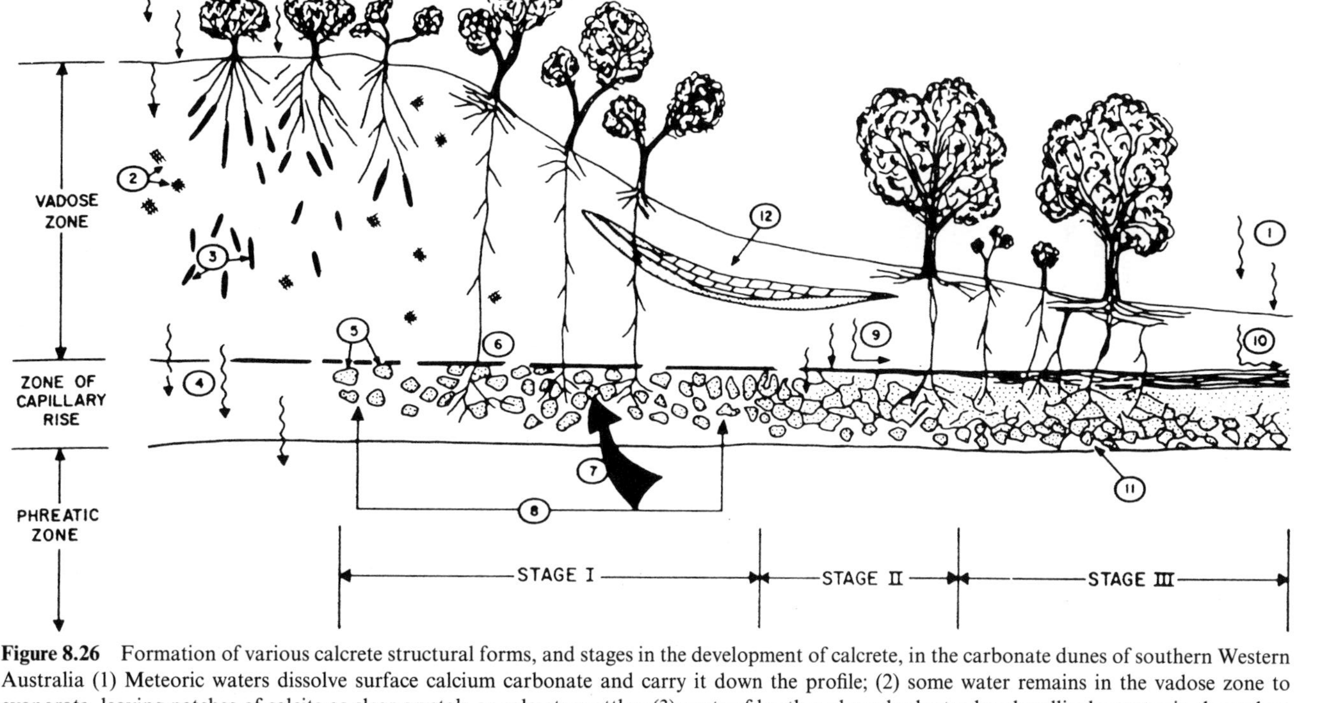

Figure 8.26 Formation of various calcrete structural forms, and stages in the development of calcrete, in the carbonate dunes of southern Western Australia (1) Meteoric waters dissolve surface calcium carbonate and carry it down the profile; (2) some water remains in the vadose zone to evaporate, leaving patches of calcite as clear crystals or calcrete mottles; (3) roots of heath and scrub plants absorb pellicular water in the vadose zone, while calcite is precipitated around the roots as rhizoconcretions; (4) excess meteoric water gravitates to the water table; (5) evaporation in the zone of capillary rise leaves an interstitial precipitate of calcite in the parent sand which aggregates to form calcrete mottles; (6) plants (phreatophytes) which utilize ground water precipitate calcite around their roots, initially as calcrete mottles; (7) ground water supplies water to the woodland vegetation during the summer months; (8) calcrete mottles gradually coalesce to form a massive calcrete sheet; (9) once a massive calcrete sheet has been formed, vertical recharge is hindered, and water tends to flow laterally or is locally ponded; (10) a stage is reached where much of the percolating meteoric water does not penetrate the massive calcrete, and evaporation of perched water forms a thin layer of laminar calcrete; (11) roots which penetrate the massive calcrete layer continue to utilize ground water, and calcrete continues to be precipitated below the massive calcrete; (12) soil horizons above the main phreatic zone may locally develop a perched moisture zone, and calcrete sheets may form above the main sheet. (After Semeniuk & Meagher 1981).

8.6.4 Karstification of aeolianites

Karst development can occur simultaneously with near-surface cementation, or may post-date it by some considerable period of time. Seasonal desiccation can cause cracks in surface or subsurface calcrete horizons, which subsequently act as foci for water penetration and dissolution. If calcrete becomes exposed at the surface it may partially disintegrate, forming a surface breccia or boulder calcrete horizon. Small-scale karst features such as lapies, karren, sink holes, and pipes, may also form (Day 1928, Coetzee 1975/6b).

During periods of Pleistocene low sea level, aeolianites and other lithified carbonate rocks in some areas, including Bermuda, experienced pronounced karstification which contributed to the formation of cave systems and large collapse features (Bretz 1960, Land *et al.* 1967).

8.6.5 Relationship between aeolianites and red soils

In many parts of the world, including Bermuda and southeast India, aeolianites are locally overlain by red soils (*terra rossa*) or unconsolidated red sands. The genetic relationship between these deposits and the aeolianite has been much discussed but has not been fully resolved. Blackburn & Taylor (1969) concluded that the well developed red soils in northern Bermuda formed by weathering of volcanic impurities in the underlying limestones. However, the possible importance of weathering of deposited dust transported from north Africa was not considered by these authors. In southeast India, aeolianite is overlain by red sands which have been reworked by aeolian action (Gardner 1981, 1983a), but it is unclear whether they formed by weathering and decalcification of the aeolianite or whether they represent a later aeolian sedimentation episode with different mineralogical composition. In the Burdekin Delta area of North Queensland, localized impure aeolianites have unquestionably weathered to form surficial red soils (Pye 1984). In this seasonally humid tropical area, formation of a partially decalcified red surface soil appears to have occurred simultaneously with calcite cementation of the underlying C-horizon sands.

In Natal, the extensive Berea Red Sands Formation has been widely interpreted as a decalcified, locally reworked weathering product of Pleistocene aeolianites [e.g. Maud (1968)]. However, as pointed out by Yaalon (1982), the field evidence is consistent with the alternative hypothesis that the original carbonate content of the older Pleistocene, now reddened, dune sand cordons was too low (<25%) to allow aeolianite formation, whereas the initial carbonate content of the two latest Pleistocene dune cordons exceeded 30% and was sufficient to allow carbonate cementation to proceed.

8.6.6 Regressive diagenesis of aeolianites

Muller & Tietz (1975) presented evidence from Fuerteventura, Canary Islands, that red algal clasts replaced by low-Mg calcite in the subaerial environment show partial reconstitution of their original high-Mg calcite mineralogy when they are subsequently exposed to sea water. In addition, high-Mg calcite cement may form locally where aeolianites are exposed to sea water or spray in the intertidal zone and in coastal cliffs.

8.7 Early diagenetic cementation by evaporite minerals

In hot desert environments, where saline lake and groundwaters evaporate in interdune depressions, interdune and basal dune deposits may become cemented by evaporites. Cementation by halite or gypsum is most common, and occurs principally in the capillary fringe zone. Schenk & Fryberger (1988) observed that, in the gypsiferous deposits at White Sands, New Mexico, crystallographically controlled dissolution of gypsum framework grains takes place in the capillary fringe and phreatic zones. Overgrowth development is widespread, and results in filling of some intergranular pores. By contrast, etching of framework grains in the vadose zone was found to be minor, with weak cementation taking place by the formation of meniscus-type gypsum cements. Evidence was found of complex diagenetic histories in the capillary fringe and phreatic zones, reflected by multiple overgrowths and several episodes of dissolution on single grains. These fluctuations probably reflect changes in saturation related to alternating dilution by meteoric recharge and subsequent evaporative concentration.

The surface sand layers of desert dunes often become cemented by thin salt crusts following evaporation of rain water, spray, dew, or fog which contains dissolved salts. This phenomenon is particularly common in coastal deserts and around saline lakes. The resulting salt-cemented crusts, termed *salcretes* (Yasso 1966), are usually ephemeral, but may persist for months or years if not dissolved by rain or eroded by saltating grains (Pye 1980b, Watson 1983a, p. 169, Pye & Tsoar 1987). In arid climates they may be preserved if they are buried under a layer of freshly deposited sand. Halite (NaCl) is the most common cement, but instances of gypsum cementation have been recorded [e.g. Watson (1983b, p. 141)].

9 Management and human use of sand dune environments

Vegetation plays a key role in the morphological development of many dune landscapes. In turn, the establishment and evolution of vegetation are in part influenced by physical and chemical properties of the sand which include thermal regime, moisture retention characteristics, and nutrient availability. This interaction between vegetation and sand substrate is of central importance in the management, human usage, and conservation of sand dunes.

9.1 Thermal properties of sand, moisture regime, and vegetation growth

9.1.1 Thermal properties

The thermal properties of dune sand exert a significant influence on the suitability of sand as a substrate for vegetation growth. The surface 1–2 mm of sand absorbs heat from incoming short-wave solar radiation (Wijk & de Vries 1963). This heat energy is then transferred to the underlying sand layers by conduction. The rate of heat flux through the sand (q) is given by (Tsoar 1990b)

$$q = Q/A = -K(\delta T/\delta z) \tag{9.1}$$

where q is measured in watts per unit area ($\mathrm{W\,m^{-2}}$), Q is the rate of heat flow across an area A, K is the *thermal conductivity* of the material ($\mathrm{W\,m^{-1}\,K^{-1}}$), and $\delta T/\delta z$ is the temperature gradient below the surface. The minus sign in Equation 9.1 signifies that the heat flux is in the direction of decreasing temperature (T).

The thermal conductivity of quartz is about $6\,\mathrm{W\,m^{-1}\,K^{-1}}$. However, 35–40% of the bulk volume of dune sand is typically occupied by air-filled pore spaces. Since air is a relatively poor conductor of heat, the thermal conductivity of bulk sand is typically about one-tenth that of quartz (i.e. $0.6\,\mathrm{W\,m^{-1}\,K^{-1}}$) (Janza 1975).

The ratio of the total amount of heat needed to increase the temperature of a unit mass of sand by 1K to the total amount of heat required to raise the temperature of the same mass of water at 15°C by 1K is known as the *specific heat* (C) (measured in units of $\mathrm{J\,kg^{-1}\,K^{-1}}$). Pure quartz has a C value of $740\,\mathrm{J\,kg^{-1}\,K^{-1}}$, whereas that of sandy soil is typically about $100\,\mathrm{J\,kg^{-1}\,K^{-1}}$ (Janza 1975). The specific heat of water at 20°C is about four times larger than that of air at the same temperature, and water therefore requires a much larger input or output of energy than air to change its temperature. Consequently, dry sand is much more sensitive to thermal changes than wet sand (Satoh 1967).

The rate of temperature change within a sand body is directly proportional to its thermal conductivity, but inversely proportional to its specific heat (Janza 1975):

$$k = K/C\gamma \tag{9.2}$$

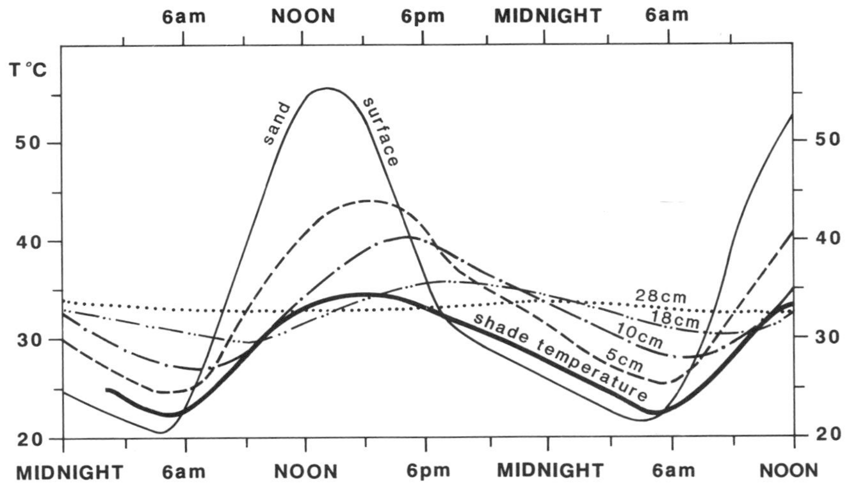

Figure 9.1 Variation in temperature with depth in a sand patch on a flat wadi bed near Cairo on 8–9th August 1923. (Data from Williams 1954).

where k is a measure of temperature change, known as the *thermal diffusivity* ($m^2 s^{-1}$), and γ is the bulk density of the sand (typically 1.6×10^3 kg m^{-3}). The thermal diffusivity of typical sandy soil is relatively low (about $3.75 \times 10^{-7} m^2 s^{-1} K$), but at a depth of 28 cm the diurnal temperature variation may be less than 3 °C (Fig. 9.1) (Sinclair 1922, Salisbury 1952, p. 192, Williams 1954, Willis *et al.* 1959b, Cloudsley-Thompson & Chadwick 1964, p. 21, Boerboom 1964, Satoh 1967, Dincer *et al.* 1974, Lettau 1978).

Since the albedo of bare sand is relatively high (List 1949, p. 442), the net radiation absorbed is not as large as might be expected. The maximum net radiation absorbed at the surface in the Mojave Desert, California, at noon in June was reported to be only 620 W m^{-2} (Vehrencamp 1953), which is comparable to higher latitude values (McNaughton & Black 1973, Ripley & Redmann 1976). If the sand is covered by dark organic debris, the albedo is reduced, more heat is absorbed, and higher surface temperatures can be attained (Otterman *et al.* 1975, Salisbury 1933, 1952, p. 194).

The daytime surface sand temperature is often as much as 20 °C higher than that of the air only a few centimetres above the surface (Vehrencamp 1953, Satoh 1967, Baldwin & Maun 1983). This steep lapse rate may give rise to local up-rushes of hot, light air which develop into whirlwinds (Fig. 2.6). At night the sand surface may cool to a temperature lower than that of the air immediately above, giving rise to a near-surface inversion which suppresses air motion near the surface.

The steep temperature gradient which develops in the uppermost 30 cm of sand at night favours the upward movement of soil moisture vapour and its distillation in the cooler near-surface sand layers, leading to the formation of sub-surface dew (Salisbury 1952, p. 183, Willis *et al.* 1959b, Migahid 1961). During the summer, nocturnal sub-surface dew may increase the moisture content of the near-surface sand by as much as 0.9 ml per 100 ml of soil (Salisbury 1952, p. 187), which is probably sufficient to meet the transpiration needs of shallow-rooted annual plants growing on British coastal dunes. Dew may also form on the surface when moisture-laden air condenses at night onto cold surfaces such as plant stems and pebbles. Dews of this type are of particular biological significance in coastal deserts such as the Namib, where precipitation from coastal fog is also important (Seely & Louw 1980, Seely 1984).

Since the presence of moisture increases the specific heat of a bulk soil, well drained sandy soils which have a low moisture retention capacity are warmer at depth than wet loamy or clayey soils which have a higher moisture retention capacity. Hence, in some subtropical areas, sandy soils provide a favourable substrate for the growth and ripening of crops during the winter season (Satoh 1967, Tsoar & Zohar 1985). However, temperatures at the sand surface may fall close to, or below, freezing owing to the low thermal diffusivity of the sand. Natural or artificial mulches may therefore be required to prevent frost damage and increase the rate of ripening (Satoh 1967).

9.1.2 Sand moisture regime

Since moisture is able to percolate relatively freely through well sorted dune sands, the amount of retained moisture available to plants is relatively low. The maximum quantity of water which can be retained by surface tension and granular absorption against the pull of gravity is defined as the *field capacity*. The field capacity of active dune sands which contain <1% fines and <1% organic matter varies from about 4% to 10% (Noy-Meir 1973, Brady 1974, p. 174). Stabilized dune sands which contain more fines and organic matter can retain up to 35% moisture at field capacity (Ranwell 1959, Brady 1974, p. 196, Kutiel & Danin 1987). In England, Salisbury (1952, p. 152) found the average moisture retention capacity to be 7% in young coastal dune sands and 33% in the near-surface layers of older dunes containing a few per cent of organic matter.

Plant roots absorb water from the soil by exerting osmotic pressure. The point at which no further moisture can be extracted by the roots is referred to as the *wilting point*. Chepil (1956) found that dune sand had an average moisture content of 1.28% at the wilting point, although the precise value depends on the plant species.

An advantage of dune sand for plant growth arises from the fact that even a small amount of rainfall can infiltrate to a relatively large depth, and thereby becomes available to plant roots, whereas the wetting depth is limited in finer textured soils. Experiments in the Negev and elsewhere have shown that 1 mm of rainfall penetrates to a depth of 7 mm on medium-grained sand and to 20 mm on coarse sand (Dincer *et al.* 1974), but reaches only 5 mm in silty soils (Orev 1984) and 2 mm in clayey soils (Walter 1973, p. 100). However, because of the relatively small difference between the field capacity and wilting point in fresh dune sands, frequent wetting events are required to sustain the growth of plants which depend on vadose zone soil moisture.

The rate at which water flows through a permeable medium is defined as the permeability or *hydraulic conductivity*. In unweathered dune sands water is able to percolate downwards relatively easily towards the groundwater table under the influence of gravity. The hydraulic conductivity is, however, dependent on a number of factors including the size distribution of the sand grains and intergranular pores, the nature of any internal structures and textural inhomogeneities, the presence of entrapped air, and the rate at which moisture is taken up by plant root systems (Bagnold 1938b, Smith *et al.* 1968, Prill 1968).

Goldschmidt & Jacobs (1956) reported an average hydraulic conductivity of about 13 m day^{-1} on coastal dunes in the Haifa–Akko region of Israel, whereas in New Mexico Hennessy *et al.* (1985) measured average hydraulic conductivities of 19 and 20 m day^{-1} at depths of 15 and 30.5 cm, respectively. Goldschmidt & Jacobs (1956) developed the following empirical equation for the groundwater recharge in the Haifa dunes:

$$D = 0.233P + 17.5 \tag{9.3}$$

where D is the groundwater recharge by infiltration (mm yr^{-1}) and P is the annual precipitation (mm). An alternative equation, also developed for dunes in the Coastal Plain of Israel, is (Lerner *et al.* 1990):

$$D = 0.81(P - 94) \qquad (9.4)$$

where $P - 94$ mm is a threshold value below which no infiltrating water reaches the water table. The amount of precipitation required for groundwater recharge also varies with grain size. According to Dincer *et al.* (1974), the minimum mean annual precipitation for recharge must be 70 mm on Israeli dunes with a mean grain size of 0.3 – 0.4 mm and 150 mm on dunes with a mean grain size of 0.2 mm.

The intensity of rainfall events is an additional important factor which determines the degree of penetration of moisture into dune sands. Light rains keep the surface sand moist but produce little recharge to the groundwater table since much of the moisture is returned to the atmosphere by evapotranspiration. Heavy showers, by contrast, produce greater depth of penetration and proportionally smaller evaporative losses. In the Negev, rainstorms of less than 5 mm, which wet the sand to an average depth of less than 50 mm, account for 40% of the total precipitation (Lerner *et al.* 1990). This is typical of many arid lands in which there are only 10–50 rainy days each year, generally occurring in 3–15 rain events, of which less than 6 are of sufficient magnitude to affect the vegetation (Noy-Meir 1973).

Following rainfall, the upper sand layers gradually dry out owing to evaporation and transpiration. In arid climates the uppermost 5–10 cm of sand is generally dry 5–25 days after rain, but on unvegetated dunes it may take many weeks or months for the sand at greater depth to dry out by evaporation. Below 30 cm surface temperature variations have little effect and there is virtually no direct evaporative loss (Bagnold 1954b, Prill 1968, Noy-Meir 1973). Capillary forces are of limited importance in sands compared with finer sediments owing to the relatively larger size and smaller number of the pores. Most upward moisture movement in sand therefore occurs in the vapour phase (Gupta 1979). Even vapour movement below the uppermost 50–60 cm is restricted owing to the relatively low thermal gradient.

In silts and clays, capillary forces are more effective, with the result that evaporation at the surface causes the sediment to dry out more quickly and to a greater depth.

On vegetated dunes, a significant proportion of the moisture which infiltrates to a depth of more than 30 cm is returned to the atmosphere by plant transpiration. The presence of root systems associated with perennial dune vegetation often results in moisture depletion at a depth of 30–180 cm below the surface (Prill 1968, Gupta 1979). For this reason, during dry periods, patches of bare sand on dunes are often moister below the surface than are neighbouring vegetated areas. In some cases, uptake of vadose moisture by plant roots may prevent any groundwater recharge. Lerner *et al.* (1990) considered that the absence of a perched water table in parts of the western Negev, which receives 100–150 mm of precipitation, is due to high transpiration losses from dune vegetation. Lerner *et al.* (1990) therefore proposed a modified empirical equation for groundwater recharge which takes account of the transpiration loss factor:

$$D = 0.6P - T_r \qquad (9.5)$$

where T_r is the annual transpiration by vegetation.

If the moisture content of the upper 30–60 cm of sand falls below the wilting point during the summer months, shallow-rooted grasses and other species die back, but sufficient vadose moisture may be present at greater depths to allow species with deeper root systems to continue to grow (Migahid 1961, Dincer *et al.* 1974, Tsoar & Zohar 1985). Plants whose root systems extend down to the groundwater table (*phreatophytes*) are also able to grow throughout the year. Sand within about 30 cm of the water table is kept moist by capillary rise, and many dune plant species have root systems which, at least on small dunes, are able to reach this zone (Salisbury 1952, Ranwell 1959, 1972, Willis *et al.* 1959b, Maun 1985). Deep root systems also provide good anchorage under conditions of shifting sand. Desert plants which do not have deep root systems, such as cacti, cannot survive on shifting sand (Ash & Wasson 1983, Bowers 1982, 1986).

Since the evaporation rate from sand is lower than that from fine-textured soils, and retained moisture is often present in sand below 30 cm depth, desert sands often support more vigorous growth of perennial mesophytic shrubs than adjacent clay and loam soils (Chadwick & Dalke 1965, Alizai & Hulbert 1970). According to Le Houerou (1986), the biomass resulting from 1 mm of rainfall on sandy soils is 2.5 times higher than that produced by the same rainfall on fine-textured soils. Correspondingly, the minimum amount of rainfall required to initiate growth of annual vegetation on sandy soils is about 10–15 mm, whereas an event of this magnitude has no effect on silty and clayey soils (Le Houerou 1986). Fine-textured soils only become more advantageous than sands for plant growth when the precipitation reaches 300–500 mm (Noy-Meir 1973). Under these circumstances the advantage of low evaporative loss from sands is offset by the relatively high seepage loss to the groundwater table.

Where a sand surface is being eroded, the depth of evaporative drying progressively increases, eventually causing the moisture content in the plant root zone to fall below the wilting point, or even exposing the roots at the surface. On many dunes vegetation is absent from the windward slopes and crests where sand is being deflated, but grows vigorously on the lee slopes or dune plinths where there is little net change in surface level or where sand is slowly accreting (Figs 9.2 & 9.3). The plinth and interdune areas also usually contain more fines, thereby encouraging moisture retention and providing a more abundant source of plant nutrients (Zobeck & Fryrear 1986b). Plants growing in these sites are also more likely to be able to reach ground water.

Within individual dunefields, many vegetation species have a sharply defined distribution in relation to soil moisture regime. Some species are confined to relatively wet

Figure 9.2 View of the windward slope of a transverse dune in the southern part of the Coastal Plain of Israel (average annual rainfall 400 mm). Note the bare windward slope while the crest is vegetated mostly by *Ammophila arenaria*.

Figure 9.3 A seif dune in the Sinai desert (average annual rainfall 100 mm). Vegetation is present only on the dune plinth and in the interdune area.

areas which experience periodic flooding, whereas others occur only on sites which never suffer waterlogging. This type of hydrological zonation is well displayed in humid coastal dune systems (Salisbury 1952, Willis & Jefferies 1963, Ranwell 1959, 1960, 1972, Boorman 1977, Willis 1985, 1990).

The water table often forms a dome in humid-region dune sand accumulations, the summit of which may be several metres or tens of metres above the level outside the dunefield (Willis *et al.* 1959a, Laycock 1975a, 1978). In desert areas, the groundwater table, if present, also usually shows a relationship to the surface topography.

The salinity of the ground water is a further important factor which influences vegetation growth and distribution. In parts of some desert basins the salinity of ground water is high enough to preclude plant growth altogether.

9.1.3 Other factors which influence dune vegetation

At the regional scale, the nature and density of vegetation cover on sand dunes is influenced by wind energy which controls the degree of sand mobility. In areas of high wind velocity and abundant sand, transverse dunes devoid of vegetation occur even under humid conditions, as along the coasts of Oregon (Cooper 1958, Hunter *et al.* 1983) and southwestern South Africa (Illenberger & Rust 1988, Illenberger 1988). In arid lands where wind energy is moderate and rates of sand transport are relatively low, vegetation may be able to grow on the flanks of linear and parabolic dunes even when the annual rainfall is less than 50 mm (Hagedorn *et al.* 1977, p. 156, Danin 1983, p. 50). The relationship between dune mobility, rainfall, and wind energy is discussed in more detail in Section 5.6.

The nutrient regime of the sands is also an important factor which influences the vigour of vegetation growth. In many young sand-dune systems, shortages of nitrogen, phosphorus, and other essential elements such as potassium limit plant growth (e.g. Wilson 1960, Willis & Yemm 1961, Willis 1963, 1985, 1990). If the sands become

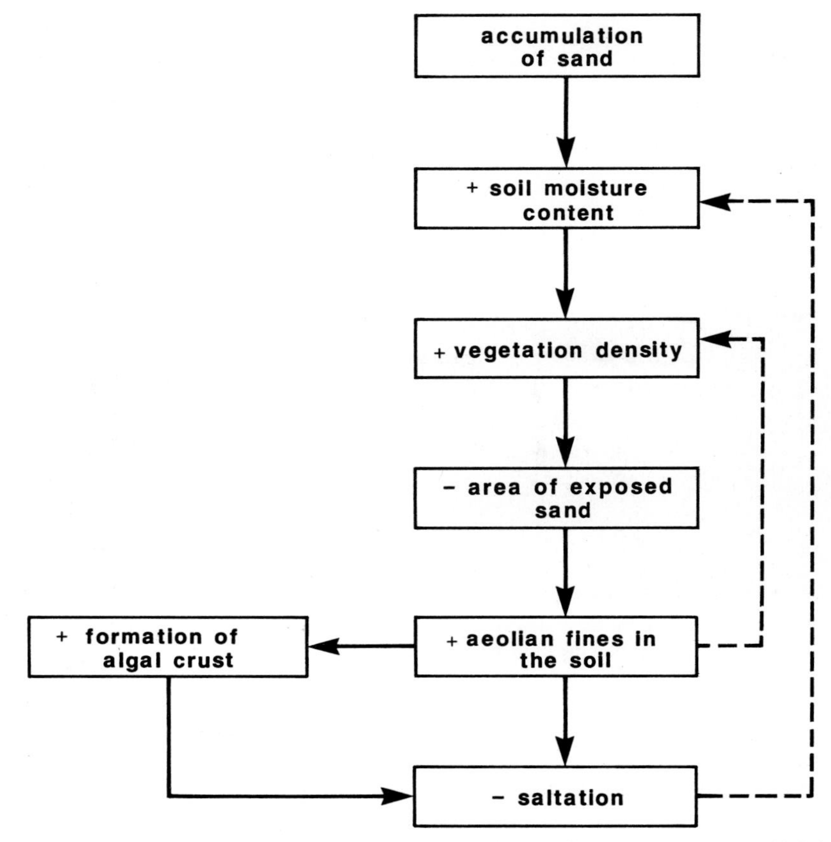

Figure 9.4 Flow chart showing the inter-relationship between the natural processes which lead to dune stabilization. Positive and negative signs represent increase and decrease of a process, respectively. Positive feedback mechanisms are indicated by broken lines. (After Tsoar & Møller 1986).

stabilized, levels of these nutrients commonly increase owing to mineral weathering and addition of airborne material. Once some plants have become established, they encourage further sand stabilization by reducing surface wind velocities and trapping windblown fines (Fig. 9.4). The presence of decaying organic matter also improves the moisture retention capacity of the sand (Lehotsky 1941, Ranwell 1959, Le Houerou 1977a), thereby increasing mineral weathering rates. The accumulation of fines encourages the formation of a surface crust which is enhanced by fungal, algal, and lichen colonization (Danin 1978, 1987, MacEntee & Bold 1978, van den Ancker *et al.* 1985). Algae and lichens subsequently become an important source of nitrogen which supports the growth of higher plants (Shields *et al.* 1957). By this system of positive feedback, the soil–vegetation system progressively becomes more complex.

The important role played by airborne dust in dune stabilization is illustrated by the fact that coastal dunes in southern Israel, which receive 230 mm of annual rainfall, possess a richer dune flora than dunes in northern Israel where the annual rainfall is higher (590 mm) but rates of dust deposition are much lower (Tear 1925, Yaalon & Ganor 1975). Formerly mobile desert dunes in the northern Negev near Beer Sheva, in an area receiving 150 mm of annual rainfall, became completely stabilized within 10

Figure 9.5 Natural regrowth of *Artemesia monosperma* vegetation cover on a sand sheet a few years after the construction of a fence prevented grazing by nomadic herds, northern Sinai.

years of the cessation of human interference (Danin 1987), (see also Fig. 9.5). Rapid natural recovery of dune vegetation has also been reported from parts of North Africa which receive 80–120 mm of annual rainfall (Le Houerou 1968, 1977a, Hagedorn *et al.* 1977, p. 156).

9.2 Water courses in dune areas

Additional water is introduced into some dunefields by watercourses whose headwaters lie outside the dune area. Because of the highly permeable character of dune sand, desert streams lose much of their water while crossing a dunefield. However, small dunes which cover the surface of a flood channel may be washed away during major flood events. In addition to the rate of seepage loss, three factors determine whether a watercourse will succeed in maintaining an open channel across a dunefield: (a) the width of the channel (w), (b) the rate of advance of dunes across the channel (c), and (c) the frequency of major floods (f). If floods are sufficiently frequent, the channel sufficiently wide, and the rate of dune advance sufficiently low, an open channel will be maintained. An example is provided by Wadi El Arish, which crosses the dunefields of northern Sinai to reach the Mediterranean coast (Sneh 1982). In this case $w = 1500\text{–}3000\,\mathrm{m}$, $c = 15\,\mathrm{m\,yr^{-1}}$, and $f \approx 1/20\,\mathrm{yr^{-1}}$. In other cases, where the frequency and/or magnitude of floods are too low, or the rate of movement of obstructing dunes is too high, the channel will become blocked. Ephemeral flood waters are then temporarily ponded within interdune depressions (Fig. 9.6) (Smith 1969, p. 71), and eventually dissipate by evaporation and infiltration into the underlying sands. In the Namib Desert, for example, several westward-draining streams are unable to cross the dunefields and terminate in shallow pans known as *vleis* (Seely & Sandelowsky 1974, Lancaster 1984, Ward 1988) (Fig. 9.7). This is an additional important way in which fine sediment is introduced into some dune areas.

Figure 9.6 Water-filling an interdune depression after heavy rain and flood, northern Sinai.

9.3 Control of windblown sand

Drifting sand and migrating dunes present a major threat to agriculture, forestry, roads, railways, and other communication and distribution systems in many parts of the world. The problems are particularly acute in subtropical desert regions (e.g. Khalaf 1989b, Al-Nakshabandi & El Robee 1988), but are also significant along humid coasts and in some humid temperate inland areas where sandy soils are cultivated (e.g. Møller 1985, 1986, Castel 1988). The negative effects of sand drift and dune migration include erosion of soils, abrasion damage to crops and paintwork, blocking of roads, canals and railways, infilling of wells and reservoirs, and burial of buildings and industrial installations (Clements *et al.* 1963, Hidore & Albokhair 1982, Watson 1985).

Attempts to limit the damage caused by blowing sand have a long history. Planting of dune grasses has been undertaken on European coastal dunes since the Middle Ages (Ashton 1909, Case 1914, van Dieren 1934), and sand fences were used to stop drifting sand at least as early as the 16th century (Norrman 1981). During the late 18th century, complex schemes for sand stabilization were implemented in Gascony by French engineers (Bremontier 1833, Harlé 1914, Fenley 1948). In the United States, stabilization measures were initiated on the Massachusetts coast in the early 18th century (Westgate 1904, Kucinski & Eisenmenger 1943), and have subsequently been widely applied (Woodhouse & Hanes 1967, Savage & Woodhouse 1969, Manohar & Bruun 1970, Woodhouse 1978, Knutson 1980). Measures to control the movement of desert sand have been employed in the Nile Valley and parts of northern China for millennia (Stallings 1953, Zhenda *et al.* 1987). However, the need for more effective measures of sand control has become more apparent in the past 40 years owing to economic growth and urban development related to exploitation of oil and gas resources in arid areas such as the Middle East (Kerr & Nigra 1952, Cooke *et al.* 1982, Watson 1985, Jones *et al.*

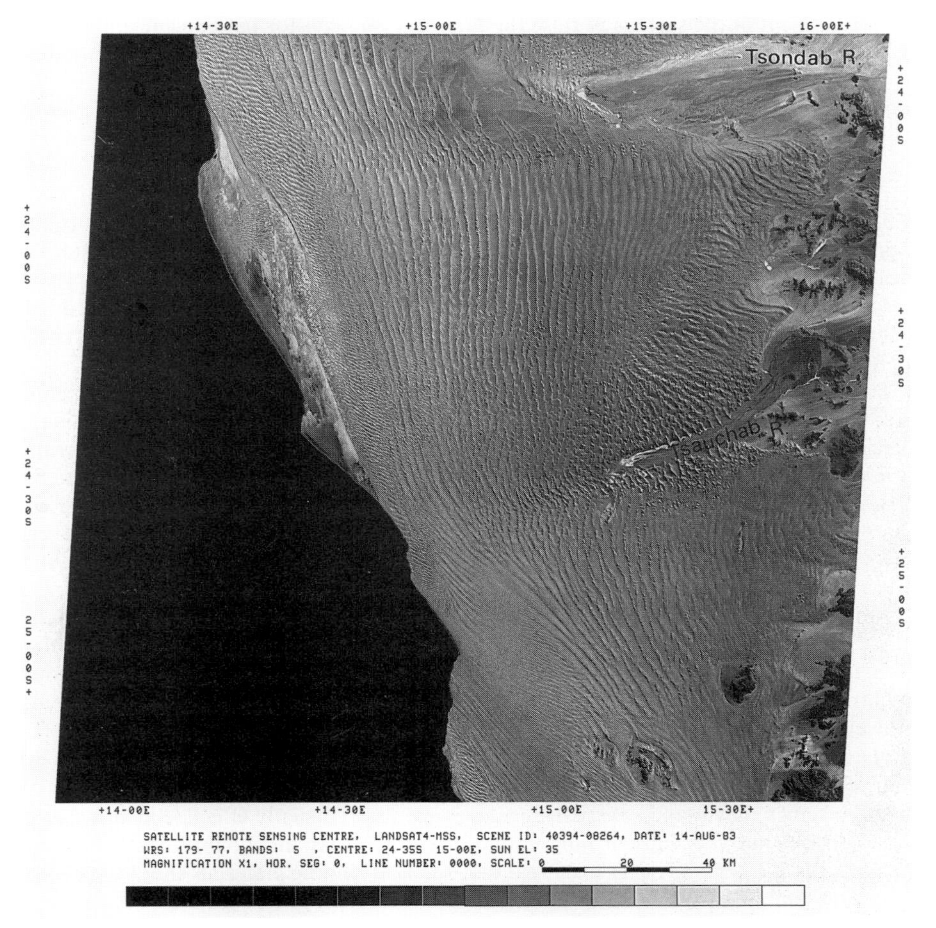

Figure 9.7 Landsat photograph of the Namib Sand Sea, showing blocking of the Tsondab and Tsauchab Rivers by linear megadunes.

1986), and to increased concern about the human and environmental consequences of desertification (Rapp, 1974, Le Houerou 1975, 1977a, b, 1986, Hagedorn *et al.* 1977).

Five main approaches have been employed to avoid or reduce the problems associated with sand encroachment [see Busche *et al.* (1984) for an annotated bibliography of older work]:

(a) reduction of sand supply and elimination of sand sources;
(b) enhancement of sand transport;
(c) deflection of moving sand;
(d) enhancement of sand deposition;
(e) stabilization, dissipation, or removal of moving dunes.

9.3.1 Reduction of sand supply

When blowing sand originates from a relatively restricted geographical area, it is often most cost effective to attempt to stabilize the source area rather than to trap the sand in motion over a wide area. Three main methods of source area control are employed:

(a) treatment of the surface to raise the threshold velocity for particle entrainment;
(b) reduction of the wind velocity over the surface area by the construction of shelter belts, fences, or other protective obstructions;
(c) exclusion of human, animal, and vehicular traffic from the source area in order to minimize disturbance to the surface and to encourage formation of a wind-stable surface layer.

9.3.1.1 Surface stabilization by mulches A wide variety of materials have traditionally been used in the past to protect surface sand from erosion, including stones, dead corn stalks, peat, brushwood, straw, and domestic refuse. In more recent years these materials have been widely replaced by oil, bitumen, lime, resin, latex, and other commercially formulated chemicals (Armbrust & Dickerson 1971, Barr & McKenzie 1976).

Surface layers of natural gravel or other coarse material, such as crushed rock, concrete, or brick, form a wind-stable pavement which can almost completely eliminate deflation. However, the protective layer needs to be at least 50 mm thick, since disruption of thinner layers is followed by rapid scouring of the underlying sediments (Logie 1981, Watson 1985). Natural gravels and crushed rock are more durable and effective than corn stalks and refuse, and are readily available in most arid regions. However, their negative visual impact makes them generally unsuitable for use in coastal dune areas which are used for recreation.

Moisture is also effective in raising the threshold for sand entrainment, and a technique which is widely used in quarries, mines, and towns in arid areas involves regular spraying of road surfaces and other open spaces. Saline water can be used advantageously for this purpose since on crystallization the salts act as an additional binding agent. However, spraying suffers from the disadvantages of being labour intensive and relatively costly. The salt crusts which form are also ephemeral, being easily destroyed by rain, scouring by strong winds, pedestrians and vehicles.

Oil, bitumen, latex, and other chemical treatments are usually effective in stabilizing sand surfaces but vary in their long-term cost effectiveness and environmental impact. Armbrust & Dickerson (1971) suggested that suitable binding materials should have the following qualities: (a) low cost, (b) resistance to wind and water erosion for at least 2 months, (c) be permeable to surface water, (d) be non-toxic and have no other adverse effect on plant growth, and (e) be easy to apply. Of 34 commercially available materials tested, only six were found to meet these requirements, with liquid polymers being considered to provide the best compromise.

When it is not intended to introduce vegetation after stabilization of the surface, where the visual impact is unimportant, and where pedestrian and vehicular traffic is not required to cross the treated areas, thick coatings of oil or bitumen usually achieve the desired result at minimum cost.

The longevity of any chemical treatment increases with the resilience and thickness of the material. However, materials which harden to form an impervious surface crust may create problems with runoff following heavy rainfall, and may be subject to wind scouring and collapse at the margins of the treated area. Oil and bitumen crusts oxidize over time, and may crack owing to alternate expansion and shrinkage in response to temperature changes. Highly viscous waxy oils do not form a brittle crust subject to cracking and deflational scouring, but are unsuitable where vehicles or personnel must cross the treated area (Watson 1985).

Since most chemical treatments affect only the uppermost few millimetres of sand, scouring of the underlying unconsolidated sand can present a serious problem if the surface crust is breached. Pressure injection techniques can be employed to achieve deeper penetration, but the higher cost is normally unwarranted except in special circumstances (e.g. around military installations and civilian arifields).

If chemical treatments are used only with the intention of stabilizing the sand surface long enough for vegetation to become established, the long-term durability of the coating is relatively unimportant. Haas & Steers (1964) reported the successful use of latex sprays as an aid to revegetation of British coastal dunes. However, Adriani & Terwindt (1974) concluded, based on field trials in The Netherlands, that bitumen and synthetic latex emulsions had a generally negative effect on plant growth. Ranwell & Boar (1986) were also of the opinion that the value of synthetic coatings as an aid to sand stabilization on British coastal dunes is unproven.

9.3.1.2 Physical barriers to airflow Physical obstacles to the wind create regions of low flow velocity both in front of and behind the barrier. Barriers built upwind can thus be used to reduce the friction velocity over a potential sand source area. Artificial barriers can be constructed of any material sturdy enough to withstand strong winds. Reeds, palm fronds, brushwood, slat-fencing, planks, stakes, and synthetic nylon mesh have all been used. Alternatively, a living shelter belt can be created by planting rows of trees or shrubs perpendicular to the dominant sand flow direction (Raheja 1963, Miller *et al.* 1975). Since the main use of sand fences is to trap moving sand, they are discussed more fully in Section 9.3.4.

9.3.1.3 Restriction of human activity in potential sand source areas The emission of windblown sand and dust from many terrain types is enhanced by human disturbance (Clements *et al.* 1963, Wilshire 1980, Tsoar & Møller 1986, Jones *et al.* 1986). Trampling by pedestrians and livestock, overgrazing, and destruction of surface crusts by trail-bikes, four-wheel-drive vehicles, and tanks are major factors leading to enhanced sand blowing in deserts (Fig. 9.5) and some coastal dunefields. Restriction of access, or banning of such activities altogether, can often reduce or eliminate the need for other sand control measures.

9.3.2 Enhancement of sand transport

In some instances where the supply of sand cannot be controlled, or where it is not possible or desirable to trap moving sand, it may be preferable to enhance the transport rate of sand across an area, and to reduce the amount which is deposited, by increasing the sand-transporting capacity of the wind or by reducing the ability of the surface to trap moving grains. Both objectives can be accomplished by streamlining the land surface. In order to prevent sand accumulation on an embanked road or railway, for example, the windward slope should have a gradient of less than 1 in 6. Steeper embankments are likely to cause sand to accumulate on the windward side. The change in slope at the crest of the embankment should also be gradual, since an abrupt change will lead to flow separation and accumulation of sand on the lee side (Watson 1985). Rates of sand transport across an area can also be locally enhanced by covering the surface with gravel or bitumen, thereby increasing the efficiency of saltation.

Where coastal dunes act as natural sea defences, coastal engineers may wish to enhance the rate of sand transport from the beach in order to rebuild foredunes following a damaging storm. By ploughing the surface of the upper beach, drying of the sand is enhanced, salt crusts are broken up, and more grains are exposed to relatively high wind velocities at the crests of the ridges (Fig. 9.8). Aeolian activity can also be enhanced by beach nourishment (e.g. Draga 1983).

Figure 9.8 Ploughed strip on the upper beach, Southport, Merseyside, UK, used for racehorse training. Aeolian entrainment of sand from the ploughed strip is enhanced compared with the adjacent part of the beach on the left which has been compacted by vehicle tyres.

Figure 9.9 Multiple wooden slat fences on a coastal dune blowout, Long Island, New York.

9.3.3 Diversion of moving sand

Sandflow can be diverted around individual structures using fences and tree belts. Two types of fence alignment have been used. In the first, the sand fences are slanted at an angle of 45° to the direction of sand drift. In the second, the fences form a V-shaped barrier pointing into the sandstream. To be most effective, and to maximize their useful life, the fences should have low permeability, thereby limiting the amount of sand deposited in their lee. In almost all cases, however, mechanical removal of accumulated sand is periodically required in order to maintain their effectiveness.

9.3.4 Enhancement of sand deposition

Enhancement of sand deposition can be achieved by construction of sand fences, by planting of vegetation, or by a combination of methods.

9.3.4.1 Sand fences Although the range of fence designs used is large, they all operate on the principle that areas of low wind velocity are created both in front of and behind the fence (Fig. 9.9). The amount of sand trapped is dependent on the fence height, its porosity, the shape and arrangement of the gaps through which the air can pass, and the wind velocity (Savage & Woodhouse 1969, Manohar & Bruun 1970, Willetts & Phillips 1978, Phillips & Willetts 1978, 1979). The alignment of the fences with respect to the sand drift, and with the number and spacing of fences, are also important (Kerr & Nigra 1952, Mulhearn & Bradley 1977).

A fence with less than 20% porosity behaves almost as a solid obstacle to the flow, inducing upstream and downstream separation bubbles (Pande *et al.* 1980) (see Section 6.3.2). Initially, most sand deposition occurs on the upstream side, forming an echo dune (Fig. 9.10). Owing to the large difference in static pressure between the two sides of the fence, it may be blown over, or it may be undermined by eddies on the windward side.

Fences with >30% porosity do not usually give rise to significant reverse flow on the windward side (Castro 1971), so that the zone of reduced wind velocity is wider than in the case of a non-porous fence of similar height. The maximum reduction in surface shear velocities, resulting in maximum sand deposition, is achieved by fences with 36–40% porosity, regardless of structure (Savage & Woodhouse 1969, Phillips & Willetts 1978, 1979). The maximum reduction in surface shear stress occurs some distance downwind of a porous fence, with the result that sand accumulation is initially greater on the leeward side than on the windward side (Fig. 9.10).

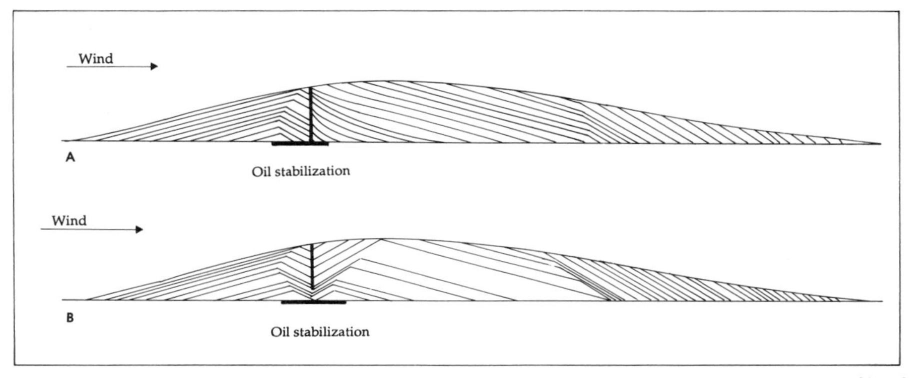

Figure 9.10 Sections showing the idealized incremental growth and ultimate steady-state profile of sand accumulations around (A) a solid fence and (B) a porous fence erected perpendicular to the wind. (After Kerr & Nigra 1951).

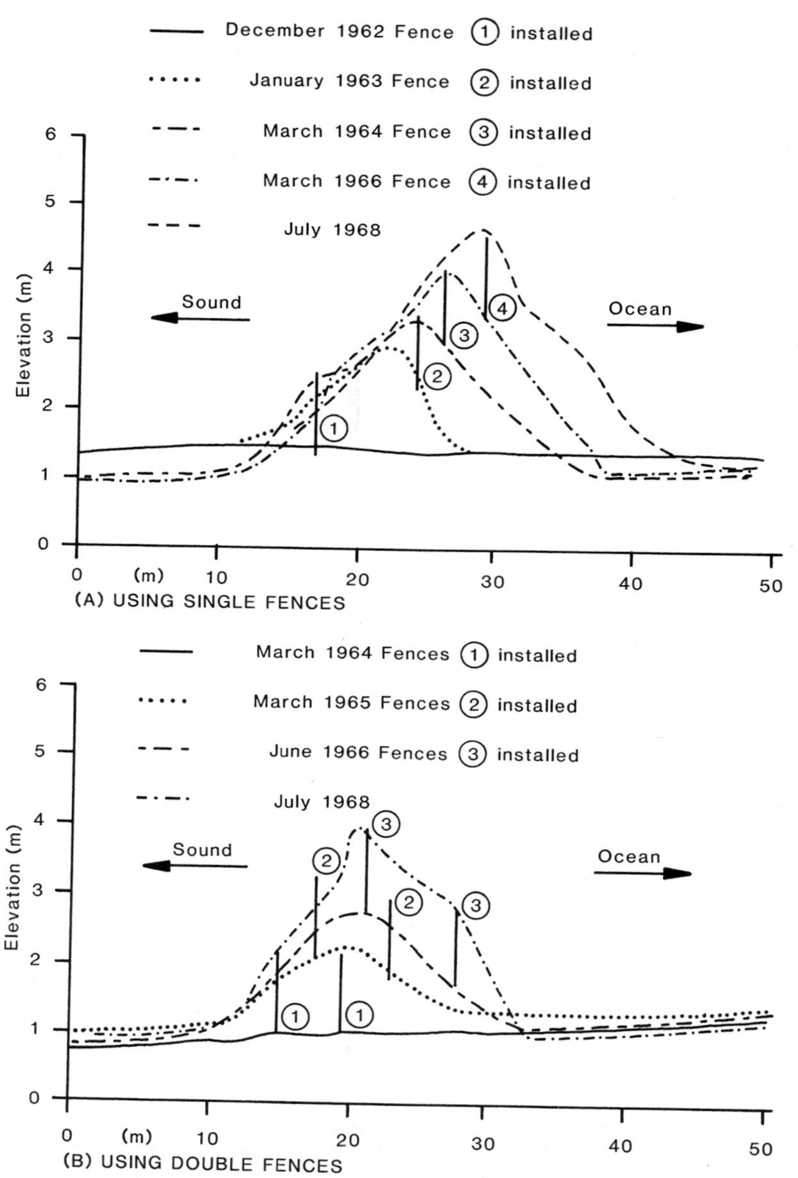

Figure 9.11 Dune construction on the coast of North Carolina, USA, (A) using single fences and (B) using double fences. (After Savage & Woodhouse 1969).

Build-up of sand gradually buries the fence, and the accumulating sand tends to assume an equilibrium form which is similar for both porous and non-porous fences. Once an equilibrium form has been established, sand passes over it unhindered. In order to trap further sand, new fences must be erected on the surface of the dune. The final dune form is strongly influenced by the sequential positioning of the fences (Kerr & Nigra 1952, Savage 1963, Blumenthal 1964, Jagschitz & Bell 1966) (Fig. 9.11).

Field tests have shown that a wooden slat fence with 50% porosity is more effective

Figure 9.12 Accumulation of blown sand behind a single brushwood fence, Formby, Merseyside, UK.

Figure 9.13 Transverse slat-type fence and spur fences constructed of chestnut palings, Formby, Merseyside, UK.

than a framed fibre fence with any degree of porosity (Savage & Woodhouse 1969). Wooden slat fences also have the advantage that they are more easily lifted for re-use after being partially buried by sand (Benito & Le Roux 1976, CERC 1977). However, the percentage of transported sand which is trapped by a slat fence varies with the wind velocity. Manohar & Bruun (1970) found that the sand trapping efficiency of a single slat fence was 60% at 10 $m\,s^{-1}$ but only 16% at 18 $m\,s^{-1}$. At velocities above 18 $m\,s^{-1}$ almost no sand was trapped by a single fence with 50% porosity. A double row of fences was found able to trap about 30% of the moving sand at such velocities but additional rows had a negligible effect. The optimum distance between the two fences was found to be about four times the fence height (Manohar & Bruun 1970).

The volume of sand trapped by a fence varies according to the square of its height. Hence the 'effective life' of a 2 m-high fence is four times than of a 1 m-high fence (Kerr & Nigra 1952). On many coasts where sand movement is appreciable a 1.2 m-high fence can easily be buried in a single year (Woodhouse 1978).

Nylon mesh or brushwood is sometimes used as an alternative to wooden slat fencing on grounds of cost or material availability (Fig. 9.12).

Whatever their construction, fences are normally orientated transverse to the dominant sand flow direction. Where the wind direction is variable, or where the sand flow is commonly parallel or highly oblique to a line of frontal dunes, spur fences may be constructed in preference to, or in addition to, transverse fences (Figs 9.8 & 9.13). Spur fences are normally 10–15 m long and spaced at intervals of 10–20 m (Brooks 1979, Ranwell & Boar 1986). Zig-zag arrangements of fences have also been employed, but the greater amount of sand trapped has been considered insufficient to justify the additional cost of their construction (Savage 1963). A further alternative form of fencing involves driving rows of regularly spaced stakes into the ground to form a regular grid pattern. In Michigan, 46 cm-long stakes were driven into the ground to a depth of 15 cm, forming a grid pattern of side length 4 m (Lehotsky 1941). Slat-type fences are also sometimes erected to form such a boxwork pattern (Fig. 9.14).

Lines of junk cars, empty oil drums and old tyres have also been used to trap mobile

Figure 9.14 Grid arrangement of chestnut paling fences, Sefton Coast, Merseyside, UK.

sand. However, use of these materials is not recommended on grounds that they are less effective, more environmentally damaging, and even more expensive than fencing (Gage 1970, CERC 1977)

9.3.4.2 Sand ditches Ditches with a minimum width of 3–4 m can be used to trap sand in saltation (Watson 1985), but they need to be deep or cleared regularly in order to remain effective. In areas such as the Jafurah Sand Sea in eastern Saudi Arabia, where measured sand drift rates exceed 12.8 m^3 m-$w^{-1}yr^{-1}$ (Fryberger *et al.* 1983), a ditch 4 m wide and 3 m deep would be filled with sand in 1 year. Under such conditions, fencing offers a better solution from both engineering and economic points of view.

9.3.4.3 Vegetation planting Vegetation can be used to trap sand in several different ways. First, belts of trees or shrubs can be planted to act as self-renewing fence systems (Brown & Hafenrichter 1962, Watson 1985, Kebin & Kaiguo 1989). Suitable trees must be able to cope with the expected rate of sand accumulation and have a bushy shape which in turn promotes sand deposition. *Tamarix* spp. and *Eucalyptus* spp. have proved to be amongst the most successful shelter-belt species in the Middle East (Stevens 1974, Hidore & Albokhair 1982). In northern China, *Populus* spp., *Fraxinus* spp., and *Ulmus* spp. are the most common tree species used in shelter belts, often being planted in combination with shrubby species such as *Tamarix chinensis* and *Amorpha fruticosa* to improve their effectiveness (Kebin & Kaiguo 1989).

The death rate among newly planted shelter belt trees and shrubs is often high, especially where rates of sand movement are high (Watson 1985). Initial protection by sand fences, possibly coupled with short-term irrigation, is often required to improve survival rates. In addition to retarding sand movement, fences cast shadows which can lower the adjacent sub-surface sand temperature significantly. Stepanov (1971) reported that in the Caspian Sea region the midday temperature of dune sand shaded by wormwood fences was reduced by 1.5–2.5°C at a depth of 15 cm on south-facing slopes and by 5.5°C on north-facing slopes. The lower temperatures, and consequent greater moisture retention, in the sands protected by fences created an environment more favourable for plant growth.

Vegetation can also be planted to trap sand over large areas rather than in discrete belts (Fig. 9.15). Selection of plant types for sand stabilization is undertaken principally by botanists and foresters who have a specialist knowledge of the growth requirements of different species. However, since dunes are dynamic bedforms, the pattern of sand transport and form evolution must be properly evaluated from a geomorphological perspective before planting (Tsuriell 1974a). Shifting sand presents a hostile environment for many seedlings and young plants, and it is therefore important to identify areas of net sand erosion and accretion in order to plan the correct pre- and post-planting management measures.

Planting can be carried out either on unmodified dune topography or after partial levelling with bulldozers (Jagschitz & Bell 1966). On unmodified active dunes, it is difficult for plants to establish themselves on the crests and upper windward slopes where wind velocities are highest and sand is continually being eroded (Fig. 9.16). Survival of plants in these mobile areas may be enhanced, however, if the surface is covered with brush matting, protected by fences, or sprayed with a binding agent.

In coastal dune areas, planting or seeding may be undertaken along the backshore, in order to encourage the formation of a new foredune, on existing frontal dunes which have been disrupted by blowouts, or on degraded dunes some distance inland. Since these environments have different conditions of salinity, chlorinity, wind exposure, pH, nutrient availability, and soil moisture regime, appropriate plants must be chosen for

Figure 9.15 Marram plantings protected by brushwood fencing, Formby, Merseyside, UK.

Figure 9.16 A seif dune in the southern coastal plain of Israel stabilized, 5 years before the picture was taken, by *Acacia cyanophylla.* Note that the plants on the crest did not thrive because of the severe erosion there. However, the originally sharp crest of the seif has become rounded and is ready at this stage for stabilization by further planting.

Figure 9.17 *Ammophila arenaria* (Link), the major dune-building grass on the coasts of Western Europe.

each location. On the coasts of Western Europe, marram grass (*Ammophila arenaria*) or sea lyme grass (*Elymus arenarius*) are the species most commonly used to stabilize sand in areas not subject to tidal inundation, excessive salt spray, or grazing (Fig. 9.17). Sea lyme is considerably more salt tolerant than marram, but in areas which are regularly affected by salt, sea couch grass (*Agropyron junceiforme*) usually grows more successfully. These grasses can tolerate inundation by up to 10–30 cm of sand each year. However, when planted on a surface which is subjected to wind scour, initial protection must be provided until the plants become established. This is usually done by building low brushwood fences around the planted areas and by thatching the surface with brushwood before planting (Fig. 9.15). In all cases, planting is best undertaken during the cooler, wetter months of the year (March/April being optimum in Western Europe). Careful handling and planting techniques are required to achieve the best results (Brooks 1979, Quinn 1977, Ranwell & Boar 1986, van der Putten & van Gulik 1987). Following planting, periodic dressings of nitrogenous fertilizer can be applied to aid establishment.

Marram and couch grass also grow well on coastal dunes in southern Europe and the eastern Mediterranean. In Israel, *Lotus creticus* and *Retama raetam* are additional useful non-irrigation requiring species used for sand stabilization (Tsuriell 1974b).

Stabilization by grasses alone is not sufficient to ensure continued protection against shifting sand (Saltiel 1963). Permanent stabilization is achieved in both humid and semi-arid areas by planting of trees and shrubs. Commonly used species in the United Kingdom include sea buckthorn (*Hippophae rhamnoides*), white poplar (*Populus alba*), privet (*Ligustrum vulgare*), and willows (*Salix* spp.). Extensive stands of conifers have also been planted on many European and Mediterranean coastal dunes, mainly pine species such as *Pinus maritima*, *Pinus nigra*, and *Pinus sylvestris* (Gooch 1947, Fenley 1948, Ovington 1950, 1951, MacDonald 1954, Thaarup 1954, Ranwell 1973). *Pinus halepensis* has given excellent results on coastal dunes in Algeria and Tunisia (Tear 1925).

In eastern Australia, spinifex grass (*Spinifex hirsutus*) is the most widely used sand-trapping plant in areas with high potential sand accumulation rates (Barr & McKenzie 1977), although European marram, *Festuca* spp., and the horsetail she-oak (*Casuarina equisetifolia*) have also been used on frontal dunes (Hesp 1979, Barr & McKenzie 1976, 1977). Areas behind the frontal dunes are commonly planted with shrubs and trees such as the coastal tea tree (*Leptospermum laevigatum*), coastal wattle (*Acacia sophorae*) and coastal banksia (*Banksia integrifolia*).

In North America, the species used most frequently on backshores and foredunes include American beach grass (*Ammophila brevigulata*) along the mid- and upper-Atlantic coast and in the Great Lakes region (Jagschitz & Bell 1966, Woodhouse 1978, Knutson 1980), European marram (*Ammophila arenaria*) along the Pacific Northwest and California coasts (Cooper 1958, 1967), sea oats (*Uniola paniculata*), and panic beach grass (*Panicum amarum*) along the southern Atlantic and Gulf coasts (Dahl *et al.* 1973, Woodhouse *et al.* 1976, Knutson 1977). These grasses are all excellent sand trappers which are able to withstand burial by shifting sand, moderate exposure to salt spray, and seasonal droughts (Dahl *et al.* 1973, Moreno-Casasola 1986). As in Europe, various species of pines have been extensively planted in many areas following initial stabilization using grass species (Kroodsma 1937, Lehotsky 1941, 1972, Kucinski & Eisenmenger 1943, Brown & Hafenrichter 1962).

In arid areas, plants used for sand stabilization must cope with prolonged droughts and very high summer temperatures in addition to sand mobility during periods of high winds. In the Middle East, successful sand arrestation has been reported using *Tamarix aphylla* (Stevens 1974, Danin 1978), and in Egypt mixed stands of *Acacia cyanophylla* and *Tamarix articulata* in the proportions 4:1 have been recommended (Tag El Din 1986). These species, together with *Ricinus communis* and *Tamarix gallica*, have also been used successfully on dunes in the semi-arid parts of southern Israel and northern Sinai. *Tamarix articulata* plays an important role in foredune formation along arid Mediterranean shorelines (Fig. 9.18) (Tear 1925, Weitz 1932, Sale 1948) since it tolerates the severest conditions of exposure to salt spray and high winds and can tolerate occasional flooding by salt water (Messines 1952, Adriani & Terwindt 1974).

Acacia cyanophylla is a valuable species for stabilizing sand in sub-humid and semi-arid areas of the Near East on account of its dense root mat and dense foliage. The leaf fall of fully established *Acacia cyanophylla* completely covers the surface with litter and encourages the formation of humus (Tear 1927, Weitz 1932, Messines 1952).

In order to reduce the surface sand movement immediately after planting, and to prevent grazing by goats, camels, and other livestock, it is usually necessary to protect planted areas by fencing. In Somalia, *Commiphora* cuttings were planted in a chequerboard pattern to provide protection for several different types of seedlings used in stabilization trials (Zollner 1986). *Prosopis juliflora* proved to be the most successful tree species planted in the experimental plots, the largest trees having branches 3 m long after 15 months and extensive root systems which could tap retained moisture below 30 cm depth. The importance of a capacity for rapid root and stem growth for the survival of

Figure 9.18 Foredune ridge in southern Israel formed by planting of *Tamarix articulata*.

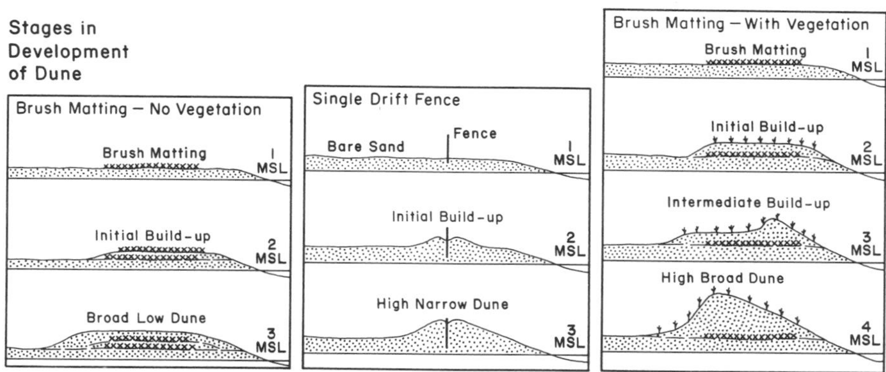

Figure 9.19 Effect of different sand accumulators on resultant dune form. (After Gale & Barr 1977).

plants in areas of shifting sand has also been demonstrated by work in North American and Tunisian deserts (Bowers 1982, Bendali *et al.* 1990).

9.3.4.4 Combined stabilization methods Successful long-term sand stabilization often requires the use of a combination of methods. Initial sand accumulation can be accomplished either by sand fences or brush matting, depending on whether the objective is to create a broad, low dune or a steep high dune (Fig. 9.19). If a broad, high dune is required, an initial period of brush matting accumulation can be followed by fence construction or vegetation planting (Fig. 9.19). Although fences and matting are very effective in the short term, they suffer from the disadvantage that the barrier ceases to function as soon as it has been buried by sand. To remain effective, artificial barriers need to be constantly maintained and modified. In the longer term, therefore, growing vegetation provides the most effective, inexpensive method of building and stabilizing dunes.

9.3.5 Control of moving dunes

The problems posed by moving sand dunes can be tackled in three ways (Watson 1985):

(a) removal;
(b) dissipation;
(c) immobilization.

The most appropriate course of action depends on the type of structure or installation being threatened, the distance of the dunes away from the structure, and the size of the dunes.

Dunes can be mechanically excavated and the sand transported to a different location. However, such action is expensive, and usually cannot be justified unless the dunes are small or the sand can be used locally for construction purposes. An alternative is to attempt to dissipate the dune so that the sand moves as individual grains rather than as a single body. This can be done by levelling or re-shaping the dune profile using bulldozers, or by selectively treating parts of the dune surface with surface coatings. Once again, these approaches are only practical if the dunes are relatively small. Where dunes are very large, or pose an immediate threat, immobilization techniques provide the best answer. Several methods can be used singly or in combination, including lowering and re-shaping of the dune crestal area, spraying the sand surface with oil or bitumen, armouring of the surface with coarse aggregate, erection of sand fences, and planting of vegetation to cut off the sand supply from upwind.

9.4 Human use of sand dune areas

Dune systems, both active and stabilized, have long attracted the attention of man. In arid and semi-arid regions, stabilized dune areas often carry a richer vegetation than other terrain types and are therefore attractive for grazing by nomadic herds. Soils on stabilized dunes are also more easily cultivated and are less prone to salinization than heavier soils in low-lying areas. In many areas, including the Sahel, overgrazing, cutting of woodland for fuel, and poor management of cultivated areas during droughts have led to extensive reactivation and erosion of formerly stabilized dunes (Barth 1982, Ellis 1987).

Coastal dunes have also been used extensively for grazing, agriculture, and forestry. Additionally, coastal dunes, particularly those near populated areas, provide useful sources of silica, building, and heavy mineral sands. Some contain large reserves of ground water used in urban water supplies; others act as important natural sea defences and sites for residential development or provide environments attractive for a wide range of recreational activities. The value of coastal dune areas as sites of special ecological significance is also being increasingly recognized.

Few coastal dune areas in Europe and North America remain untouched by human activities and, despite the greater attention given to conservation and management in recent decades, coastal dunes remain a diminishing resource (Ranwell 1972, p. 215, Gares *et al.* 1979, Olson & van der Maarel 1989, Piotrowska 1989, Westhoff 1989, Doody 1989, Hewett 1989, Martins 1989).

9.4.1 Cultivation on desert sand

Dune sand has for many years been considered to be almost useless for agriculture, although the value of pasture on stabilized dunes has long been recognized (Dainelli 1931). Cultivation on dune sand in humid areas is generally more difficult than on neighbouring fine-textured soils, even using modern techniques, owing to the tendency for nutrients to be rapidly leached from sandy soils in high-rainfall areas (Shreve 1938, Satoh 1967). However, this problem is much less severe in arid climates, where agriculture on sand can be successful if sufficient water is provided. For hundreds of years, a method of simple horticulture, termed *mawasi* (suction in Arabic), has been practised on coastal dunes in Gaza and northern Sinai. Vegetable and fruit crops are grown in interdune depressions, some natural and others artificially excavated, where a lens of fresh water approaches within a few tens of centimetres of the surface (Fig. 9.20). The water naturally contains relatively high levels of nitrogen and phosphorus, but additional fertilizer is added in the form of animal manure and green manure. Sand which is cleared from the levelled plots is heaped into marginal ramparts which give protection from saltating sand (Fig. 9.21). Palm and Tamarix trees planted around the depressions provide additional shelter from the wind and shifting sand (Tsoar & Zohar 1985).

Although traditional *mawasi* horticulture continues, modern intensive methods are being increasingly used in southern Israel and Gaza to grow out-of-season vegetables. The supply of water to the growing plants is strictly controlled using a trickle-irrigation system (Shoji 1977, Fujiyama & Nagai 1986, 1987). Using computer-controlled systems, the amount, rate, and frequency of irrigation can be closely monitored. This prevents wastage of water due to seepage losses and evaporation, and also avoids excessive leaching of nutrients. Soluble fertilizer can be supplied simultaneously with the irrigation water. In arid regions this method of cultivation can be performed in the open, but in more humid areas rainfall would leach the nutrients and substantially increase the cost of fertilizer unless the crops are protected by glass or polythene sheets.

The requirement for only small quantities of water means that low rainfall is no longer a factor precluding intensive agriculture in many deserts (Richmond 1985, Tsoar & Zohar 1985). Most of the water available in deserts is brackish, but owing to the relatively high permeability and ease with which salts are leached from dune sands it is possible to use water with a higher chlorinity than would be possible on finer textured soils (Shoji 1977, Ben-Asher 1987). In southern Israel, irrigation water with a chlorinity of 1300 mg l^{-1} is used routinely.

Another advantage of desert sand, as discussed in Section 9.1, is its tendency to experience relatively small diurnal changes in temperature at shallow depth. This

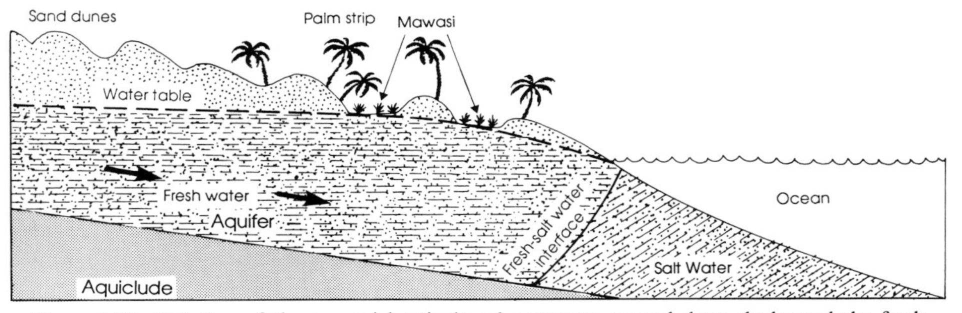

Figure 9.20 Relation of the mawasi horticultural system to coastal dune slacks and the fresh groundwater table. (Adapted from Tsoar & Zohar 1985).

Figure 9.21 Typical mawasi horticultural plot, southern Israel.

Figure 9.22 Intensive cultivation of tomatoes using trickle-irrigation in greenhouses on sand dunes, southern Israel.

tendency can be enhanced with the aid of mulches, thereby hastening the ripening of crops. However, quick winter ripening of many vegetables is still possible only when protection against low night temperatures is provided in glasshouses or by polythene sheeting (Fig. 9.22).

Perhaps the major disadvantage associated with intensive agriculture on desert sand is one of high capital cost. A single hectare of vegetables requires ca 5500 m of trickle-irrigation pipe at a cost in excess of $5000 (1988 prices). The lifetime of irrigation pipework used for vegetable production is 4–5 years, but is longer in the case of fruit orchards which require also only 1000 m of irrigation pipe per hectare. Therefore, trickle irrigation is economic only when crops can be sold at relatively high prices.

A second problem is presented by saltating sand, which can cause physical damage to plants and encourage the spread of disease (Rempel 1936). This problem is most serious when cultivation takes place on fine dune sand. Coarser sands are entrained by the wind less frequently and, when they are set in motion, tend to move by surface creep, which is less damaging than saltation. Bimodal sand sheets, which cover vast areas of the Sahara and other major deserts, could potentially provide a highly suitable substrate for trickle-irrigation agriculture, provided that supplies of either fresh or brackish water can be made available.

9.4.2 *Cultivation and grazing on coastal dunes*

Cultivation is not a major activity in most recent humid coastal dunefields, principally because better quality land is normally available nearby. However, in some areas, weathered and degraded dunes of early Holocene or Pleistocene age support mature soils which are used for agriculture. Examples are provided by the red terra rossa-type soils which overlie weathered aeolianite in the Coastal Plain of Israel, Bermuda, and Natal.

In northwest Scotland, some of the more inland areas of coastal blown sand, which generally support a cover of short grassland, known as *machair* (Ritchie 1976), are still ploughed and used principally for cereal cultivation (Knox 1974). The remainder of the

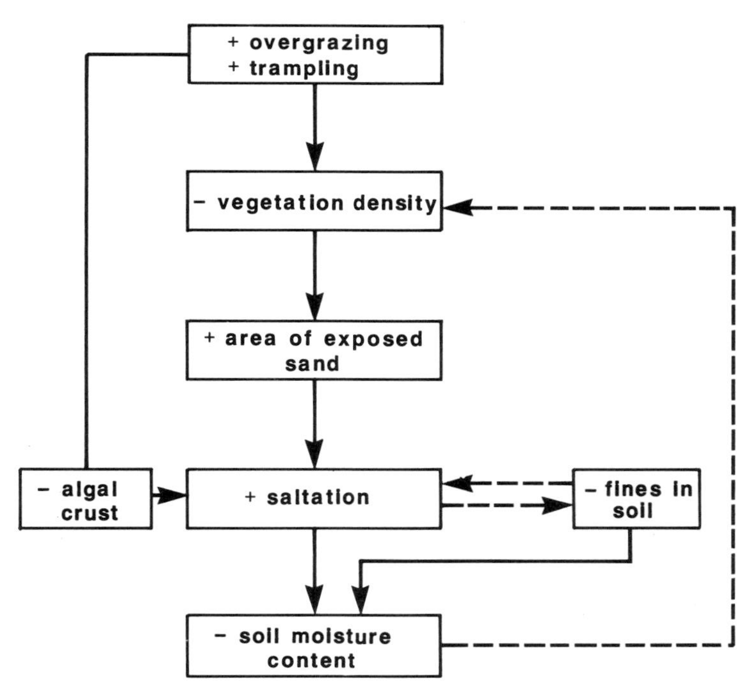

Figure 9.23 Flow chart showing the sequence of events following destruction of vegetation on desert sand dunes. Increases or decreases in process intensity are indicated by + and −, respectively. Positive feedback mechanisms are indicated by broken lines. Compare with Figure 9.4. (After Tsoar & Møller 1986).

Figure 9.24 Residential development on coastal dunes near Monterey, California.

machair is used for stock grazing (sheep and some cattle). Sheep, cattle, and rabbit grazing has also taken place in many other European coastal dune systems at least since the Middle Ages (Boorman 1989). In most cases, the present dune vegetation has established an equilibrium with the grazing regime, and blowouts represent only a localized problem.

Stabilized dunes in arid and semi-arid areas are much more sensitive to damage by cultivation and overgrazing. These activities can rapidly destroy the relatively thin vegetation cover and stabilizing surface crust (Thomas 1921, Hefley & Sidwell 1945). In turn, this leads to an increase in surface sand movement, loss of fine material in suspension, and reduction in soil moisture, causing a further negative effect on the vegetation (Le Houerou 1975, Tsoar & Møller 1986, Danin 1987) (Fig. 9.23). Such a sequence of events occurred in parts of the United States during the 1930s (Whitfield 1937, Oosting & Billings 1942, Hefley & Sidwell 1945), and more recently has affected large tracts of the Sahel and parts of the Middle East (Barth 1982, Ellis 1987, Niknam & Ahranjani 1975).

9.4.3 Urban development and recreational activities

Coastal dunes and adjacent broad sandy beaches provide a setting which is attractive for permanent residential development, holiday homes, caravan parks, and camping sites (Fig. 9.24). Such developments not only have a direct physical impact arising from the levelling of dunes and extraction of sand for construction, but also have a number of indirect effects which arise when large numbers of residents or visitors are attracted to the area (Nordstrom & McCluskey 1985). Damaging effects associated with large numbers or pedestrians include picking and trampling of vegetation, increased risk of fire, physical erosion of sand by the passage of feet, and initiation of wind funnelling along trackways, leading to the development of blowouts (Trew 1973, Carter 1980, Liddle & Greig-Smith 1975a, Boorman & Fuller 1977, Carter 1980, Hylgaard & Liddle 1981, Carter & Stone 1989, Pye 1990). Pressure is particularly severe around beach access points, and great care must be taken to ensure that public facilities such as toilets and car parks are properly sited and cordoned off from adjoining sensitive areas (Barr & Watt 1969, Lundberg 1984).

Other indirect effects, including large-scale sliding, slumping, and flowage of wet sand, may arise from changes in the hydrological regime of the dunes following phases of house and road construction (e.g. Castro & Vicuña 1986).

The passage of motorized vehicles poses a particularly serious damage risk. At Cape Cod, Massachusetts, *Ammophila brevigulata* took 3 years to recover from damage caused by 100 of off-road vehicles passes (Brodhead & Godfrey 1977). Other types of dune vegetation are more easily damaged and may take much longer to recover (Godfrey *et al.* 1978, Leatherman & Godfrey 1979, Anders & Leatherman 1987). In addition to direct crushing of the aerial parts and roots of plants, the passage of vehicle tyres can compact the sand and increase its water repellence by up to 100%, leading to erosion by surface wash, reduced infiltration, and increased drought damage to the vegetation (Liddle & Greig-Smith 1975a, b).

In desert areas, even low pressure from off-road vehicles may have a serious effect on the fragile dune vegetation. Luckenback & Bury (1983) found that in parts of the Sonoran Desert where off-road vehicle traffic was intense, virtually no native plants or wildlife remained.

Desert dune areas which are used for military manoeuvres, such as parts of New Mexico and the Negev, also suffer serious damage (Clements *et al.* 1963, Marston 1986). In Libya and southern Tunisia, tracks left by tanks in 1941–3 were still visible 30 years later (Le Houerou 1975, 1977b). Some coastal dune systems have also been significantly affected by tank movements, shelling, and bombing practice [e.g. Braunton Burrows and Camber in the United Kingdom; Kidson & Carr (1960), Pizzey (1975)].

9.4.4 *Sand mining*

Sand is extracted from dune systems for several different purposes. Dune sands which consist of high-purity silica are mined for glass and ceramic manufacture, as at Cape Flattery on the east coast of Cape York Peninsula, Queensland (Anon. 1987) (Fig. 9.25). Lower quality siliceous sands can also be useful as glass sands after beneficiation (Carter *et al.* 1964), but are more widely used for building purposes and as foundry sands. Some dune sands also constitute significant resources of feldspar (Willman 1942) and heavy minerals. In southern Queensland and northern New South Wales, for example, zircon, ilmenite, rutile, and monazite are the principal minerals of economic interest (Coaldrake 1962, Morley 1981). Unconsolidated carbonate dune sands have been dug for centuries to improve the fertility of neighbouring acid peat soils in some humid areas (e.g. northwest Scotland and western Ireland). Lithified aeolianite has also been quarried for centuries and used as a building stone in semi-arid regions (e.g. northwest India and the Arabian Gulf).

The environmental impact of sand mining varies, depending on whether the bulk of the sand is removed, as in silica sand mining, or only a selected part is extracted, as in heavy mineral mining. During silica sand mining, whole dunes may be completely removed, leaving a bare, flat, or gently undulating surface. Heavy mineral mining, on the other hand, can have a less dramatic impact. In eastern Australia, the topsoil, which contains most of the nutrients, is often carefully removed and stored before heavy mineral mining begins. Heavy minerals are then extracted from the underlying sands and the washed residue, which often represents more than 95% of the original sand volume, replaced and recontoured to form surface relief which approximates the original. Finally, the topsoil layer is replaced and vegetation colonization is encouraged by planting or seeding (Rogers 1977). Brush matting, sand fences, and surface coatings are widely used to aid re-establishment of the vegetation (Sless 1958, Barr & Atkinson 1970).

Figure 9.25 Silica sand mining at Cape Flattery, North Queensland.

In southeast Queensland, Thatcher & Westman (1975) observed that a shrub layer of leguminous species developed within a few years of the cessation of mining, but estimated that it might take 100–250 years to develop a vegetation cover similar to that removed.

9.4.5 Dunes and water supply

Sand dune systems in humid areas often contain large lenses of fresh water which form a potential source of water supply for domestic, industrial, and agricultural purposes. On North Stradbroke Island, southeast Queensland, for example, the surface of the water table roughly follows the dune topography, reaching a maximum elevation of 100 m above sea level in the centre of the Island (Laycock, 1975a, b, 1978) (Fig. 9.26). These dune sands, which were partly formed during times of lower Pleistocene sea level, rest on bedrock basement ca. 50–60 m below sea level, and are saturated with fresh water almost throughout. The permeability of the dune sands is sufficiently high for much of the annual rainfall (1650 mm) to percolate rapidly down to the water table, even though the annual potential evaporation is also high (1522 mm). Allowing for surface losses (mainly by runoff), annual infiltration for the whole island was estimated by Laycock

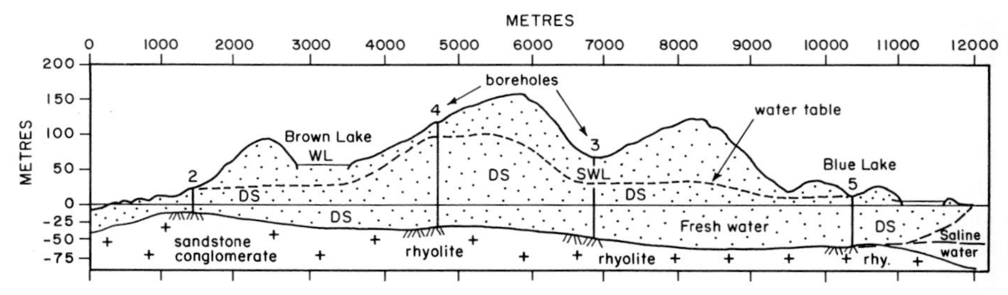

Figure 9.26 Section across North Stradbroke Island, southern Queensland, showing variations in water table level. (Based on Laycock 1975a, 1978).

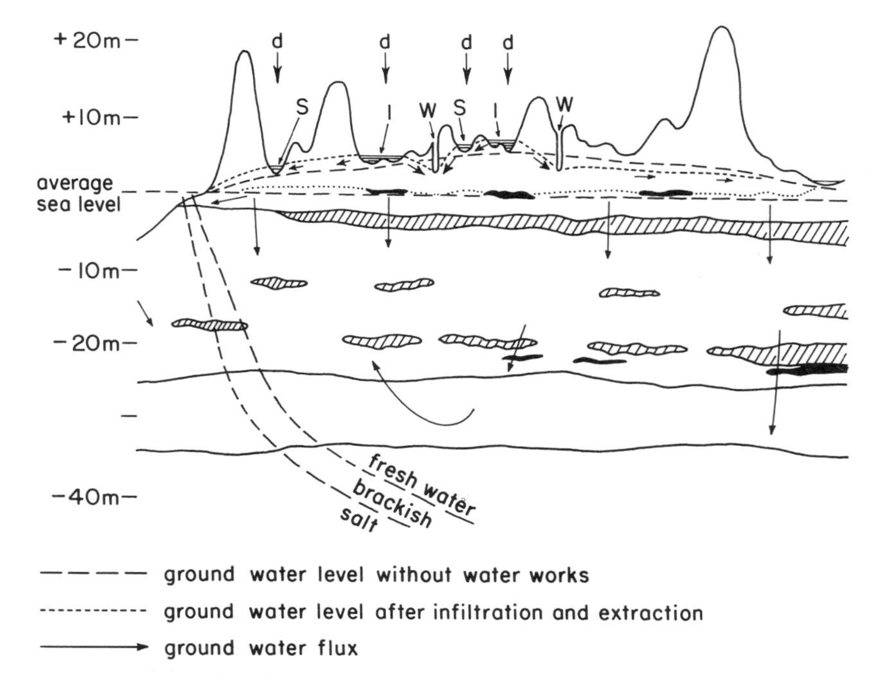

Figure 9.27 Cross-section of part of the Berkheide dunes, The Netherlands, showing the effect of water works on the groundwater table. d = depression; s = seepage pool; w = well; i = infiltration pool. (After van Dijk 1989).

(1975a, b) to be 1.66×10^8 m^3, while the total amount of water in storage was estimated to be 3.6×10^9 m^3. Although this reserve has been recognized for more than a century as a potential source of drinking water for Brisbane, no significant exploitation has yet taken place.

Dutch coastal dune areas have been used for public water supply purposes for more than a century, and for this reason they have been largely protected from urban development, the effects of mass tourism, waste disposal, and pollution which have seriously affected many other dune areas in Europe (van der Maarel 1979, van Dijk 1989). The total dune infiltration capacity was about 202×10^6 m^3 in 1975, and has subsequently been increased by the construction of additional infiltration ponds.

Until about 1955, recharge of the dune aquifer was dependent on rainfall and seepage from a relatively small number of artificially created infiltration ponds fed by rainfall and local runoff. Progressive water extraction caused a gradual fall in average water table levels in many of the dune systems, bringing about serious ecological consequences for the dune slack environments (Fig. 9.27). There was also some incursion of salt and brackish water in areas near the shore. Consequently, since 1955 recharge has been enhanced using water from the Rhine and Meuse rivers. The water is purified before being punped into seepage ponds in the dunes. However, rising water levels and changes in water quality have had further notable effects on the dune ecology (van der Meulen 1982, van Dijk 1989).

9.4.6 Coastal dunes as natural sea defences

On low-lying coasts, littoral dunes are often important as natural sea defences. In many parts of the western Netherlands, for example, where extensive areas of agricultural land

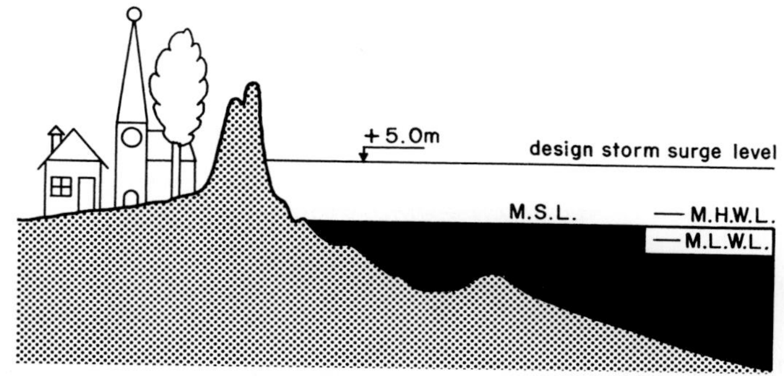

Figure 9.28 Schematic representation of a typical Dutch coastal profile. (After Vellinga 1978).

and urban development lie at or below mean high water level, and up to 5 m below storm surge level (Fig. 9.28), the primary sea defence system consists of sandy beaches and dunes (Vellinga 1978, 1982, van de Graaf 1986, van der Meulen & van der Maarel 1989).

A continuous belt of coastal dunes acts as a barrier to wave overwash and flooding of the area landward of it. In this respect it performs the same function as a sea wall. However, in many situations coastal dunes have a number of advantages over sea walls: (a) they are considerably less expensive to construct and maintain; (b) dunes act as a sand store which can release sand to the beach during periods of storm wave attack, thereby helping to dissipate wave energy (Leatherman 1979); (c) they are more flexible than sea walls and can adjust to changing conditions, such as a long-term fall in beach levels or a natural tendency for shoreline retreat; and (d) they have a less damaging visual impact on the coastline. Additionally, although dunes reflect some wave energy during storms, the degree of wave reflection and consequent beach scouring is less severe than that induced by sea walls (Leatherman 1979). However, in some instances only construction of a 'hard' sea wall can provide an adequate level of protection. This is the case, for example, in highly built-up areas where any change in shoreline position would have grave financial consequences, or along shores which are too muddy, or experience too little onshore wind to form sizable coastal dunes.

Model studies and field observations have clearly shown that the presence of a high foredune will reduce foreshore lowering and shoreline recession during a storm (van der Meulen & Gourlay 1968, Edelman 1968, 1973, van de Graaf 1977). During the initial stages of wave attack, the average level of the foreshore is lowered and any berm present is removed as sand is moved offshore (Fig. 9.29). Especially during storm surges, waves may begin to break close to, or even against, the duneline. Deflection of water along the dune foot often erodes a slot, thereby making the seaward dune slope unstable and causing it to slide into the sea, forming a cliff (Fig. 9.30). Slumping of the dune cliff is also encouraged when standing water saturates the sand at the base of the dune during surges. The rate of recession of the duneline varies inversely with the dune height, and the rate of dune recession can be predicted as a function of dune height assuming that the sand released is spread uniformly over the foreshore and is not immediately lost offshore (Edelman 1968) (Fig. 9.31).

Wave run-up during severe storms may pass right over the crest of a low dune (< 5 m high), resulting in transfer of sand from the seaward to the landward side. This process of overwash is important in the landward migration of low dune-capped barriers in the eastern and southern United States (Leatherman 1976, 1979, Leatherman *et al.* 1977, Leatherman & Zaremba 1987). The principal factors which determine the height of wave run-up are the slope of the beach, the slope of the seaward dune face, beach width,

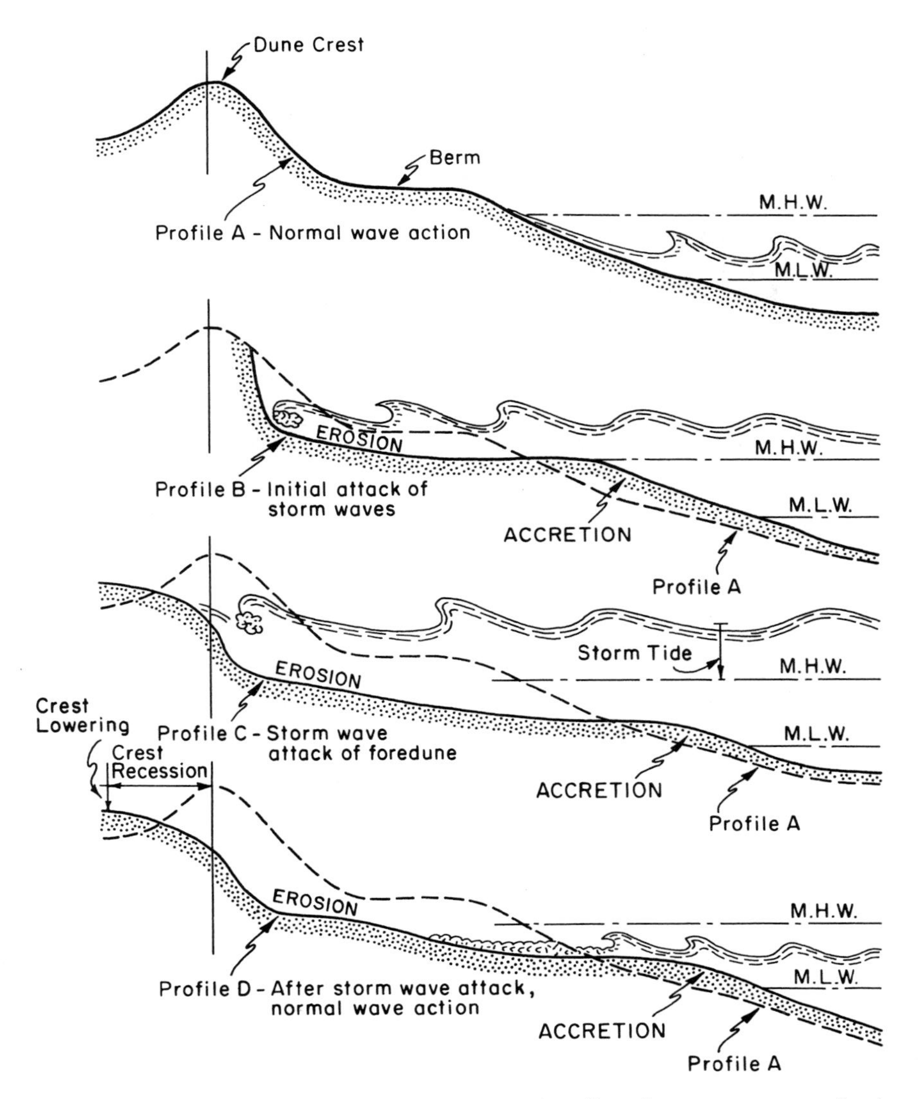

Figure 9.29 Schematic diagram showing the progressive effect of storm waves on a foredune/beach system. (After CERC 1977).

deepwater wave height, wave period, and any increase in water level due to storm surge [see Gares *et al.* (1979), pp.18–24, for dune run-up calculation procedures].

The use of coastal dunes for sea defence usually requires that they be kept as stable as possible. As a result, conflicts of interest frequently arise with nature conservation and leisure groups, since dune stability tends to reduce ecological diversity and restrictions usually need to be placed on recreational activities (van der Meulen & van der Maarel 1989). A central task facing coastal managers and planners is, therefore, the reconciliation of these conflicting interests [see Carter (1988) and papers in van der Meulen *et al.* (1989) for further discussion of management and conservation aspects].

Figure 9.30 Frontal dune cliffed by storm waves, Sefton Coast, Merseyside, UK.

Figure 9.31 Calculated relationship between dune height and dune erosion rate, assuming that all of the eroded sand is distributed evenly across the foreshore. (After Edelman 1968).

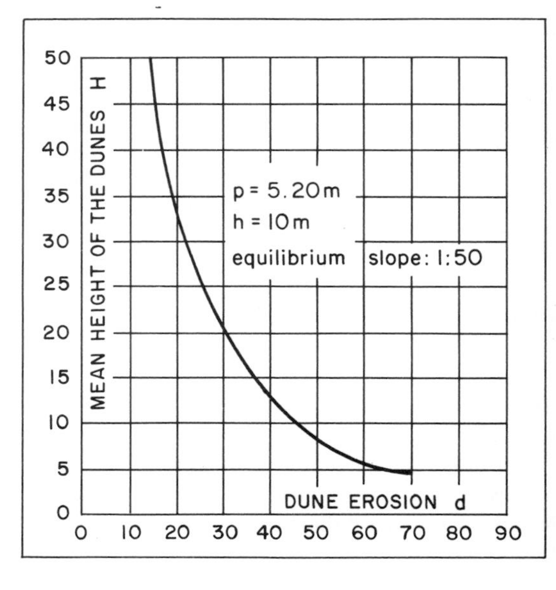

10
Aeolian research techniques

A very wide range of investigative approaches have been employed in aeolian studies and only a few of the more important techniques used by geomorphologists and sedimentologists can be considered here.

10.1 Wind tunnel studies

A major advantage of wind tunnel studies is that they make it possible to conduct experiments under scientifically controlled conditions. It is easier to control the number of variables operating at any one time, compared with typical field situations, and conditions can be held constant long enough for experiments to be completed and repeated. The main drawbacks relate to problems of scaling, but these need not undermine the value of modelling work if appropriate precautions are taken.

Most laboratory wind tunnel studies of aeolian processes have been concerned with the threshold for particle entrainment (Bagnold 1941, p. 32, Iversen & White 1982, Iversen *et al.* 1987), the nature of particle trajectories, and grain-bed interactions (Bagnold 1936, 1937a, White 1985, Willetts & Rice 1988). However, airflow over model dunes, the effect of obstacles on sand deposition, and the formation of micro-dunes under simulated Venusian atmospheric conditions have also been studied in laboratory wind tunnels (Greeley *et al.* 1974b, Tsoar 1983b, Tsoar *et al.* 1985, Greeley *et al.* 1984).

Two distinct types of laboratory wind tunnel have been used. The first is an open-circuit tunnel which usually has three parts: entrance cone, test section, and diffuser (Fig. 10.1). Air is sucked in through the bell-shaped entrance by a fan located at the end of the diffuser. In this way, disturbances to the airflow inside the tunnel, which can be a serious problem in tunnels equipped with blower fans, are kept to a minimum (Bagnold 1941, p. 25). Since the open-circuit tunnel is sensitive to outside winds, it is normally housed inside a building.

The second type is a closed-circuit tunnel which, as the name implies, is sealed and has a continuously circulating flow of air. This type of tunnel has been used principally for aeolian abrasion experiments and simulation of sediment transport at high pressures (e.g. Kuenen 1960, Seppala & Lindé 1978, Greeley *et al.* 1984).

Open-floored wind tunnels have also been employed in the field, mainly to determine threshold velocities and the relationship between shear velocity and particle flux for a range of natural soils and sediments (Chepil & Milne 1939, 1941, Malina 1941, Zingg 1951, Gillette 1978, Gillette *et al.* 1980, 1982, Nickling & Gillies 1989) (Fig. 10.2).

It is important to ensure that conditions in the wind tunnel are representative of natural full-scale conditions (Cermak 1971).

Care should therefore be taken to ensure that:

(a) any model dune or obstruction to the flow is geometrically identical with the original;
(b) the vertical wind velocity profile and level of turbulence are representative of natural flow conditions;

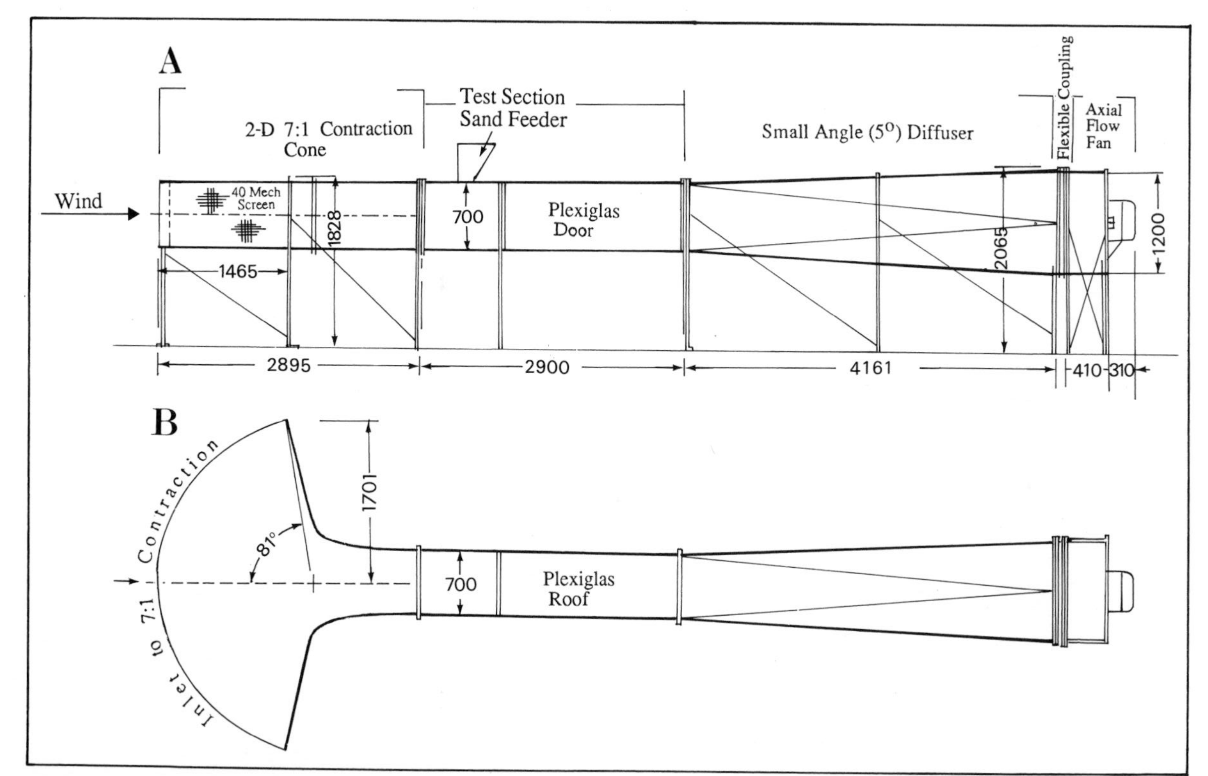

Figure 10.1 Schematic diagram of an open-circuit wind tunnel used at Ben-Gurion University of the Negev (designed by Prof. M. Cohen). Dimensions are in mm. A, side view; B, plan view.

Figure 10.2 Open-floored wind tunnel used in the field to determine the threshold shear velocities of natural sediments and soils. (Photograph by W. G. Nickling).

(c) the test section is sufficiently long to allow sand transport to attain a uniform rate; and
(d) dynamic similitude is maintained by appropriate scaling of all the variables which affect sediment transport (see Table 10.1).

Jensen (1958) first recognized that true scaling of the roughness criterion is necessary if the boundary layer is to be appropriately modelled:

$$z_{0m}/z_{0p} = L_m/L_p \qquad (10.1)$$

Table 10.1 Variables involved in wind tunnel simulation work on aeolian processes. Dimensions of each variable are indicated in paretheses. L = length, M = mass, T = time.

u	velocity in the boundary layer ($L\ T^{-1}$)
u_∞	velocity outside the boundary layer ($L\ T^{-1}$)
u_*	friction velocity ($L\ T^{-1}$)
u_{*t}	threshold friction velocity ($L\ T^{-1}$)
w_f	settling velocity of grains ($L\ T^{-1}$)
g	acceleration due to gravity ($L\ T^{-2}$)
ρ_f	fluid density ($M\ L^{-3}$)
ρ_s	grains density ($M\ L^{-3}$)
γ	bulk density of sand ($M\ L^{-3}$)
μ	dynamic viscosity of fluid ($M\ L^{-1}\ T^{-1}$)
ν	kinematic viscosity of fluid ($L^2\ T^{-1}$)
L	size of model as specific to some characteristic dimension (L)
z_o	surface roughess (L)
d	grain diameter (L)
δ	boundary-layer thickness (L)

Table 10.2 Set of variables and their dimensions.

			n		
r	ρ_f^a	u^b	d^c	μ	ρ_s
M	1	0	0	1	1
L	−3	1	1	−1	−3
T	0	−1	0	−1	0

where m and p denote the model and the full-size prototype, respectively, and z_0 and L are as defined in Table 10.1.

A thick wind tunnel boundary layer grows naturally if the floor of the tunnel over a considerable distance upwind of the test section is roughened. This requires a very long wind tunnel with a long fetch ahead of the test section. Such a naturally grown boundary layer with sufficient depth shows excellent agreement with atmospheric data (Cook 1978). With short wind tunnels, however, a thick boundary layer can be created by placing a velocity profile generator upwind of the test section. This may consist of a grid with small apertures near the floor and larger apertures towards the roof of the tunnel (Cowdrey 1967). Alternatively, a bent wire-mesh screen, oriented vertically close to the roof and horizontally near the floor, will serve the same purpose (Elder 1959).

Another method of artificially controlling the growth of the turbulent boundary layer involves using longitudinal vortex generators (Counihan 1969). In most cases a boundary layer is allowed to develop by a combination of a long fetch of floor roughness and vortex generators (Cook 1978). Exact replication of natural boundary layer conditions, including flow field temperature stratification, is not needed for simulation of sand movement.

Establishment of an equilibrium boundary layer under conditions of active sand transport occurs more slowly than that under sand-free conditions. The length of test section required depends to a marked degree on the type of sand used in the experiments. With dune sand, equilibrium conditions can be attained at the downwind end of a test section which is 10–15 m long (Bagnold 1941, p. 26). The tunnel length can be judged to be adequate if, in the absence of obstructions in the tunnel, an initially flat bed of sand does not change its depth along the test section.

Having a knowledge of the variables which affect the air and sand movement in the wind tunnel, it is possible to combine the 12 variables listed in Table 10.1 into a smaller number of dimensionless groups (Kline 1965, p. 17). The *Buckingham pi theorem* provides one method of identifying a competent number of dimensionless groups to define a problem (Buckingham 1914). In non-compressible fluid mechanics, including low-speed airflow in a wind tunnel, there are three primary dimensions, L, M, and T (Table 10.1). Some variables selected for dimensionless analysis are shown in Table 10.2. The number of variables is indicated by n and the number of the primary dimensions by r. A number of $(n–r)$ equations relating n and r can be set up by combining parameters selected from Table 10.2 to form dimensionless groups of π. From Table 10.2, one dimensionless group is

$$\pi_1 = \rho_f{}^a u^b d^c \mu \tag{10.2}$$

Therefore

$$\pi_1 = (M/L^3)^a (L/T)^b L^c M/LT = 0 \tag{10.3}$$

The exponents of the variables are equated from Table 10.2:

M: $a+1=0$

L: $-3a+b+c-1=0$

T: $-b-1=0$

The solution is $a=-1$, $c=-1$, $b=-1$. A dimensionless π_1 group can be set up as

$$\pi_1 = \mu/\rho_f\, d\, u_{\text{model}} = \mu/\rho_f\, d\, u_{\text{prototype}} \tag{10.4}$$

The second dimensionless group (according to Table 10.2) is

$$\pi_2 = \rho_f{}^a u^b d^c \rho_s \tag{10.5}$$

Therefore,

$$\pi_2 = (M/L^3)^a (L/T)^b L^c M/L^3 = 0 \tag{10.6}$$

The exponents of the variables are equated from Table 10.2:

M: $a+1=0$

L: $-3a+b+c-3=0$

T: $-b=0$

The solution is $a=-1$, $b=0$, $c=0$. A dimensionless π_2 group can then be set up as

$$\pi_2 = \rho_s/\rho_{f\ \text{model}} = \rho_s/\rho_{f\ \text{prototype}} \tag{10.7}$$

It is often impossible or impractical to reach a satisfactory similarity of all the parameters simultaneously. For example, the Reynolds number (Eqns 2.9 & 10.4) and Froude number (u^2/Lg) can never both be matched over a range of velocities. When static models of dunes are used, it is impossible to achieve an exact match between model and full-size scale parameters because L of the model has to be scaled down to between 1:50 and 1:200 (Kind 1976, Kind & Murray 1982). It is more important, however, to obtain a fully developed turbulent flow, and a strict matching of Reynolds number is not always necessary for dynamic similarity (Mironer 1979, p. 330). A fully developed turbulent flow requires that the roughness Reynolds number ($u_* z_0/\nu$) be > 2.5 (Isyumov & Tanaka 1980). This may be difficult to achieve if the scale of the roughness parameter in the model (Eqn 10.1) is adhered to. The level of turbulence has an important influence on the mode of particle movement (Section 4.2.1) and also on the separation and reattachment characteristics of the flow (Jensen 1958).

There are some physical limitations to dynamic similarity because not all variables can be changed without a drastic concomitant change in the characteristics of particle movement. For instance, if the grain size or density is reduced to such low values that the particle might be affected by the vertical component of the turbulence, it may well be transported in suspension instead of saltation.

In conclusion, maintaining exact simulation conditions for small-scale models of sand transport is sometimes problematic because of the large number of variables involved. Since it is almost impossible to match all of the model's dimensional parameters with the full-scale original, there is a small penalty to pay in terms of the accuracy of modelling. However, good experimental practice should identify any such 'scale effects' and allow them be taken into account when the results are interpreted.

10.2 Measurement of sand movement using sand traps

The rate of sand transport can be measured either in the wind tunnel or in the field using sand traps. Traps for field use are usually larger than those used in wind tunnel studies.

The results obtained from any sand trap can only be regarded as approximate, since interference with the air flow is unavoidable (Bagnold 1938a). The airflow immediately in front of a vertical sand trap is impeded, leading to the development of stagnation pressure which diverts the flow around the trap. Some of the grains follow this deflected flow and do not enter the trap. The trap may also generate vortices, leading to localized areas of enhanced bed erosion or deposition around the trap. To minimize these problems, a trap should be as narrow as possible and have a streamlined shape (Jones & Willetts 1979, Illenberger & Rust 1986).

Horizontal sand traps do not disturb the flow to the same degree as vertical traps, but at high wind velocities they need to be very long to trap grains with flattened saltation trajectories (Horikawa & Shen 1960).

Trap efficiency can be defined as the relative ratio of trapped sand to the actual quantity of blown sand (Chepil & Milne 1941). Efficiency determinations of this kind are usually carried out in wind tunnels. Whereas it is relatively easy to assess the relative efficiencies of different trap designs, absolute efficiencies are more difficult to establish.

10.2.1 Horizontal sand traps

A simple horizontal sand trap designed by Owens (1927) is shown in Figure 10.3. It consists of a box, 67 cm long and 34 cm high, which is half-buried in the sand. The box has vents on the upwind and downwind sides to allow entry and exit of air, and internal baffles to cause sand deposition.

A horizontal trap 91 cm long and 15 cm wide and fitted with transverse riffles was used by O'Brien & Rindlaub (1936). It was buried in the sand with its long axis pointing

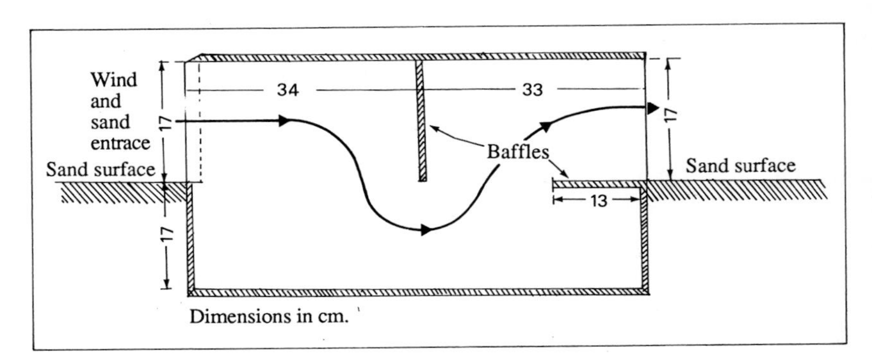

Figure 10.3 Low box-type sand trap. (After Owens 1927).

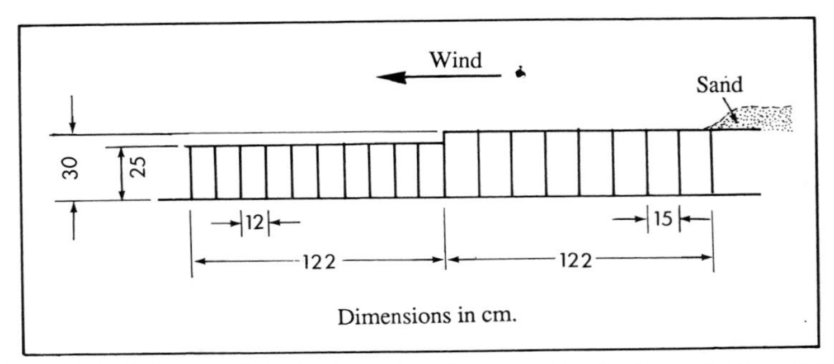

Figure 10.4 Side view of the horizontal sand trap used by Horikawa & Shen (1960).

parallel to the wind direction. Horikawa & Shen (1960) and Belly (1964) also employed a rectangular box, 2.4 m long and with 18 internal compartments, which was orientated with its long axis parallel to the wind direction (Fig. 10.4). This type of trap has a high collection efficiency when the wind direction is constant, but narrow traps do not cope well when the wind direction fluctuates to a significant extent.

10.2.2 Vertical sand traps

Several devices have been used to collect blown sand at different heights above the ground surface. Sharp (1964) used a vertical array of metal tubes, bent through an angle of 90° at the downwind end. Bagnold (1938a) designed a vertical collector which was

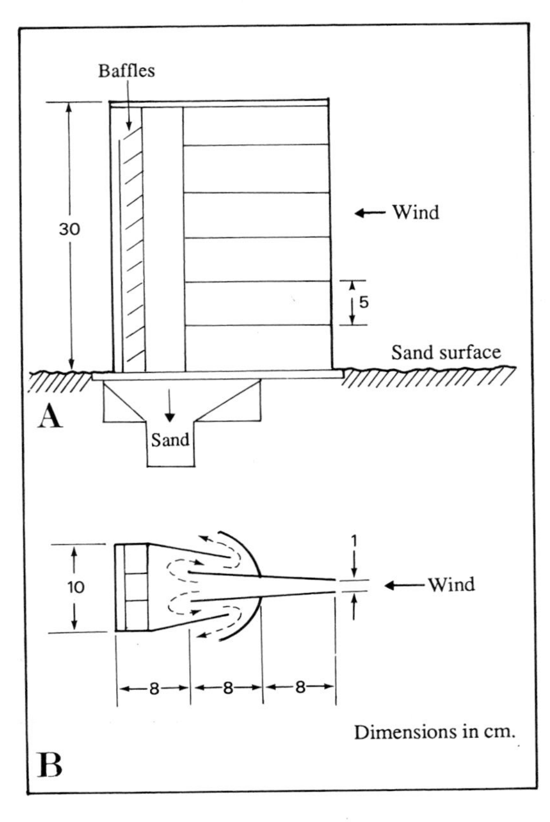

Figure 10.5 Improved Bagnold vertical sand trap as used by Horikawa & Shen (1960). A, side view; B, plan view showing airflow exits.

Figure 10.6 Modified Bagnold vertical trap equipped with a fin which allows the trap to rotate and always face into the wind.

76 cm high and 1.3 cm wide to minimize interference with the airflow. The collector contained baffle plates inserted at an angle of 40° to prevent grains bouncing out of the collector. Sand entering the trap was collected in a bin buried in the sand below the collector. Horikawa & Shen (1960) improved this basic design by providing an exit port for the airflow, thereby reducing the stagnation pressure in front of the collector (Fig. 10.5). The collection efficiency of modified Bagnold collectors is relatively low, generally ranging from 20 to 40% (Knott & Warren 1981).

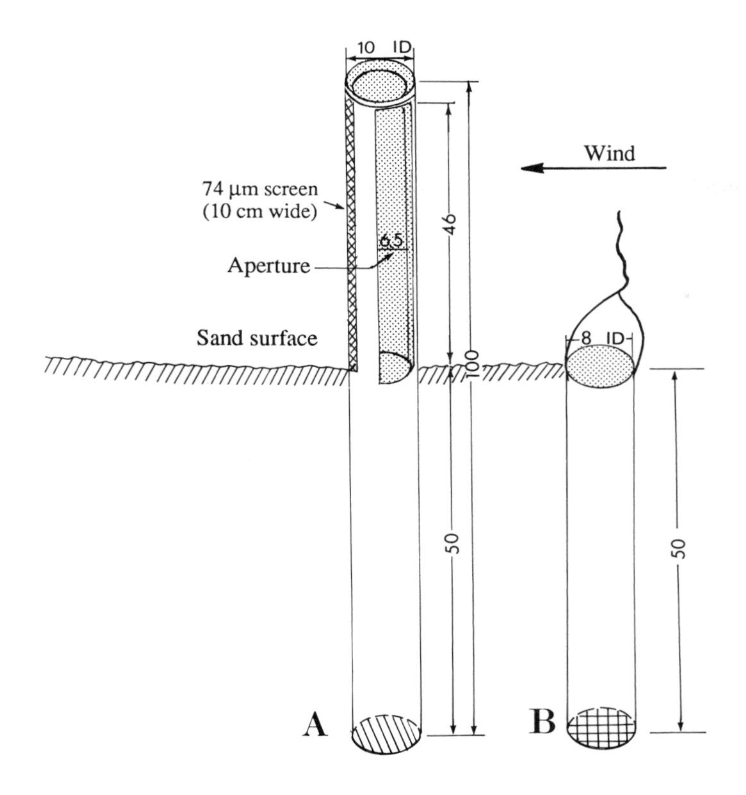

Figure 10.7 Modified Leatherman trap. A, vertical PVC tube; B, inner liner for collecting the sand. (As used by Goldsmith *et al.* 1988).

The Bagnold-type trap can be equipped with a fin to enable it to rotate about an axial pole so that it always faces into the wind (Fig. 10.6). A trap of this type can be used in conjunction with several underground containers to separate sand blown from different directions (Fryberger *et al.* 1984). The collecting canister can also be connected to a pressure transducer which enables the weight of trapped sand to be recorded at regular intervals on a data-logger (Fryberger *et al.* 1979).

A very simple and inexpensive vertical trap was developed by Leatherman (1978) and modified and enlarged by Rosen (1979). A version of Leatherman's trap, used in Israel (Goldsmith *et al.* 1988), is shown in Figure 10.7. It consists of a 10 cm diameter, 100 cm long PVC tube which is half-buried in the sand. The exposed 50 cm length has two slits which extend 46 cm down from the top. Air enters the front slit (6.5 cm wide) and leaves through the back slit (10 cm wide), which is covered by 60-μm mesh to trap the sand grains. Sand is collected in a removable tube liner which has a fine mesh base to allow free drainage of water. Several such traps can be positioned to face in different directions. Stagnation pressure is reduced by the screened exit port, but scouring often occurs around the base of the trap (Fig. 10.8). Also, since the trap is only 50 cm tall, it may not intercept all of the moving sand during severe wind storms.

Illenberger & Rust (1986) designed a vertical sand trap with the aim of minimizing the stagnation pressure both inside and outside the trap. This was achieved by attaching a

Figure 10.8 Scouring around the base of a modified Leatherman trap. Note that the collector is full to the brim and that scouring has occurred from a wind direction not directly facing the trap's aperture. (Photograph by D. Blumberg).

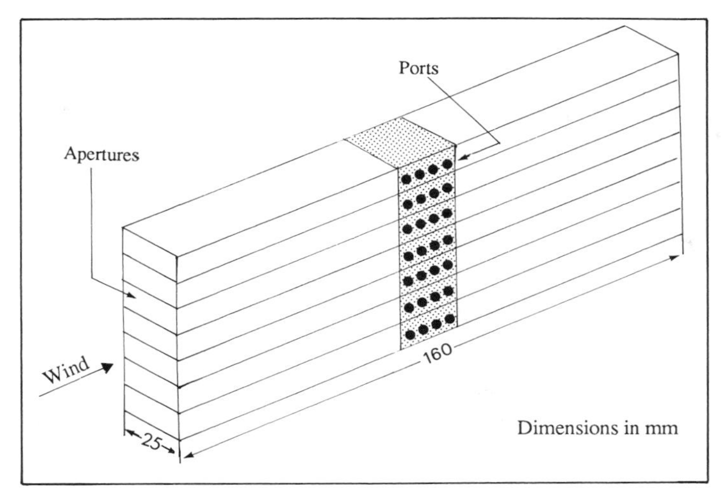

Figure 10.9 Seven-chamber saltation trap used to sample sand simultaneously at different heights. (After Jensen *et al.* 1984).

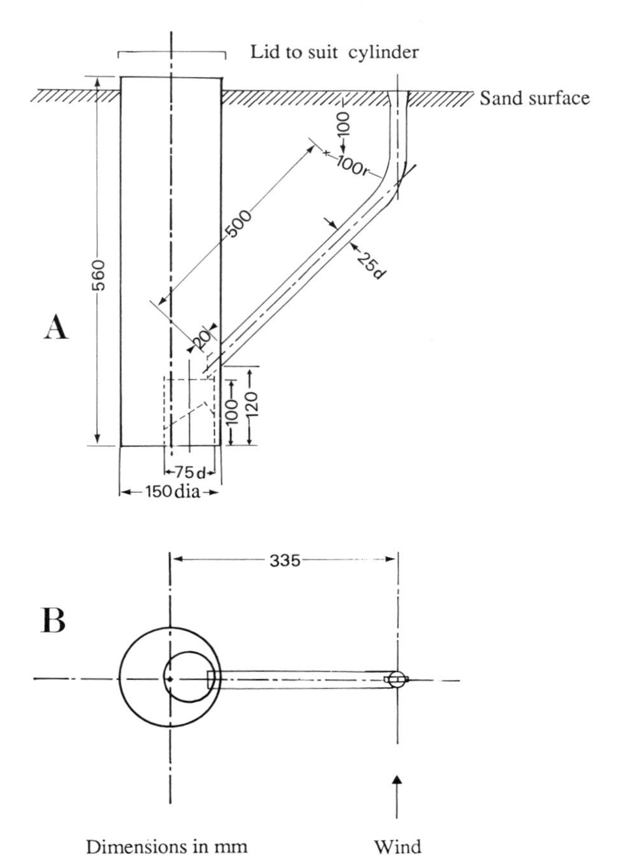

Figure 10.10 Improved Bagnold creep collector. A, side view; B, plan view. (After Ross *et al.* 1984).

venturi vacuum generator to the exit port and by providing a streamlined wedge-shaped entrance. However, scouring around the trap was not completely eliminated.

Jensen *et al.* (1984) designed a trap to collect sand at different heights which consists of a stack of seven chambers, each with a rectangular aperture 25 mm wide and 14 mm high. The chambers have a length of 160 mm (Fig. 10.9). Small mesh-covered ports along the sides of the chamber act to reduce the stagnation pressure.

Owing to the small capacity of all the sand traps discussed, they can be filled within a few minutes or at most a few hours during major sand storms. Unless frequently emptied, they are of little value in terms of providing long-term data about sand transport rates.

10.2.3 Surface creep traps

Bagnold (1937b) measured surface creep in wind tunnel experiments by means of a narrow transverse slot in the bed. This simple concept was developed for field applications by Ross *et al.* (1984). Their creep collector had an orifice 8 mm wide and 30 mm long, installed flush with the surface and connected by a 25 mm diameter tube to a plastic container placed in the bottom of a buried aluminium cylinder (Fig. 10.10). Although such traps are effective in trapping virtually all the grains moving in true creep, they inevitably trap a proportion of grains travelling in saltation.

10.3 Sand tracer techniques

Tracer techniques have been widely used in fluvial and beach sediment transport studies (e.g. Jolliffe 1963, Ingle 1966, Crickmore 1967, Kennedy & Kouba 1970, Lavelle *et al.* 1978, Hung & Shen 1979), but relatively few authors have used them to monitor aeolian transport in the field (Berg 1983, Tsoar & Yaalon 1983) or in wind tunnels (Willetts & Rice 1985b, Barndorff-Nielsen *et al.* 1983, 1985, 1988, Sørensen 1988).

Sand grains can be labelled using coloured dyes, fluorescent dyes, or radioactive tracers such as ^{198}Au. Radioactive tracers have the advantage that it is possible to locate grains buried beneath the surface using a scintillation counter, and they have proved useful in wind tunnel studies of sand movement. ^{198}Au has a suitable half-life of 2.7 days and it emits gamma rays of sufficient energy to allow detection. Since the gold layer is abraded relatively easily during saltation, the labelled grains are first coated with chromium to improve adhesion of the gold layer (Barndorff-Nielsen *et al.* 1985a).

For field studies, the coloured dye method is more practical and less expensive. Dyed grains are carefully released at some point on a dune and subsequently the sand surface is sampled at regular intervals along transects downwind of the insertion point. Sampling is most easily done using a vaseline-covered card mounted on the end of a pole (Ingle 1966, p. 21). The concentration of dyed grains is subsequently determined in the laboratory. Fluorescent dyes, which allow treated grains to be identified under ultra-violet light, may produce less sampling bias than coloured dyes, which are more obvious to the naked eye. However, where visual effect is important, coloured dyes give the best result. Several different colours can also be used for different grain sizes. It is possible to determine the distribution of fluorescent grains at night using a portable ultraviolet lamp, but this is generally less convenient than working during the day. By choosing organic dyes which fade within a few days and which are non-toxic, the use of coloured dyes need have no long-term effects. Large quantities of sand can be treated with coloured or fluorescent dyes relatively easily, and the aerodynamic properties of grains are not significantly affected (Yasso 1966b, Tsoar & Yaalon 1983).

Great care needs to be taken when the labelled grains are released to ensure that the surface is disturbed as little as possible, and that no artificial relief is created. Berg (1983) mixed fluorescent sand with an equal amount of natural sand before placing the mixture in a shallow trench, 8 cm wide, 0.5 cm deep, and 5 m long, orientated perpendicular to the prevailing wind.

Data obtained from sand-coated sampling cards can be analysed either qualitatively or quantitatively. The former may be adequate to give information about sand transport directions and areas of net erosion and deposition, whereas the latter allows the sand transport rate to be estimated by determining the centroid of the tracer particles (Crickmore 1967). The *velocity of tracer centroid* (u_{tc}) is calculated according to the equation (Berg 1983):

$$u_{tc} = \left[\left(\frac{\sum_{o}^{n} C_i X_i}{\sum_{o}^{n} C_i}\right)_{t_2} - \left(\frac{\sum_{o}^{n} C_i X_i}{\sum_{o}^{n} C_i}\right)_{t_1}\right][1/(t_2 - t_1)] \quad (10.8)$$

where C_i is the concentration of tracer grains (grains per unit area), n is the number of sampling stations, X_i is the distance from the ith sampling station to the point at which the grains were released, and t is the time between release and sampling. Berg (1983) found that rates of sand transport in a natural dune environment calculated using Equation 10.8 were as much as one order of magnitude lower than rates calculated using Equations 4.34 and 4.41.

An accurate evaluation of sand transport rates using tracers is difficult in dune terrain where the pattern of wind velocities varies considerably in response to the topography. It is also difficult to relate the dispersion pattern of dyed sand, which usually has a restricted size range, to that of the natural sand which may show wide spatial variations in grain size. Sand tracers are therefore most useful in providing a qualitative indication of sand movement and deposition on dunes (e.g. Fig. 10.11).

10.4 Methods of sample collection for grain size and mineralogical analysis

A number of investigators (e.g. Apfel 1938, Otto 1938, Ehrlich 1964) have stressed the importance of sampling individual sedimentary units deposited under uniform environmental conditions. White & Williams (1967) suggested that the ideal sampling unit was a single grain layer. However, the most appropriate scale of sampling depends on the purpose to which the results are to be put. Sampling of individual laminae may be appropriate where the objective is to understand the relationship between flow conditions, sediment transport, and depositional processes but is not appropriate where the aim is to characterize the grain size of a whole dune or a whole dunefield.

Most aeolian grain size studies have used samples collected from a depth interval of 0–10 cm below the surface, usually after removing surface ripples which are regarded as lag sediments (e.g. Folk 1971a, Warren 1972, Vincent 1984, Watson 1986). A scoop, trowel, or spoon has typically been used, and samples have even been ‘grabbed’ by hand. Such samples inevitably represent a mixture of grains from many different laminae.

However, it is possible to collect samples from a thin surface layer in order to compare, for example, the grain size of ripple crests with that of ripple troughs. Tsoar (1975, 1990a) used a quick-setting spray adhesive for this purpose (Fig. 10.12) and found that it was possible to take samples from layers not more than two grains thick. In the laboratory the adhesive is dissolved with a water-soluble solvent such as chloroform. The grains are then dispersed ultrasonically, washed, and dried ready for size analysis.

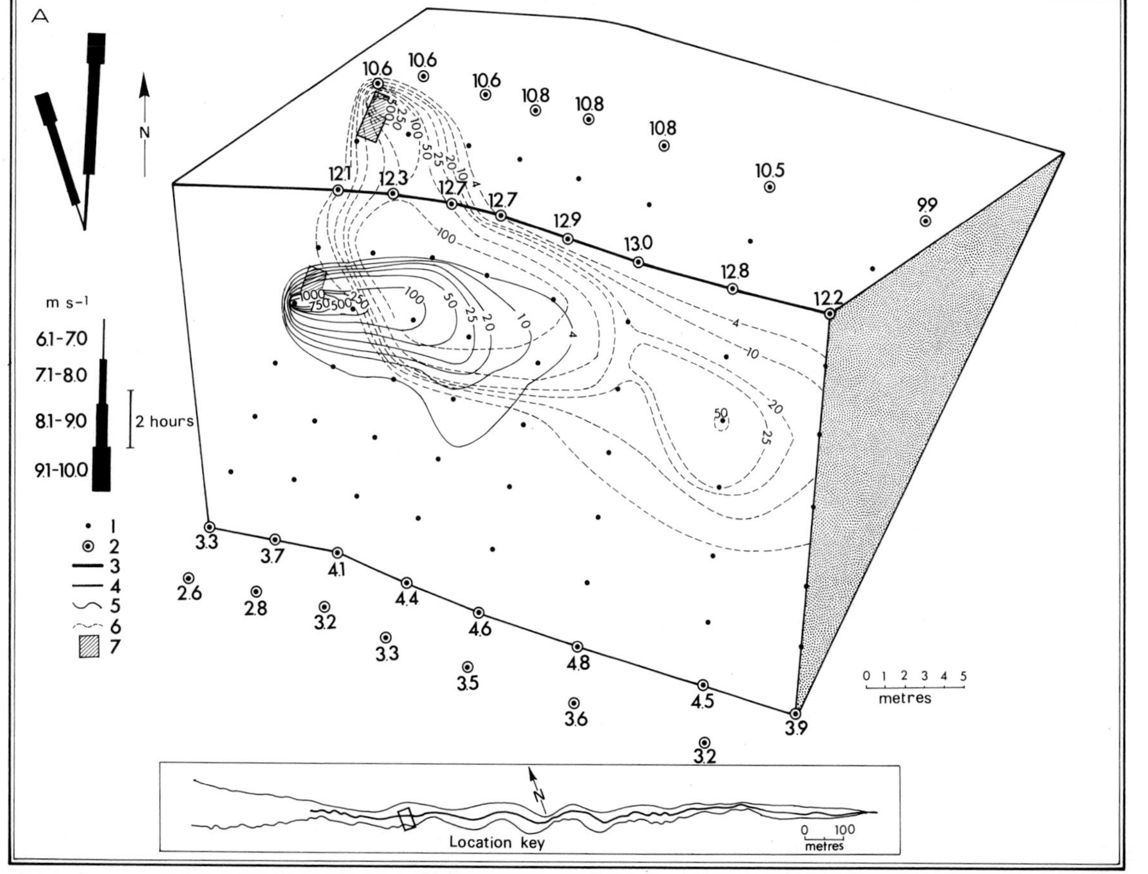

Figure 10.11 (A) Diagram showing concentration isograms of dyed grains (grains per 56 cm^2) released from sources on each flank of a seif dune. Different colours of sand were released on each flank. The wind rose indicates the direction, duration, and magnitude of wind measured at an elevation 11 m above the ground during the tracer experiment. The location key shows the

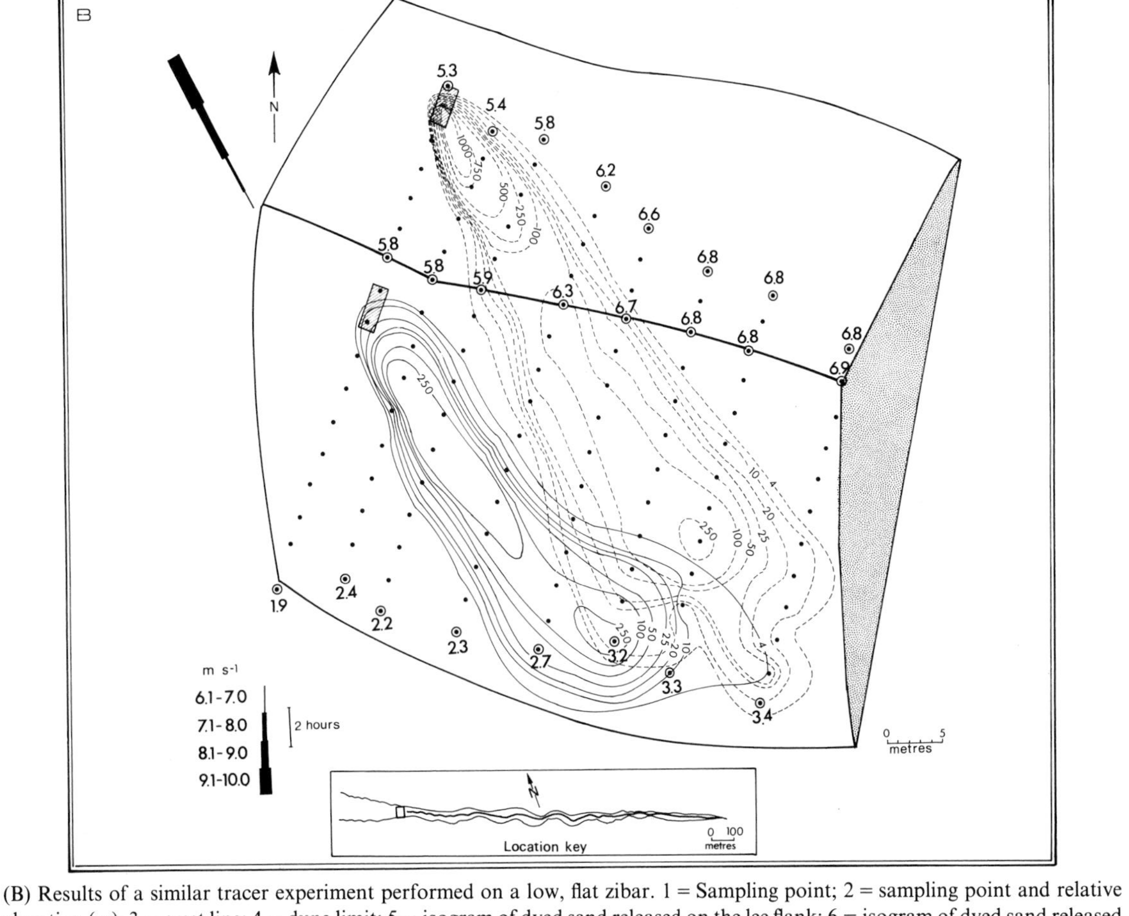

(B) Results of a similar tracer experiment performed on a low, flat zibar. 1 = Sampling point; 2 = sampling point and relative elevation (m); 3 = crest line; 4 = dune limit; 5 = isogram of dyed sand released on the lee flank; 6 = isogram of dyed sand released on the windward flank; 7 = sites of dyed sand release. (After Tsoar & Yaalon 1983).

(a)

(b)

Figure 10.12 (a) Surface of a megaripple showing armoured layer of coarse grains on its windward slope and crest; (b) the same megaripple after removal of the coarse surface layer using spray adhesive.

Figure 10.13 Subsurface sediment sampling using a powered sand auger.

The spray technique works well with dry sand but is less efficient if the sand is moist. With care it is possible to collect samples from several successive layers in order to investigate changes in grain size with depth in the uppermost 1–2 cm. Adhesive sprays can also be used on vertical sections to make sediment peels (Yasso & Hartman 1972).

Variations in grain size over greater depth intervals are normally investigated by collecting samples using a hand-auger or power-auger (Fig. 10.13). A typical shell auger suitable for sands can collect up to 30 cm of sand at a time (Fig. 10.14). The sample is inevitably disturbed by the action of augering, but this method has proved to be extremely useful for studying general subsurface trends in grain size to a depth of 10 m or more (e.g. Pye 1982b).

10.5 Methods of determining the grain size of sands

The grain size distribution of sands is currently determined using one of four main methods: (a) sieving (either wet or dry), (b) settling tube analysis, (c) electro-optical techniques, including Coulter Counter analysis and laser granulometry, and (d) computerized image analysis. The choice of the most appropriate method is governed largely by the amount of fine material in the sample and the use to which the data are to be put. Samples which contain only very small amounts of fines are most readily analysed by

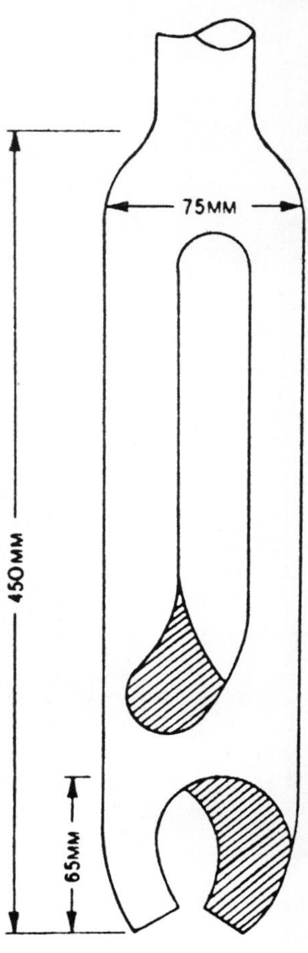

Figure 10.14 Sampling head of a typical sand auger.

dry-sieving or settling tube analysis, whereas Coulter Counter analysis or laser granulometry may be more convenient (although not necessarily as accurate) if the sample contains significant amounts of silt and clay. Image analysis is normally employed only where both size and shape information are required, or where the samples are very small.

10.5.1 Sieving

10.5.1.1 Sample pretreatment Samples from beaches, foredunes, and playa-margins in deserts may contain significant amounts of salt which should normally be removed before sieving. However, some playa-margin dune deposits contain wind-transported grains of halite which should properly be included in the analysis. In such cases removal of salts by washing with distilled water may not be appropriate, and it is advisable first to examine such samples with an optical microscope to establish whether detrital halite grains (or other water-soluble salts) are present. In most cases, however, salt is present as intergranular cement which, if not removed, will introduce errors into the analysis (Ingram 1971, McManus 1988). A sand sample which contains no silt and clay can be simply washed two or three times with distilled or deionized water. This can be done by placing the sample in a glass beaker containing the water and agitating it periodically for 15 min. The supernatant liquid is then decanted and the process repeated twice before the beaker is placed in an oven to dry at 105 °C.

Samples which contain significant amounts (> 2%) of silt and clay are more difficult to process. Where possible, it is preferable to separate the fine fraction from the sands before drying the sample, since during drying the silt and clay form aggregates and surface crusts which are subsequently difficult to disperse and to treat chemically (McManus 1988). However, many aeolian sediment samples which contain fines are already dry and may be crusted. It therefore simplifies the procedure to oven-dry the sample and to obtain its dry weight before proceeding with the analysis. Dispersion of fine particles in the dried sample can usually be achieved by shaking overnight in a dilute solution of sodium hexametaphosphate, followed by ultrasonic vibration for 20 min. Separation of the sand and fine fractions is then performed by washing through a 63-μm sieve. Care must be taken to retain all of the liquid passing through the sieve for subsequent analysis.

The process of wet sieving normally removes any salt present in the sand fraction, which can then be oven-dried and analysed by dry sieving. Low concentrations of dissolved salts present in the silt and clay suspension need not be removed, but if large amounts of salt are present they should be removed by repeated washing and centrifugation or by dialysis techniques (McManus 1988).

10.5.1.2 Dry sieving Dry sieving is undertaken using a stack of successively finer sieves which are mounted on an electrically powered shaker (Fig. 10.15). Some shakers have a simple vibrating action whereas others also rotate, tilt, or have a hammer action. Each sieve consists of a stainless-steel, brass, phosphor-bronze, or nylon mesh, the composition depending on the mesh size and manufacturer's specification. The number of square apertures per unit length defines the mesh number and the diagonal distance

Figure 10.15 Endecotts Octagon 200 sieve shaker and nest of sieves.

Table 10.3 List of maximum permissible sieve loadings. (After BS 1377:1975).

BS sieve mesh (mm)	Maximum weight (kg)	Sieve diameter (mm)
20	2.0	300
14	1.5	300
10	1.0	300
6.3	0.75	300
5	0.5	300
3.35	0.3	200
2.0	0.200	200
1.18	0.100	200
0.600	0.075	200
0.425	0.075	200
0.300	0.050	200
0.212	0.050	200
0.150	0.040	200
0.063	0.025	200

between the corners of the aperture defines the nominal size of the mesh. Most sedimentologists have used nests of sieves which have aperture dimensions at quarter-phi or half-phi intervals (see Table. 3.2).

The optimum size of sample used for dry sieving depends on the number of sieves and the dimensions of the mesh apertures. If too much sample is used the sieves will become overloaded and the size distribution will appear coarser than it actually is. Permissible sieve loadings recommended by the British Standards Institution (BS 1377:1975) are shown in Table 10.3. In the analysis of typical aeolian sands, 30–40 g of dry sand are commonly placed on a nest of eight sieves spanning the range 1–4 phi. However, if data are required at quarter-phi intervals, or the sample has a wider size range, only half of the required sieves may fit on the shaker at one time. In this instance the pan fraction from the first sieve run is emptied into the top sieve of the second set of finer sieves.

The optimum sieving time also varies with the number of sieves and the size range of the particles. A longer sieving time is required when much fine material is present, because fine material must pass through a greater number of sieves before reaching its final position, and because it takes a longer time for a grain to pass through a sieve with a small aperture size (Mizutani 1963, Dalsgaard & Jensen 1985). For aeolian sands, a sieving time of 20 min is adequate for each nest of sieves when the size of sample used is 40 g or less. Longer sieving times do not significantly improve the accuracy or reproducibility of the results (Dalsgaard & Jensen 1985), and may cause excessive grain breakage in some weathered sands (Pye & Sperling 1983).

The material retained in each sieve is carefully emptied onto a sheet of A3 size paper and any grains trapped in the mesh are removed by gentle brushing. The grains are then tipped carefully into a pre-weighed container and weighed to two decimal places on a suitable balance.

10.5.1.3 Wet sieving Samples which contain 2–10% fines are often most conveniently analysed by wet sieving. Several manufacturers provide sieves in the silt-size range which can be used to separate particles larger than 20 μm. The remaining 'pan' fraction is then rendered sufficiently small to allow the calculation of grain size parameters without the need for a separate pipette analysis of the fine fraction. The disadvantage of this technique is that it may be necessary to pass large amounts of water through the nest of sieves to achieve a full size separation, and it is often difficult to concentrate the finest material suspended in a large volume of water.

10.5.2 Settling tube analysis

Settling tube analysis is based on the settling velocity of grains and is applicable to silts, clays and sands. One of the first widely used settling tubes (Emery 1938) consists of a broad tube which narrows into a smaller diameter tube at the base. The height of sediment accumulation at known time periods after introduction of the sediment at the top of the column was measured with an optical micrometer. Particle sizes were then calculated from these figures using a derivation of Stokes' Law. In a later development, the Woods Hole Rapid Sediment Analyzer (Ziegler *et al.* 1960) used pressure transducers to measure the weight of water column above specific levels. A further variant of the sedimentation tube uses a balance pan near the bottom of the tube to measure the weight of accumulated grains (Sengupta & Veenstra 1968, Rigler *et al.* 1981).

All settling tube methods involve releasing the grains simultaneously into the top of the water column. This is normally done by holding the sample on the platten using a wetting agent and lowering it evenly into the water. The sample must be small (typically < 5 g) to satisfy the requirement of unhindered settling.

Most settling tubes used for sands are several tens of centimetres wide and 2 m or more long. Special measures may be required to ensure thermal stability and to prevent water convection within the tube.

Many workers have reported that settling tube analyses give good reproducibility, and they have the advantage that a continuous size distribution can be derived (Hartmann & Christiansen 1988).

10.5.3 Electro-optical methods of size analysis

Coulter Counter analysis and laser granulometry can also be used to analyse the distribution of samples containing sand, silt, and clay. In Coulter Counter analysis, the sample is suspended in an electrolyte which is drawn through an aperture across which an electrical current is passed. The presence of particles causes fluctuations in the electrical conductivity which are recorded and translated into volumes and numbers of particles per unit volume of suspension (McAllister 1981). Samples which contain a wide range of particle sizes may require the use of two or more apertures, in which case the data are merged using a special computer program. Although procedures for Coulter Counter analysis of sandy sediments have been established (McCave & Jarvis 1973), the method has not been widely used for the analysis of aeolian sands.

Laser granulometry is based on the principle that there is a direct relationship between the size of particles and the degree to which they diffract light. In the case of the Malvern Instruments Laser Particle Sizer Type 3600E, a beam of monochromatic light (wavelength 633 nm) is passed through a cell containing the sample in suspension and the diffracted light is focused onto a detector which senses the angular distribution of scattered light energy (McCave *et al.* 1986). The size range detected depends on the focal length of the focusing lens, which is placed between the sample cell and the detector. Grains are kept in suspension by a mechanical stirring device. Three lenses are available, each of which divides the distribution into 15 size classes. The 300 mm focal length lens has a range of 5.8–560 μm and is therefore most appropriate for sands. However, the coarsest of the 15 class intervals has a very wide range (261–564 μm). Although they are relatively simple to use, laser analysers in general have been judged to have poor accuracy and reproducibility compared with other methods (Dodge 1984, Cooper *et al.* 1984, McCave *et al.* 1986).

10.5.4 Direct measurement of grain size by image analysis

A wide variety of computer-based image analysers are now commercially available for the quantitative description of particle size and shape (Jongerius 1974, Slater & Ralph 1976, Jones 1977, Allen 1981, Serra 1982, Joyce-Loebl 1985). Their major advantage lies in their ability to provide simultaneous shape and size information, both for individual grains and for bulk samples. The principal limitation is that most rely on two-dimensional projected images and are thus unable to differentiate, for example, between a disc, a sphere, and an up-ended cylinder of equal radius.

10.6 Characterization of airflow

10.6.1 Wind velocity measurements

A wide range of instruments are available for measurement of wind velocity and direction [see Knott & Warren (1981) and WMO (1983) for more detailed reviews]. The choice of measuring system depends on the nature of the data required and the funds available. Different instruments are suited for the determination of mean flow velocity and short-term velocity fluctuations due to turbulence. Mean flow velocities, which are normally averaged over a period of a few minutes to 1 h or longer, are frequently measured using a vertical axis rotating cup anemometer. These instruments range in size from large anemometers with a radius of 50 cm, which are designed to be tower mounted at a standard height of 10 m above the ground, to small micro-anemometers, which are employed in vertical arrays to document the wind velocity profile near the ground. Most anemometers have three cups, although five-cup versions are also available.

All cup anemometers display a delayed response following changes in windspeed. A hysteresis effect exists, the response being faster following an increase in windspeed than following a decrease (Kaganov & Yaglom 1976). The high inertia of some instruments can result in the mean windspeed being overestimated by as much as 15% (Knott & Warren 1981). However, cup anemometers show an almost linear relationship between the rate of rotation and windspeed, and their calibration is unaffected by variations in air temperature or density.

Portable air meters, in which up to eight vanes rotate about a horizontal axis, can also be used to determine mean wind velocity and total wind run over a period ranging from a few minutes to a few hours. As with conventional cup anemometers, they display a lag response and hysteresis effect following sudden changes in windspeed.

Short-term fluctuations in wind velocity are often measured using hot-wire anemometers. These instruments operate on the principle that the electrical resistance of a wire varies with its temperature, which in turn is dependent on the flux of air past the wire (i.e. windspeed). Owing to their small size, arrays of closely spaced hot-wire anemometers can be employed in the field and in wind tunnels. However, they are easily damaged by saltating sand and are relatively expensive to repair.

Short-term gusts can also be measured using strain-gauge anemometers. A number of different designs have been described (Chepil & Siddoway 1959, Knott & Warren 1981), all of which are reported to be efficient provided that they are properly calibrated.

10.6.2 Flow visualization

Flow visualization techniques provide qualitative information about the nature of wind flow around obstacles, including the formation of eddies in the lee of dunes. Tape streamers and neutrally buoyant helium-filled balloons have been used for this purpose, but smoke candles which generate a trail of dense smoke are generally more effective in the field (e.g. Tsoar 1978) (Fig. 6.29). For wind tunnel experiments, smoke can be introduced into the flow at different levels above the bed through a small pipe.

10.7 Methods of monitoring changes in sand dune terrain

10.7.1 Field surveys

The most detailed information about changes in dune form is obtained by repeated theodolite and plane table surveys (e.g. Sharp 1966, Hastenrath 1967, Tsoar 1978, Warren & Kay 1987). Relevant introductions to surveying techniques were given by Jackson (1972) and Pugh (1975). However, since such surveys are time consuming, they are impractical over very large areas and cannot be repeated very frequently. An alternative method of monitoring changes in sand level or dune position, which is less comprehensive but much quicker, is to make regular measurements relative to fixed posts or pegs (e.g. Cooper 1958, Inman *et al.* 1966, Tsoar 1978, Pye 1980a, Warren & Kay 1987, Sarre 1989, Jungerius & van der Meulen 1989, Livingstone 1989b).

10.7.2 Remote sensing

A larger and more representative sample of morphological changes can be obtained by comparison of air photographs taken at different dates (e.g. McKee 1966, Clos-Arceduc 1969, 1969–70, Pye 1980a). Photographs taken at scales larger than 1:10 000 are most useful in this respect. Stereo-coverage is required if contour maps of dune terrain are to be made and changes in surface level documented using photogrammetric techniques. Smaller scale air photographs and satellite images are also useful for qualitative evaluation and morphometric measurements of large-scale aeolian features (e.g. Mainguet & Callot 1974, Breed & Grow 1979, Lancaster 1981a, Walker 1986), and can be used to monitor changes in vegetation cover at the regional scale (e.g. Tsoar & Møller 1986). The greatest problem facing all dune morphometric work, however, lies in deciding precisely what features to measure (Knott & Warren 1981). This problem is particularly acute when dunes have no regular geometry, no slip face, or a complex morphology with several slip faces. A number of morphometric parameters which can be used to describe and compare simple dune forms, such as barchans, are shown in Fig. 6.18.

A wide range of remote sensing information is now available for use in aeolian studies [see Sabins (1987) for a general review of remote sensing techniques]. These include Landsat Multispectral Scanner and Thematic Mapper images (e.g. Breed *et al.* 1987), SPOT satellite imagery, airborne radar and infrared imagery (Blom & Elachi 1981, Stembridge 1978, McCauley *et al.* 1982), and large-format camera photography taken by the Skylab and space shuttle astronauts (McKee *et al.* 1977). SPOT images have a better spatial resolution than Landsat Thematic Mapper images and may be obtained as stereo pairs. However, SPOT imagery is more expensive and lacks the spectral range of Landsat.

10.7.3 Sand dating methods

Long-term rates of dune movement, and the timing of phases of large-scale aeolian sand encroachment, can only be determined by dating. A minimum age can be provided by archaeological artifacts, such as stone tools, hearths, and shell middens, which are found on stabilized dune surfaces. In this case the age of a particular tool culture must be known fairly precisely. The age of charcoal, bone, and shell material associated with archaeological sites can be determined by radiocarbon dating. In certain circumstances soil carbonate nodules, humus layers, and woody material such as tree stumps can also be usefully dated. However, great care needs to be taken in the selection and pretreatment of material for dating in order to ensure that the relationship between the dated material and the host sediment is known (Goh & Molloy 1979, Worsley 1981, Pye & Switsur 1981, Geyh *et al.* 1983, Bradley 1985, pp. 47–69). Radiocarbon dating is routinely capable of providing ages up to a maximum of 40 000 years, although some workers have claimed that the range may ultimately be extended to 75 000 years and beyond using thermal enrichment (Stuiver *et al.* 1978) and accelerator mass spectrometric techniques (Stuiver 1978, Hedges 1981).

The past decade has also seen an increasing use of luminescence techniques to date dune sediments directly. Luminescence dating is based on the principle that light energy is emitted from a mineral crystal which has been exposed to ionizing radiation if it is heated (thermoluminescence) (Wintle & Huntley 1982) or subjected to light from a laser source (optical luminescence) (Huntley *et al.* 1985). The natural decay of radioisotopes in the surrounding sediment produces free electrons in the crystal which become trapped within lattice defects. The amount of stored light energy shows an asymptotic relationship with the length of exposure to ionizing radiation, and eventually all of the electron traps become filled. Dating is based on the assumption that exposure to sunlight during grain transport empties the electron traps, effectively 'zeroing' the luminescence clock. Following burial the electron traps progressively fill up again with time. The detailed procedures for determining luminescence ages were reviewed by Aitken (1985) and Berger (1988).

Some of the earliest thermoluminescence dating studies of sand dunes attempted to date fine quartz and feldspar grains (4–11 μm) present in the sands (e.g. Singhvi *et al.* 1982). However, such material often represents infiltrated airborne dust, which may post-date the host sand by several thousand years (Pye 1982f). It is therefore preferable to date the sand grains themselves. Several laboratories have recently reported thermoluminescence dates obtained from aeolian sand grains which show reasonable agreement with ages suggested by other evidence (Prescott 1983, Hutton *et al.* 1984, Bluszcz & Pazdur 1985, Poupeau *et al.* 1985, Singhvi *et al.* 1986, Gardner *et al.* 1987, Lundquist & Mejdahl 1987, Readhead 1988, Dijkmans *et al.* 1988). However, many difficulties remain to be resolved before luminescence dating of sand can be regarded as entirely reliable.

Appendix 1
SI units and c.g.s. equivalents

Quantity	Dimensions	SI	c.g.s.
Basic			
Length	L	1 m	$= 10^2$cm
Mass	M	1 kg	$= 10^3$ g
Time	T	1 s (or min,h,day,yr)	= 1 s
Temperature	θ	1 K (or 1 °C)	= 1 K (or 1 °C)
Derived			
Area	L^2	1 m^2	$= 10^4\ cm^2$
Volume	L^3	1 m^3	$= 10^6\ cm^3$
Density	ML^{-3}	1 $kg\,m^{-3}$	$= 10^{-3}\ g\,cm^{-3}$
Velocity	LT^{-1}	1 $m\,s^{-1}$	$= 10^2\ cm\,s^{-1}$
Acceleration	LT^{-2}	1 $m\,s^{-2}$	$= 10^2\ cm\,s^{-2}$
Force	MLT^{-2}	1 $kg\,m\,s^{-2}$ = 1 N (newton)	$= 10^5\ g\,cm\,s^{-2} = 10^5$ dyn
Pressure	$ML^{-1}T^{-2}$	1 $kg\,m^{-1}\,s^{-2}$ = 1 Pa(pascal)	$= 10\ g\,cm^{-1}\,s^{-2} = 10^{-2}$ mb
Work, energy	ML^2T^{-2}	1 $kg\,m^2\,s^{-2}$ = 1 J (joule)	$= 10^7\ g\,cm^2\,s^{-2} = 10^7$ erg
Power	ML^2T^{-3}	1 $kg\,m^2\,s^{-3}$ = 1 W (watt)	$= 10^7\ g\,cm^2\,s^{-3} = 10^7\ erg\,s^{-1}$
Heat, energy	Q (or ML^2T^{-2})	1 J	= 0.2388 cal
Heat flux	QT^{-1}	1 W	$= 0.2388\ cal\,s^{-1}$
Heat flux density	$QL^{-2}T^{-1}$	1 $W\,m^{-2}$	$= 2.388 \times 10^{-5}\ cal\,cm^{-2}\,s^{-1}$
Latent heat	QM^{-1}	1 $J\,kg^{-1}$	$= 2.388 \times 10^{-4}\ cal\,g^{-1}$
Specific heat	$QM^{-1}\theta^{-1}$	1 $J\,kg^{-1}\,°C^{-1}$	$= 2.388 \times 10^{-4}\ cal\,g^{-1}\,°C^{-1}$
Thermal conductivity	$QL^{-1}\theta^{-1}T^{-1}$	1 $W\,m^{-1}\,°C^{-1}$	$= 2.388 \times 10^{-3}\ cal\,cm^{-1}\,s^{-1}\,°C^{-1}$
Thermal diffusivity	L^2T^{-1}	1 $m^2\,s^{-1}$	$= 10^4\ cm^2\,s^{-1}$
Viscosity dynamic	$ML^{-1}T^{-1}$	1 $kg\,m^{-1}s^{-1}$ = 1 $N\,s\,m^{-2}$	= 10 poise
Viscosity kinematic	L^2T^{-1}	1 m^2s^{-1}	$= 10^4 cm^2s^{-1}$

References

Adetunji, J., J. McGregor and C. K. Ong 1979. Harmattan haze. *Weather* **34**, 430–6.

Adriani, M. J. and J. H. J. Terwindt 1974. *Sand stabilization and dune building*. Rijkswaterstaat Communications, No. 19. The Hague: Government Publishing Office.

Agassiz, A. 1895. A visit to the Bermudas in March 1894. *Harvard Coll. Mus. Comp. Zool. Bull.* **26**, 205–81.

Ahlbrandt, T. S. 1973. *Sand dunes, geomorphology and geology, Killpecker Creek area, northern Sweetwater County, Wyoming*. Unpublished PhD Thesis, Univ. Wyoming.

Ahlbrandt, T. S. 1974. The source of sand for the Killpecker sand dune field, southwestern Wyoming. *Sediment. Geol.* **11**, 39–57.

Ahlbrandt, T. S. 1975. Comparison of textures and structures to distinguish eolian environments, Killpecker dunefield, Wyoming. *Mt. Geol.* **12**, 61–73.

Ahlbrandt, T. S. 1979. Textural parameters of eolian deposits. In *A study of global sand seas*, E. D. McKee (ed.), 21–51. Prof. Pap. US Geol. Surv., No. 1052.

Ahlbrandt, T. S. & S. Andrews 1978. Distinctive sedimentary features of cold climate eolian environments, North Park, Colorado. *Palaeogeogr. Palaeoclimatol. Palaeoecol.* **25**, 327–51.

Ahlbrandt, T. S. & S. G. Fryberger 1980. *Eolian deposits in the Nebraska Sand Hills*. Prof. Pap. US Geol. Surv., No. 1120–A.

Ahlbrandt, T. S. & S. G. Fryberger 1981. Sedimentary features and significance of interdune deposits. In *Recent and ancient non-marine depositional environments: models for exploration*, F. G. Etheridge & R. M. Flores (eds), 293–314. Tulsa: Soc. Econ. Palaeontol. Mineral. Spec. Publ. No. 31.

Ahlbrandt, T. S., S. Andrews & D. T. Gwynne 1978. Bioturbation in eolian deposits. *J. Sediment. Petrol.* **48**, 839–48.

Aitken, M. J. 1985. *Thermoluminescence dating*. London: Academic Press.

Alizai, H. U. & L. C. Hulbert 1970. Effects of soil texture on evaporative loss and available water in semi-arid climates. *Soil Sci.* **110**, 328–32.

Al-Janabi, K. Z., A. J. Ali, F. H. Al-Taie & F. J. Jack 1988. Origin and nature of sand dunes in the alluvial plain of southern Iraq. *J. Arid Environ.* **14**, 27–34.

Allchin, B., A. S. Goudie & K. T. M. Hegde 1978. *The prehistory of the Great Indian Desert*. London: Academic Press.

Allen, J. R. L. 1963. Asymmetrical ripple marks and the origin of water-laid co-sets of cross-strata. *Geol. J.* **3**, 187–236.

Allen, J. R. L. 1968. *Current ripples*. Amsterdam: North-Holland.

Allen, J. R. L. 1970a. The avalanching of granular solids on dune and similar slopes. *J. Geol.* **78**, 326–51.

Allen, J. R. L. 1970b. *Physical processes of sedimentation*. London: Allen and Unwin.

Allen, J. R. L. 1973. A classification of climbing ripple cross-lamination. *J. Geol. Soc. Lond.* **129**, 537–41.

Allen, J. R. L. 1982. *Sedimentary structures*. Amsterdam: Elsevier.

Allen, J. R. L. 1985. *Principles of physical sedimentology*. London: Allen and Unwin.

Allen, T. 1981. *Particle size measurement*, 3rd edn. London: Chapman and Hall.

Allison, R. J. 1988. Sediment types and sources in the Wahiba Sands, Oman. In *The scientific results of the Oman Wahiba Sands Project 1985–1987*, R. W. Dutton (ed.), 161–8. J. Oman Stud. Spec. Rep. 3.

Almagor, G. 1979. Relict sandstones of Pleistocene age on the continental shelf of northern Sinai and Israel. *Isr. J. Earth Sci.* **28**, 70–6.

Al-Nakshabandi, G. A. & F. T. El Robee 1988. Aeolian deposits in relation to climatic conditions, soil characteristics and vegetative cover in the Kuwait desert. *J. Arid Environ.* **15**, 229–43.

Amiel, A. J. 1975. Progressive pedogenesis of eolianite sandstone. *J. Sediment. Petrol.* **45**, 513–19.

Ancker, J. A. M. van den, P. D. Jungerius & L. R. Mur 1985. The role of algae in the stabilization of coastal dune blowouts. *Earth Surf. Proc. Landf.* **10**, 189–92.
Anders, F. J. & S. P. Leatherman 1987. Effects of off-road vehicles on coastal foredunes at Fire Island, New York, USA. *Environ. Management* **11**, 45–52.
Anderson, H. A., M. L. Berrow, V. C. Farmer, A. Hepburn, J. D. Russell & A. D. Walker 1982. A reassessment of podzol formation processes. *J. Soil Sci.* **33**, 125–36.
Anderson, R. S. 1987a. Eolian sediment transport as a stochastic process: the effects of a fluctuating wind on particle trajectories. *J. Geol.* **95**, 497–512.
Anderson, R. S. 1987b. A theoretical model for aeolian impact ripples. *Sedimentology* **34**, 943–56.
Anderson, R. S. 1988. The pattern of grainfall deposition in the lee of aeolian dunes. *Sedimentology* **35**, 175–88.
Anderson, R. S. & P. K. Haff 1988. Simulation of eolian saltation. *Science* **241**, 820–3.
Anderson, R. S. and B. Hallet 1986. Sediment transport by wind: toward a general model. *Bull. Geol. Soc. Am.* **97**, 523–35.
Andrews, S. 1981. Sedimentology of Great Sand Dunes, Colorado. In *Recent and ancient non-marine depositional environments: models for exploration*, F. G. Etheridge and R. M. Flores (eds), 279–91. Tulsa: Soc. Econ. Paleontol. Mineral. Spec. Pub. 31.
Andriesse, J. P. 1969/70. The development of the podzol morphology in the tropical lowlands of Sarawak (Malaysia). *Geoderma* **3**, 261–79.
Angell, J. K. 1971. Helical circulation in the planetary boundary layer. *J. Atmos. Sci.* **28**, 135–8.
Angell, J. K., D. H. Pack & C. R. Dickson 1968. A Lagrangian study of helical circulations in the planetary boundary layer. *J. Atmos. Sci.* **25**, 707–17.
Anon. 1987. Australia's minerals: silica minerals. *Ind. Min.* December, 21–2.
Anton, D. & F. Ince 1986. A study of sand color and maturity in Saudi Arabia. *Z. Geomorph. N.F.* **30**, 339–56.
Anton, D. & P. Vincent 1986. Parabolic dunes of the Jafurah Desert, Eastern Province, Saudi Arabia. *J. Arid Environ.* **11**, 187–98.
Apfel, E. T. 1938. Phase sampling of sediments. *J. Sediment. Petrol.* **8**, 67–8.
Armbrust, D. V. & J. D. Dickerson 1971. Temporary wind erosion control: cost and effectiveness of 34 commercial materials. *J. Soil Water Conserv.* **26**, 152–7.
Ash, J. E. & R. J. Wasson 1983. Vegetation and sand mobility in the Australian desert dunefield. *Z. Geomorph Suppl. Bd.* **45**, 7–25.
Ashley, G. M. 1985. Proglacial eolian environment. In *Glacial sedimentary environments*, G. M. Ashley, J. Shaw & N. D. Smith (eds), 217–32, SEPM Short Course Notes 16. Tulsa: Soc. Econ. Paleontol. Mineral.
Ashour, M. M. 1985. Textural properties of Qatar dune sands. *J. Arid Environ.* **8**, 1–14.
Ashton, W. 1909. *The battle of land and sea on the Lancashire, Cheshire and North Wales coasts, and the origin of the Lancashire sandhills.* Southport: W. Ashton.
Atkinson, W. J. 1974. Problems arising from the intensive use of coastal dunes in New South Wales, Australia. *Int. J. Biometeorol.* **18**, 94–100.
Aufrère, L. 1930. L'orientation des dunes continentales. *Proc. 12th Int. Geogr. Congr., Cambridge*, 220–31.
Aufrère, L. 1931. Le cycle morphologique de dunes. *Geogr. Ann.* **40**, 362–85.
Aufrère, L. 1933. Classification des dunes. *Int. Geogr. Congr. Paris C.R., Actes* **2**, 699–711.
Aufrère, L. 1935. Essai sur les dunes du Sahara algerien. *Geogr. Ann.* **17**, 481–500.
Avis, A. M. & R. A. Lubke 1985. The effect of wind-borne sand and salt spray on the growth of *Scirpus nodosus* in a mobile dune system. *S. Afr. J. Bot.* **51**, 100–10.
Ayyad, M. A. 1973. Vegetation and environment of the western Mediterranean coastal land of Egypt. 1. The habitat of sand dunes. *J. Ecol.* **61**, 509–23.
Azizov, A. 1977. The influence of soil moisture on the resistance of soil to wind erosion. *Soviet Soil Sci.* **9**, 105–8.

Baba, J. & P. D. Komar, 1981. Measurements and analysis of settling velocities of natural quartz sand grains. *J. Sediment. Petrol.* **51**, 631–40.
Bagnold, R. A. 1931. Journeys in the Libyan Desert, 1929 and 1930. *Geogr. J.* **78**, 13–39, 524–33.
Bagnold, R. A. 1933. A further journey in the Libyan Desert. *Geogr. J.* **82**, 103–29, 211–35.
Bagnold, R. A. 1935a. *Libyan sands.* London: Hodder and Stoughton.
Bagnold, R. A. 1935b. The movement of desert sand. *Geogr. J.* **85**, 342–69.
Bagnold, R. A. 1936. The movement of desert sand. *Proc. R. Soc. London, Ser. A* **157**, 594–620.
Bagnold, R. A. 1937a. The size-grading of sand by wind. *Proc. R. Soc. London, Ser. A* **163**, 250–64.
Bagnold, R. A. 1937b. The transport of sand by wind. *Geogr. J.* **89**, 409–38.
Bagnold, R. A. 1938a. The measurement of sand storms. *Proc. R. Soc. London, Ser. A* **167**, 282–91.

Bagnold, R. A. 1938b. Grain structure of sand dunes and its relation to their water content. *Nature* **142**, 403–4.
Bagnold, R. A. 1941. *The physics of blown sand and desert dunes.* London: Methuen.
Bagnold, R. A. 1951. Sand formation in southern Arabia. *Geogr. J.* **117**, 78–86.
Bagnold, R. A. 1953a. Forme des dunes de sable et régime des vents. In *Actions Éoliennes.* CNRS Colloq. Int., Vol. 35, 23–32.
Bagnold, R. A. 1953b. The surface movement of blown sand in relation to meteorology. In *Desert research*, 89–96. Jerusalem: Research Council of Israel, Special Publication 2.
Bagnold, R. A. 1954a. Experiments on a gravity-free dispersion of large solid spheres in a Newtonian fluid under shear. *Proc. R. Soc. London, Ser. A* **225**, 49–63.
Bagnold, R. A. 1954b. The physical aspects of dry deserts. In *Biology of deserts*, J. L. Cloudsley-Thompson (ed.), 7–12. London: Institute of Biology.
Bagnold, R. A. 1956. Flow of cohesionless grains in fluids. *Philos. Trans. R. Soc. London, Ser. A* **249**, 235–97.
Bagnold, R. A. 1960. The re-entrainment of settled dusts. *Int. J. Air Pollut.* **2**, 357–63.
Bagnold, R. A. 1973. The nature of saltation and "bed-load" transport in water. *Proc. R. Soc. London, Ser. A* **332**, 473–504.
Bagnold, R. A. 1979. Sediment transport by wind and water. *Nordic Hydrol.* **10**, 309–22.
Bagnold, R. A. & O. E. Barndorff-Nielsen 1980. The pattern of natural grain size distributions. *Sedimentology* **27**, 199–207.
Baldwin, K. A. & M. A. Maun 1983. Microenvironment of Lake Huron sand dunes. *Can. J. Bot.* **61**, 241–55.
Ball, M. M. 1967. Carbonate sand bodies of Florida and the Bahamas. *J. Sediment. Petrol.* **37**, 556–91.
Ballantyne, C. K. & G. Whittington 1987. Niveo-aeolian sand deposits on An Teallach, Wester Ross, Scotland. *Trans. R. Soc. Edinb. Earth Sci.* **78**, 51–63.
Barclay, W. S. 1917. Sand dunes in the Peruvian Desert. *Geogr. J.* **49**, 53–6.
Barndorff-Nielsen, O. E. 1977. Exponentially decreasing distributions for the logarithm of particle size. *Proc. R. Soc. London, Ser. A* **353**, 401–19.
Barndorff-Nielsen, O. E. 1986. Sand, wind and statistics: some recent investigations. *Acta Mech.* **64**, 1–18.
Barndorff-Nielsen, O. E. & C. Christiansen 1986. *Erosion, deposition and size distributions of sand.* Dept. Theoretical Statistics, Institute of Mathematics, Univ. Aarhus, Res. Rep. 149.
Barndorff-Nielsen, O. E. & C. Christiansen 1988. Erosion, deposition and size distributions of sand. *Proc. R. Soc. London, Ser. A* **417**, 335–52.
Barndorff-Nielsen, O. E., K. Dalsgaard, C. Halgreen, H. Kuhlman, J. T. Møller & G. Schou 1982. Variation in particle size over a small dune. *Sedimentology* **29**, 53–65.
Barndorff-Nielsen, O. E., J. L. Jensen & M. Sørensen 1983. On the relation between size and distance travelled for wind-driven sand grains — results and discussion of a pilot experiment using coloured sand. In *Mechanics of sediment transport*, B. M. Sumer & A. Muller (eds), 55–64. Rotterdam: Balkema.
Barndorff-Nielsen, O. E., J. L. Jensen, H. L. Nielsen, K. R. Rasmussen & M. Sørensen 1985a. Wind tunnel tracer studies of grain progress. In *Proceedings of international workshop on the physics of blown sand*, O. E. Barndorff-Nielsen, J. T. Møller, K. R. Rasmussen & B. B. Willetts (eds), 243–52. Dept. Theoretical Statistics, Institute of Mathematics, Univ. Aarhus, Mem. 8.
Barndorff-Nielsen, O. E., P. Blaesild, J. L. Jensen & M. Sørensen 1985b. The fascination of sand. In *A celebration of statistics — the ISI centenary volume*, A. C. Atkinson and S. E. Fienberg (eds), 57–87. New York: Springer.
Barndorff-Nielsen, O. E., J. L. Jensen, H. L. Nielsen, K. R. Rasmussen & M. Sørensen 1986. *Sand transport studies in a wind tunnel using radioactive grain. Report on a pilot experiment.* Dept. Theoretical Statistics, Institute of Mathematics, Univ. Aarhus, Res. Rep. 140.
Barndorff-Nielsen, O. E., J. L. Jensen, H. L. Nielsen, K. R. Rasmussen & M. Sørensen 1988. *Sand transport studies in a wind tunnel using radioactive grains: report on a pilot experiment.* Dept. Theoretical Statistics, Institute of Mathematics, Univ. Aarhus, Res. Rep. 140.
Barr, D. A. & J. B. McKenzie 1976. Dune stabilization in Queensland, Australia, using vegetation and mulches. *Int. J. Biometeorol.* **20**, 1–8.
Barr, D. A. & J. B. McKenzie 1977. Progress in coastal and sand dune stabilization and management experiments on South Stradbroke Island, Queensland. *Proc. 3rd Aust. Conf. Coastal and Ocean Eng., Melbourne 18–21 April 1977*, 207–13.
Barr, D. A. & B. G. Watt 1969. Pedestrian access to beaches. *J. Soil Conserv. N.S.W.* **25**, 286–94.
Barrett, P. J. 1980. The shape of rock particles, a critical review. *Sedimentology* **27**, 291–304.
Barth, H. K. 1982. Accelerated erosion of fossil dunes in the Gourma region (Mali) as a manifestation of desertification. *Catena Suppl.* **1**, 212–19.

Basu, A. 1976. Petrology of Holocene fluvial sand derived from plutonic source rocks: implications to paleoclimatic interpretation. *J. Sediment. Petrol.* **46**, 649–709.
Basu, A. 1985. Influence of climate and relief on compositions of sands released at source areas. In *Provenance of arenites*, G. G. Zuffa (ed.), 1–15. Dordrecht: Reidel.
Bathurst, R. G. C. 1975. *Carbonate sediments and their diagenesis*, 2nd edn. Amsterdam: Elsevier.
Battan, L. J. 1961. *The nature of violent storms*. New York: Anchor Books.
Beadnell, H. J. L. 1909. Desert sand dunes. *Cairo Sci. J.* **3**, 171–2.
Beadnell, H. J. L. 1910. The sand dunes of the Libyan desert. *Geogr. J.* **35**, 379–95.
Beal, M. A. & F. P. Shepard 1956. A use of roundness to determine depositional environments. *J. Sediment. Petrol.* **26**, 49–60.
Beard, D. C. & P. K. Weyl 1973. Influence of texture on porosity and permeability of unconsolidated sand. *Bull. Am. Assoc. Petrol. Geol.* **51**, 349–69.
Beheiry, S. A. 1967. Sand forms in the Coachella Valley, Southern California. *Ann. Am. Assoc. Geogr.* **57**, 25–48.
Beier, J. A. 1987. Petrographic and geochemical analysis of caliche profiles in a Bahamian Pleistocene dune. *Sedimentology* **34**, 991–8.
Belly, P. Y. 1964. *Sand movement by wind.* US Army Corps of Engineers, Coastal Engineering Research Center, Tech. Memo., No. 1.
Ben-Asher, J. 1987. Irrigation with saline water *GeoJournal* **15**, 267–72.
Bendali, F., C. Floret, E. Le Floc'h & R. Pontanier 1990. The dynamics of vegetation and sand mobility in arid regions of Tunisia. *J. Arid Environ.* **18**, 21–32.
Benito, G. A. de, & P. J. Le Roux 1976. Stabilization of sand dunes in the Western Sahara. *S. Afr. For. J.* **97**, 36–43.
Berg, N. H. 1983. Field evaluation of some sand transport models. *Earth Surf. Proc. Landf.* **8**, 101–14.
Berger, G. W. 1988. Dating Quaternary events by luminescence. *Geol. Soc. Am. Spec. Pap.* No. 227, 13–50.
Berner, R. A. 1978. Rate control of mineral dissolution under earth surface conditions. *Am. J. Sci.* **278**, 1235–51.
Besler, H. 1980. Die dunen-Namib: Entstehung und dynamik eines ergs. *Stuttgarter Geographische. Studien Band 96*. Stuttgart: Geographisches Institut der Universität.
Besler, H. 1983. The response diagram: distinction between aeolian mobility and stability of sands and aeolian residuals by grain size parameters. *Z. Geomorph. Suppl. Bd.* **45**, 287–302.
Besler, H. 1984. The development of the Namib dune field according to sedimentological and geomorphological evidence. In *Late Cainozoic palaeoclimates of the southern hemisphere*, J. C. Vogel (ed.), 445–53. Rotterdam: Balkema.
Bettenay, E. 1962. The salt lake systems and their associated aeolian features in the semi-arid regions of Western Australia. *J. Soil Sci.* **13**, 10–17.
Bigarella, J. J. 1972. Eolian environments — their characteristics, recognition and importance. In *Recognition of ancient sedimentary environments*, J. K. Rigby & W. K. Hamblin (eds), 12–62. Tulsa: Soc. Econ. Paleontol. Mineral., Spec. Publ. No. 16.
Bigarella, J. J. 1975a. Lagoa dunefield, State of Santa Catarina, Brazil — a model of eolian and pluvial activity. *Bol. Paran. Geocien.* **33**, 133–67.
Bigarella, J. J. 1975b. Structures developed by dissipation of dune and beach ridge deposits. *Catena* **2**, 107–52.
Bigarella, J. J. & I. Salamuni 1961. Early Mesozoic wind patterns as suggested by dune bedding in the Botucatu Sandstone of Brazil and Uruguay. *Geol. Soc. Am. Bull.* **72**, 1089–105.
Bigarella, J. J., A. H. Alessi, R. D. Becker & G. M. Duarte, 1969a. Textural characteristics of the coastal dune, sand ridge, and beach sediments, *Bol. Paran. Geocien.* **27**, 15–89.
Bigarella, J. J., R. D. Becker & G. M. Duarte 1969b. Coastal dune structures from Paran, Brazil. *Mar. Geol.* **7**, 5–55.
Binda, P. L. & P. R. Hildred 1973. Bimodal grain-size distributions of some Kalahari-type sands from Zambia. *Sediment. Geol.* **10**, 233–7.
Bird, E. C. F. 1965. The formation of coastal dunes in the humid tropics: some evidence from North Queensland. *Aust. J. Sci.* **27**, 258–9.
Bird, E. C. F. 1974. Dune stability on Fraser Island. *Queensl. Nat.* **21**, 15–21.
Bisal, F. & J. Hsieh 1966. Influence of moisture on erodibility of soil by wind. *Soil Sci.* **102**, 143–6.
Bisal, F. & K. F. Nielsen 1962. Movement of soil particles in saltation. *Can. J. Soil Sci.* **42**, 81–6.
Blackburn, G. & R. M. Taylor 1969. Limestones and red soils of Bermuda. *Geol. Soc. Am. Bull.* **80**, 1595–8.
Blake, W. P. 1855. On the grooving and polishing of hard rocks and minerals by dry sand. *Am. J. Sci.* **20**, 178–81.

Blakey, R. C. 1988. Superscoops: their significance as elements of eolian architecture. *Geology* **16**, 483–7.

Blakey, R. C. & L. T. Middleton 1983. Permian shoreline eolian complex in Central Arizona: dune changes in response to cyclic sea level changes. In *Eolian sediments and processes*, M. E. Brookfield & T. S. Ahlbrandt (eds), 551–81. Amsterdam: Elsevier.

Blanford, W. T. 1876. On the physical geography of the Great Indian Desert with special reference to the existence of the sea in the Indus valley; and on the origin and mode of formation of the sand hills. *J. Asiatic Soc. Bengal* **45**, 86–103.

Blatt, H. 1967. Original characteristics of clastic quartz grains. *J. Sediment. Petrol.* **37**, 401–24.

Blatt, H. 1970. Determination of mean sediment thickness in the crust: a sedimentologic method. *Bull. Geol. Soc. Am.* **81**, 255–62.

Blom, J. & L. Wartena 1969. The influence of changes in surface roughness on the development of the turbulent boundary layer in the lower layer of the atmosphere. *J. Atmos. Sci.* **26**, 255–65.

Blom, R. & C. Elachi 1981. Spaceborne and airborne imaging radar observations of sand dunes. *J. Geophys. Res.* **86**, 3061–73.

Blumenthal, K. P. 1964. The construction of a drift sand dyke on the island of Rottumerplaat. *Proc. 9th Coastal Engng. Conf., Lisbon*, 346–67.

Bluszcz, A. & M. F. Pazdur 1985. Comparison of TL and ^{14}C dates of young eolian sediments — a check of the zeroing assumption. *Nucl. Tracks* **10**, 703–10.

Boerboom, J. H. A. 1964. Microklimatologische waarnemingen in de Wassenaarse duinen. *Meded. Landbouwhogesch. Wageningen* **64**, 9–28.

Bond, R. D. 1964. The influence of the microflora on the physical properties of soils. II. Field studies on water repellent sands. *Aust. J. Soil Res.* **2**, 123–31.

Bond, R. D. & J. R. Harris 1964. The influence of the microflora on the physical properties of soils. I. Effects associated with filamentous algae and fungi. *Aust. J. Soil Res.* **2**, 111–22.

Bond, W. J. 1976. Illuvial band formation in a laboratory column of sand. *Soil Sci. Soc. Am. J.* **50**, 265–7.

Boorman, L. A. 1977. Sand dunes. In *The Coastline*, R. S. K. Barnes (ed.), 161–97. New York: Wiley.

Boorman, L. A. 1989. The influence of grazing on British sand dunes. In *Perspectives in coastal dune management*, F. van der Meulen, P. D. Jungerius & J. Visser (eds), 121–24. The Hague: SPB Academic Publishing.

Boorman, L. A. & R. M. Fuller 1977. Studies on the impact of paths on the dune vegetation at Winterton, Norfolk, England. *Biol. Conserv.* **12**, 203–16.

Borowka, R. K. 1980. Present day processes and dune morphology on the Leba Barrier, Polish coast of the Baltic. *Geogr. Ann.* **62A**, 75–82.

Bosworth, T. O. 1922. *Geology of the Tertiary and Quaternary periods in the north-west part of Peru.* London: Macmillan.

Boulaine, J. 1954. La zebkha de Ben Ziane et sa 'lunette' ou bourrelet. *Rév. Géomorph. Dyn.* **5**, 102–23.

Boulaine, J. 1956. Les lunettes des bassess plaines oranaises; formation éolienne argileuses lies a l'extension des sols salins; La Sebkha de Ben Ziane; la depression de Chantrit. *Proc. 4th INQUA Conf.* **1**, 143–50.

Bourman, R. P. 1986. Aeolian sand transport along beaches. *Aust. Geogr.* **17**, 30–4.

Bowen, A. J. & D. Lindley 1977. A wind tunnel investigation of the wind speed and turbulence characteristics close to the ground over various escarpment shapes. *Boundary-Layer Meteorol.* **12**, 259–71.

Bowers, J. E. 1982. The plant ecology of inland dunes in western North-America. *J. Arid Environ.* **5**, 199–220.

Bowers, J. E. 1986. *Seasons of the wind.* Flagstaff: Northland Press.

Bowler, J. M. 1973. Clay dunes; their occurrence, formation and environmental significance. *Earth Sci. Rev.* **9**, 315–38.

Bowler, J. M. 1978. Glacial-age aeolian events at high and low latitudes. In *Antarctic glacial history and world palaeoenvironments*, E. M. Van Zinderen-Bakker (ed.), 149–72. Rotterdam: Balkema.

Bowler, J. M., G. S. Hope, J. N. Jennings, G. Singh & D. Walker 1976. Late Quaternary climates of Australia and New Guinea. *Quat. Res.* **6**, 359–94.

Bradley, E. F. 1980. An experimental study of the profiles of wind speed, shearing stress and turbulence at the crest of a large hill. *Q. J. R. Meteorol. Soc.* **106**, 101–23.

Bradley, E. F. 1983. The influence of thermal stability and angle of incidence on the acceleration of wind up a slope. *J. Wind Eng. Ind. Aerodyn.* **15**, 231–42.

Bradley, R. S. 1985. *Quaternary paleoclimatology.* Boston: Allen and Unwin.

Brady, N. C. 1974. *The nature and properties of soils*, 8th edn. New York: Macmillan.

Branner, J. C. 1890. The aeolian sandstone of Fernando de Noronha. *Am. J. Sci.* **39**, 247–57.
Brazel, A. & S. Hsu 1981. The climatology of hazardous Arizona dust storms. In *Desert dust: origin, characteristics, and effect on man*, T. L. Péwé (ed.), 293–303. Geol. Soc. Am. Spec. Pap., No. 186.
Breed, C. S. & W. J. Breed 1979. Dunes and other windforms of Central Australia (and a comparison with linear dunes on the Moenkopi Plateau, Arizona). In: *Apollo–Soyuz test project summary science report, Vol. 2. Earth observations and photography*, F. El-Baz and D. M. Warner (eds), 319–58. Washington, DC: NASA, SP-412.
Breed, C. S. & T. Grow 1979. Morphology and distribution of dunes in sand seas observed by remote sensing. In *A study of global sand seas*, E. D. McKee (ed.), 253–302. Prof. Pap. US Geol. Surv., No. 1052.
Breed, C. S., S. G. Fryberger, S. Andrews, C. McCauley, F. Lennartz, D. Gebel & K. Horstman 1979. Regional studies of sand seas, using Landsat (ERTS) imagery. In *A study of global sand seas*, E. D. McKee (ed.), 305–97. Prof. Pap. US Geol. Surv., No. 1052.
Breed, C. S., J. F. McCauley, W. J. Breed, C. K. McCauley & A. S. Cotera 1984. Eolian (wind-formed) landscapes. In *Landscapes of Arizona, the geological story*, P. L. Smiley, J. D. Nations, T. L. Péwé and J. P. Schafer (eds), 359–413. Lanham: Univ. Press of America.
Breed, C. S., J. F. McCauley & P. A. Davis 1987. Sand sheets of the eastern Sahara and ripple blankets on Mars. In *Desert sediments: ancient and modern*, L. E. Frostick and I. Reid (eds), 337–59. Geol. Soc. Spec. Publ., No. 35. Oxford: Blackwell.
Breed, C. S., J. F. McCauley & M. I. Whitney 1989. Wind erosion forms. In *Arid zone geomorphology*, D. S. G. Thomas (ed.), 284–307. London: Belhaven Press.
Bremontier, N. T. 1833. Mémoire sur les dunes. *Annales des Ponts et Chaussées* **1**, 145–224.
Bressolier, C. & Y. F. Thomas 1977. Studies on wind and plant interactions on the French Atlantic coastal dunes. *J. Sediment. Petrol.* **47**, 331–8.
Bretz, J. H. 1960. Bermuda, a partially drowned, late mature, Pleistocene karst. *Bull. Geol. Soc. Am.* **71**, 1729–54.
Brewer, R. 1964. *Fabric and mineral analysis of soils.* New York: Wiley.
Bridge, B. J. & P. J. Ross 1983. Water erosion in vegetated sand dunes at Cooloola, southeast Queensland. *Z. Geomorph. Suppl. Bd.* **45**, 227–44.
Briggs, L. I., D. S. McCulloch, & F. Moser, 1962. The hydraulic shape of sand particles. *J. Sediment. Petrol.* **32**, 645–56.
British Standards Institution 1975. *BS 1377. Methods of test for soils for civil engineering purposes.* London: HMSO.
Britter, R. E., J. C. R. Hunt & K. J. Richards 1981. Air flow over a two-dimensional hill, studies of velocity speed-up, roughness effects and turbulence. *Q. J. R. Meteorol. Soc.* **107**, 91–110.
Brodhead, J. M. & P. J. Godfrey 1977. Off road vehicle impact in Cape Cod National Seashore: disruption and recovery of dune vegetation. *Int. J. Biometeorol.* **21**, 299–306.
Brookfield, M. 1970. Dune trends and wind regime in central Australia. *Z. Geomorph. Suppl. Bd.* **10**, 121–53.
Brookfield, M. E. 1977. The origin of bounding surfaces in ancient aeolian sandstones. *Sedimentology* **24**, 303–32.
Brookfield, M. E. & T. S. Ahlbrandt (eds) 1983. *Eolian sediments and processes.* Amsterdam: Elsevier.
Brooks, A. 1979. *Coastlands.* London: British Trust for Conservation Volunteers.
Brooks, C. E. P. & N. Carruthers 1953. *Handbook of statistical methods in meteorology.* London: Air Ministry, Meteorological Office.
Brooks, C. E. P. & S. T. A. Mirrlees 1932. *A study of the atmospheric circulation over tropical Africa.* Geophys. Mem. No. 55. London: Meteorological Office.
Brooks, H. B. 1960. Rotation of dust devils. *J. Meteorol.* **17**, 84–6.
Brown, R. A. 1983. The flow in the planetary boundary layer. In *Eolian sediments and processes*, M. E. Brookfield and T. S. Ahlbrandt (eds), 291–319. Amsterdam: Elsevier.
Brown, R. A. & A. L. Hafenrichter 1962. *Stabilizing sand dunes on the Pacific coast with woody plants.* US Dept. Agric., Soil Conserv. Serv., Misc. Publ. 1892.
Brugmans, F. 1983. Wind ripples in an active drift sand area in the Netherlands: a preliminary report. *Earth Surf. Proc. Landf.* **8**, 527–34.
Brunt, D. 1939. *Physical and dynamical meteorology.* Cambridge: Cambridge Univ. Press.
Bruun, P. 1962. Sea level rise as a cause of shore erosion. *Proc. Am. Soc. Civ. Eng., Waterways Harbours Div.* **88**, 1117–30.
Bryan, K. 1932. Characteristic forms of dune fields, *Geogr. Rev.* **22**, 325–7.
Bucher, W. H. 1919. On ripples and related surface sedimentary forms. *Am. J. Sci.* **47**, 149–210, 241–69.

Buckingham, E. 1914. On physically similar systems; illustrations of the use of dimensional equations. *Phys. Rev.* **4**, 345–76.
Buckland, P. C. 1982. The coversands of north Lincolnshire and the Vale of York. In *Papers in earth studies, Lovatt Lectures, Worcester*, B. H. Adlam, C. R. Fenn and L. Morris (eds), 143–77. Norwich: Geobooks.
Buckley, R. 1981. Central Australian sandridges. *J. Arid Environ.* **4**, 91–101.
Buckley, R. 1987. The effect of sparse vegetation on the transport of dune sand by wind. *Nature* **325**, 426–28.
Buckley, R. 1989. Grain-size characteristics of linear dunes in central Australia. *J. Arid Environ.* **16**, 23–8.
Bui, E. N., J. M. Mazzullo & L. P. Wilding 1990. Using quartz grain size and shape analysis to distinguish between aeolian and fluvial deposits in the Dallol Bosso of Niger (West Africa). *Earth Surf. Proc. Landf.* **14**, 157–66.
Bull, W. B. 1975. Allometric change of landforms. *Bull. Geol. Soc. Am.* **86**, 1489–98.
Busche, D., M. Draga & H. Hagedorn 1984. *Les sables éoliens, modelés et dynamique, la menace éolienne et son contrôlé, bibliographie annotée*. Eschborn: Gesellschaft fur Technische Zusammenarbeit.
Byers, H. R. 1959. *General meteorology*. New York: McGraw-Hill.
Byrne, R. J. 1968. Aerodynamic roughness criteria in aeolian sand transport. *J. Geophys. Res.* **73**, 541–7.

Cailleux, A. 1967. Periglacial of McMurdo Strait (Antarctica). *Biul. Periglac.* **17**, 57–90.
Cailleux, A. 1969. Quaternary periglacial wind-worn sand grains in USSR. In *The periglacial environment*, T. L. Péwé (ed.), 285–301. Montreal: McGill–Queens Univ. Press.
Cailleux, A. 1973. Éolienisations périglaciaires quaternaires au Canada. *Biul. Periglac.* **22**, 81–115.
Cailleux, A. 1978. Niveo-eolian deposits. In *The Encyclopedia of Sedimentology*, R. W. Fairbridge and J. Bourgeois (eds), 501–3. Stroudsberg: Dowden Hutchinson Ross.
Calder, K. L. 1949. Eddy diffusion and evaporation in flow over aerodynamically smooth and rough surfaces. A treatment based on laboratory laws of turbulent flow with special reference to conditions in the lower atmosphere. *Q. J. Mech. Appl. Math.* **2**, 153–76.
Calkin, P. E. & R. H. Rutford 1974. The sand dunes of Victoria Valley, Antarctica. *Geogr. Rev.* **64**, 189–216.
Calvet, F., F. Plana & A. Traveria 1980. La tendencia mineralogica de las eolianitas del Pleistoceno de Mallorca, mediante la aplicacias del metodo de Chung. *Acta Geol. Hispan.* **15**, (2), 39–44.
Calvet, F., L. Pomar, & M. Esteban 1975. Las rhizoconcretions del Pleistoceno de Mallorca. *Inst. Invest. Geol., Univ. Barcelona* **30**, 35–60.
Campbell, E. M. 1968. Lunettes in southern South Australia. *Trans. R. Soc. S. Austr.* 92–109.
Capot-Rey, R. 1945. Dry and humid morphology in the Western Erg. *Geogr. Rev.* **35**, 391–407.
Carrigy, M. A. 1970. Experiments on the angles of repose of granular material. *Sedimentology* **14**, 147–58.
Carroll, D. 1939. The movement of sand by wind. *Geol. Mag.* **76**, 6–22.
Carroll, D. 1970. *Rock weathering*. New York: Plenum.
Carson, M. A. & P. A. MacLean 1985a. Storm-controlled oblique dunes of the Oregon coast: discussion. *Bull. Geol. Soc. Am.* **96**, 409–10.
Carson, M. A. & P. A. MacLean 1985b. Hybrid eolian dunes of William River dune field, Northern Saskatchewan, Canada. *Bull. Am. Assoc. Petrol. Geol.* **69**, 242–3.
Carson, M. A. & P. A. MacLean 1986. Development of hybrid aeolian dunes: The William River dunefield, Northwest Saskatchewan, Canada. *Can. J. Earth Sci.* **23**, 1974–90.
Carter, G., H. M. Harris & K. S. Strandsberg 1964. *Beneficiation studies of the Oregon coastal dune sands for use as glass sand.* US Bureau of Mines, Rep. Invest. 6484.
Carter, L. D. 1981. A Pleistocene sand sea on the Alaskan Arctic coastal plain. *Science* **211**, 381–3.
Carter, R. W. G. 1976. Formation, maintenance and geomorphological significance of an eolian shell pavement. *J. Sedim. Petrol.* **46**, 418–29.
Carter, R. W. G. 1980. Human activities and geomorphic processes: the example of recreation pressure on the Northern Ireland coast. *Z. Geomorph.* **34**, 155–64.
Carter, R. W. G. 1982. Some problems associated with the analysis and interpretation of mixed carbonate and quartz beach sands, illustrated by examples from north-west Ireland. *Sediment. Geol.* **33**, 35–56.
Carter, R. W. G. 1988. *Coastal environments*. London: Academic Press.
Carter, R. W. G. & G. W. Stone 1989. Mechanisms associated with the erosion of sand dune cliffs, Magilligan, Northern Ireland. *Earth Surf. Proc. Landf.* **14**, 1–10.

Carver, R. E. & G. A. Brook. 1989. Late Pleistocene paleowind directions, Atlantic Coastal Plain, U.S.A. *Palaeogeogr. Palaeoclimatol. Palaeoecol.* **74**, 205–16.

Case, G. O. 1914. *Coast sand dunes, sand spits and sand wastes.* London: St. Brides Press.

Castel, I. I. Y. 1988. A simulation model of wind erosion and sedimentation as a basis for management of a drift sand area in the Netherlands. *Earth Surf. Proc. Landf.* **13**, 501–9.

Castel, L., E. Koster & R. Slotboom 1989. Morphogenetic aspects and age of Late Holocene eolian drift sands in Northwest Europe. *Z. Geomorph.* **33**, 1–26.

Castro, C. A. & P. V. Vicuña 1986. Man's impact on coastal dunes in Central Chile (32°–34°S). *Thalassas* **4**, 17–25.

Castro, I. P. 1971. Wake characteristics of two dimensional perforated plates normal to an air stream. *J. Fluid Mech.* **46**, 599–609.

Catt, J. A. 1977. Loess and coversands. In *British Quaternary studies: recent advances*, F. W. Shotton (ed.), 22–9. Oxford: OUP.

Catt, J. A. 1986. *Soils and Quaternary geology: a handbook for field scientists*, Oxford: OUP.

CERC 1977. *Shore Protection Manual*, 3rd edn. Fort Belvoir, VA: U.S. Army Coastal Engineering Research Centre.

Cermak, J. E. 1971. Laboratory simulation of the atmospheric boundary layer. *Am. Inst. Aeron. Astron. J.* **9**, 1746–54.

Chadwick, H. W. & P. D. Dalke 1965. Plant succession on dune sands in Fremont County, Idaho. *Ecology* **46**, 765–80.

Chan, M. A. & G. Kocurek 1988. Complexities in eolian and marine interactions: processes and eustatic controls on erg development. *Sediment. Geol.* **56**, 283–300.

Chandler, M. A., G. Kocurek, Goggin, D. J. & L. W. Lake 1989. Effects of stratigraphic heterogeneity on permeability in eolian sandstone sequence, Page Sandstone, Northern Arizona. *Am. Assoc. Petrol. Geol. Bull.* **73**, 658–68.

Chang, P. K. 1976. *Control of flow separation.* New York: McGraw-Hill.

Chapman, F. 1900. Notes on the consolidated aeolian sands of Kathiawar. *Q. J. Geol. Soc. Lond.* **56**, 584–8.

Charba, J. 1974. Application of gravity current models to analysis of squall-line gust front. *Month. Weather Rev.* **102**, 140–56.

Chaudhri, R. S. & H. M. M. Kahn, 1981. Textural parameters of desert sediments — Thar Desert (India). *Sediment. Geol.* **28**, 43–62.

Chepil, W. S. 1941. Relation of wind to the dry aggregate structure of a soil. *Sci. Agric.* **21**, 488–507.

Chepil, W. S. 1945a. Dynamics of wind erosion: I. Nature of movement of soil by wind. *Soil Sci.* **60**, 305–20.

Chepil, W. S. 1945b. Dynamics of wind erosion: II. Initiation of soil movement. *Soil Sci.* **60**, 397–411.

Chepil, W. S. 1951. Properties of soil which influence wind erosion: IV. State of dry aggregate structure. *Soil Sci.* **72**, 387–401.

Chepil, W. S. 1956. Influence of moisture on erodibility of soil by wind. *Proc. Soil Sci. Soc. Am.* **20**, 288–92.

Chepil, W. S. 1957. Sedimentary characteristics of dust storms: III. Composition of suspended dust. *Am. J. Sci.* **255**, 206–13.

Chepil, W. S. 1958a. *Soil conditions that influence wind erosion.* US Dept. Agri. Tech. Bull., No. 1185.

Chepil, W. S. 1958b. The use of evenly spaced hemispheres to evaluate aerodynamic forces on a soil surface. *Trans. Am. Geophys. Union* **39**, 397–403.

Chepil, W. S. 1959. Equilibrium of soil grains at the threshold movement by wind. *Proc. Soil Sci. Soc. Am.* **23**, 422–8.

Chepil, W. S. 1961. The use of spheres to measure lift and drag on wind-eroded soil grains. *Proc. Soil Sci. Soc. Am.* **25**, 343–5.

Chepil, W. S. & R. A. Milne 1939. Comparative study of soil drifting in the field and in a wind tunnel. *Sci. Agric.* **19**, 249–57.

Chepil, W. S. & R. A. Milne 1941. Wind erosion of soils in relation to size and nature of the exposed area. *Sci. Agric.* **21**, 479–87.

Chepil, W. S. & F. H. Siddoway 1959. Strain-gauge anemometer for analyzing various characteristics of wind turbulence. *J. Meteorol.* **16**, 411–18.

Chepil, W. S. & N. P. Woodruff 1957. Sedimentary characteristics of dust storms. II. Visibility and dust concentration. *Am. J. Sci.* **255**, 104–14.

Chepil, W. S. & N. P. Woodruff 1963. The physics of wind erosion and its control. *Adv. Agron.* **15**, 211–302.

Chepil, W. S., F. H. Siddoway & D. V. Armbrust 1962. Climatic factor for estimating wind erodibility of farm fields. *J. Soil Water Conserv.* **17**, 162–5.

Chepil, W. S., F. H. Siddoway & D. V. Armbrust 1963. Climatic index of wind erosion conditions in the Great Plains. *Proc. Soil Sci. Soc. Am.* **27**, 449–52.

Chiu, T. Y. 1972. *Sand transport by wind.* Univ. Florida (Gainesville), Dept. Coastal Oceanogr. Eng., Tech. Rep., TR-040.

Chorley, R. J. & B. A. Kennedy 1971. *Physical geography, a systems approach.* London: Prentice-Hall.

Christiansen, C. 1984. A comparison of sediment parameters from log-probability plots and log–log plots of the same sediments. *Dept. Geol., Univ. Aarhuus, Geoskrifter* 20.

Christiansen, C. & D. Hartmann 1988a. SAHARA: a package of PC-computer programs for estimating both log-hyperbolic grain size parameters and standard moments. *Comput. Geosci.* **14**, 557–625.

Christiansen, C. & D. Hartmann 1988b. On using the log-hyperbolic distribution to describe the textural characteristics of aeolian sediments: a discussion. *J. Sediment. Petrol.* **58**, 159–60.

Christiansen, C., P. Blaesild & K. Dalsgaard 1984. Re-interpreting 'segmented' grain-size curves. *Geol. Mag.* **121**, 47–51.

Chudeau, R. 1920. Étude sur les dunes Sahariennes. *Ann. Geogr.* **29**, 334–451.

Clapp, F. G. 1920. Along and across the Great Wall of China. *Geogr. Rev.* **9**, 221–49.

Clark, J. R. 1977. *Coastal ecosystem management*, New York: Wiley.

Clark, M. W. 1987. Image analysis of clastic particles. In *Clastic particles*, J. R. Marshall (ed.), 256–66. New York: Van Nostrand Reinhold.

Clarke, R. H. & C. H. B. Priestley 1970. The asymmetry of Australian desert sand ridges. *Search* **1**, 77–8.

Cleary, W. J. & J. R. Conolly 1971. Embayed quartz grains in soils and their significance. *J. Sediment. Petrol.* **42**, 899–904.

Clements, T. F. 1952. Windblown rocks and trails on the Little Bonnie Claire Playa, Nye County, Nevada. *J. Sediment. Petrol.* **22**, 182–6.

Clements, T., R. O. Stone, J. F. Mann, Jr, & J. L. Eymann 1963. *A study of windborne sand and dust in desert areas.* Natick: US Army Natick Laboratory, Report ES-8.

Clemmensen, L. B. 1986. Storm-generated eolian sand shadows and their sedimentary structures, Vejers Strand, Denmark. *J. Sediment. Petrol.* **56**, 520–7.

Clemmensen, L. B. 1987. Complex star dunes and associated aeolian bedforms, Hopeman Sandstone (Permo-Triassic), Moray Firth Basin, Scotland. In *Desert sediments ancient and modern*, L. E. Frostick & I. Reid (eds), 213–31. Geol. Soc. Lond. Spec. Publ. No. 35. Oxford: Blackwell.

Clemmensen, L. B. 1989. Preservation of interdraa and plinth deposits by the lateral migration of large linear draas (Lower Permian Yellow Sands, northeast England). *Sediment. Geol.* **65**, 139–51.

Clemmensen, L. B. & K. Abrahamsen 1983. Aeolian stratification and facies associations in desert sediments, Arran Basin (Permian), Scotland. *Sedimentology* **30**, 31–9.

Clemmensen, L. B. & R. C. Blakey 1989. Erg deposits in the Lower Jurassic Wingate Sandstone, northeastern Arizona: oblique dune sedimentation. *Sedimentology* **36**, 449–70.

Clos-Arceduc, A. 1966. Le róle déterminant des ondes aeriennes stationnaires dans la structure des ergs sahariennes et les formes d'erosion aroisantes. *C. R. Acad. Sci. Paris* D **262**, 2673–6.

Clos-Arceduc, A. 1967. La direction des dunes et ses rapports avec celle du vent. *C. R. Acad. Sci., Paris Ser. D* **264**, 1393–6.

Clos-Arceduc, A. 1969. Essai d'explication de formes dunaires Sahariennes. *Études de Photo-Interpretation No. 4.* Paris: Imprimerie de l'Institut Geog. National.

Clos-Arceduc, A. 1969–70. The use of aerial photographs in the study of the lengthening of sand dunes in a direction close to that of the wind in the Sahara. *Photogrammetria* **25**, 189–99.

Cloudsley-Thompson, J. L. & M. J. Chadwick 1964. *Life in deserts.* London: Foulis.

Coaldrake, J. E. 1962. The coastal dunes of southern Queensland. *Proc. R. Soc. Qld.* **2**, 101–15.

Coetzee, F. 1975/6a. Coastal aeolianites at Black Rock, northern Zululand. *Trans. Geol. Soc. S. Afr.* **78**, 313–22.

Coetzee, F. 1975/6b. Solution pipes in coastal aeolianites of Zululand and Mozambique. *Trans. Geol. Soc. S. Afr.* **78**, 323–34.

Coffey, G. N. 1909. Clay Dunes. *J. Geol.* **17**, 754–5.

Cook, N. J. 1978. Wind-tunnel simulation of the adiabatic atmospheric boundary layer by roughness, barrier and mixing device methods. *J. Ind. Aerodyn.* **3**, 157–76.

Cooke, R. U. & A. Warren 1973. *Geomorphology in Deserts.* London: Batsford.

Cooke, R. U., D. Brunsden, J. C. Doornkamp & D. K. C. Jones 1982. *Urban geomorphology in drylands.* Oxford: Oxford University Press.

Cooper, L. R., R. L. Haverland, D. M. Hendricks & W. G. Knisel 1984. Microtrac particle size analysis: an alternative particle-size determination method for sediments and soils. *Soil Sci.* **138**, 138–46.
Cooper, W. S. 1944. Development and maintenance of the mature profile of a transverse dune ridge. *Am. Philos. Soc. Year Book 1944*, 150–3.
Cooper, W. S. 1958. Coastal sand dunes of Oregon and Washington. *Mem. Geol. Soc. Am.* No. 72.
Cooper, W. S. 1967. Coastal dunes of California. *Mem. Geol. Soc. Am.* No. 104.
Corn, M. 1966. Adhesion of particles. In *Aerosol Science*, C. N. Davies (ed.), 389–92. New York: Academic Press.
Cornish, V. 1897. On the formation of sand-dunes. *Geogr. J.* **9**, 278–309.
Cornish, V. 1900. On desert sand dunes bordering the Nile delta. *Geogr. J.* **15**, 1–30.
Cornish, V. 1914. *Waves of sand and snow.* London: Unwin.
Cornish, V. 1928. Limits of form and magnitude in desert dunes. *Nature* **121**, 620–2.
Counihan, J. 1969. An improved method of simulating a neutral atmospheric boundary layer in a wind tunnel. *Atmos. Environ.* **3**, 197–214.
Cowdrey, C. F. 1967. *A simple method for the design of wind tunnel velocity-profile grids.* Aerodynamics Division, National Physical Laboratory, Aero Note 1055.
Cowie, J. D. 1968. Pedology of soils from wind-blown sand in the Manawatu district. *N.Z. J. Sci.* **11**, 459–87.
Cressey, G. B. 1928. *The Indiana sand dunes and shore lines of the Lake Michigan basin.* Chicago: Univ. Chicago Press.
Crickmore, M. H. 1967. Measurement of sand transport in rivers with special reference to tracer methods. *Sedimentology* **8**, 175–228.
Crook, K. A. W. 1968. Weathering and roundness of quartz sand grains. *Sedimentology* **11**, 171–82.
Cui, B., P. D. Komar & J. Baba 1983. Settling velocities of natural sand grains in air. *J. Sediment. Petrol.* **53**, 1205–11.
Culver, S. J., P. A. Bull, S. Campbell, R. A. Shakesby & W. B. Whalley 1983. Environmental discrimination based on quartz grain surface textures: a statistical investigation. *Sedimentology* **30**, 129–36.
Curran, P. 1987. Life on loess. *Geogr. Mag.* **59**, 381–4.

Dahl, B. E., B. A. Fall & L. C. Otteni 1973. Vegetation for creation and stabilization of foredunes, Texas coast. In *Estuarine research*, Vol. II, *Geology and engineering*, L. E. Cronin (ed.), 457–70. New York: Academic Press.
Dainelli, G. 1931. The agricultural possibilities of Italian Somalia. *Geogr. Rev.* **21**, 56–69.
Dalsgaard, K. & J. L. Jensen 1985. A methodological study of the sieving of small sand samples. In *Proceedings of the International Workshop on the Physics of Blown Sand, Aarhus, May 28–31 1985*, O. E. Barndorff-Nielsen, J. T. Møller, K. R. Rasmussen & B. B. Willetts (eds), 609–32. Dept. Theoretical Statistics, Institute of Mathematics, Univ. Aarhus, Mem. 8.
Dan, J., D. H. Yaalon & H. Koyumdjisky 1968/9. Catenary soil relationship in Israel. 1. The Netanya catena on coastal dunes of the Sharon. *Geoderma* **2**, 95–120.
Dana, J. D. & C. S. Harlbut 1959. *Manual of mineralogy*, 7th edn. New York: Wiley.
Dangavs, N. V. 1979. Presencia de dunas de arcilla fosiles en La Pampa Deprimida. *Rev. Assoc. Geol. Argent.* **34**, 31–5.
Danin, A. 1978. Plant species diversity and plant succession in a sandy area in the Northern Negev. *Flora* **167**, 409–22.
Danin, A. 1983. *Desert vegetation of Israel and Sinai.* Jerusalem: Cana.
Danin, A. 1987. Impact of man on biological components of desert in Israel. *Proc. Annual Meeting of the Israeli Bot. Soc., Beer Sheva*, 6–7.
Danin, A. & D. H. Yaalon 1982. Silt plus clay sedimentation and decalcification during plant succession in sands of the Mediterranean Coastal Plain. *Isr. J. Earth Sci.* **31**, 101–9.
Dann, J. 1939. Sandgebirge in Alagschan. *Z. Geomorph.* **11**, 28–51.
David, P. 1977. *Sand dune occurrences in Canada.* National Parks Branch, Can., Dept. Indian and Northern Affairs, Contract 74–230, Report.
David, P. 1981. Stabilized dune ridges in northern Saskatchewan. *Can. J. Earth Sci.* **18**, 286–310.
Davies, J. L. 1974. The coastal sediment compartment. *Aust. Geogr. Stud.* **12**, 139–51.
Davies, P. J., B. Bubela & J. Ferguson 1978. The formation of ooids. *Sedimentology* **25**, 703–30.
Day, A. E. 1928. Pipes in the coast sandstone of Syria. *Geol. Mag.* **65**, 412–15.
Deacon, E. L. 1953. Vertical profiles of mean wind in the surface layers of the atmosphere. *Geophys. Mem.* **91**, 1–68.
De Coninck, F. 1980. Major mechanisms in the formation of spodic horizons. *Geoderma* **24**, 101–28.

De Ploey, J. 1980. Some field measurements and experimental data on wind blown sands. In *Assessment of erosion*, M. de Boodt & D. Gabrieli (eds), 541–52. Chichester: Wiley.

Deynoux, M., G. Kocurek & J. N. Proust 1989. Late Proterozoic periglacial aeolian deposits on the West African platform, Taoudeni Basin, western Mali. *Sedimentology* **36**, 531–49.

Dieren van, J. W. 1934. *Organogene Dunenbildung*. The Hague: Martinus Nijhoff.

Dijk, van H. W. J. 1989. Ecological impact of drinking water production in Dutch coastal dunes. In *Perspectives in coastal dune management*, F. van der Meulen, P. D. Jungerius & J. Visser (eds), 163–72. The Hague: SPB Academic Publishing.

Dijkerman, J. C., M. G. Cline & G. W. Olson 1967. Properties and genesis of textural subsoil lamellae. *Soil Sci.* **104**, 7–16.

Dijkmans, J. W. A. 1990. Niveo-eolian sedimentation and resulting sedimentary structures; examples from the Sondre Stromfjiord area, W. Greenland. In *Permafrost and Periglacial Processes* (in press).

Dijkmans, J. W. A. & H. J. Mucher 1989. Niveo-aeolian sedimentation of loess and sand, and experimental and micromorphological approach. *Earth Surf. Proc. Landf.* **14**, 303–15.

Dijkmans, J. W. A., E. A. Koster, J. P. Galloway & W. G. Mook 1986. Characteristics and origin of calcretes in a subarctic environment, Great Kobuk Sand Dunes, northwestern Alaska, USA. *Arctic Alpine Res.* **18**, 377–87.

Dijkmans, J. W. A., A. G., Wintle & V. Mejdahl 1988. Some thermoluminescence properties and dating of eolian sands from the Netherlands. *Quat. Sci. Rev.* **7**, 349–55.

Dincer, T., A. Al-Mugrim & U. Zimmerman 1974. Study of the infiltration and recharge through the sand dunes in arid zones with special reference to the stable isotopes and thermonuclear tritium. *J. Hydrol.* **23**, 79–109.

Dodge, L. G. 1984. Calibration of the Malvern particle size analyzer. *Appl. Opt.* **23**, 2415–19.

Doe, T. W. & R. H. Dott, Jr 1980. Genetic significance of deformed cross-bedding — with examples from the Navajo and Weber sandstones of Utah. *J. Sediment. Petrol.* **50**, 793–812.

Doody, J. P. 1989. Conservation and development of the coastal dunes in Great Britain. In *Perspectives in coastal dune management*, F. van der Meulen, P. D. Jungerius & J. Visser (eds), 53–68. The Hague: SPB Academic Publishing.

Douglass, A. E. 1909. The crescentic dunes of Peru. *Appalachia* **12**, 34–45.

Draga, M. 1983. Eolian activity as a consequence of beach nourishment — observations at Westerland (Sylt), German North Sea coast. *Z. Geomorph. Suppl. Bd.* **45**, 303–19.

Dubief, J. 1952. Le vent et le deplacement du sable au Sahara. *Trav. Inst. Rech. Sahara* **8**, 123–62.

Dubief, J. 1979. Review of the North African climate with particular emphasis on the production of eolian dust in the Sahel Zone and in the Sahara. In *Saharan dust*, C. Morales (ed.), 27–48. Chichester: Wiley.

Dubief, J. & P. Queney 1935. Les grands traits du climate du Sahara Algerien. *La Meteorologie* **11**, 80–91.

Durst, C. S. 1935. Dust in the atmosphere. *Q. J. R. Met. Soc.* **61**, 81–7.

Durward, J. 1931. Rotation of dust devils, *Nature* **128**, 412–13.

Durward, J. 1936. Weather changes on the west African air route. *Meteorol. Mag.* **71**, 227–9.

Dyer, K. 1986. *Coastal and estuarine sediment dynamics*. Chichester: Wiley.

Eardley, A. J. & B. Stringham 1952. Selenite crystals in the clays of Great Salt Lake. *J. Sediment Petrol.* **22**, 234–8.

Eaton, K. J. 1981. Building and tropical windstorms. *Overseas Building Notes*, No. 188.

Edelman, T. 1968. Dune erosion during storm conditions. *Proc. 11th Conf. Coastal Eng., London*, Vol. I, 719–22.

Edelman, T. 1973. Dune erosion during storm conditions. *Proc. 14th Conf. Coastal Eng.* 1305–12.

Ehlers, J. 1988. *The morphodynamics of the Wadden Sea*. Rotterdam: Balkema.

Ehrenberg, C. G. 1847. The Sirocco dust that fell at Genoa on the 16th May 1846. *Q. J. Geol. Soc. Lond.* **3**, 25–6.

Ehrlich, R. 1964. The role of the homogeneous unit in sampling plans for sediments. *J. Sediment. Petrol.* **34**, 437–9.

Ehrlich, R. & B. Weinberg 1970. An exact method for characterization of grain shape. *J. Sediment. Petrol.* **40**, 205–12.

Ehrlich, R., S. K. Kennedy & C. D. Brotherhood 1987. Respective roles of Fourier and SEM techniques in analyzing sedimentary quartz. In *Clastic particles*, J. R. Marshall (ed.), 292–301. New York: Van Nostrand Reinhold.

Ehrlich, R., J. Orzeck & B. Weinberg 1974. Detrital quartz as a natural tracer — Fourier grain shape analysis. *J. Sediment. Petrol.* **44**, 145–50.

Ehrlich, R., P. J. Brown, J. M. Yarus & R. S. Przygocki 1980. The origin of shape frequency distributions and the relationship between size and shape. *J. Sediment. Petrol.* **50**, 475–83.
Einstein, H. A. 1950. The bed-load function for sediment transportation in open channel flows. *US Department of Agriculture Tech. Bull.* **1026**.
Einstein, H. A. & E. A. El-Samni 1949. Hydrodynamic forces on a rough wall. *Rev. Mod. Phys.* **21**, 520–4.
El-Baz, F. 1986. The formation and motion of dunes and sand seas. In *Physics of desertification*, F. El-Baz and M. H. A. Hassan (eds), 70–93. Dordrecht: Martinus Nijhoff.
El-Baz, F. & M. H. A. Hassan (eds) 1986. *Physics of desertification*. Dordrecht: Martinus Nijhoff.
Elder, J. W. 1959. Steady flow through non-uniform gauzes of arbitrary shape. *J. Fluid Mech.* **5**, 355–68.
Eldridge, F. R. 1980. *Wind machines*, 2nd edn. New York: Van Nostrand Reinhold.
El-Fandy, M. G. 1940. The formation of depressions of the khamsin type. *Q. J. R. Meteorol. Soc.* **66**, 323–35.
Elliott, W. P. 1958. The growth of the atmospheric internal boundary layer. *Trans. Am. Geophys. Union* **39**, 1048–54.
Ellis, W. S. 1987. Africa's Sahel: the stricken land. *Nat. Geogr.* **172**, 141–79.
Ellwood, J. M., P. D. Evans & I. G. Wilson 1975. Small-scale aeolian bedforms. *J. Sediment. Petrol.* **45**, 554–61.
Elzenga, W., J. Schwan, T. A. Baumfalk, J. Vendenberghe & L. Krook 1987. Grain surface characteristics of periglacial aeolian and fluvial sands. *Geol. Mijn.* **65**, 273–86.
Embabi, N. S. 1982. Barchans of the Kharga depression. In *Desert landforms of southwest Egypt: a basis for comparison with Mars*, F. El-Baz and T. A. Maxwell (eds), 141–55. Washington, DC: NASA CR-3611.
Emery, K. O. 1938. Rapid method of mechanical analysis of sands. *J. Sediment. Petrol.* **8**, 105–11.
Engelhardt, W. F. von, 1977. *The origin of sediments and sedimentary rocks. Sedimentary petrology, Part III*. New York: Wiley.
Enquist, F. 1932. The relation between dune form and wind direction. *Geol. Foren. Stockh. Forhand.* **54**, 19–59.
Eschner, T. B. & G. Kocurek 1986. Marine destruction of eolian sand seas: origin of mass flows. *J. Sediment. Petrol.* **56**, 401–11.
Eschner, T. B. & G. Kocurek 1988. Origins of relief along contacts between eolian and overlying marine strata. *Am. Assoc. Petrol Geol. Bull.* **72**, 932–49.
Esteban, M. 1976. Vadose pisolite and caliche. *Am. Assoc. Petrol. Geol. Bull.* **60**, 2048–57.
Evans, J. R. 1962. Falling and climbing sand dunes in the Cronese ("Cat") Mountain area, San Bernardino County, California. *J. Geol.* **70**, 107–13.
Evans, J. W. 1900. Mechanically-formed limestones from Junagarh (Kathiawar) and other localities. *Q. J. Geol. Soc. Lond.* **56**, 559–83, 588–9.

Fairbridge, R. W. 1950. The geology and geomorphology of Point Peron, Western Australia. *J. R. Soc. West. Austr.* **34**, (3), 35–72.
Fairbridge, R. W. & D. L. Johnson 1978. Eolianites. In *The Encyclopedia of Sedimentology*, R. W. Fairbridge & J. Bourgeois (eds), 279–82. Stroudsberg, Dowden Hutchinson and Ross.
Fairbridge, R. W. & C. Teichert 1953. Soil horizons and marine bands in the coastal limestones of Western Australia. *J. Proc. R. Soc. New South Wales* **86**, 68–86.
Faller, A. J. 1965. Large eddies in the atmospheric boundary layer and their possible role in the formation of cloud rows. *J. Atmos. Sci.* **22**, 176–84.
Faller, A. J. & A. H. Woodcock 1964. The spacing of windrows of Sargassum in the ocean. *J. Mar. Res.* **22**, 22–9.
Farquharson, J. S. 1937. Haboobs and instability in the Sudan. *Q. J. R. Meteorol. Soc.* **63**, 393–414.
Feniak, M. W. 1944. Grain sizes and shapes of various minerals in igneous rocks. *Am. Mineral.* **29**, 415–21.
Fenley, J. M. 1948. Sand dune control in Les Landes, France. *J. For.* **46**, 514–20.
Ferguson, J. B., B. Bubela & P. J. Davies 1978. Synthesis and possible mechanism of formation of radial carbonate ooids. *Chem. Geol.* **22**, 285–308.
Filion, L. 1984. A relationship between dunes, fire and climate recorded in the Holocene deposits of Quebec. *Nature* **309**, 543–6.
Filion, L. & P. Morisset 1983. Eolian landforms along the eastern coast of Hudson Bay, Northern Quebec. *Nordicana* **47**, 73–94.
Finkel, H. J. 1959. The barchans of southern Peru. *J. Geol.* **67**, 614–47.
Fisher, P. & P. Galdies 1988. Computer model for barchan dune movement. *Comput. Geosci.* **14**, 229–53.

Flint, R. F. & G. Bond 1968. Pleistocene sand ridges and pans in Western Rhodesia. *Bull. Geol. Soc. Am.* **79**, 299–313.
Flohn, H. 1969. Local wind systems. In *World survey of climatology*, Vol. 2, H. Flohn (ed.), 139–71. Amsterdam: Elsevier.
Flower, W. D. 1936. *Sand devils*. Air Ministry, Meteorological Office, London, Prof. Notes 71.
Folk, R. L. 1955. Student operator error in determination of roundness, sphericity and grain size. *J. Sediment. Petrol.* **25**, 297–301.
Folk, R. L. 1968. Bimodal supermature sandstone: product of the desert floor *Proc. 23rd Int. Geol. Congr.* **8**, 9–32.
Folk, R. L. 1971a. Longitudinal dunes of the northwestern edge of the Simpson Desert, Northern Territory, Australia, 1. Geomorphology and grain size relationships. *Sedimentology* **16**, 5–54.
Folk, R. L. 1971b. Genesis of longitudinal and oghurd dunes elucidated by rolling upon grease. *Bull. Geol. Soc. Am.* **82**, 3461–8.
Folk, R. L. 1976a. Rollers and ripples in sand, streams and sky: rhythmic alteration of transverse and longitudinal vortices in three orders. *Sedimentology* **23**, 649–69.
Folk, R. L. 1976b. Reddening of desert sands: Simpson desert, Northern Territory, Australia. *J. Sediment. Petrol.* **46**, 604–15.
Folk, R. L. 1977a. Longitudinal ridges with turning fork junctions in the laminated interval of flysch beds: Evidence for low order helicoidal flow in turbidities. *Sediment. Geol.* **19**, 1–6.
Folk, R. L. 1977b. Folk's bedform theory — reply. *Sedimentology* **24**, 864–74.
Folk, R. L. 1978. Angularity and silica coatings of Simpson Desert sand grains, Northern Territory. *J. Sediment. Petrol.* **52**, 93–101.
Folk, R. L. & W. C. Ward 1957. Brazos River bar: a study in the significance of grain size parameters. *J. Sediment. Petrol.* **27**, 3–26.
Francis, J. R. D. 1973. Experiments on the motion of solitary grains along the bed of a water stream. *Proc. R. Soc. London, Ser. A* **332**, 413–71.
Free, E. E. 1911. The movement of soil material by the wind. *USDA Bur. Soils Bull.* No. 68.
Frere, H. B. E. 1870. Notes on the Rann of Cutch. *J. R. Geogr. Soc.* **40**, 181–207.
Friedman, G. M. 1961. Distinction between dune, beach and river sands from their textural characteristics. *J. Sediment. Petrol.* **31**, 514–29.
Friedman, G. M. 1967. By name processes and statistical parameters compared for size frequency distribution of beach and river sands. *J. Sediment. Petrol.* **37**, 327–54.
Friedman, G. M. & J. E. Sanders 1978. *Principles of sedimentology*. New York: Wiley.
Frostick, L. E. & I. Reid (eds) 1987. *Desert sediments ancient and modern*. Geol. Soc. Lond. Spec. Publ. No. 35. Oxford: Blackwell.
Fryberger, S. G. & T. S. Ahlbrandt 1979. Mechanisms for the formation of eolian sand seas. *Z. Geomorph.* **23**, 440–60.
Fryberger, S. G. & G. Dean 1979. Dune forms and wind regime. In *A study of global sand seas*, E. D. McKee (ed.), 137–69. Prof. Pap. US Geol. Surv. No. 1052.
Fryberger, S. G. & A. S. Goudie 1981. Arid geomorphology. *Prog. Phys. Geogr.* **5**, 420–8.
Fryberger, S. G. & C. J. Shenk 1981. Wind sedimentation tunnel experiments on the origins of aeolian strata. *Sedimentology* **28**, 805–22.
Fryberger, S. G. & C. J. Shenk 1988. Pin stripe lamination: a distinctive feature of modern and ancient eolian sediments. *Sediment. Geol.* **55**, 1–16.
Fryberger, S. G., T. S. Ahlbrandt & S. Andrews 1979. Origin, sedimentary features, and significance of low-angle eolian "sand sheet" deposits, Great Sand Dunes National Monument and vicinity, Colorado. *J. Sediment. Petrol.* **49**, 733–46.
Fryberger, S. G., A. M. Al-Sari & T. J. Clisham 1983. Eolian dune, interdune, sand sheet, and siliciclastic sabkha sediments of an offshore prograding sand sea, Dhahran area, Saudi Arabia. *Bull. Am. Assoc. Petrol. Geol.* **67**, 280–312.
Fryberger, S. G., A. M. Al-Sari, T. J. Clisham, S. A. R. Rizvi & K. G. Al-Hinai 1984. Wind sedimentation in the Jafurah Sand Sea, Saudi Arabia. *Sedimentology* **31**, 413–31.
Fryberger, S. G., L. F. Krystinik & C. J. Schenk 1990. Tidally flooded back barrier dunefield, Guerrero Negro area, Baja California, Mexico. *Sedimentology* **37**, 23–43.
Fryberger, S. G., C. J. Shenk & L. F. Krystinik 1988. Stokes surfaces and the effects of near-surface groundwater table on aeolian deposition. *Sedimentology* **35**, 21–41.
Fujiyama, H. & T. Nagai 1986. Studies of improvement of nutrient and water supply in crop cultivation on sand dune soil. 1. Comparison of irrigation methods. *Soil Sci. Plant Nutr.* **32**, 511–21.
Fujiyama, H. & T. Nagai 1987. Studies of improvement of nutrient and water supply in crop cultivation on sand dune soil. 2. Effect of fertilizer placement and irrigation method of growth and nutrient uptake of tomatoes. *Soil Sci. Plant Nutr.* **33**, 461–70.

Full, W., R. Ehrlich, & J. E. Klovan 1981. Extended Q-Model in objective definition of external members in the analysis of mixtures. *J. Math. Geol.* **13**, 331–44.
Full, W., R. Ehrlich, & S. K. Kennedy 1984. Optimal configuration and information content of sample data generally displayed as histograms or frequency plots. *J. Sediment. Petrol.* **54**, 117–26.

Gad-el-Hak, M., D. Pierce, A. Howard & J. B. Morton 1976. *The interaction of unidirectional winds with an isolated barchan sand dune*. School of Engineering and Applied Science, Univ. Virgina, Report No. UVN 528035/ESS76/102.
Gage, M. O. 1970. *Experimental dunes on the Texas coast*. Fort Belvoir, VA: Coastal Engineering Research Centre, US Army Corps of Engineers, MP 1–70.
Gale, R. W. & D. A. Barr 1977. Vegetation and coastal sand dunes. *Beach Conserv*. 28. Brisbane: Beach Protection Authority of Queensland.
Galloway, J. P., E. A. Koster & T. D. Hamilton 1985. Comments on cemented horizon in subarctic Alaskan sand dunes. *Am. J. Sci.* **285**, 186–90.
Gardner, G. J., A. J. Mortlock, D. M. Price, M. L. Readhead & R. J. Wasson 1987. Thermoluminescence and radiocarbon dating of Australian desert dunes. *Aust. J. Earth Sci.* **34**, 343–57.
Gardner, R. A. M. 1981. Reddening of dune sands — evidence from southeast India. *Earth Surf. Proc.* **6**, 459–68.
Gardner, R. A. M. 1983a. Reddening of tropical coastal dune sands. In *Residual deposits*, R. C. L. Wilson (ed.), 103–15. Oxford: Blackwell.
Gardner, R. A. M. 1983b. Aeolianite. In *Chemical sediments and geomorphology*, A. S. Goudie and K. Pye (eds), 265–300. London: Academic Press.
Gardner, R. A. M. & K. Pye 1981. Nature, origin and palaeoenvironmental significance of red coastal and desert dune sands. *Prog. Phys. Geogr.* **5**, 514–34.
Gares, P. A., K. F. Nordstrom & N. P. Psuty 1979. *Coastal dunes: their function, delineation and management*. Center for Coastal and Environmental Studies, Rutgers University, and New Jersey Department of Environmental Protection, Division of Coastal Resources.
Gares, P. A., K. F. Nordstrom & N. P. Psuty 1980. Delineation and implementation of a dune management district. *Proc. Conf. Coastal Zone 80* 1269–88. Hollywood, FL: ASCE.
Gautier, E. F. 1935. *Sahara, the great desert*. London: Frank Cass.
Gavish, E. & G. M. Friedman 1969. Progressive diagenesis in Quaternary to late Tertiary carbonate sediments: sequence and time scale. *J. Sediment. Petrol.* **39**, 980–1006.
Gaylord, D. R. & P. J. Dawson 1987. Airflow terrain interactions through a mountain gap, with an example of eolian activity beneath an atmospheric hydraulic jump. *Geology*, **15**, 789–92.
Gees, R. A. & A. K. Lyall 1969. Erosional sand columns in dune sands, Cape Sable Island, Nova Scotia, Canada. *Can. J. Earth Sci.* **6**, 344–7.
Gemmel, A. R., P. Greig-Smith & C. H. Gimingham 1953. A note on the behaviour of *Ammophila arenaria* (L.) Link in relation to sand dune formation. *Trans. Bot. Soc. Edinb.* **36**, 132–6.
Gerety, K. M. 1984. *A wind-tunnel study of the saltation of heterogeneous (size, density) sands*. Ph.D. Thesis, Pennsylvania State Univ.
Gerety, K. M. 1985. Problems with determination of u_* from wind velocity profiles measured in experiments with saltation. In *Proceedings of international workshop on the physics of blown sand*, O. E. Barndorff-Nielsen, J. T. Møller, K. R. Rasmussen & B. B. Willetts (eds), 271–300. Dept. Theoretical Statistics, Institute of Mathematics, Univ. Aarhus, Mem. 8.
Gerety, K. M. & R. Slingerland 1983. Nature of the saltating population in wind tunnel experiments with heterogeneous size-density sands. In *Eolian sediments and processes*, M. E. Brookfield & T. S. Ahlbrandt (eds), 115–31. Amsterdam: Elsevier.
Geyh, M. A., G. Roeschmann, T. A. Wiimstra & A. A. Middeldorp 1983. The unreliability of ^{14}C dates obtained from buried sandy podzols. *Radiocarbon* **25**, 409–16.
Gile, L. H. 1966. Coppice dunes and the Rotorua Soil. *Proc. Soil Sci. Soc. Am.* **30**, 657–60.
Gillette, D. A. 1974. On the production of soil wind erosion aerosols having the potential for long range transport. *J. Réch. Atmos.* **8**, 735–44.
Gillette, D. A. 1978. Tests with a portable wind tunnel for determining wind erosion threshold velocities. *Atmos. Environ.* **12**, 2309–13.
Gillette, D. A. 1979. Environmental factors affecting dust emission by wind erosion. In *Saharan dust*, C. Morales (ed.), 71–91. Chichester: Wiley.
Gillette, D. A. 1981. Production of dust that may be carried great distances. In *Desert dust*, T. L. Péwé (ed.), 11–26. Geol. Soc. Am. Spec. Pap. 186.
Gillette, D. A. & P. A. Goodwin 1974. Microscale transport of sand-sized aggregates eroded by wind. *J. Geophys. Res.* **79**, 4080–4.
Gillette, D. A. & T. R. Walker 1977. Characteristics of airborne particles produced by wind erosion of sandy soil, High Plains of West Texas, *Soil Sci.* **123**, 97–110.

Gillette, D. A., I. H. Blifford & D. W. Fryrear 1974. The influence of wind velocity on the size distributions of aerosols generated by the wind erosion of soils. *J. Geophys. Res.* **79**, 4068–75.

Gillette, D. A., J. Adams, L. Endo & D. Smith 1980. Threshold velocities for input of soil particles into the air by desert soils. *J. Geophys. Res.* C **85**, 5621–30.

Gillette, D. A., J. Adams, D. Muhs & R. Kihl 1982. Threshold friction velocities and rupture moduli for crusted desert soil for input of soil particles into the air. *J. Geophys. Res.* **87**, 9003–15.

Glassford, D. K. & L. P. Killigrew 1976. Evidence for Quaternary westward extension of the Australian Desert into southwestern Australia. *Search* **7**, 394–6.

Glennie, K. W. 1970. *Desert sedimentary environments.* Developments in Sedimentology 14. Amsterdam: Elsevier.

Glennie, K. W. 1972. Permian Rotliegendes of Northwest Europe interpreted in light of modern desert sedimentation studies. *Bull. Am. Assoc. Petrol. Geol.* **56**, 1048–71.

Glennie, K. W. 1983a. Lower Permian Rotliegend desert sedimentation in the North Sea area. In *Eolian sediments and processes*, M. E. Brookfield & T. S. Ahlbrandt (eds), 521–41. Amsterdam: Elsevier.

Glennie, K. W. 1983b. Early Permian (Rotliegendes) palaeowinds of the North Sea. *Sediment Geol.* **34**, 245–65.

Glennie, K. W. 1985. Early Permian (Rotliegendes) palaeowinds of the North Sea — Reply. *Sediment Geol.* **45**, 297–313.

Glennie, K. W. 1987. Desert sedimentary environments, present and past – a summary. *Sediment Geol.* **50**, 135–65.

Glennie, K. W. & A. T. Buller 1983. The Permian Wiessliegendes of N. W. Europe: the partial deformation of aeolian dune sands caused by the Zechstein transgression. *Sediment Geol.* **35**, 43–81.

Glennie, K. W. and B. D. Evamy 1968. Dikaka: plants and plant root structures associated with eolian sand. *Palaeogeogr. Palaeoclimatol. Palaeoecol.* **23**, 77–87.

Glennie, K. W., G. C. Mudd & P. J. C. Nagtegaal 1978. Depositional environment and diagenesis of Permian Rotliegendes sandstones in Leman Bank and Sole Pit areas of the U.K. southern North Sea. *J. Geol. Soc. Lond.* **135**, 25–34.

Godfrey, P. J., S. P. Leatherman & P. A. Buckley 1978. Impact of off-road vehicle on coastal ecosystems. *Proc. Symp. on Technical, Environmental, Socioeconomic and Regulatory Aspects of Coastal Zone Planning and Management, San Francisco, California.* 581–600.

Goff, R. C. 1976. Vertical structure of thunderstorm outflows. *Month. Weather Rev.* **104**, 1429–40.

Goggin, D. J., M. A. Chandler, G. A. Kocurek & L. W. Lake 1986. *Patterns of permeability in eolian deposits.* Paper presented at the SPE/DOE Fifth Symposium on Enhanced Oil Recovery, Tulsa, April 20–23, 1986.

Goh, K. M. & B. P. J. Molloy 1979. Contaminants in charcoals used for radiocarbon dating. *N. Z. J. Sci.* **22**, 39–47.

Goldschmidt, M. J. & M. Jacobs 1956. Underground water in the Haifa–Acco sand dunes and its replenishment. *State of Israel Hydrological Service, Jerusalem, Hydrological Paper*, No. 2.

Goldsmith, V. 1973. Internal geometry and origin of vegetated coastal sand dunes. *J. Sediment. Petrol.* **43**, 1128–42.

Goldsmith, V. 1985. Coastal dunes. In *Coastal sedimentary environments*, 2nd edn, R. A. Davis, Jr (ed.), 303–78. New York: Springer.

Goldsmith, V., H. F. Hennigar & A. L. Gutman 1977. The "VAMP" coastal dune classification. In *Coastal processes and resulting forms of sediment accumulation, Currituck Spit, Virginia/North Carolina*, V. Goldsmith (ed.), 26–1 – 26–20. Gloucester Point, Virginia: Virginia Institute of Marine Science, SRAMSOE No. 143.

Goldsmith, V., P. Rosen & Y. Gertner 1988. *Eolian sediments transport on the Israeli coast.* Final Report, US–Israel BSF. Haifa: National Oceanographic Institute.

Goldstein, S. (ed.) 1938. *Modern developments in fluid dynamics*, Vol. II. Oxford: Clarendon Press.

Gooch, W. L. 1947. Present conditions in the pine forests of Les Landes, France. *J. For.* **45**, 263–4.

Good, T. R. & I. D. Bryant 1985. Fluvio-aeolian sedimentation — an example from Banks Island, N.W.T., Canada. *Geogr. Ann.* **67A**, 33–46.

Gooding, J. L. 1982. Petrology of dune sand derived from basalt on the Ka'U Desert, Hawaii. *J. Geol.* **90**, 97–108.

Goossens, D., 1985. The granulometrical characteristics of a slowly-moving dust cloud. *Earth Surf. Proc. Landf.* **10**, 353–62.

Gorycki, M. A. 1973. Sheetflood structure: mechanism of beach cusp formation and related phenomena. *J. Geol.* **81**, 109–17.

Goudie, A. S. 1969. Statistical laws and dune ridges in South Africa. *Geogr. J.* **135**, 404–6.

Goudie, A. S. 1970. Notes on some major dune types in Southern Africa. *S. Afr. Geogr. J.* **52**, 93–101.

Goudie, A. S. 1974. Further experimental investigation of rock weathering by salt and other mechanical processes. *Z. Geomorph. Suppl. Bd.* **21**, 1–12.
Goudie, A. S. 1983a. Dust storms in space and time. *Prog. Phys. Geogr.* **7**, 502–30.
Goudie, A. S. 1983b. The arid earth. In *Megageomorphology*, R. Gardner & H. Scoging (eds), 152–71. Oxford: Oxford Univ. Press.
Goudie, A. S. 1985. Salt weathering. *School Geogr., Oxford Univ., Res. Pap.* No. 33.
Goudie, A. S. & C. H. B. Sperling 1977. Long distance transport of foraminiferal tests by wind in the Thar Desert, northwestern India. *J. Sediment. Petrol.* **47**, 630–3.
Goudie, A. S. & D. S. G. Thomas 1986. Lunette dunes in southern Africa. *J. Arid Environ.* **10**, 1–12.
Goudie, A. S. & A. Watson 1981. The shape of desert sand dune grains *J. Arid Environ.* **4**, 185–90.
Goudie, A. S., R. U. Cooke & I. Evans 1970. Experimental investigation of rock weathering by salts. *Area* **4**, 42–8.
Goudie, A. S., B. Allchin & K. T. M. Hegde 1973. The former extensions of the Great Indian Sand Desert. *J. Geogr.* **139**, 243–57.
Goudie, A. S., R. U. Cooke & J. C. Doornkamp 1979. The formation of silt from quartz dune sand by salt weathering processes in deserts. *J. Arid Environ.* **2**, 105–12.
Goudie, A. S., A. Warren, D. K. C. Jones & R. U. Cooke 1987. The character and possible origins of the aeolian sediments of the Wahiba Sand Sea. *Geogr. J.* **153**, 231–56.
Graaf, J. van de 1977. Dune erosion during a storm surge. *Coastal Eng.* **1**, 99–134.
Graaf, J. van de 1986. Probabilistic design of dunes; an example from the Netherlands. *Coastal Eng.* **9**, 479–500.
Grass, A. J. 1971. Structural features of turbulent flow over smooth and rough boundaries. *J. Fluid Mech.* **50**, 233–55.
Graton, L. C. & H. J. Fraser 1935. Systematic packing of spheres with particular relation to porosity and permeability. *J. Geol.* **43**, 785–909.
Greeley, R. 1986. Aeolian landforms: laboratory simulations and field studies. In *Aeolian geomorphology*, W. G. Nickling (ed.), 195–211. Boston: Allen and Unwin.
Greeley, R. & J. D. Iversen 1985. *Wind as a geological process.* Cambridge: Cambridge Univ. Press.
Greeley, R. & J. D. Iversen 1987. Measurements of wind friction speeds over lava surfaces and assessment of sediment transport. *Geophys. Res. Lett.* **14**, 925–8.
Greeley, R. & R. Leach 1978. A preliminary assessment of the effects of electrostatics on aeolian processes. *Reports Planetary Geology Program 1977–78*, 236–7. NASA TM 79729.
Greeley, R. & A. R. Peterfreund 1981. Aeolian "megaripples": examples from Mono Craters, California and Northern Iceland. *Geol. Soc. Am. Nat. Conf. Abstr.* **13**, 463.
Greeley, R., J. D. Iversen, J. B. Pollack, N. Udovich & B. White 1974a. Wind tunnel studies of Martian aeolian processes. *Proc. R. Soc. London, Ser. A* **341**, 331–60.
Greeley, R., J. D. Iversen, J. B. Pollack, N. Udovich & B. White 1974b. Wind tunnel simulations of light and dark streaks on Mars. *Science* **183**, 847–9.
Greeley, R., J. R. Marshall & R. N. Leach 1984. Microdunes and other aeolian bedforms on Venus: wind tunnel simulation. *Icarus* **60**, 152–60.
Greeley, R., B. R. White, J. B. Pollack, J. D. Iversen & R. N. Leach 1981. Dust storms on Mars: consideration and simulations. In *Desert dust: origin, characteristics and effect on man*, T. L. Péwé (ed.), 101–21. Spec. Pap. Geol. Soc. Am., 186.
Green, H. L. & W. R. Lane 1964. *Particulate clouds: dusts, smokes and mists*, 2nd edn. London: Spon.
Griffiths, J. F. & K. H. Soliman 1972. The northern desert (Sahara). In *Climates of Africa*, J. F. Griffiths (ed.), 75–131. World survey of climatology, 10. Amsterdam: Elsevier.
Grolier, M. J., G. E. Ericksen, J. F. McCauley & E. C. Morris 1974. The desert landforms of Peru; a preliminary photographic atlas. *Interagency Report US Geol. Surv.: Astrogeology* 57.
Gross, M. C. 1964. Variations in the $^{18}O/^{16}O$ and $^{13}C/^{12}C$ ratios of diagenetically altered limestones in the Bermuda Islands. *J. Geol.* **72**, 170–94.
Grove, A. T. 1958. The ancient erg of Hausaland, and similar formations on the south side of the Sahara. *J. Geogr.* **124**, 528–33.
Grove, A. T. 1969. Landforms and climatic change in the Kalahari and Ngamiland. *Geogr. J.* **135**, 191–212.
Grove, A. T. & A. Warren 1968. Quaternary landforms and climate on the south side of the Sahara. *J. Geogr.* **134**, 194–208.
Guilcher, A. & F. Joly 1954. Récherches sur la morphologie de la Côte Atlantique du maroc. *Trav. Inst. Sci. Chérif., Sér. Géol. Géogr. Phys.* 2.
Gupta, J. P. 1979. Some observations on the periodic variations of moisture in stabilized and unstabilized sand dunes in the Indian desert. *J. Hydrol.* **41**, 153–6.

Haas, J. A. & J. A. Steers 1964. An aid to stabilization of sand dunes: experiments at Scolt Head Island. *Geogr. J.* **130**, 265–7.

Haberle, R. M. 1986. The climate of Mars. *Sci. Am.* **254** (5), 54–62.

Hack, J. T. 1941. Dunes of the western Navajo country. *Geogr. Rev.* **31**, 240–63.

Haff, P. K. & D. E. Presti 1984. *Barchan dunes of the Salton Sea region, California*. California Inst. Tech. Brown Bag Preprint Series in Basic and Applied Science, BB-16.

Hagedorn, H., K. Giessner, O. Weise, A. Busche & G. Grunet 1977. *Dune stabilization*. Eschborn: German Agency for Technical Cooperation.

Hamilton, P. A. & J. W. Archbold 1945. Meteorology of Nigeria and adjacent territory. *Q. J. R. Meteorol. Soc.* **71**, 231–65.

Hand, B. M. 1967. Differentiation of beach and dune sands using settling velocities of light and heavy minerals. *J. Sediment. Petrol.* **37**, 514–20.

Hanna, S. R. 1969. The formation of longitudinal sand dunes by large helical eddies in the atmosphere. *J. Appl. Meteorol.* **8**, 874–83.

Hardisty, J. & K. J. S. Whitehouse 1988. Evidence for a new sand transport process from experiments on Saharan dunes. *Nature* **332**, 532–4.

Harlé, E. 1914. La fixation des dunes de Gasgogne. *Bull. Sect. Geogr.* **29**, 181–224.

Hartline, B. K. 1980. Coastal upwelling: Physical factors feed fish. *Science* **208**, 38–40.

Hartmann, D. 1988. The goodness-of-fit to ideal Gauss and Rosin distributions: a new grain size parameter — discussion. *J. Sediment. Petrol.* **58**, 913–17.

Hartmann, D. & C. Christiansen 1988. Settling velocity distributions and sorting processes on a longitudinal dune: a case study. *Earth Surf. Proc. Landf.* **13**, 649–56.

Hastenrath, S. L. 1967. The barchans of the Arequipa region, southern Peru *Z. Geomorph.* **11**, 300–31.

Hastenrath, S. L. 1978. Mapping and surveying dune shape and multiannual displacement. In *Exploring the world's driest climate*, H. H. Lettau & K. Lettau (eds), 74–88. Madison: Univ. Wisconsin, IES Report 101.

Hastenrath, S. L. 1987. The barchan dunes of southern Peru revisited. *Z. Geomorph.* **31**, 167–78.

Hastings, J. D. 1971. Sand streets. *Meteorol. Mag.* **100**, 155–9.

Havholm, K. & G. Kocurek 1988. A preliminary study of the dynamics of a modern draa, Algodones, southeastern California, U.S.A. *Sedimentology* **35**, 649–69.

Haynes, C. V. 1982. Great Sand Sea and Selima Sand Sheet, eastern Sahara: geochronology of desertification. *Science* **217**, 629–33.

Haynes, C. V., Jr 1989. Bagnold's barchan: a 57 year record of dune movement in the eastern Sahara and implications for dune origin and paleoclimate since Neolithic times. *Quat. Res.* **32**, 153–67.

Hedges, R. M. 1981. Radiocarbon dating with an accelerator: review and preview. *Archaeometry* **23**, 3–18.

Hefley, H. M. & R. Sidwell 1945. Geological and ecological observations of some High Plains dunes. *Am. J. Sci.* **243**, 361–76.

Heine, K. 1982. The main stages of the Quaternary evolution of the Kalahari region, southern Africa. *Palaeoecol. Afr.* **15**, 53–76.

Hendry, D. A. 1987. Silica and calcium carbonate replacement of plant roots in tropical dune sands, S. E. India. In *Desert sediments ancient and modern*, L. E. Frostick & I. Reid (eds), 309–19. Oxford: Blackwell.

Hennessy, J. T., R. P. Gibbens, J. M. Tromble & M. Cardenas 1985. Mesquite (*Prosopis glandulosa* Torr.) dunes and interdunes in southern New Mexico: a study of soil properties and soil water relations. *J. Arid Environ.* **9**, 27–38.

Hesp, H. 1979. Sand trapping ability of culms of marram grass (*Ammophila arenaria*). *J. Soil Conserv. Serv. N.S.W.*. **35**, 156–60.

Hesp, P. A. 1981. The formation of shadow dunes. *J. Sediment. Petrol.* **51**, 101–12.

Hesp, P. A. 1983. Morphodynamics of incipient foredunes in New South Wales, Australia. In *Eolian sediments and processes*, M. E. Brookfield & T. S. Ahlbrandt (eds), 325–40. Amsterdam: Elsevier.

Hesp, P. A. 1988. Morphology, dynamics and internal stratification of some established foredunes in southeast Australia. *Sediment Geol.* **55**, 17–41.

Hesp, P. A. & S. G. Fryberger (eds). 1988. *Special Issue on Eolian Sediments Sediment. Geol.*, 55.

Hesp, P. A., R. Hyde, V. Hesp & Q. Zhengyu 1989. Longitudinal dunes can move sideways. *Earth Surf. Proc. Landf.* **14**, 447–51.

Hewett, D. G. 1989. Dunes and dune management on the Atlantic coast of Europe. In *Perspectives in coastal dune management*, F. van der Meulen, P. D. Jungerius & J. Visser (eds), 69–80. The Hague: SPB Academic Publishing.

Hidore, J. J. & Y. Albokhair 1982. Sand Encroachment in Al-Hasa Oasis. *Geogr. Rev.* **72**, 350–6.
Higgins, G. M., S. Baig & R. Brinkman 1974. The sands of Thal: wind regimes and sand ridge formations *Z. Geomorph. N.F.* **18**, 272–90.
Hills, E. S. 1940. The lunette; a new landform of aeolian origin. *Aust. Geogr.* **3**, 15–21.
Hjulstrom, F. 1935. Studies of the morphological activity of rivers as illustrated by the River Fyris. *Bull. Geol. Inst. Univ. Uppsala* **25**, 221–527.
Hjulstrom, F. 1939. Transportation of detritus by moving water. In *Recent marine sediments*, P. D. Trask (ed.), 5–31. Tulsa: Am. Assoc. Petrol. Geol.
Hobbs, W. H. 1943. The glacial anticyclones and the European continental glacier. *Am. J. Sci.* **241**, 333–6.
Hobday, D. K. 1977. Late Quaternary history of Inhaca Island, Mozambique. *Trans. Geol. Soc. S. Afr.* **80**, 189–91.
Hobday, D. K. & G. R. Orme 1975. The Port Durnford Formation: a major Pleistocene barrier–lagoon complex along the Zululand coast. *Trans. Geol. Soc. S. Afr.* **77**, 171–249.
Hogbom, I. 1923. Ancient inland dunes of northern and middle Europe. *Geogr. Ann.* **5**, 113–243.
Holm, D. A. 1953. Dome-shaped dunes of central Nejd, Saudi Arabia. *C. R. 19th Int. Geol. Cong. Algiers, 1952*, 107–12.
Holm, D. A. 1960. Desert geomorphology in the Arabian peninsula. *Science* **132**, 1369–79.
Holm, D. A. 1968. Sand dunes. In *The encyclopedia of geomorphology*, R. W. Fairbridge (ed.), 973–9. New York: Reinhold.
Holton, J. R. 1979. *An introduction to dynamic meteorology* New York: Academic Press.
Horikawa, K. & H. W. Shen 1960. *Sand movement by wind action.* US Army, Corps of Engineers, Beach Erosion Board, Tech. Memo. 119.
Horikawa, K., S. Hotta & S. Kubota 1982. Experimental study of blown sand on a wetted sand surface. *Coastal Eng. Jpn.* **25**, 177–95.
Horikawa, K., S. Hotta, S. Kubota & S. Katori 1984. Field measurement of blown sand transport rate by trench trap. *Coastal Eng. Jpn.* **27**, 214–32.
Horikawa, K., S. Hotta & N. C. Kraus 1986. Literature review of sand transport by wind on a dry sand surface. *Coastal Eng.* **9**, 503–26.
Horowitz, D. H. 1982. Geometry and origin of large-scale deformation structures in some ancient windblown sand deposits. *Sedimentology* **29**, 155–80.
Horton, R. E. 1945. Erosional development of streams and their drainage basins; hydrophysical approach to quantitative morphology. *Bull. Geol. Soc. Am.* **56**, 275–370.
Hotta, S., S. Kubota, S. Katori & K. Horikawa 1985. Sand transport by wind on a wet sand surface. *Proc. 19th Coastal Eng. Conf. Houston*, 1265–81.
Houghton, J. T. 1986. *The physics of atmospheres*, 2nd edn. Cambridge: Cambridge Univ. Press.
Howard, A. D. 1977. Effect of slope on the threshold of motion and its application to orientation of wind ripples. *Geol. Soc. Am. Bull.* **88**, 853–6.
Howard, A. D. 1985. Interaction of sand transport with topography and local winds in the northern Peruvian coastal desert. In *Proceedings of the international workshop on the physics of blown sand*, O. E. Barndorff-Nielsen, J. T. Møller, K. R. Rasmussen & B. B. Willetts (eds), 511–43. Dept. Theoretical Statistics, Institute of Mathematics, Univ. Aarhus, Mem. 8.
Howard, A. D. & J. L. Walmsley 1985. Simulation model of isolated sand sculpture by wind. In *Proceedings of the international workshop on the physics of blown sand*, O. E. Barndorff-Nielsen, J. T. Møller, K. R. Rasmussen & B. B. Willetts (eds), 377–9. Dept. Theoretical Statistics, Institute of Mathematics, Univ. Aarhus, Mem. 8.
Howard, A. D., J. B. Morton, M. Gad-el-Hak & D. B. Pierce 1978. Sand transport model of barchan dune equilibrium. *Sedimentology* **25**, 307–38.
Hoyt, J. H. 1966. Air and sand movements to the lee of dunes. *Sedimentology* **7**, 137–43.
Hsu, S. A. 1971a. Measurement of shear stress and roughness length on a beach *J. Geophys. Res.* **76**, 2880–5.
Hsu, S. A. 1971b. Wind stress criteria in aeolian sand transport. *J. Geophys. Res.* **76**, 8684–6.
Hsu, S. A. 1973. Computing eolian sand transport from shear velocity measurements. *J. Geol.* **81**, 739–43.
Hsu, S. A. 1974. Computing eolian sand transport from routine weather data. *Proc. 14th Coastal Eng. Conf.*, Copenhagen Vol. II, 1619–26.
Huffman, G. G. & W. A. Price 1949. Clay dune formation near Corpus Christi, Texas. *J. Sediment. Petrol.* **19**, 118–27.
Hummel, G. & G. Kocurek 1984. Interdune areas of the Back-Island dune field, North Padre Island, Texas. *Sediment. Geol.* **39**, 1–26.
Hung, C. S. & H. W. Shen 1979. Statistical analysis of sediment motions on dunes. *J. Hyd. Div. Am. Soc. Civ. Eng.* **105**, 213–27.

Hunt, J. C. R. 1980. Wind over hills. In *Workshop on planetary boundary layer*, J. C. Wyngaard (ed.), 107–44. Boston: American Meteorological Society.

Hunt, J. C. R. & P. Nalpanis 1985. Saltating and suspended particles over flat and sloping surfaces I. Modelling concepts. In *Proceedings of the international workshop on the physics of blown sand*, O. E. Barndorff-Nielsen, J. J. Møller, K. R. Rassmussen & B. B. Willetts (eds), 9–36. Dept. Theoretical Statistics, Institute of Mathematics, Univ. Aarhus, Mem. 8.

Hunt, J. C. R., W. H. Snyder & R. Lawson 1978. *Flow structure and turbulent diffusion around a three-dimensional hill. Part 1. Flow Structure*. EPA Report 60/4–78041.

Hunt, J. C. R., S. Leibovich & K. J. Richards 1988a. Turbulent shear flow over low hills. *Q. J. R. Meteorol. Soc.* **114**, 1435–70.

Hunt, J. C. R., K. J. Richards & P. W. M. Brighton 1988b. Stratified shear flow over low hills. *Q. J. R. Meteorol. Soc.* **114**, 859–86.

Hunter, R. E. 1969. Eolian microridges on modern beaches and a possible ancient example. *J. Sediment. Petrol.* **39**, 1573–78.

Hunter, R. E. 1973. Pseudo-cross lamination formed by climbing adhesion ripples. *J. Sediment. Petrol.* **43**, 1125–7.

Hunter, R. E. 1977a. Basic types of stratification in small eolian dunes. *Sedimentology* **24**, 361–87.

Hunter, R. E. 1977b. Terminology of cross-stratified sedimentary layers and climbing ripple structures. *J. Sediment. Petrol.* **47**, 697–706.

Hunter, R. E. 1980. Quasi-planar adhesion stratification — an eolian structure formed in wet sand. *J. Sediment. Petrol.* **50**, 263–6.

Hunter, R. E. & B. M. Richmond 1988. Daily cycles in coastal dunes. *Sediment. Geol.*. **55**, 43–67.

Hunter, R. E., B. M. Richmond & T. R. Alpha 1983. Storm-controlled oblique dunes of the Oregon coast. *Bull. Geol. Soc. Am.* **94**, 1450–65.

Hunter, R. E., B. M. Richmond & T. R. Alpha 1985. Storm-controlled oblique dunes of the Oregon coast: reply. *Bull. Geol. Soc. Am.* **96**, 410.

Huntley, D. J., D. I. Godfrey-Smith & M. L. W. Thewalt 1985. Optical dating of sediments. *Nature* **313**, 105–7.

Hutton, J. T., J. R. Prescott & C. R. Twidale 1984. Thermoluminescence dating of coastal dune sand related to a higher stand of Lake Woods, Northern Territory. *Aust. J. Soil Res.* **22**, 15–21.

Hyde, R. & R. J. Wasson 1983. Radiative and meteorological control on the movement of sand at Lake Mungo. In *Eolian sediments and processes*, M. E. Brookfield & T. S. Ahlbrandt (eds), 311–23. Amsterdam: Elsevier.

Hylgaard, T. & M. J. Liddle 1981. The effect of human trampling on a sand dune ecosystem dominated by *Empetrum nigrum*. *J. Appl. Ecol.* **18**, 559–69.

Idso, S. B. 1973. Haboobs in Arizona. *Weather* **28**, 154–5.

Idso, S. B. 1974. Thunderstorm outflows: different perspectives over arid and mesic terrain. *Month. Weather Rev.* **102**, 603–4.

Idso, S. B. 1976. Dust storms. *Sci. Am.* **235**, 108–14.

Idso, S. B., R. S. Ingram & J. M. Pritchard 1972. An American haboob. *Bull. Am. Meteorol. Soc.* **53**, 930–5.

Iler, R. K. 1979. *The geochemistry of silica*. New York: Wiley.

Illenberger, W. K. 1988. The dunes of the Alexandria coastal dunefield, Algoa Bay, South Africa, *S. Afr. J. Geol.* **91**, 381–90.

Illenberger, W. K. & I. C. Rust 1986. Venturi-compensated eolian sand trap for field use. *J. Sediment. Petrol.* **56**, 541–3.

Illenberger, W. K. & I. C. Rust 1988. A sand budget for the Alexandria coastal dunefield, South Africa. *Sedimentology* **35**, 513–21.

Ingle, J. C. 1966. *The movement of beach sand.* Developments in Sedimentology 5. Amsterdam: Elsevier.

Ingram, R. L. 1971. Sieve analysis. In *Procedures in sedimentary petrology*, R. E. Carver (ed.) 49–67. New York: Wiley-Interscience.

Inman, D. L., G. C. Ewing & J. B. Corliss 1966. Coastal sand dunes of Guerro Negro, Baja California, Mexico. *Bull. Geol. Soc. Am.* **77**, 787–802.

Isyumov, N. & H. Tanaka 1980. Wind tunnel modelling of stack gas dispersion — difficulties and approximations. In *Wind engineering*, J. E. Cermak (ed.), 987–1001. Oxford: Pergamon Press.

Iversen, J. D. 1985. Aeolian threshold: effect of density ratio. In *Proceedings of the international workshop on the physics of blown sand*, O. E. Barndorff-Nielsen, J. J. Møller, K. R. Rasmussen & B. B. Willetts (eds), 67–82. Dept. Theoretical Statistics, Institute of Mathematics, Univ. Aarhus, Mem. 8.

Iversen, J. D. & B. R. White 1982. Saltation threshold on Earth, Mars and Venus. *Sedimentology* **29**, 111–19.
Iversen, J. D., R. Greeley & J. B. Pollack 1976a. Windblown dust on Earth, Mars and Venus. *J. Atmos. Sci.* **33**, 2425–9.
Iversen, J. D., J. B. Pollack, R. Greeley & B. R. White 1976b. Saltation threshold on Mars: the effect of interparticle force surface roughness and low atmospheric density. *Icarus* **29**, 381–93.
Iversen, J. D., R. Greeley, J. R. Marshall & J. B. Pollack 1987. Aeolian saltation threshold: the effect of density ratio. *Sedimentology* **34**, 699–706.
Ives, R. L. 1947. Behavior of dust devils. *Bull. Am. Meteorol. Soc.* **28**, 168–74.

Jackson, J. E. 1972. *Plane and geodetic surveying by the Late David Clark*. Volume 1. *Plane survey*. London: Constable.
Jackson, M. L., D. A. Gillette, E. F. Danielsen, J. H. Blifford, R. A. Bryson & S. K. Syers 1973. Global dustfall during the Quaternary as related to environments. *Soil Sci.* **116**, 135–45.
Jackson, P. S. 1976. A theory for flow over escarpments. In *Wind effects on buildings and structures*, K. J. Eaton (ed.), 33–40. Cambridge: Cambridge Univ. Press.
Jackson, P. S. 1977. Aspects of surface wind behaviour. *Wind Eng.* **1**, 1–14.
Jackson, P. S. & J. C. R. Hunt 1975. Wind flow over a low hill. *Q. J. R. Meteorol. Soc.* **101**, 929–55.
Jagschitz, J. A. & R. S. Bell 1966. Restoration and retention of coastal dunes with fences and vegetation. *Bull. Rhode Island Agric. Exp. Station* No. 404.
Jagschitz, J. A. & R. C. Wakefield 1971. How to build and save beaches and dunes. *Bull. Rhode Island Agric. Exp. Station* No. 382.
James, N. P. & P. W. Choquette 1984. Diagenesis 9 — Limestones — the meteoric diagenetic environment. *Geosci. Can.* **11**, 161–93.
James, W. C., G. H. Mack & L. J. Suttner 1981. Relative alteration of microcline and sodic plagioclase in semi-arid and humid climates. *J. Sediment. Petrol.* **51**, 151–64.
Janza, F. J. 1975. Interaction mechanisms. In *Manual of remote sensing*, Vol. I, R. G. Reeves, A. Anson & D. Landen (eds), 75–179. Falls Church: American Society of Photogrammetry.
Jeffreys, H. 1929. On the transport of sediment by streams. *Proc. Camb. Philos. Soc.* **25**, 272–6.
Jeffreys, H. 1934. Additional notes. In *Ocean waves and kindred geophysical phenomena*, V. Cornish (ed.), 121–59. Cambridge: Cambridge Univ. Press.
Jenkin, C. F. 1931. The pressure exerted by granular material: an application of the principles of dilatancy. *Proc. R. Soc. London* **131**, 53–89.
Jenkin, C. F. 1933. The pressure on retaining walls. *Minutes Proc. Inst. Civ. Eng.* **234**, 103–54.
Jennings, J. N. 1957. On the orientation of parabolic or U-dunes. *Geogr. J.* **123**, 474–80.
Jennings, J. N. 1964. The question of coastal dunes in tropical humid climates. *Z. Geomorph.* **8**, 150–4.
Jennings, J. N. 1965. Further discussion of factors affecting coastal dune formation in the tropics. *Aust. J. Sci.* **28**, 166–7.
Jennings, J. N. 1967. Cliff-top dunes. *Aust. Geogr. Stud.* **5**, 40–9.
Jennings, J. N. 1968. A revised map of the desert dunes of Australia. *Aust. Geogr.* **10**, 408–9.
Jensen, J. L. 1988. Maximum likelihood estimation of the log-hyperbolic parameters from grouped observations. *Comput. Geosci.* **14**, 389–408.
Jensen, J. L. & M. Sørensen 1983. On the mathematical modeling of aeolian saltation. In *Mechanics of sediment transport*, B. M. Sumer & A. Muller (eds), 65–72. Rotterdam: Balkema.
Jensen, J. L. & M. Sørensen 1986. Estimation of some eolian saltation transport parameters: a re-analysis of Williams' data. *Sedimentology* **33**, 547–55.
Jensen, J. L., K. R. Rasmussen, M. Sørensen & B. B. Willetts 1984. *The Hanstholm experiment 1982. Sand grain saltation on a beach.* Dept. Theoretical Statistics, Institute of Mathematics, Univ. Aarhus, Res. Rep. 125.
Jensen, M. 1958. The model-law phenomena in natural wind. *Ingenioren (Int. Ed.)* **2**, 121–8.
Jensen, N. O. 1983. Escarpment induced flow perturbations, a comparison of measurements and theory. *J. Wind Eng. Ind. Aerodyn.* **15**, 243–51.
Jensen, N. O. & O. Zeman 1985. Perturbation in mean wind and turbulence in flow over topographic forms. In *Proceedings of the international workshop of the physics of blown sand*, O. E. Barndorff-Nielsen, J. T. Møller, K. R. Rasmussen & B. B. Willetts (eds), Vol. 2, 351–68. Dept. Theoretical Statistics, Institute of Mathematics, Univ. Aarhus, Mem. 8.
Johnson, J. W. 1965. Sand movement on coastal dunes. *Proc. Federal Inter-Agency Sedimentation Conference*, 747–55. US Dept. Agric. Misc. Publ. No. 970.
Johnson, R. B. 1967. The Great Sand Dunes of southern Colorado. *US Geol. Surv. Prof. Pap.* 575C, C177–C183.
Joly, J. 1904. Formation of sand ripples. *Sci. Proc. R. Dublin Soc.* N.S. **10**, 328–30.

Jolliffe, I. P. 1963. A study of sand movements on the Lowestoft sandbank using fluorescent tracers. *Geogr. J.* **129**, 480–93.

Jones, D. J. 1938. Gypsum–Oolite dunes, Great Salt Lake Desert, Utah. *Bull. Am. Assoc. Petrol. Geol.* **37**, 2530–8.

Jones, D. K. C., R. U. Cooke & A. Warren 1986. Geomorphological investigation for engineering purposes, of blowing sand and dust hazard. *Q. J. Eng. Geol.* **19**, 251–70.

Jones, J. R. & B. B. Willetts 1979. Errors in measuring uniform aeolian sand flow by means of an adjustable trap. *Sedimentology* **26**, 463–8.

Jones, M. P. 1977. Automatic image analysis. In *Physical methods in determinative mineralogy*, J. Zussman (ed.), 167–200. London: Academic Press.

Jongerius, A. 1974. Recent developments in soil micromorphology. In *Soil microscopy*, G. K. Rutherford (ed.), 67–83. Kingston, Ontario: Limestone Press.

Jordan, W. M. 1965. Prevalence of sand dune types in the Sahara desert. *Spec. Pap. Geol. Soc. Am.* **82**, 104–5.

Joyce-Loebl 1985. *Image analysis — principles and practice*. Gateshead: Joyce-Loebl.

Junge, C. 1979. The importance of mineral dust as an atmospheric constituent. In *Saharan dust*, C. Morales (ed.), 49–60. Chichester: Wiley.

Jungerius, P. D. & F. van der Meulen 1988. Erosion processes in a dune landscape along the Dutch coast. *Catena* **15**, 217–28.

Jungerius, P. D. & F. van der Meulen 1989. The development of dune blowouts, as measured with erosion pins and sequential air photos. *Catena* **16**, 369–76.

Jungerius, P. D. & J. H. de Jong 1989. Variability in water repellence in the dunes along the Dutch coast. *Catena* **16**, 491–7.

Jutson, J. T. 1920. Dust whirls in Western Australia. *Proc. R. Soc. Vict.* **32**, 314–22.

Kadar, L. 1934. A study of the sand sea in the Libyan desert. *Geogr. J.* **83**, 470–8.

Kadib, A. A. 1965. *A function of sand movement by wind.* Hydraulic Eng. Lab. Tech. Rep. HEL–2–12. Berkeley: Univ. California.

Kaganov, E. I. & A. M. Yaglom 1976. Errors in wind speed measurements by rotation anemometers. *Boundary Layer Meteorol.* **10**, 229–44.

Kahle, C. F. 1974. Ooids from Great Salt Lake, Utah, as an anologue for the genesis and diagenesis of ooids in marine limestones. *J. Sediment. Petrol.* **44**, 30–9.

Kaldi, J., D. G. Krinsley & D. Lawson 1978. Experimentally produced eolian surface textures on quartz sand grains from various environments. In *Scanning electron microscopy in the study of sediments*, W. B. Whalley (ed.), 261–74. Norwich: Geo Abstracts.

Kalinske, A. A. 1947. Movement of sediment as bedload in rivers. *Trans. Am. Geophys. Union* **28**, 615–20.

Kalu, A. E. 1979. The African dust plume: its characteristics and propagation across West Africa in winter. In *Saharan dust*, C. Morales (ed.), 95–118. Chichester: Wiley.

Kar, A. 1987. Origin and transformation of longitudinal sand dunes in the Indian Desert. *Z. Geomorph.* **33**, 311–37.

Kar, A. 1990. Megabarchanoids of the Thar; their environment, morphology and relationship with longitudinal dunes. *Geogr. J.* **156**, 51–61.

Kármán, T. von, 1934. Turbulence and skin friction. *J. Aeron. Sci.* **1**, 1–20.

Kármán, T. von, 1935. Some aspects of the turbulence problem. *Proc. 4th Int. Congr. Appl. Mech. Cambridge*, 54–91.

Kármán, T. von, 1937. Turbulence *J. R. Aeron. Soc.* **41**, 1109–43.

Kármán, T. von, 1947. Sand ripples in the desert. *Technion Yearbook*, 52–4.

Kasting, J. F., O. B. Toon & J. B. Pollack 1988. How climate evolved on the terrestrial planets. *Sci. Am.* **256** (2), 90–7.

Kawamura, R. 1964. *Study of sand movement by wind.* Hydraulic Eng. Lab. Tech. Rep. HEL–2–8, 99–108. Berkeley: Univ. California.

Kebin, Z. & Z. Kaiguo 1989. Afforestation for sand fixation in China. *J. Arid Environ.* **16**, 3–10.

Kelly, R. D. 1984. Horizontal roll and boundary-layer interrelationships observed over Lake Michigan. *J. Atmos. Sci.* **41**, 1816–26.

Kennedy, S. K., T. P. Mèloy & T. E. Gurney 1985. Sieve data — size and shape information. *J. Sediment. Petrol.* **55**, 356–60.

Kennedy, V. C. & D. L. Kouba 1970. Fluorescent sand as a tracer of fluvial sediments. *US Geol. Surv. Prof. Pap.* 562E, El–E13.

Kerr, R. C. & J. O. Nigra 1951. *Analysis of eolian sand control.* New York: Arabian American Oil.

Kerr, R. C. & J. O. Nigra 1952. Eolian sand control. *Bull. Am. Assoc. Petrol. Geol.* **36**, 1541–73.

Khalaf, F. I. 1989a. Textural characteristics and genesis of the aeolian sediments in the Kuwaiti desert. *Sedimentology* **36**, 253–71.
Khalaf, F. I. 1989b. Desertification and aeolian processes in the Kuwait desert. *J. Arid Environ.* **16**, 125–45.
Khalaf, F. I. & I. M. Gharib 1985. Roundness parameters of quartz sand grains of recent aeolian sand deposits in Kuwait. *Sediment. Geol.* **45**, 147–58.
Kidson, C. & A. P. Carr 1960. Dune reclamation at Braunton Burrows, Devon. *Chart. Surv.* December, 3–8.
Kind, R. J. 1976. A critical examination of the requirements for model simulation of wind-induced erosion/deposition phenomena such as snow drifting. *Atmos. Environ.* **10**, 219–27.
Kind, R. J. & S. B. Murray 1982. Saltation flow measurements relating to modeling of snowdrifting. *J. Wind Eng. Ind. Aerodyn.* **10**, 89–102.
Kindle, E. M. 1923. A note on rhizoconcretions. *J. Geol.* **33**, 744–6.
King, D. 1960. The sand ridge deserts of South Australia and related aeolian landforms of the Quaternary arid cycles. *Trans. R. Soc. S. Austr.* **83**, 99–109.
King, L. C. 1963. *South African scenery*, 3rd edn. New York: Hafner.
King. R. C. & K. K. Kindel 1969. One-step control of coastal sand dunes in Chile. *J. For.* **67**, 810–12.
King, W. H. J. 1916. The nature and formation of sand ripples and dunes. *Geogr. J.* **47**, 189–209.
King, W. H. J. 1918. Study of a dune belt. *Geogr. J.* **51**, 16–33.
Klappa, C. F. 1978. Biolithogenesis of *Microdium*: elucidation. *Sedimentology* **25**, 489–522.
Klappa, C. F. 1979. Calcified filaments in Quaternary calcretes: organo–mineral interactions in the subaerial vadose environment. *J. Sediment. Petrol.* **49**, 955–68.
Klappa, C. F. 1980. Rhizoliths in terrestrial carbonates: classification, recognition, genesis and significance. *Sedimentology* **27**, 613–29.
Kline, S. J. 1965. *Similitude and approximation theory*. New York: McGraw-Hill.
Klovan, J. E. 1966. The use of factor analysis in determining depositional environments from grain size distributions. *J. Sediment. Petrol.* **36**, 115–25.
Knott, P. 1979. *The structure and pattern of dune-forming winds*. Unpub. PhD Thesis, Univ. London.
Knott, P. & A. Warren 1981. Aeolian processes. In *Techniques in geomorphology*, A. S. Goudie (ed.), 226–46. London: Allen and Unwin.
Knotternus, D. F. C. 1980. Relative humidity of the air and critical wind velocity in relation to erosion. In *Assessment of erosion*, M. de Boodt & D. Gabriels (eds), 531–40. Chichester: Wiley.
Knox, A. J. 1974. Agricultural use of machair. In *Sand dune machair*, D. S. Ranwell (ed.), 19. Monks Wood Experimental Station: Institute of Terrestrial Ecology.
Knutson, P. L. 1977. Planting guidelines for dune creation and stabilization. *US Army Corps Eng., Coastal Eng. Res. Center, CETA 77–4*, Fort Belvoir, Virginia.
Knutson, P. L. 1980. Experimental dune restoration and stabilization, Nauset beach, Cape Cod, Massachusetts. *US Army Corps Eng., Coastal Eng. Res. Center, TP 80–5*, Fort Belvoir, Virginia.
Kocurek, G. 1981a. Significance of interdune deposits and bounding surfaces in aeolian dune sands. *Sedimentology* **28**, 753–80.
Kocurek, G. 1981b. Erg reconstruction: the Entrada Sandstone (Jurassic) of northern Utah and Colorado. *Palaeogeogr., Palaeoclimatol., Palaeoecol.* **36**, 125–53.
Kocurek, G. 1984. Origin of first order bounding surfaces in eolian sandstones — reply. *Sedimentology* **31**, 125–7.
Kocurek, G. 1986. Origins of low-angle stratification in aeolian deposits. In *Aeolian geomorphology*, W. G. Nickling (ed.), 177–94. Boston: Allen and Unwin.
Kocurek, G. (ed.) 1988a. Special Issue: Late Paleozoic and Mesozoic eolian deposits of the western interior of the United States. *Sediment. Geol.* 56.
Kocurek, G. 1988b. First order and super bounding surfaces in eolian sequences — bounding surfaces revisited. *Sediment. Geol.* **56**, 193–206.
Kocurek, G. & R. H. Dott, Jr 1981. Distinctions and uses of stratification types in the interpretation of eolian sand. *J. Sediment. Petrol.* **51**, 579–95.
Kocurek, G. & G. Fielder 1982. Adhesion structures. *J. Sediment. Petrol.* **52**, 1229–41.
Kocurek, G. & J. Nielson 1986. Conditions favourable for the formation of warm-climate aeolian sand sheets. *Sedimentology* **33**, 795–816.
Kolbuszewski, J. 1953. Note on factors governing the porosity of wind deposited sands. *Geol. Mag.* **90**, 48–56.
Kolbuszewski, J., L. Nadolski & Z. Dydacki 1950. Porosity of wind deposited sands. *Geol. Mag.* **87**, 433–5.
Kolstrup, E. 1983. Cover sands in southern Jutland (Denmark). *Proc. Fourth Int. Permafrost Conf., Washington, DC*, 639–44.

Kolstrup, E. & J. B. Jorgensen 1982. Older and younger coversand in southern Jutland (Denmark). *Bull. Geol. Soc. Denmark* **30**, 71–8.

Komar, P. D. & B. Cui 1984. The analysis of grain size measurements by sieving and settling tube techniques. *J. Sediment. Petrol.* **54**, 603–14.

Komar, P. D. & C. E. Reimers 1978. Grain shape effects on settling rates. *J. Geol.* **86**, 193–209.

Konischev, V. N. 1982. Characteristics of cryogenic weathering in the permafrost zone of the European USSR. *Arctic Alpine Res.* **14**, 261–5.

Koster, E. A. 1982. Terminology and lithostratigraphic division of (surficial) sandy eolian deposits in the Netherlands: an evaluation. *Geol. Mijn.* **61**, 121–9.

Koster, E. A. 1988. Ancient and modern cold-climate aeolian sand deposition: a review. *J. Quat. Sci.* **3**, 69–83.

Koster, E. A. & Dijkmans, J. W. A. 1988. Niveo-aeolian deposits and denivation forms, with special reference to the Great Kobuk Sand Dunes, northwestern Alaska. *Earth Surf. Proc. Landf.* **13**, 153–70.

Krinsley, D. H. & J. C. Doornkamp 1973. *An atlas of quartz sand grain surface textures*, Cambridge: Cambridge Univ. Press.

Krinsley, D. H. & F. McCoy 1978. Eolian quartz sand and silt. In *Scanning electron microscopy in the study of sediments*, W. B. Whalley (ed.), 249–60. Norwich: Geo Abstracts.

Krinsley, D. H. & P. Trusty 1985. Environmental interpretation of quartz grain surface textures. In *Provenance of arenites*, G. G. Zuffa (ed.), 213–29. Dordrecht: Reidel.

Krinsley, D. H. & W. Wellendorf 1980. Wind velocities determined from surface textures of sand grains. *Nature* **282**, 372–3.

Krinsley, D. H., P. F. Friend & R. Klimentides 1976. Eolian transport textures on the surfaces of sand grains of Early Triassic age. *Bull. Geol. Soc. Am.* **87**, 130–2

Kroodsma, R. P. 1937. Permanent fixation of sand dunes in Michigan. *J. For.* **35**, 365–71.

Krumbein, W. C. 1934. Size frequency distributions of sediments. *J. Sediment. Petrol.* **4**, 65–77.

Krumbein, W. C. & G. D. Monk 1942. Permeability as a function of the size parameters of unconsolidated sand. *Am. Inst. Min. Metall. Eng. Tech. Pub.* 1492.

Krumbein, W. C. & L. L. Sloss 1963. *Stratigraphy and sedimentation*, 2nd edn. San Francisco: Freeman.

Kucinski, K. J. & W. S. Eisenmenger 1943. Sand dune stabilization on Cape Cod. *Econ. Geogr.* **19**, 206–14.

Kuenen, P. H. 1960. Experimental abrasion. 4. Eolian action. *J. Geol.* **68**, 427–49.

Kuenen, P. H. & W. G. Perdok 1962. Experimental abrasion. 5. Frosting and defrosting of quartz grains. *J. Geol.* **70**, 648–58.

Kuettner, J. 1959. The band structure of the atmosphere. *Tellus* **11**, 267–94.

Kuhlman, H. 1960. The terminology of the geo-aeolian environment. *Geogr. Tidsskr.* **59**, 70–88.

Kutiel, P. & A. Danin 1987. Annual-species diversity and above ground phytomass in relation to some soil properties in the sand dunes of the northern Sharon plains, Israel. *Vegetatio* **70**, 45–9.

Lai, R. J. & J. Wu 1978. *Wind erosion and deposition along a coastal sand dune*. Sea Grant Program, Univ. Delaware, Newark, Delaware.

Laing, C. 1954. *The ecological life history of the marram grass community on Lake Michigan dunes.* Unpubl. PhD Thesis, Univ. Chicago.

Lancaster, N. 1978. The pans of the southern Kalahari. *J. Geogr.* **144**, 80–98.

Lancaster, N. 1980. The formation of seif dunes from barchans—supporting evidence for Bagnold's model from the Namib Desert. *Z. Geomorph. N. F.* **24**, 160–7.

Lancaster, N. 1981a. Aspects of the morphometry of linear dunes of the Namib desert. *S. Afr. J. Sci.* **77**, 366–8.

Lancaster, N. 1981b. Grain size characteristics of Namib Desert linear dunes. *Sedimentology* **28**, 115–22.

Lancaster, N. 1981c. Paleoenvironmental implications of fixed dune systems in Southern Africa. *Palaeogeogr. Palaeoclimatol. Palaeoecol.* **33**, 327–46.

Lancaster, N. 1982a. Dunes on the Skeleton Coast, Namibia (South West Africa): geomorphology and grain size relationships. *Earth Surf. Proc. Landf.* **7**, 575–87.

Lancaster, N. 1982b. Linear dunes. *Prog. Phys. Geogr.* **6**, 475–504.

Lancaster, N. 1983a. Controls of dune morphology in the Namib sand sea. In *Eolian sediments and processes*, M. E. Brookfield & T. S. Ahlbrandt (eds), 261–89. Amsterdam: Elsevier.

Lancaster, N. 1983b. Linear dunes of the Namib sand sea. *Z. Geomorph. Suppl. Bd.* **45**, 27–49.

Lancaster, N. 1984. Paleoenvironments in the Tsondab Valley, central Namib Desert. *Palaeoecol. Afr.* **16**, 411–19.

Lancaster, N. 1985a. Variations in wind velocity and sand transport on the windward flanks of desert sand dunes. *Sedimentology* **32**, 581–93.
Lancaster, N. 1985b. Wind and sand movements in the Namib Sand Sea. *Earth Surf. Proc. Landf.* **10**, 607–19.
Lancaster, N. 1986. Grain size characteristics of linear dunes in the southwestern Kalahari. *J. Sediment. Petrol.* **56**, 395–499.
Lancaster, I. N. 1987. Reply (to comments by A. Watson). *Sedimentology* **34**, 516–20.
Lancaster, N. 1988a. Development of linear dunes in the southwestern Kalahari, Southern Africa. *J. Arid Environ.* **14**, 233–44.
Lancaster, N. 1988b. The development of large aeolian bedforms. *Sediment. Geol.* **55**, 69–89.
Lancaster, N. 1988c. On desert sand seas. *Episodes* **11**, 12–17.
Lancaster, N. 1988d. *A bibliography of dunes: Earth, Mars and Venus.* NASA Contractor Report **4149**.
Lancaster, I. N. 1989a. Star dunes. *Prog. Phys. Geogr.* **13**, 67–91.
Lancaster, N. 1989b. The dynamics of star dunes: an example from the Gran Desierto, Mexico, *Sedimentology* **36**, 273–89.
Lancaster, N. 1989c. *The Namib Sand Sea.* Rotterdam: Balkema.
Lancaster, N. 1989d. Late Quaternary palaeoenvironments in the southwestern Kalahari. *Palaeogeogr. Palaeoclimatol. Palaeoecol.* **70**, 367–76.
Lancaster, N. & J. R. Hallward 1984. *A bibliography of desert dunes.* Cape Town: Department of Environmental and Geographical Sciences, Univ. Cape Town.
Lancaster, N. & C. D. Ollier 1983. Sources of sand for the Namib Sand Sea. *Z. Geomorph.* **45**, 71–83.
Lancaster, N. & J. T. Teller 1988. Interdune deposits of the Namib Sand Sea. *Sediment. Geol.* **55**, 91–108.
Lancaster, N., R. Greeley & P. R. Christensen 1987. Dunes of the Gran Desierto sand-sea, Sonora, Mexico. *Earth Surf. Proc. Landf.* **12**, 277–88.
Land, L. S. 1964. Eolian cross-bedding in the beach dune environment, Sapelo Island, Georgia. *J. Sediment. Petrol.* **34**, 389–94.
Land, L. S. 1967. Diagenesis of skeletal carbonates. *J. Sediment. Petrol.* **37**, 914–30.
Land, L. S. 1970. Phreatic versus vadose meteoric diagenesis of limestones: evidence from a fossil watertable. *Sedimentology* **14**, 175–85.
Land, L. S., F. T. Mackenzie & S. J. Gould 1967. Pleistocene history of Bermuda. *Geol. Soc. Am. Bull.* **78**, 993–1006.
Landsberg, H. 1942. The structure of the wind over a sand-dune. *Trans. Am. Geophys. Union* **23**, 237–9.
Landsberg, H. & N. A. Riley 1943. Wind influences on transportation of sand over a Michigan sand dune. *Bull. Univ. Iowa Stud. Eng.* **27**, 342–52.
Landsberg, S. Y. 1956. The orientation of dunes in Britain and Denmark in relation to wind. *Geogr. J.* **122**, 176–89.
Langford, R. P. 1989. Fluvio-aeolian interactions. Part I. Modern systems. *Sedimentology* **36**, 1023–35.
Langford, R. P. & M. A. Chan 1989. Fluvio-aeolian interactions. Part II. Ancient systems. *Sedimentology* **36**, 1037–51.
Langford-Smith, T. 1982. The geomorphic history of the Australian deserts. *Striae* **17**, 4–19.
Langmuir, I. 1938. Surface motion of water induced by wind. *Science* **87**, 119–23.
Larsen, F. D. 1969. Eolian sand transport on Plum Island, Massachusetts. In *Coastal environments of northeastern Massachusetts and New Hampshire. Field trip guidebook*, M. O. Hayes (ed.), 356–67. Contribution No. 1, Coastal Research Group, Geology Dept., Univ. Massachusetts.
Lavelle, J. W., D. J. P. Swift, P. E. Gadd, W. L. Stubblefield, F. N. Case, H. R. Brashear & K. W. Haff 1978. Fair weather and storm sand transport on the Long Island, New York, inner shelf. *Sedimentology* **25**, 823–42.
Lawson, T. J. 1971. Haboob structure at Khartoum. *Weather* **26**, 105–12.
Laycock, J. W. 1975a. Hydrogeology of North Stradbroke Island. *Proc. R. Soc. Qld.* **86**, 15–19.
Laycock, J. W. 1975b. North Stradbroke Island – hydrogeological report. *Rep. Geol. Surv. Qld.* 88.
Laycock, J. W. 1978. North Stradbroke Island. In *Handbook of recent geological studies of Moreton Bay, Brisbane River and North Stradbroke Island*, G. R. Orme and R. W. Day (eds), 89–96. Dept. Geol. Univ. Qld. Pap. 8.
Leatherman, S. P. 1976. Barrier island dynamics: overwash processes and eolian transport. *Proc. 15th Conf. Coastal Engng., Honolulu*, 1958–74.
Leatherman, S. P. 1978. A new aeolian sand trap design. *Sedimentology* **25**, 303–6.
Leatherman, S. P. 1979. Beach and dune interactions during storm conditions. *Q. J. Eng. Geol.* **12**, 281–90.

Leatherman, S. P. & P. J. Godfrey 1979. The impact of off-road vehicles on coastal ecosystems in Cape Cod National Seashore: an overview. *Univ. Massachusetts, National Park Service Cooperative Research Unit,* Rep. No. 34, Amherst, MA.

Leatherman, S. P. & R. E. Zaremba 1987. Overwash and aeolian processes on a U.S. northeast coast barrier. *Sediment. Geol.* **52**, 183–206.

Leatherman, S. P., A. J. Williams & J. S. Fisher 1977. Overwash sedimentation associated with a large scale northeaster. *Mar. Geol.* **24**, 109–21.

Le Bissonnais, Y., A. Bruand & M. Jamagne 1989. Laboratory experimental study of soil crusting: relation between aggregate breakdown mechanisms and crust structure. *Catena* **16**, 377–92.

Lee, J. A. 1987. A field experiment on the role of small scale wind gustiness in aeolian sand transport. *Earth Surf. Proc. Landf.* **12**, 331–5.

Leeder, M. R. 1977. Folk's bedform theory. *Sedimentology* **24**, 863–4.

Leeder, M. R. 1982. *Sedimentology — process and product.* London: Allen and Unwin.

Lehotsky, K. 1941. Sand dune fixation in Michigan. *J. For.* **39**, 998–1004.

Lehotsky, K. 1972. Sand dune fixation in Michigan — thirty years later. *J. For.* **70**, 155–60.

Le Houerou, H. N. 1968. La désertisation du Sahara septentrional et des steppes limitrophes (Libye, Tunisie, Algérie). *Ann. Algér. Geogr.* **6**, 2–27.

Le Houerou, H. N. 1975. Deterioration of the ecological equilibrium in the arid zone of North Africa. In *Ecological research on development of arid zones (Mediterranean deserts) with winter precipitation,* 45–57. Spec. Publ., Volcani Centre, Bet-Dagan, Israel, No. 39.

Le Houerou, H. N. 1977a. Biological recovery versus desertization. *Econ. Geogr.* **53**, 413–20.

Le Houerou, H. N. 1977b. Man and desertization in the Mediterranean region. *Ambio* **6**, 363–5.

Le Houerou, H. N. 1986. The desert and arid zones of Northern Africa. In *Hot deserts and arid shrublands*, B. M. Evenari, I. Noy-Meir & D. W. Goodall (eds), 101–47. Amsterdam: Elsevier.

LeMone, M. A. 1973. The structure and dynamics of horizontal roll vortices in the planetary boundary layer. *J. Atmos. Sci.* **30**, 1077–91.

Lerner, D., A. Issar & I. Simmers 1990. *Groundwater recharge.* Hannover: Heise.

Lettau, H. H. 1978. Extremes of diurnal surface temperature ranges. In *Exploring the world's driest climate*, H. H. Lettau & K. Lettau (eds), 67–73. Madison: Center for Climatic Research, Univ. Wisconsin.

Lettau, K. & H. Lettau 1969. Bulk transport of sand by the barchan of the Pampa de La Joya in Southern Peru. *Z. Geomorph.* **13**, 182–95.

Lettau, K. & H. Lettau 1978. Experimental and micrometeorological field studies of dune migration. In *Exploring the world's driest climate*, H. H. Lettau & K. Lettau (eds.), 110–47. Madison: Center for Climatic Research, Univ. Wisconsin.

Lewis, A. D. 1936. Sand dunes of the Kalahari within the Union. *S. Afr. Geogr. J.* **19**, 22–32.

Lewis, D. W. & D. G. Titheridge 1978. Small-scale sedimentary structures resulting from foot impressions in dune sands. *J. Sediment. Petrol.* **48**, 835–8.

Li, Z. & P. D. Komar 1986. Laboratory measurements of pivoting angles for applications to selective entertainment of gravel in a current. *Sedimentology* **33**, 413–23.

Liddle, M. J. & P. Greig-Smith 1975a. A survey of tracks and paths in a dune ecosystem, I. Soils. *J. Appl. Ecol.* **12**, 893–908.

Liddle, M. J. & P. Greig-Smith 1975b. A survey of tracks and paths in a sand dune ecosystem. 2. Vegetation. *J. Appl. Ecol.* **12**, 909–30.

Lin, I. J., V. Rohrlich & A. Slatkins 1974. Surface microtextures of heavy minerals from the Mediterranean coast of Israel. *J. Sediment. Petrol.* **44**, 1281–95.

Lindé, K. 1987. Experimental aeolian abrasion of different sand-size materials: some preliminary results. In *Clastic particles*, J. R. Marshall (ed.), 242–7. New York: Van Nostrand Reinhold.

Lindé, K. & E. Mycielska–Dowgiallo 1980. Some experimentally produced microtextures on grain surfaces of quartz sand. *Geogr. Ann.* **62A**, 171–84.

Lindroos, P. 1972. On the development of late glacial and post-glacial dunes in North Karelia, Eastern Finland. *Bull. Geol. Surv. Finland.* **254**, 1–85.

Lindquist, S. J. 1988. Practical characterization of eolian reservoirs for development: Nugget Sandstone, Utah-Wyoming thrust belt. *Sediment. Geol.* **56**, 315–39.

Lindsay, H. A. 1933. A typical Australian line-squall dust storm. *Q. J. R. Meteorol Soc.* **59**, 350.

Lindsay, J. F. 1973. Reversing barchan dunes in lower Victoria Valley, Antarctica. *Bull. Geol. Soc. Am.* **84**, 1799–806.

List, R. J. 1949. *Smithsonian metereological tables*, 6th edn. Washington, DC: Smithsonian Institution.

Little, I. P. 1986. Mobile iron, aluminium and carbon in sandy coastal podzols of Fraser Island, Australia: a quantitative analysis. *J. Soil Sci.* **37**, 439–54.

Little, I. P., T. M. Armitage & R. J. Gilkes 1978. Weathering of quartz in dune sands under subtropical conditions in eastern Australia. *Geoderma* **20**, 225–37.
Livingstone, I. 1982. Dynamics of Namib linear sand dunes. *Namib Bull.* **4**, 10–11.
Livingstone, I. 1986. Geomorphological significance of wind flow patterns over a Namib linear dune. In *Aeolian geomorphology*, W. G. Nickling (ed.), 97–112. Boston: Allen and Unwin.
Livingstone, I. 1987. Grain-size variation on a 'complex' linear dune in the Namib Desert. In *Desert sediments: ancient and modern*, L. Frostick & I. Reid (eds), 281–91. Oxford: Blackwell.
Livingstone, I. 1988. New models for the formation of linear sand dunes. *Geography* **73**, 105–15.
Livingstone, I. 1989a. Monitoring surface changes on a Namib linear dune. *Earth Surf. Proc. Landf.* **14**, 317–32.
Livingstone, I. 1989b. Temporal trends in grain size measures on a linear dune. *Sedimentology* **36**, 1017–22.
Logie, M. 1981. Wind tunnel experiments on dune sands. *Earth Surf. Proc.* **6**, 364–74.
Logie, M. 1982. Influence of roughness elements and soil moisture on the resistance of sand to wind erosion. *Catena Suppl.* **1**, 161–73.
Long, J. T. & R. P. Sharp 1964. Barchan-dune movement in Imperial Valley, California. *Bull. Geol. Soc. Am.* **75**, 149–56.
Longman, M. W., T. G. Fertal, J. S. Glennie, C. G. Krazan, D. H. Suek, W. G. Toller & S. K. Wiman 1983. Description of a paraconformity between carbonate grainstones, Isla Cancun, Mexico. *J. Sediment. Petrol.* **53**, 533–42.
Loope, D. B. 1984. Origin of extensive bedding planes in aeolian sandstones: a defence of Stokes' hypothesis. *Sedimentology* **31**, 123–5.
Loope, D. B. 1985a. Episodic desposition and preservation of eolian sands: a late Paleozoic example from southeastern Utah. *Geology* **13**, 73–6.
Loope, D. B. 1985b. Rhizoliths in ancient eolianites. *Sediment. Geol.* **56**, 301–14.
Lorenz, E. N. 1970. The nature of the global circulation of the atmosphere: a present view. In *The global circulation of the atmosphere*, G. A. Corby (ed.), 3–23. London: Royal Meteorological Society.
Loughnan, F. C. 1969. *Chemical weathering of silicate minerals*. New York: Elsevier.
Luckenback, R. A. & R. B. Bury 1983. Effects of off-road vehicles on the biota of the Algodones dunes, Imperial County, California. *J. Appl. Ecol.* **20**, 265–86.
Lumley, J. L. & H. A. Panofsky 1964. *The structure of atmospheric turbulence*. New York: Wiley.
Lundberg, A. 1984. A controversy between recreation and ecosystem protection in the sand dune areas on Karmoy, Southwestern Norway. *GeoJournal* **8**, 147–57.
Lundquist, J. & V. Mejdahl 1987. Thermoluminescence dating of eolian sediments in central Sweden. *Geol. Foren. Stock. Forhand.* **109**, 147–58.
Lunson, E. A. 1950. *Sandstorms on the northern coasts of Libya and Egypt*. London: Meteorological Office, Prof. Notes 102.
Lupe, R. & T. S. Ahlbrandt 1979. Sediments in ancient eolian environments — reservoir inhomogeniety. In *A study of global sand seas*, E. D. McKee (ed.), 241–51. Prof. Pap. US Geol. Surv. No. 1052.
Lutz, H. J. 1941. The nature and origin of layers of fine-textured material in sand dunes. *J. Sediment. Petrol.* **11**, 105–23.
Lyles, L. & R. K. Krauss 1971. Threshold velocities and initial particle motion as influenced by air turbulence. *Trans. Am. Soc. Agric. Eng. Abs.* **14**, 563–66.

Maarel, F. van der 1979. Environmental management of coastal dunes in the Netherlands. In *Ecological processes in coastal environments*, R. L. Jefferies & A. J. Davies (eds), 543–70. Oxford: Oxford Univ. Press.
Maarleveld, G. C. 1960. Wind directions and cover sands in the Netherlands. *Biul. Peryglac.* **8**, 49–58.
Mabbutt, J. A. 1968. Aeolian landforms in Central Australia. *Aust. Geogr. Stud.* **6**, 139–50.
Mabbutt, J. A. 1977. *Desert landforms,* Cambridge: MIT Press.
Mabbutt, J. A. & M. E. Sullivan 1968. The formation of longitudinal dunes: evidence from the Simpson Desert. *Aust. Geogr.* **10**, 483–7.
Mabbutt, J. A. & R. A. Wooding 1983. Analysis of longitudinal dune patterns in the northwestern Simpson Desert, central Australia. *Z. Geomorph. Suppl. Bd.* **45**, 51–69.
Mabbutt, J. A., R. A. Wooding & J. N. Jennings 1969. The asymmetry of Australian desert sand ridges. *Aust. J. Sci.* **32**, 159–60.
MacCarthy, G. R. 1935. The rounding of beach sands, a comparison. *Am. J. Sci. Ser. 5,* **25**, 204–24.
MacCarthy, G. R. & J. W. Huddle 1938. Shape sorting of sand grains by wind action. *Am. J. Sci. Ser. 5,* **35**, 64–73.

MacDonald, J. 1954. Tree planting on coastal sand dunes in Great Britain. *Adv. Sci.* **11**, 33–7.
MacEntee, F. J. & H. C. Bold 1978. Some microalgae from sand. *Texas J. Sci.* **30**, 167–73.
Mader, D. 1982. Aeolian sands in continental red beds of the Bundsandstein (Lower Triassic) at the western margin of the German Basin. *Sediment. Geol.* **31**, 191–230.
Mader, D. & M. J. Yardley 1985. Migration, modification and merging in aeolian systems and the significance of the depositional mechanisms in Permian and Triassic dune sands of Europe and North America. *Sediment. Geol.* **43**, 85–218.
Madigan, C. T. 1936. The Australian sand-ridge deserts. *Geogr. Rev.* **26**, 205–27.
Madigan, C. T. 1946. The Simpson Desert Expedition, 1939 scientific reports: No. 6, Geology — the sand formations. *Trans. R. Soc. S. Aust.* **70**, 45–63.
Maegley W. J. 1976. Saltation and Martian sandstorms: *Rev. Geophys. Space Phys.* **14**, 135–42.
Magaritz, M., E. Gavish, N. Bakler & U. Kafri 1979. Carbon and oxygen isotope composition – indicators of cementation environment in Recent, Holocene and Pleistocene sediments along the coast of Israel. *J. Sediment Petrol.* **49**, 401–12.
Mainguet, M. 1978. The influence of trade winds, local air-masses and topographic obstacles on the aeolian movement of sand particles and the origin and distribution of dunes in the Sahara and Australia. *Geoforum* **9**, 17–28.
Mainguet, M. 1983. Dunes vivés, dunes fixées, dunes vetués: une classification selon le bilan d'alimentation, le regime éolien et la dynamique des edifices sableux. *Z. Geomorph. Suppl. Bd.* **45**, 265–85.
Mainguet, M. 1984a. Cordons longitudinaux dunes allongées a ne plus confondre avec les sifs, autres dunes linéaires. *Trav. Inst. Geogr. Reims,* **59–60**, 61–83.
Mainguet, M. 1984b. A classification of dunes based on aeolian dynamics and the sand budget. In *Deserts and arid lands*, F. El.-Baz (ed.), 31–58. The Hague: Martinus Nijhoff.
Mainguet, M. 1986. The wind and desertification processes in the Saharo-Sahelian and Sahelian regions. In *Physics of desertification*, F. El-Baz & M. H. A. Hassan (eds), 210–40. Dordrecht: Martinus Nijhoff.
Mainguet, M. & Y. Callot 1974. Air photo study of typology and interrelations between the texture and structure of dune patterns in the Fachi-Bilma Erg, Sahara. *Z. Geomorph. Suppl. Bd.* **20**, 62–8.
Mainguet, M. & Y. Callot 1978. L'erg de Fachi-Bilma. *Memoires et Documents* 18. Paris: Centre National de la Recherche Scientifique.
Mainguet, M. & M.-C. Chemin 1983. Sand seas of the Sahara and Sahel: an explanation of their thickness and sand dune type by the sand budget principle. In *Eolian sediments and processes*, M. E. Brookfield & T. S. Ahlbrandt (eds), 353–63. Amsterdam: Elsevier.
Mainguet, M. & L. Cossus 1980. Sand circulation in the Sahara: geomorphological relation between the Sahara Desert and its margin. In *Sahara and surrounding seas*, M. Sarnthein, E. Seibold & P. Rognon (eds), 69–78. Rotterdam: Balkema.
Mainguet, M., Y. Callot & M. Guy 1974. Taxonomic classification of dunes in Fachi-Bilma erg, diagnosis of wind direction. *Rev. Photo-Interpretation* **74–1**, 47–53.
Mainguet, M., J.-M. Borde & M.-C. Chemin 1984. Sedimentation éolienne au Sahara et sur ses marges. *Trav. Inst. Geogr. Reims* **59–60**, 15–27.
Malina, F. J. 1941. Recent developments in the dynamics of wind erosion. *Trans. Am. Geophys. Union* **22**, 262–84.
Manohar, M. & P. Bruun 1970. Mechanics of dune growth by sand fences. *Dock. Harb. Auth.* **51**, 243–52.
Margolis, S. V. & D. H. Krinsley 1971. Submicroscopic frosting on eolian and subaqueous quartz sand grains. *Bull. Geol. Soc. Am.* **82**, 3395–406.
Marsh, W. M. & B. D. Marsh 1987. Wind erosion and sand dune formation on high Lake Superior bluffs. *Geog. Ann.* **69A**, 379–91.
Marsland, P. S. & J. G. Woodruff 1937. A study of the effect of wind transportation on grains of several minerals. *J. Sediment Petrol.* **7**, 18–30.
Marston, R. A. 1986. Maneuver-caused wind erosion impacts, south-central New Mexico. In *Aeolian geomorphology*, W. G. Nickling (ed.), 273–306. Boston: Allen and Unwin.
Martins, F. 1989. Morphology and management of dunes at Leiria District, Portugal. In *Perspectives in coastal dune management*, F. van der Meulen, P. D. Jungerius & J. Visser (eds), 1–9. The Hague: SPB Academic Publishing.
Marzolf, J. E. 1988. Controls on late Paleozoic and early Mesozoic eolian deposition of the western United States. *Sediment. Geol.* **56**, 167–92.
Mason, C. C. & R. L. Folk 1958. Differentiation of beach, dune and aeolian flat environments by size analysis, Mustang Island, Texas. *J. Sediment Petrol.* **28**, 211–26.
Mason, P. J. & J. C. King 1985. Measurements and predictions of flow and turbulence over an isolated hill of moderate slope. *Q. J. R. Meteorol. Soc.* **111**, 617–40.

Mason, P. J. & R. I. Sykes 1979. Flow over an isolated hill of moderate slope. *Q. J. R. Meteorol. Soc.* **105**, 383–95.
Mather, K. B. 1969. The pattern of surface wind flow in Antarctica. *Geofis. Pura Appl.* **75**, 332–54.
Mathews, B. 1970. Age and origin of aeolian sand in the Vale of York. *Nature* **227**, 1234–6.
Mattox, R. B. 1955. Eolian shape sorting. *J. Sediment. Petrol.* **25**, 111–14.
Maud, R. R. 1968. Quaternary geomorphology and soil formation in coastal Natal. *Z. Geomorph. Suppl. Bd.* **7**, 155–99.
Maun, M. A. 1985. Population biology of *Ammophila breviligulata* and *Calamovilfa longifolia* on Lake Huron sand dunes. 1. Habitat, growth form, reproduction, and establishment. *Can. J. Bot.* **63**, 113–24.
Mawson, D. 1930. *The home of the blizzard.* London: Hodder and Stoughton.
Maxwell, T. A. 1982. Sand sheet and lag deposits in the Southwestern Desert. In *Desert landforms of southwest Egypt: a basis for comparison with Mars*, F. El-Baz & T. A. Maxwell (eds), 157–73. Washington, DC: NASA CR–3611.
Maxwell, T. A. & C. V. Haynes 1989. Large-scale, low amplitude bedforms (chevrons) in the Selima sand sheet, Egypt. *Science* **243**, 1179–82.
Mazzullo, J., D. Sims & D. Cunningham 1986. The effects of eolian sorting and abrasion upon the shapes of fine quartz sand grains *J. Sediment. Petrol.* **56**, 45–56.
McAllister, J. 1981. Particle size. In *Geomorphological techniques*, A. S. Goudie (ed.), 81–5. London: Allen and Unwin.
McArthur, D. S. 1987. Distinctions between grain size distributions of accretion and encroachment deposits in an inland dune. *Sediment. Geol.* **54**, 147–63.
McBride, E. F. 1971. Mathematical treatment of grain distribution data. In *Procedures in sedimentary petrology*, R. C. Carver (ed.), 109–27. New York: Wiley-Interscience.
McBride, E. F. & M. O. Hayes 1962. Dune cross bedding on Mustang Island, Texas. *Am. Assoc. Petrol. Geol. Bull.* **46**, 546–51.
McCarthy, M. J. 1967. Stratigraphical and sedimentological evidence from the Durban region of major sea level movements since the late Tertiary. *Trans. Geol. Soc. S. Afr.* **70**, 135–66.
McCauley, J. F., C. S. Breed, G. G. Schaber, W. P. McHugh, B. Issawi, C. V. Haynes, M. J. Grolier & A. El Kilani 1986. Paleodrainages of the Eastern Sahara – the radar rivers revisited (SIR-A/B implications for Mid-Tertiary Trans African drainage system). *IEEE Trans. Geosci. Remote Sensing* **24**, 624–48.
McCauley, J. F., G. G. Schaber, C. S. Breed, M. J. Grolier, C. V. Haynes, C. V. Issawi, C. Elachi & R. Blom 1982. Subsurface valleys and geochronology of the eastern Sahara revealed by Shuttle radar. *Science* **218**, 1004–19.
McCave, I. N. & J. Jarvis 1973. Use of the Model T Coulter Counter in size analysis of fine to coarse sand. *Sedimentology* **20**, 305–15.
McCave, I. N., R. J. Bryant, H. F. Cooke & C. A. Coughanowr 1986. Evaluation of a laser-diffraction-size analyzer for use with natural sediments. *J. Sediment. Petrol.* **56**, 561–4.
McClelland, J. E. 1950. The effect of time, temperature and particle size on the release of bases from some common soil forming minerals of different crystal structure. *Proc. Soil Sci. Soc. Am.* **15**, 301–7.
McGee, W. J. 1908. Outlines of Hydrology. *Bull. Geol. Soc. Am.* **19**, 193–220.
McKee, E. D. 1945. Small-scale structures in the Coconino sandstone of northern Arizona. *J. Geol.* **53**, 313–25.
McKee, E. D. 1957. Primary structures in some Recent sediments. *Am. Assoc. Petrol. Geol. Bull.* **41**, 1704–47.
McKee, E. D. 1966. Structures of dunes at White Sands National Monument, New Mexico. *Sedimentology* **7**, 3–69.
McKee, E. D. 1979a. Introduction to a study of global sand seas. In *A study of global sand seas*, E. D. McKee (ed.), 1–19. Prof. Pap. US Geol. Surv. No. 1052.
McKee, E. D. 1979b. Ancient sandstones considered to be eolian. In *A study of global sand seas*, E. D McKee (ed.), 187–238. Prof. Pap. US Geol. Surv. No. 1052.
McKee, E. D. 1982. Sedimentary structure in dunes of the Namib desert, South West Africa. *Spec. Pap. Geol. Soc. Am.* No. 188.
McKee, E. D. 1983. Eolian sand bodies of the world. In *Eolian sediments and processes*, M. E. Brookfield & T. S. Ahlbrandt (eds), 1–25. Amsterdam: Elsevier.
McKee, E. D. & J. J. Bigarella 1972. Deformational structures in Brazilian coastal dunes. *J. Sediment. Petrol.* **42**, 670–81.
McKee, E. D. & J. J. Bigarella 1979. Sedimentary structures in dunes. In *A study of global sand seas*, E. D. McKee (ed.), 83–134. Prof. Pap. US Geol. Surv. No. 1052.
McKee, E. D. & C. S. Breed 1974. An investigation of major sand seas in desert areas throughout

the world. In *Third Earth Resources Technology Satellite — 1. Symposium*, S.C. Freden, E.P. Mercanti & M.A. Becker (eds), 665–78. Washington, DC: NASA SP–351.

McKee, E.D. & J. Moiola 1975. Geometry and growth of the White Sands dune field, New Mexico. *J. Res. US Geol. Surv.* **3**, 59–66.

McKee, E.D. & G.C. Tibbitts 1964. Primary structures of a seif dune and associated deposits in Libya. *J. Sediment. Petrol.* **34**, 5–17.

McKee, E.D. & W.C. Ward 1983. Eolian environments. In *Carbonate depositional environments*, P.A. Scholle, D.G. Bebout & C.D. Moore (eds), 131–70. Am. Assoc. Petrol. Geol. Mem. 33.

McKee, E.D., J.R. Douglass & S. Rittenhouse 1971. Deformation of lee side laminae in eolian dunes. *Bull. Geol. Soc. Am.* **82**, 359–78.

McKee, E.D., C.S. Breed & S.G. Fryberger 1977. Desert sand seas. In *Skylab explores the Earth*, 5–48. Washington, DC: NASA SP–380.

McKenna-Neuman, C. 1989. Kinetic energy transfer through impact and its role in entrainment by wind of particles from frozen surfaces. *Sedimentology* **36**, 1007–16.

McKenna-Neuman, C. & R. Gilbert 1986. Aeolian processes and landforms in glaciofluvial environments of southeastern Baffin Island, N.W.T., Canada. In *Aeolian geomorphology*, W.C. Nickling (ed.), 213–36. Boston: Allen and Unwin.

McLachlan, A., C. Ascaray & P. du Toit 1987. Sand movement, vegetation succession and biomass spectrum in a coastal dune slack in Algoa Bay, South Africa. *J. Arid Environ.* **12**, 9–25.

McManus, D.A. 1963. A criticism of certain usage of the phi notation. *J. Sediment. Petrol.* **35**, 792–6.

McManus, J. 1988. Grain size determination and interpretation. In *Techniques in sedimentology*, M. Tucker (ed.), 63–85. Oxford: Blackwell.

McNaughton, K. & T.A. Black 1973. A study of evaporation from a Douglas fir forest using the energy balance approach. *Water Res.* **9**, 1957–90.

McTainsh, G. 1984. The nature and origin of the aeolian mantles of central northern Nigeria. *Geoderma*, **33**, 13–37.

McTainsh, G.H., & P.H. Walker 1982. Nature and distribution of Harmattan dust. *Z. Geomorph.* **26**, 417–35.

Medlicott, H.B. & W.T. Blanford 1879. *A manual of the geology of India, Part I: Peninsular area.* London: Trubner.

Melton, F.A. 1940. A tentative classification of sand dunes — its application to dune history in the southern High Plains. *J. Geol.* **48**, 113–74.

Merk, G.P. 1960. Great sand dunes of Colorado. In *Guide to the geology of Colorado*, Geol. Soc. Am., Rocky Mtn. Assoc. Geologists, and Colorado Sci. Soc., 127–9.

Messines, J. 1952. Sand-dune fixation and afforestation in Libya. *Unasylva* **6**, 51–8.

Meteorological Office 1962. *Weather in the Mediterranean I. General meteorology*, 2nd edn. M.O.391. London: HMSO.

Meulen, F. van der 1982. Vegetation changes and water catchment in a Dutch coastal dune area. *Biol. Conserv.* **24**, 305–16.

Meulen, F. van der & M.R. Gourlay 1968. Beach and dune erosion tests. *Proc. 11th Int. Coastal Engng. Conf., London*, 701–7.

Meulen, F. van der & E. van der Maarel 1989. Coastal defence alternatives and nature development perspectives. In *Perspectives in coastal dune management*, F. van der Meulen, P.D. Jungerius & J. Visser (eds), 183–96. The Hague: SPB Academic Publishing.

Meulen, F. van der, P.D. Jungerius & J. Visser (eds) 1989. *Perspectives in coastal dune management.* The Hague: SPB Academic Publishing.

Middleton, G.V. 1970. Experimental studies related to problems of flysch sedimentation. In *Flysch sedimentology of North America*, J. Lajoie (ed.), 253–72. Geol. Soc. Canada Spec. Publ. No. 7.

Middleton, G.V. 1976. Hydraulic interpretation of sand size distributions, *J. Geol.* **84**, 405–26.

Middleton, G.V. & J.B. Southard 1978. *Mechanics of sediment movement.* Binghamton, S.E.P.M. short course No. 3.

Middleton, N.J. 1985. Effect of drought on dust production in the Sahel. *Nature,* **316**, 431–4.

Migahid, A.A. 1961. The drought resistance of Egyptian desert plants. *Arid Zone Res.* **16**, 213–33.

Miller, D.R., N.J. Rosenberg & W.T. Bagley 1975. Wind reduction by highly permeable tree shelterbelt. *Agric. Meteorol.* **14**, 327–33.

Miller, M.C. & P.D. Komar 1977. The development of sediment threshold curves for unusual environments (Mars) and for inadequately studied materials (foram sands). *Sedimentology* **24**, 709–21.

Miller, M.C., I.N. McCave & P.D. Komar 1977. Threshold of sediment motion under unidirectional currents. *Sedimentology* **24**, 507–27.

Milnes, A.R. & N.H. Ludbrook 1986. Provenance of microfossils in aeolian calcarenites and calcretes in southern South Australia. *Aust. J. Earth Sci.* **33**, 145–59.

Mironer, A. 1979. *Engineering fluid mechanics*. New York: McGraw-Hill.
Mitha, S., M. Q. Tran, B. T. Werner & P. K. Haff 1986. The grain-bed impact process in aeolian saltation. *Acta Mech.* **63**, 267–78.
Mizutani, S. 1963. A theoretical and experimental consideration on the accuracy of sieving analysis. *J. Earth Sci. Nagoya Univ.* **11**, 1–27.
Moberley, R., L. D. Bauer & A. Morrison 1965. Source and variation of Hawaiian littoral sand. *J. Sediment. Petrol.* **35**, 589–98.
Moiola, R. J. & A. B. Spencer 1979. Differentiation of eolian deposits by discriminant analysis. *U.S. Geol. Surv. Prof. Pap.* 1052, 53–B.
Moiola, R. J. & D. Weiser 1968. Textural parameters: an evaluation. *J. Sediment. Petrol.* **38**, 45–53.
Moiola, R. J., A. B. Spencer & D. Weiser 1974. Differentiation of modern sand bodies by linear discriminant analysis. *Trans. Gulf Coast Assoc. Geol. Socs.* **24**, 321–6.
Møller, J. T. 1985. *Soil and sand drift in Denmark*. Geologisk Institut Aarhus Universitet Geoskrifter No. 22.
Møller, J. T. 1986. Soil degradation in a North European region. In R. Fantechi & N. S. Margaris (eds), *Desertification in Europe*. Dordrecht: Reidel, 214–30.
Monin, A. S. & A. M. Yaglom 1965. *Statistical fluid mechanics: mechanics of turbulence*, Vol. I. Cambridge, MA: MIT Press.
Monod, T. 1958. Majabat al-Koubra. Contribution a l'étude de l'Empty Quarter ouest-Saharien. *Mem. Inst. Fr. Afr. Noire* 52.
Monteith, J. L. 1973. *Principles of environmental physics*. London: Edward Arnold.
Moomen, S. E. & C. W. Barney 1981. A modern technique to halt desertification in the Libyan Jamahiriya. *Agric. Meteorol.* **23**, 131–6.
Moore, D. J. 1979. Offshore wind data. In *Proc. First BWEA Wind Energy Workshop, April 1979*, 199–207. London: Multi-Science Publishers.
Moreno-Casasola, P. 1986. Sand movement as a factor in the distribution of plant communities in a coastal dune system. *Vegetatio* **65**, 67–76.
Morey, G. W., Fournier, R. O. & J. J. Rowe 1962. The solubility of quartz in water in the temperature interval from 25 °C to 300 °C. *Geochim. Cosmochim. Acta* 22, 1029–43.
Morley, I. W. 1981. *Black sands: a history of the mining industry in eastern Australia*. Brisbane: University of Queensland Press.
Moss, A. J. 1966. Origin, shaping and significance of quartz sand grains. *J. Geol. Soc. Aust.* **13**, 97–136.
Moss, A. J. 1972. Initial fluviatile fragmentation of granitic quartz. *J. Sediment. Petrol* **42**, 905–16.
Moss, A. J., P. H. Walker & J. Hutka 1973. Fragmentation of granitic quartz in water. *Sedimentology* **20**, 489–511.
Moss, A. J. & P. Green 1975. Sand and silt grains: predetermination of their formation and properties by microfractures in quartz. *J. Geol. Soc. Aust.* **22**, 485–95.
Moss, A. J., P. Green & J. Hutka 1981. Static breakage of granitic detritus by ice and water in comparison with breakage by flowing water. *Sedimentology* **28**, 261–72.
Mulhearn, P. J. & E. F. Bradley 1977. Secondary flow in the lee of porous shelterbelts. *Boundary-Layer Meteorol.* **12**, 75–92.
Muller, G. & G. Tietz 1975. Regressive diagenesis in Pleistocene eolianites from Fuerteventura, Canary Islands. *Sedimentology* **22**, 485–96.
Mulligan, K. R. 1985. *The movement of transverse coastal dunes, Pismo Beach, California, 1982–83*. Unpubl. MA Thesis, Dept. Geogr., Univ. California, Los Angeles.
Mulligan, K. R. 1987. Velocity profiles measured on the windward slope of a transverse dune. *Earth Surf. Proc. Landf.* **13**, 573–82.

Nalivkin, D. V. 1982. *Hurricanes, storms and tornadoes: geographic characteristics and geological activity*. New Delhi: Amerind.
Nalpanis, P. 1985. Saltating and suspended particles over flat and sloping surfaces. II. Experiments and numerical simulations. In *Proceedings of the international workshop on the physics of blown sand*, O. E. Barndorff-Nielsen, J. T. Møller, K. R. Rasmussen & B. B. Willetts (eds), 37–66. Dept. Theoretical Statistics, Institute of Mathematics, Univ. Aarhus, Mem. 8.
Newell, N. D. & D. W. Boyd 1955. Extraordinarily coarse eolian sand of the Ica desert, Peru. *J. Sediment. Petrol.* **25**, 226–8.
Newell, R. E., S. Gould-Stewart & J. C. Chung 1981. A possible interpretation of paleoclimatic reconstructions for 18,000 B.P. for the region 60°N to 60°S, 60°W to 100°E. *Palaeoecol. Afr.* **13**, 1–19.
Nicholson, I. A. 1952. *A study of Agropyron junceum (Beauv.) in relation to the stabilization of coastal sand and the development of sand dunes*. Unpubl. MSc Thesis, Univ. Durham.

Nicholson, S. E. & H. Flohn 1980. African environmental and climatic changes and the general atmospheric circulation in Late Pleistocene and Holocene. *Climatic Change* **2**, 313–48.

Nickling W. G. 1978. Eolian sediment transport during dust storms: Slims River Valley, Yukon Territory. *Can. J. Earth Sci.* **15**, 1069–84.

Nickling, W. G. 1983. Grain-size characteristics of sediment transported during dust storms. *J. Sediment. Petrol.* **53**, 1011–24.

Nickling, W. G. 1984. The stabilizing role of bonding agents on the entrainment of sediment by wind. *Sedimentology* **31**, 111–17.

Nickling, W. G. (ed.) 1986 *Aeolian geomorphology*. Boston: Allen & Unwin.

Nickling, W. G. 1988. The initiation of particle movement by wind. *Sedimentology* **35**, 499–511.

Nickling, W. G. & M. Ecclestone 1981. The effects of soluble salts on the threshold shear velocity of fine sand. *Sedimentology* **28**, 505–10.

Nickling, W. G. & J. A. Gillies 1989. Emission of fine-grained particulates from desert soils. In *Paleoclimatology and paleometeorology: modern and past processes of global atmospheric transport*, 133–65. NATO ASI Series No. 282. Dordrecht: Kluwer.

Nielson, J. & G. Kocurek 1986. Climbing zibars of the Algodones. *Sediment Geol.* **48**, 1–15.

Nielson, J. & G. Kocurek 1987. Surface processes, deposits, and development of star dunes: Dumont dune field, California. *Bull. Geol. Soc. Am.* **99**, 177–86.

Niessen, A. C. H. M., E. A. Koster & J. P. Galloway 1984. Periglacial sand dunes and eolian sand sheets; an annotated bibliography. *U.S. Geol. Surv. Open File Rep.* 84-167, 1–61.

Nieter, W. M. & D. H. Krinsley 1976. The production and recognition of aeolian features on sand grains by silt abrasion. *Sedimentology* **23**, 713–20.

Niknam, F. & B. Ahranjani 1975. *Dunes and development in Iran*. Imperial Government of Iran, Ministry of Agriculture & Natural Resources.

Nikuradse, J. 1933. *Laws of flow in rough pipe*. Translation of 'Stromungsgesetze in rauhen rohren': VDI-Forschungscreft 361. Beilage zu 'Forschung auf dem Gebiete des Ingenieurwesens' Ausgabe B, Band 4. NACA Tech. Mem. 1292, November 1950.

Nordstrom, K. F. & J. M. McCluskey 1985. The effects of houses and sand fences on the eolian sediment budget at Fire Island, New York. *Coastal Res.* **1**, 39–46.

Nordstrom, K. F., N. P. Psuty & R. W. G. Carter (eds) 1990. *Coastal dunes*. Chichester: Wiley.

Norris, R. M. 1956. Crescentic beach cusps and barkhan dunes. *Bull. Am. Assoc. Petrol. Geol.*, **40**, 1681–6.

Norris, R. M. 1966. Barchan dunes of Imperial Valley, California. *J. Geol.* **74**, 292–306

Norris, R. M. 1969. Dune reddening and time. *J. Sediment. Petrol.* **39**, 7–11.

Norrman, J. O. 1981. Coastal dune systems. In *Coastal dynamics and scientific sites*, E. C. F. Bird & K. Koike (eds), 119–57. Tokyo: Dept. of Geography, Komazawa Univ.

Norstrud, H. 1982. Wind flow over low arbitrary hill. *Boundary Layer Meteorol.* **23**, 115–24.

Nowaczyk, B. 1976. Eolian coversands in central-west Poland. *Quaest. Geogr.* **3**, 57–77.

Noy-Meir, I. 1973. Desert ecosystems: environment and producers. *Annu. Rev. Ecol. Systematics* **4**, 25–51.

Numerical Algorithms Group 1982. *NAG Library Manual Mark 9*. Oxford: Numerical Algorithms Group.

O'Brien, M. P. & B. D. Rindlaub 1936. The transportation of sand by wind. *Civil Eng.*, **6**, 325–7.

Oertel, G. F. & M. Larsen 1976. Developmental sequences in Georgia coastal dunes and distribution of dune plants. *Georgia Acad. Sci. Bull.* **34**, 35–48.

Oke, T. R. 1978. *Boundary layer climates*. London: Methuen.

Oliver, F. W. 1945. Dust storms in Egypt and their relation to the war period, as noted by Maryot 1939–1945. *Geogr. J.* **106**, 26–49.

Oliver, F. W. 1947. Dust storms in Egypt as noted in Maryot: a supplement. *Geogr. J.* **108**, 221–6.

Olsen, H., P. Due & L. B. Clemmensen 1989. Morphology and genesis of asymmetric adhesion warts – a new adhesion surface structure. *Sediment Geol.* **61**, 277–85.

Olson, J. S. 1958a. Lake Michigan dune development 1. Wind velocity profiles. *J. Geol.* **66**, 254–63.

Olson, J. S. 1958b. Lake Michigan dune development 2. Plants as agents and tools in geomorphology. *J. Geol.* **66**, 345–51.

Olson, J. S. 1958c. Rates of succession and soil changes on southern Lake Michigan sand dunes. *Bot. Gaz.* **119**, 125–70.

Olson, J. S. & E. van der Maarel 1989. Coastal dunes in Europe: a global view. In *Perspectives in coastal dune management*, F. van der Meulen, P. D. Jungerius & J. Visser (eds), 3–32. The Hague: SPB Academic Publishing.

Oosting, H. J. & W. D. Billings 1942. Factors effecting vegetation zonation on coastal dunes. *Ecology* **23**, 131–42.

Orev, Y. 1984. Sand is greener. *Teva-va-Aretz* **26**, 15–16 (in Hebrew).
Orford, J. D. 1981. Particle form. In *Geomorphological techniques*, A. S. Goudie (ed.), 86–9. London: Unwin Hyman.
Orford, J. D. & W. B. Whalley 1983. The use of fractal dimensions to quantify the morphology of irregular-shaped particles. *Sedimentology* **30**, 655–68.
Orford, J. D. & W. B. Whalley 1987. The quantitative description of highly irregular sedimentary particles: the use of the fractal dimension. In *Clastic particles*, J. R. Marshall (ed.). 267–80. New York: Van Nostrand Reinhold.
Orme, G. R. & V. P. Tchakerian 1986. Quaternary dunes of the Pacific Coast of the Californias. In *Aeolian geomorphology*, W. G. Nickling (ed.), 149–75. Boston: Allen and Unwin.
Otterman, J., Y. Waisel & E. Rosenberg 1975. Western Negev and Sinai ecosystems: comparative study of vegetation, albedo, and temperatures. *Agro-Ecosystems* **2**, 47–59.
Otto, G. H. 1938. The sedimentation unit and its use in field sampling. *J. Geol.* **46**, 569–82.
Ovington, J. D. 1950. The afforestation of the Culbin Sands. *J. Ecol.* **38**, 303–19.
Ovington, J. D. 1951. The afforestation of Tentsmuir Sands. *J. Ecol.* **39**, 363–75.
Owen, P. R. 1960. Dust deposition from a turbulent airstream. *Int. J. Air Pollut.* **3**, 8–25.
Owen, P. R. 1964. Saltation of uniform grains in air. *J. Fluid Mech.* **20**, 225–42.
Owen, P. R. 1980. *The physics of sand movement.* Lecture Notes, Workshop on physics of flow in deserts. Trieste: International Centre for Theoretical Physics.
Owens, J. S. 1908. Experiments on the transporting power of sea currents. *Geogr. J.* **31**, 415–25.
Owens, J. S. 1918. Discussion on Mr. Harding King's paper "Study of a dune belt". *Georg. J.* **51**, 254–6.
Owens, J. S. 1927. The movement of sand by wind. *Engineer* **143**, 377.
Ower, E. & R. C. Pankhurst 1977. *The measurement of air flow*, 5th edn. Oxford: Pergamon Press.

Paepe, R. & R. Vanhoorne 1967. The stratigraphy and palaeobotany of the late Pleistocene in Belgium. *Mém. Explic. Cartes Géol. Min. Belg.* 8.
Page, H. G. 1955. Phi–millimetre conversion tables. *J. Sediment. Petrol.* **25**, 285–92.
Pande, P. K., R. Prakash & M. L. Agrawal 1980. Flow past fence in turbulent boundary layer. *J. Hydraul. Div. Am. Soc. Civ. Eng.* **106**, 191–207.
Panofsky, H. A. 1982. The atmosphere. In *Engineering meteorology*, E. J. Plate (ed.), 1–32. Amsterdam: Elsevier.
Parkin, D. W. 1974. Trade-winds during the glacial cycles. *Proc. R. Soc. London, Ser. A* **337**, 73–100.
Parkin, D. W. & R. C. Padgham 1975. Further studies on trade winds during the glacial cycles. *Proc. R. Soc. London, Ser. A* **346**, 245–60.
Parkin, D. W. & N. J. Shackleton 1973. Trade wind and temperature correlations down a deep sea core off the Sahara coast. *Nature* **245**, 455–7.
Parrish, J. T. & F. Peterson 1988. Wind directions predicted from global circulation models and wind directions determined from eolian sandstones of the western United States—a comparison. *Sediment. Geol.* **56**, 261–82.
Pasquill, F. 1974. *Atmospheric diffusion*, 2nd edn. Chichester: Ellis Horwood.
Paton, T., R. Mitchell, P. B. Adamson, D. Buchanan & G. M. Bowman 1976. Speed of podzolization. *Nature* **260**, 601–2.
Paul, K. 1944. Morphologie und vegetation der Kurische Nehrung. *Nova Acta Leopoldina Carol. NF* **13**, 217–378.
Pearse, J. R. 1982. Wind flow over conical hills in a simulated atmospheric boundary layer. *J. Wind Eng. Ind. Aerodyn.* **10**, 303–13.
Pearse, J. R., D. Lindley & D. C. Stevenson 1981. Wind flow over ridges in simulated atmospheric boundary layers. *Boundary-Layer Meteorol.* **21**, 77–92.
Peeters, L. 1983. Les dunes continentales de la Belgique. *Bull. Soc. Belge Géol.* **52**, 51–62.
Peters, S. P. 1932. *Some upper air observation over Lower Egypt*. Geophysical Memoir No. 56. London: Meteorological Office.
Petrov, M. P. 1976. *Deserts of the World.* New York: Wiley.
Péwé, T. L. 1960. Multiple glaciation in the McMurdo Sound region, Antarctica—a progress report. *J. Geol.* **68**, 498–514.
Péwé T. L. 1974. Geomorphic processes in polar deserts. In *Polar deserts and modern man*, T. L. Smiley and J. H. Zuamberge (eds), 33–52. Tucson: Univ. Arizona Press.
Phillips, C. J. & B. Willetts 1978. A review of selected literature on sand stabilization. *Coast Eng.* **2**, 133–48.
Phillips, C. J. & B. Willetts 1979. Predicting sand deposition at porous fences. *J. Am. Soc. Civ. Eng., Waterways, Port, Coastal & Ocean Eng. Div.* **105**, 15–31.

Phillips, J. A. 1882. The red sands of the Arabian Desert. *Q. J. Geol. Soc. Lond.* **38**, 110–13.
Phillpot, H. R. 1985. Physical geography—climate. In *Antarctica.* W. N. Bonner & D. W. H. Walton (eds), 23–38. Oxford: Pergamon Press.
Piotrowska, H. 1989. Natural and anthropogenic changes in sand dunes and their vegetation on the southern Baltic coast. In *Perspectives on coastal dune management*, F. van der Meulen, P. D. Jungerius & J. Visser (eds), 33–40. The Hague: SPB Academic Publishing.
Pissart, A., J. S. Vincent & S. A. Edlund 1977. Dépots et phenomènés éoliens sur l'île de Banks, Térritoires du Nord-Ouest, Canada. *Can. J. Earth Sci.* **14**, 2462–80.
Pizzey, J. M. 1975. Assessment of dune stabilization at Camber, Sussex, using air photographs. *Biol. Conserv.* **7**, 275–88.
Pluis, J. L. A. & B. de Winder 1989. Spatial patterns of algae colonization of dune blowouts. *Catena* **16**, 499–506.
Porter, M. L. 1986. Sedimentary record of erg migration. *Geology* **14**, 497–500.
Porter, M. L. 1987. Sedimentology of an ancient erg margin: the Lower Jurassic Aztec Sandstone, southern Nevada and southern California *Sedimentology* **34**, 661–80.
Poupeau, G., J. H. Souza, & A. Rivera 1985. Thermoluminescence dating of Pleistocene sediments. A review of some preliminary results on sand formations from Brazil. In *Quaternary of South America and Antarctic Peninsula 3*, J. Rabassa (ed.), 9–42. Rotterdam: Balkema.
Powers, M. C. 1953. A new roundness scale for sedimentary particles. *J. Sediment. Petrol.* **23**, 117–19.
Prandtl. L. 1935. The mechanics of viscous fluids. In *Aerodynamic theory*, Vol. III, W. F. Durand (ed.), 34–208. Berlin: Springer.
Prescott, J. R. 1983. Thermoluminescence dating of sand dunes at Roonka, South Australia. *Pact* **9**, 505–12.
Prescott, J. R. V. & H. P. White 1960. Sand formations in the Niger valley between Niamey and Bourem. *Geogr. J.* **126**, 200–3.
Press, F. & R. Siever 1982. *Earth*, 3rd edn. San Francisco: Freeman.
Price, V. J. 1974. A dune is reborn. *Soil Conserv.* **39**, 4–6.
Price, W. A. 1944. Greater American deserts. *Proc. Trans. Texas Acad. Sci.* **37**, 163–70.
Price, W. A. 1950. Saharan sand dunes and the origin of the longitudinal dune: a review. *Geogr. Rev.* **40**, 462–5.
Price, W. A. 1958. Sedimentology and Quaternary geomorphology of South Texas. *Trans. Gulf Coast Assoc. Geol. Socs.* **8**, 410–75.
Prill, R. C. 1968. Movement of moisture in the unsaturated zone in a dune area, southwestern Kansas. *Prof. Pap. U.S. Geol. Surv.* 600-D, DI–D9.
Pryor, W. A. 1971. Grain shape. In *Procedures in sedimentary petrology*, R. E. Carver (ed.), 131–50. New York: Wiley-Interscience.
Pryor, W. A. 1973. Permeability–porosity patterns and variations in some Holocene sand bodies. *Am. Assoc. Petrol. Geol. Bull.* **57**, 162–89.
Psuty, N. P. (ed.) 1988. Beach-dune interaction. *J. Coastal Res. Spec. Issue* No. 3.
Pugh, J. C. 1975. *Surveying for field scientists.* London: Methuen.
Pulvertaft, T. C. R. 1985. Aeolian dune and wet interdune sedimentation in the middle Proterozoic Dala Sandstone, Sweden. *Sediment. Geol.* **44**, 93–111.
Purdie, R. 1984. *Land systems of the Simpson Desert region.* CSIRO Division of Water and Land Resources, Natural Resources Series No. 2.
Putten, W. H. van der & W. J. M. van Gulik 1987. Stimulation of vegetation growth on raised coastal fore dune ridges. *Neth. J. Agric. Sci.* **35**, 198–201.
Pye, K. 1980a. *Geomorphic evolution of coastal sand dunes in a humid tropical environment: North Queensland.* Unpubl. PhD Thesis, Univ. Cambridge.
Pye, K. 1980b. Beach salcrete and eolian sand transport: evidence from North Queensland. *J. Sediment. Petrol.* **50**, 257–61.
Pye, K. 1981. Rate of dune reddening in a humid tropical climate. *Nature* **290**, 282–4
Pye, K. 1982a. Morphological development of coastal dunes in a humid tropical environment, Cape Bedford and Cape Flattery, North Queensland. *Geogr. Ann.* **A64**, 213–27.
Pye, K. 1982b. Negatively skewed aeolian sands from a humid tropical coastal dunefield, Northern Australia. *Sediment. Geol.* **31**, 249–66.
Pye, K. 1982c. Morphology and sediments of the Ramsay Bay sand dunes, Hinchinbrook Island, North Queensland. *Proc. R. Soc. Qld.* **93**, 31–47.
Pye, K. 1982d. Characteristics and significance of some humate-cemented sands (humicretes) at Cape Flattery, North Queensland, Australia. *Geol. Mag.* **119**, 229–36.
Pye, K. 1982e. SEM observations on some sand fulgurites from Northern Australia. *J. Sediment. Petrol.* **52**, 991–8.

Pye, K. 1982f. Thermoluminescence dating of sand dunes. *Nature* **299**, 376.
Pye, K. 1983a. Formation of quartz silt during humid tropical weathering of dune sands. *Sediment. Geol.* **34**, 267–82.
Pye, K. 1983b. Formation and history of Queensland coastal dunes. *Z. Geomorph. Suppl. Bd.* **45**, 175–204.
Pye, K. 1983c. Dune formation on the humid tropical sector of the North Queensland coast, Australia. *Earth Surf. Proc. Landf.* **8**, 371–81.
Pye, K. 1983d. Coastal dunes. *Prog. Phys. Geogr.* **7**, 531–57.
Pye, K. 1983e. The coastal dune formations of northern Cape York Peninsula, Queensland. *Proc. R. Soc. Qld.* **94**, 33–9.
Pye, K. 1983f. Post-depositional modification of aeolian dune sands. In *Eolian sediments and processes*, M. E. Brookfield & T. S. Ahlbrandt (eds) 197–221. Amsterdam: Elsevier.
Pye, K. 1983g. Post-depositional reddening of late Quaternary coastal dune sands, north-eastern Australia. In *Residual deposits*, R. C. L. Wilson (ed.), 117–29. Oxford: Blackwell.
Pye, K. 1983h. Red beds. In *Chemical sediments and geomorphology*, A. S. Goudie & K. Pye (eds), 227–63. London: Academic Press.
Pye, K. 1984. Models of transgressive coastal dune building episodes and their relationship to Quaternary sea level changes: a discussion with reference to evidence from eastern Australia. In *Coastal research: U.K. perspectives*, M. Clark (ed.), 81–104. Norwich: Geo Books.
Pye, K. 1985a. Granular disintegration of gneiss and migmatites. *Catena* **12**, 191–9.
Pye, K. 1985b. Controls on fluid threshold velocity, rates of aeolian sand transport and dune grain size parameters along the Queensland coast. In *Proceeding of the international conference on the physics of blown sand,* O. E. Barndorff-Nielsen, J. T. Møller, K. R. Rasmussen & B. B. Willetts (eds), Vol. 3, 483–510. Dept. Theoretical Statistics, Institute of Mathematics, Univ. Aarhus, Mem. 8.
Pye, K. 1987. *Aeolian dust and dust deposits.* London: Academic Press.
Pye, K. 1990. Physical and human influences on coastal dune development between the Ribble and Mersey estuaries, northwest England. In *Coastal dunes: processes and morphology*, K. F. Nordstrom, N. P. Psuty & R. W. G. Carter (eds), 339–59. Chichester: Wiley, (in press).
Pye, K. & G. M. Bowman 1984. The Holocene marine transgression as a forcing function in episodic dune activity on the eastern Australian coast. In *Coastal geomorphology in Australia*, B. G. Thom (ed.), 179–96. Sydney: Academic Press.
Pye, K. & B. Jackes 1981. Vegetation of the coastal dunes at Cape Bedford and Cape Flattery, North Queensland. *Proc. R. Soc. Qld.* **92**, 37–42.
Pye, K. & D. H. Krinsley 1986. Diagenetic carbonate and evaporite minerals in Rotliegend aeolian sandstones of the southern North Sea: their nature and relationship to secondary porosity development. *Clay Min.* **21**, 441–57.
Pye, K. & A. D. M. Paine 1984. Nature and source of aeolian deposits near the summit of Ben Arkle, northwest Scotland. *Geol. Mijn.* **63**, 13–18.
Pye, K. & C. H. B. Sperling 1983. Experimental investigation of silt formation by static breakage processes: the effect of temperature, moisture and salt on quartz dune sand and granitic regolith. *Sedimentology* **30**, 49–62.
Pye, K. & V. R. Switsur 1981. Radiocarbon dates from the Cape Bedford–Cape Flattery dunefield, North Queensland. *Search* **12**, 225–6.
Pye, K. & E. G. Rhodes 1985. Holocene development of an episodic transgressive dune barrier, Ramsay Bay, North Queensland, Australia. *Marine Geol.* **64**, 189–202.
Pye, K. & H. Tsoar 1987. The mechanics and geological implications of dust transport and deposition in deserts with particular reference to loess formation and dune sand diagenesis in the northern Negev, Israel. In *Desert sediments: ancient and modern*, L. Frostick & I. Reid (eds), 139–56. Oxford: Blackwell.
Pye, K. & L-P. Zhou 1989. Late Pleistocene and Holocene aeolian dust deposition in North China and the northwest Pacific Ocean. *Palaeogeogr. Palaeoclimatol. Palaeoecol.* **73**, 11–23.
Pye, K., A. S. Goudie & A. Watson 1985. An introduction to the physical geography of the Kora area of central Kenya. *Geogr. J.* **151**, 168–81.

Quinn, C. M. 1977. *Sand dunes. Formation, erosion and management.* Dublin: An Foras Forbartha.

Rabinowicz, E. 1965. *Friction and wear of materials.* New York: Wiley.
Rae, J. 1884. Wind sand ripples. *Nature* **29**, 357.
Raheja, P. C. 1963. Shelter belts in arid climates and special techniques for tree planting. *Ann. Arid Zone* **2**, 77–82.
Ranwell, D. S. 1958. Movement of vegetated sand dunes at Newborough Warren, Anglesey. *J. Ecol.* **46**, 83–100.

Ranwell, D. S. 1959. Newborough Warren, Anglesey. I. The dune system and dune slack habitat. *J. Ecol.* **47**, 571–601.
Ranwell, D. S. 1960. Newborough Warren, Anglesey. II. Plant associes and succession cycles of the sand dune and dune slack vegetation. *J. Ecol.* **48**, 117–41.
Ranwell, D. S. 1972. *Ecology of salt marshes and sand dunes.* London: Chapman & Hall.
Ranwell, D. S. 1973. Management of salt-marsh and coastal dune vegetation. In *Estuarine research*, Vol. II. *Geology and engineering*, L. E. Cronin (eds.), 471–83. New York: Academic Press.
Ranwell, D. S. & R. Boar 1986. *Coast dune management guide.* Monks Wood Experimental Station: Institute of Terrestrial Ecology.
Rapp, A. 1974. *A review of desertification in Africa – water vegetation and man.* Stockholm: Secretariat for International Ecology, Sweden, Report No. 1.
Rasmussen, K. R. 1990. Flow over rough terrain. *Proc. R. Soc. Edinb. Biol. Sci.* in press.
Raup, H. M. & G. W. Argus 1982. *The Lake Athabasca sand dunes of northern Saskatchewan and Alberta, Canada, 1. The land and vegetation.* Ottawa: Nat. Museums of Canada Publ. Bot., No. 12.
Read, J. F. 1974. Calcrete deposits and Quaternary sediments, Edel Province, Shark Bay, Western Australia. *Am. Assoc. Petrol. Geol. Mem.* **22**, 250–82.
Readhead, M. 1988. Thermoluminescence dating study of quartz in aeolian sediments from southeastern Australia. *Quat. Sci. Rev.* **7**, 257–64.
Reeckman, S. A. & E. D. Gill 1981. Rates of vadose diagenesis in Quaternary dune and shallow marine calcarenites, Warrnambool, Victoria, Australia. *Sediment. Geol.* **30**, 157–72.
Reid, D. G. 1985. Wind statistics and the shape of sand dunes. In *Proceedings of the international workshop on the physics of blown sand,* O. E. Barndorff-Nielsen, J. T. Møller, K. R. Rasmsussen & B. B. Willetts (eds), 393–419. Dept. Theoretical Statistics, Institute of Mathematics, Univ. Aarhus, Mem. 8.
Reineck, H. E. 1955. Haftrippeln und haftwarzen, ablagerungsformen von flugsand. *Senckenbergiana Lethaea* **38**, 347–57.
Rempel, P. 1936. The crescentic dunes of the Salton Sea and their relation to vegetation. *Ecology* **17**, 347–58.
Rettger, R. E. 1935. Experiments on soft-rock deformation. *Am. Assoc. Petrol. Geol. Bull.* **19**, 271–92.
Richardson, J. G., J. B. Sangree & R. M. Sneider 1988. Aeolian dunes. *J. Petrol. Techol.*, Jan., 11–12.
Richmond, A. 1985. Desert agriculture – past and future. In *Desert development,* Y. Gradus (ed.), 167–83. Dordrecht: Reidel.
Richthofen F. von, 1882. On the mode of origin of the loess. *Geol. Mag.* **9**, 293–305.
Rigler, J. K., M. B. Collins & S. J. Williams 1981. A high precision, digital-recording sedimentation tower for sands. *J. Sediment. Petrol.* **51**, 642–3.
Riley, M. C. 1941. Projection sphericity. *J. Sediment. Petrol.* **11**, 94–7.
Rim, M. 1951. The influence of geophysical processes on the stratification of sandy soils. *J. Soil Sci.* **2**, 188–95.
Ripley, E. A. & R. E. Redmann 1976. Grassland. In *Vegetation and the atmosphere,* Vol. 2. *Case Studies*, J. L. Monteith (ed.), 349–98. London: Academic Press.
Ritchie, W. 1976. The meaning and definition of machair. *Trans. Bot. Soc. Edinb.* **42**, 431–40.
Roberts, H. H., W. Ritchie & A. Mather 1973. Cementation in high latitude dunes. *Coastal Stud. Bull.* **7**, 95–112.
Robinson, G. H. & C. I. Rich 1960. Characteristics of the multiple yellowish-red bands common to certain soils in the southeastern United States. *Soil Sci. Soc. Am. Proc.* **24**, 226–30.
Rogers, R. W. 1977. Resources and management. *Proc. Ecol. Soc. Aust.* **9**, 296–306.
Rognon, P. & M. A. J. Williams 1977. Late Quaternary climatic changes in Australia and North Africa: a preliminary interpretation. *Palaeogeogr. Palaeoclimatol. Palaeoecol* .**21**, 285–327.
Rosen, P. S. 1979. An efficient, low cost, aeolian sampling system. *Sci. Tech. Notes, Current Res. Part A, Geol. Surv. Can., Pap.* 78-1A, 531–2.
Ross, G. M. 1983a. Proterozoic aeolian quartz arenites from the Hornby Bay Group, Northwest Territories, Canada: implication for Precambrian aeolian processes. *Precambrian Res.* **20**, 149–69.
Ross, G. M. 1983b. Bigbear erg: a Proterozoic intermontane eolian sand sea in the Hornby Bay Group, Northwest Territories, Canada. In *Eolian sediments and processes*, M. E. Brookfield & T. S. Ahlbrandt (eds), 483–519. Amsterdam: Elsevier.
Ross, P. J., B. J. Bridge, I. F. Fergus, J. R. Forth, R. E. Prebble & R. Reeve 1984. *Studies in landscape dynamics in the Cooloola–Noosa River area. Queensland. 2. Field measurement techniques.* CSIRO, Division of Soils, Rep. 74.

Rossby, C. G. 1941. The scientific basis of modern meteorology. In *Climate and man, yearbook of agriculture,* 599–655. US Dept. Agriculture.
Rouse, H. 1937. Modern conceptions of the mechanics of turbulence. *Trans. Am. Soc. Civ. Eng.* **102**, 463–543.
Rubin, D. M. 1990. Lateral migration of linear dunes in the Strzelecki Desert, Australia. *Earth Surf. Proc. Landf.* **15**, 1–14.
Rubin, D. M. & R. E. Hunter 1982. Bedform climbing in theory and nature. *Sedimentology* **29**, 121–38.
Rubin, D. M. & R. E. Hunter 1983. Reconstructing bedform assemblages from compound cross-bedding. In *Eolian sediments and processes*, M. E. Brookfield & T. S. Ahlbrandt (eds), 407–27. Amsterdam: Elsevier.
Rubin, D. M. & R. E. Hunter 1984. Origin of first-order bounding surfaces — reply. *Sedimentology* **31**, 128–32.
Rubin, D. M. & R. E. Hunter 1985. Why deposits of longitudinal dunes are rarely recognized in the rock record. *Sedimentology* **32**, 147–57.
Rubin, D. M. & R. E. Hunter 1987. Bedform alignment in directionally varying flows. *Science* **237**, 276–8.
Ruegg, G. H. J. 1983. Periglacial evenly laminated sandy deposits in the late Pleistocene of N.W. Europe, a facies unrecorded in modern sedimentological handbooks. In *Eolian sediments and processes*, M. E. Brookfield & T. S. Ahlbrandt (eds), 455–82. Amsterdam: Elsevier.
Rumpel, D. A. 1985. Successive aeolian saltation: studies of idealized collisions. *Sedimentology* **32**, 267–80.
Rutin, Y. 1983. *Erosional processes on a coastal sand dune, De Blink, Noordwijkerhout, The Netherlands*. Publicaties van het Fysisch Geografisch en Bodemkundig Lab., 35.

Sabins, F. 1987. *Remote sensing — principles and interpretation.* New York: Freeman.
Sacré, C. 1981. Strong wind structure near a sea-land roughness discontinuity. *Boundary-Layer Meteorol.* **21**, 57–76.
Sagan, C. & R. A. Bagnold 1975. Fluid transport on Earth and aeolian transport on Mars. *Icarus* **26**, 209–18.
Sakamoto-Arnold, C. M. 1981. Eolian features produced by the December 1977 windstorm, southern San Joaquin Valley, California. *J. Geol.* **89**, 129–37.
Sale, G. N. 1948. Note on sand dune fixation in Palestine. *Empire For. J.* **27**, 60–1.
Salisbury, E. J. 1922. The soils of Blakeney Point: a study of soil reaction and succession in relation to the plant covering. *Ann. Bot.* **36**, 391–431.
Salisbury, E. J. 1925. Note on the edaphic succession in some dune soils with special reference to the time factor. *J. Ecol.* **13**, 322–8.
Salisbury, E. J. 1933. On the day temperatures of sand dunes in relation to the vegetation at Blakeney Point, Norfolk. *Trans. Norfolk Norwich Nat. Soc.* **13,** 333–55.
Salisbury, E. J. 1952. *Downs and Dunes*. London: Bell.
Sallenger, A. H., Jr 1979. Inverse grading and hydraulic equivalence in grain flow deposits. *J. Sediment. Petrol.* **49**, 553–62.
Saltiel, M. 1963. *Sand dune stabilization for the protection of engineering structures*. Tahal, Water Planning for Israel, Govt. Israel, UN Special Fund, FOA, Tel Aviv Technical Report No. 2.
Samways, J. 1976. Ill wind over Africa. *Geogr. Mag.* **48**, 218–20.
Sarnthein, M. 1978. Sand deserts during glacial maximum and climatic optimum. *Nature* **272**, 43–6.
Sarnthein, M. & L. Diester-Haas 1977. Eolian-sand turbidites. *J. Sediment. Petrol.* **47**, 868–90.
Sarnthein, M., M. G. Tetzlaaf, B. Koopman, K. Wolter & U. Plaumann 1981. Glacial and interglacial wind regimes over the eastern sub-tropical Atlantic and northwest Africa. *Nature,* **293**, 193–6.
Sarre, R. D. 1987. Aeolian sand transport. *Prog. Phys. Geogr.* **11**, 157–82.
Sarre, R. D. 1988. Evaluation of aeolian sand transport equations using intertidal zone measurements, Saunton Sands, England. *Sedimentology* **35**, 671–9.
Sarre, R. D. 1989. Aeolian sand drift from the intertidal zone on a temperate beach: potential and actual rates. *Earth Surf. Proc. Landf.* **14**, 247–58.
Satoh, I. 1967. Studies on some peculiar environmental factors related to cultivation in sand dune field. *J. Fac. Agric. Tottori Univ.* **5**, 1–41.
Savage, R. P. 1963. Experimental study of dune building with sand fences. *Proc. 8th Coastal Eng. Conf., Mexico City*, 380–96.
Savage, R. P. & W. W. Woodhouse 1969. Creation and stabilization of coastal barrier dunes. *Proc. 11th Coastal Eng. Conf., London* **1**, 671–700.
Sayles, R. W. 1931. Bermuda during the Ice Age. *Proc. Am. Acad. Arts Sci.* **66**, 381–468.

Schempf, W. H. 1943. On Haboob in the Egyptian Sudan. *Bull. Am. Met. Soc.* **24**, 371–7.
Schenk, C. J. & S. G. Fryberger 1988. Early diagenesis of eolian dunes and interdune sands at White Sands, New Mexico. *Sediment. Geol.* **55**, 109–20.
Schlee, J., E. Uchupi & J. V. A. Trumbull 1965. Statistical parameters of Cape Cod beach and eolian sands. *US Geol. Surv. Prof. Pap.* 501D, 118–22.
Schofield, J. C. 1975. Sea level fluctuations cause periodic post-glacial progradation, South Kaipara barrier, North Island, New Zealand. *N.Z. J. Geol. Geophys.* **18**, 295–316.
Schroeder, J. H. 1985. Eolian dust in the coastal desert of the Sudan: aggregates cemented by evaporites. *J. Afr. Earth Sci.* **3**, 370–86.
Schwan, J. 1986. The origin of horizontal alternating bedding in Weichselian aeolian sands in northwestern Europe. *Sediment. Geol.* **49**, 73–108.
Schwan, J. 1987. Sedimentologic characteristics of a fluvial to aeolian succession in Weichselian Talsand in Emsland (F.R.G.). *Sediment. Geol.* **52**, 273–90.
Schwan, J. 1988. The structure and genesis of Weichselian to Early Holocene aeolian sand sheets in Western Europe. *Sediment. Geol.* **55**, 197–232.
Schwan, J. 1989. Grain fabrics of natural and experimental low-angle aeolian sand deposits. *Geol. Mijn.* **68**, 211–19.
Scoček, V. & A. A. Saadallah 1972. Grain size distribution, carbonate content and heavy minerals in eolian sands, southern desert, Iraq. *Sediment. Geol.* **8**, 29–46.
Seely, M. K. 1984. The Namib's place among deserts of the world. *S. Afr. J. Sci.* **80**, 155–8.
Seely, M. K. & G. N. Louw 1980. First approximation of the effect of rainfall on the ecology and energetics of a Namib desert dune ecosystem. *J. Arid Environ.* **3**, 25–54.
Seely, M. K. & B. H. Sandelowsky 1974. Dating the regression of a river's end point. *Bull. S. Afr. Archaeol.* **2**, 61–4.
Selby, M. J., R. B. Rains & R. W. P. Palmer 1974. Eolian deposits of the ice-free Victoria Land, Antarctica. *N.Z. J. Geol. Geophys.* **17**, 543–62.
Selim, A. A. 1974. Origin and lithification of the Pleistocene carbonates of the Salum area, western coastal plain of Egypt. *J. Sediment. Petrol.* **44**, 70–8.
Semeniuk, V. 1986. Holocene history of coastal southwestern Australia using calcrete as an indicator. *Palaeogeogr. Palaeoclimatol. Palaeoecol.* **53**, 289–308.
Semeniuk, V. & T. D. Meagher 1981. Calcrete in Quaternary coastal dunes in southwestern Australia: a capillary rise phenomenon associated with plants. *J. Sediment. Petrol.* **51**, 47–68.
Semeniuk, V. & D. J. Searle 1985. Distribution of calcrete in Holocene coastal sands in relationship to climate, southwestern Australia. *J. Sediment. Petrol.* **55**, 86–95.
Sengupta, S. & H. J. Veenstra 1968. On sieving and settling techniques for sand analysis. *Sedimentology* **11**, 83–98.
Seppala, M., 1972. Location morphology and orientation of inland dunes in northern Sweden. *Geogr. Ann.* **54A**, 85–104.
Seppala, M. & K. Lindé 1978. Wind tunnel studies of ripple formation. *Geogr. Ann.* **60A**, 29–42.
Serra, J. 1982. *Image analysis and mathematical morphology*. London: Academic Press.
Setlow, L. W. 1978. Age determination of reddened coastal dunes in northwest Florida, USA, by use of scanning electron microscopy. In *Scanning electron microscopy in the study of sediments*, W. B. Whalley (ed.), 283–305. Norwich: Geo Abstracts.
Sharp, R. P. 1963. Wind ripples. *J. Geol.* **71**, 617–36.
Sharp, R. P. 1964. Wind-driven sand in Coachella Valley, California. *Bull. Geol. Soc. Am.* **75**, 785–804.
Sharp, R. P. 1966. Kelso dunes, Mojave Desert, California. *Bull. Geol. Soc. Am.* **77**, 1045–73.
Sharp, R. P. 1979. Intradune flats of the Algodones Chain, Imperial Valley, California. *Bull. Geol. Soc. Am.* **90**, 908–16.
Sharp, R. P. 1980. Wind-driven sand in Coachella Valley, California: further data. *Bull. Geol. Soc. Am.* **91**, 724–30.
Sharp, R. P. & D. L. Carey, 1976. Sliding stones, Racetrack Playa, California. *Bull. Geol. Soc. Am.* **87**, 1704–17.
Shaw, N. 1936. *Manual of Meteorology Vol. 2 Comparative Meteorology*, 2nd edn. Cambridge: Cambridge Univ. Press.
Shepard, F. P. & R. Young 1961. Distinguishing between beach and dune sands. *J. Sediment. Petrol.* **31**, 196–214.
Shideler, G. L. & K. P. Smith 1984. Regional variability of beach and foredune characteristics along the Texas Gulf Coast barrier system. *J. Sediment. Petrol.* **54**, 507–26.
Shields, A. 1936. *Application of similarity principles and turbulence research to bed-load movement.* Translation of *Mitteilungen der preussischen Versuchsanstalt für Wasserbau und Schiffbau.* W. P.

Ott & J. C. van Wehelen (translators), California Inst. Technol. Hydrodynamic Lab., Publ. No. 167.
Shields, L. M., C. Mitchell & F. Drouet 1957. Algae and lichen-stabilized surface crusts as soil nitrogen sources. *Am. J. Bot.* **44**, 489–98.
Shikula, N. K. 1981. Prediction of dust storms from meteorological observations in the South Ukraine, U.S.S.R. In *Desert dust: origin, characteristics, and effect on man*, T. L. Péwé (ed.), 261–6. Geol. Soc. Am. Spec. Pap. No. 186.
Shoji, K. 1977. Drip irrigation. *Sci. Am.* **237**, November, 62–8.
Short, A. & P. Hesp 1982. Wave, beach and dune interactions in southeastern Australia. *Mar. Geol.* **48**, 259–84.
Shotton, F. W. 1937. The Lower Bunter Sandstone of north Worcestershire and east Shropshire. *Geol. Mag.* **74**, 534–53.
Shreve, F. 1938. The sandy areas of the North American desert. *Assoc. Pacific Geogr. Yearbook* **4**, 11–14.
Sidhu, P. S. 1977. Aeolian additions to soils of northwest India. *Pedologie* **27**, 323–36.
Sidwell, R. & W. F. Tanner 1939. Sand grain patterns of west Texas dunes. *Am. J. Sci.* **237**, 181–7.
Siever, R. 1962. Silica solubility 0 °C–200 °C and the diagenesis of siliceous sediments. *J. Geol.* **70**, 127–50.
Simonett, D. S. 1960. Development and grading of dunes in western Kansas. *Ann. Assoc. Am. Geogr.* **50**, 216–41.
Simons, D. B., E. V. Richardson & C. F. Nordin 1965. Sedimentary structures generated by flow in alluvial channels. In *Sedimentary structures and their hydrodynamic interpretation*, G. V. Middleton (ed.), S.E.P.M. Spec. Publ. 12, 34–52. Tulsa: Soc. Econ. Paleontol. Mineral.
Simons, F. S. 1956. A note on Pur-Pur dune, Viru Valley, Peru. *J. Geol.* **64**, 517–21.
Simpson, E. L. & D. B. Loope 1985. Amalgamated interdune deposits, White Sands, New Mexico. *J. Sediment. Petrol.* **55**, 361–6.
Sinclair, J. G. 1922. Temperature of the soil and air in a desert. *Monthly Weather Rev.* **50**, 142–4.
Sinclair, P. C. 1964. Some preliminary dust devil measurements. *Monthly Weather Rev.* **92**, 363–7.
Sinclair, P. C. 1969. General characteristics of dust devils. *J. Appl. Meteorol.* **8**, 32–45.
Singhvi, A. K., Y. P. Sharma & D. P. Agrawal 1982. Thermoluminescence dating of sand dunes in Rajasthan, India. *Nature* **195**, 313–15.
Singhvi, A. K., S. U. Deraniyagala & D. Sengupta 1986. Thermoluminescence dating of Quaternary red sand beds: a case study of coastal dunes in Sri Lanka. *Earth Planet. Sci. Lett.* **80**, 139–44.
Slater, J. & B. Ralph 1976. The determination of particle shape and size distributions using automatic image analysis techniques. *Proceedings of the 4th International Congress on Stereology, Gaithersburg*, 177–80. US Natl. Bureau of Standards Spec. Publ. 431.
Sless, J. B. 1958. Coastal sand drift. Part II. Control measures. *J. Soil Conserv. Serv. N.S.W.* **14**, 50–68.
Smedman, A. S. & H. Bergstrom 1984. Flow characteristics above a very low and gently sloping hill. *Boundary-Layer Meteorol.* **29**, 21–37.
Smith, D. M., C. R. Twidale & J. A. Bourne 1975. Kappakoola dunes — aeolian landforms induced by man. *Aust. Geogr.* **13**, 90–6.
Smith, H. T. U. 1943. The physics of blown sand and desert dunes by R. A. Bagnold, a Review. *Geogr. Rev.* **33**, 170–2.
Smith, H. T. U. 1946. Sand dunes. *Trans. New York Acad. Sci., Ser. 2* **8**, 197–9.
Smith, H. T. U. 1953. Classification of sand dunes (abstr.). *C.R. 19th Int. Geol. Congr. Algiers, 1952*, 105.
Smith H. T. U. 1954. Eolian sand on desert mountains. *Bull. Geol. Soc. Am.* **65**, 1036–7.
Smith, H. T. U. 1956. Giant composite barchans of the northern Peruvian desert *Bull. Geol. Soc. Am.* **67**, 1735.
Smith, H. T. U. 1963. *Eolian geomorphology, wind direction and climatic change in north Africa.* Final report, Bedford Geophysics Research Directorate, Air Force Cambridge Research Laboratories, AFCRL-63-443.
Smith, H. T. U. 1965. Wind-formed pebble ripples in Antarctica. *Spec. Pap. Geol. Soc. Am.* **87**, 160.
Smith, H. T. U. 1968. Nebraska dunes compared with those of North Africa and other regions. In *Loess and related eolian deposits of the world*, C. B. Schultz & J. C. Frye (eds), 29–42. Lincoln: Univ. Nebraska Press.
Smith, H. T. U. 1969. *Photo-interpretation studies of desert basins in Northern Africa.* US Air Force, Office of Aerospace Research, Cambridge Research Laboratories, Bedford, Mass., Final report, AFCRKL-68-0590.
Smith, J. D. 1970. Stability of a sand bed subjected to a shear flow of low Froude Number. *J. Geophys. Res.* **75**, 5928–40.

Smith, R. S. U. 1978. Field trip to dunes at Superstition Mountain. In *Aeolian features of southern California: a comparative planetary geology guide book*, R. Greeley, M. B. Womer, R. P. Papson & P. D. Spudis (eds), 66–71. Arizona State Univ., College of the Desert, and NASA — Ames Res. Center.

Smith, R. S. U. 1982. Sand dunes in the North American desert. In *Reference handbook on the deserts of North America*, G. L. Bender (ed.), 481–526. Westport: Greenwood Press.

Smith, W. O., H. W. Olsen, R. A. Bagnold & J. C. Rice 1968. Certain flows of air and water in sands during infiltration. *Soil Sci.* **101**, 441–9.

Sneed, E. D. & R. L. Folk 1958. Pebbles in the lower Colorado River, Texas, a study in particle morphogenesis. *J. Geol.* **66**, 114–50.

Sneh, A. 1982. Drainage systems of the Quaternary in Northern Sinai with emphasis on Wadi El-Arish *Z. Geomorph. N.F.* **26**, 179–95.

Sneh, A. 1988. Permian dune patterns in northwestern Europe challenged. *J. Sediment. Petrol.* **58**, 44–51.

Sneh, A. & T. Weissbrod 1983. Size-frequency distributions of longitudinal dune rippled flank sands compared to that of slip face sands of various dune types. *Sedimentology* **30**, 717–25.

Sokolow, N. A. 1894. *Die dunen*. Berlin: Bildung, Entwicklung und innerer Ban.

Sorby, H. C. 1877. The application of the microscope to geology. *Mon. Micros. J.* **17**, 113–36.

Sorby, H. C. 1880. On the structure and recognition of non-calcareous stratified rocks. *Proc. Geol. Soc. Lond.* **36**, 46–92.

Sørensen, M. 1988. *Radioactive tracer studies of grain progress in aeolian sand transport: a statistical analysis*. Dept. Theoretical Statistics, Institute of Mathematics, Univ. Aarhus, Res. Rep. 141.

Sperling, C. H. B. & A. S. Goudie 1975. The miliolite of western India: a discussion of the aeolian and marine hypotheses. *Sediment. Geol.* **13**, 71–5.

Stallings, J. H. 1953. *Wind erosion control*. Washington, DC: US Dept. Agriculture, Soil Conservation Service, SCS-TP-115.

Stanton, T. E. 1911. The mechanical viscosity of fluids. *Proc. R. Soc. London, Ser. A* **85**, 366–76.

Stapor, I. W., J. P. May & J. Barwis 1983. Eolian shape sorting and aerodynamic traction equivalence in the coastal dunes of Hout Bay, Republic of South Africa. In *Eolian sediments and processes*, M. E. Brookfield & T. S. Ahlbrandt (eds), 149–64. Amsterdam: Elsevier.

St. Arnaud, R. J. & E. P. Whiteside 1963. Physical breakdown in relation to soil development. *J. Soil Sci.* **14**, 267–81.

Steele, R. P. 1983. Longitudinal draa in the Permian Yellow Sands of northeast England. In *Eolian sediments and processes*, M. E. Brookfield & T. S. Ahlbrandt (eds), 543–50. Amsterdam: Elsevier.

Steele, R. P. 1985. Early Permian (Rotliegendes) palaeowinds of the North Sea — comment. *Sediment. Geol.* **45**, 293–7.

Steidtmann, J. R. 1973. Ice and snow in eolian sand dunes of southwestern Wyoming. *Science* **179**, 794–8.

Steidtmann, J. R. 1982. Structures in the moist, cold-climate sand dunes of southwestern Wyoming. *Geol. Soc. Am. Spec. Pap.* **192**, 83–7.

Steinen, R. P. 1974. Phreatic and vadose diagenetic modification of Pleistocene limestone: petrographic observations from sub-surface Barbados, West Indies. *Am. Assoc. Petrol. Geol. Bull.* **58**, 1008–24.

Stembridge, J. E. 1978. Vegetated coastal dunes: growth detected from aerial infrared photography. *Remote Sensing Environ.* **7**, 73–6.

Stepanov, A. M. 1971. The effect of shelter on the temperature and moisture of desert sands. *Sov. Geogr.* **12**, 695–701.

Stephens, C. G. & R. L. Crocker 1946. Composition and genesis of lunettes. *Trans. R. Soc. S. Austr.* **70**, 303–12.

Stevens, J. H. 1974. Sand stabilization in Saudi Arabia's Al-Hasa Oasis. *J. Soil Water Conserv.* **29**, 129–33.

Stokes, W. L. 1968. Multiple truncation bedding planes — a feature of wind deposited sandstone formations. *J. Sediment. Petrol.* **38**, 510–15.

Stone, R. O. 1967. A desert glossary. *Earth-Sci. Rev.* **3**, 211–68.

Strahler, A. N. 1952. Dynamic basis of geomorphology. *Bull. Geol. Soc. Am.* **63**, 923–38.

Striem, H. L. 1954. The seifs on the Israeli–Sinai border and the correlation of their alignment. *Bull. Res. Council. Israel.* **4**, 195–8.

Stuiver, M. 1978. Carbon-14 dating: a comparison of beta and ion counting. *Science* **202**, 881–3.

Stuiver, M., C. J. Heusser & I. C. Yang 1978. North American glacial history extended to 75,000 years ago. *Science* **200**, 16–21.

Sundborg, A. 1955. Meteorological and climatological conditions for the genesis of aeolian sediments. *Geogr. Ann.* **37**, 94–111.

Sung-Chiao, C. 1984. The sandy deserts and the Gobi of China. In *Deserts and arid lands*, F. El-Baz (ed.), 95–113. The Hague: Martinus Nijhoff.
Sutton, J. C. & B. R. Sheppard 1975. Aggregation of sand dune soil by endo-mycorrhizal fungi. *Can J. Bot.* **54**, 326–33.
Sutton, L. J. 1925. Haboobs. *Q. J. R. Meteorol. Soc.* **51**, 25–30.
Sutton, L. J. 1931. Haboobs. *Q. J. R. Meteorol. Soc.* **57**, 143–61.
Sutton, O. G. 1934. Wind structure and evaporation in a turbulent atmosphere. *Proc. R. Soc. London, Ser. A* **146**, 701–22.
Sutton, O. G. 1953. *Micrometeorology*. New York: McGraw-Hill.
Svasek, J. N. & J. H. J. Terwindt 1974. Measurements of sand transport by wind on a natural beach. *Sedimentology* **21**, 311–22.
Swan, B. 1979. Sand dunes in the humid tropics: Sri Lanka. *Z. Geomorph.* **23**, 152–71.
Sweet, M. L., J. Nielson, K. Havholm & J. Farralley 1988. Algodones dune field of southeastern California: case history of a migrating modern dune field. *Sedimentology* **35**, 939–52.
Sykes, R. I. 1980. An asymptotic theory of incompressible flow over a small hump. *J. Fluid Mech.* **101**, 647–70.

Tag El Din, S. S. 1986. Some aspects of sand stabilization in Egypt. In *Physics of desertification*, F. El-Baz & M. H. A. Hassan (eds), 118–26. Dordrecht: Martinus Nijhoff.
Talbot, M. R. 1980. Environmental responses to climate change in the West African Sahel over the past 20,000 years. In *The Sahara and the Nile*, M. A. J. Williams & H. Faure (eds), 37–62. Rotterdam: Balkema.
Talbot, M. R. 1984. Late Pleistocene dune building and rainfall in the Sahel. *Palaeoecol. Afr.* **16**, 203–14.
Talbot, M. R. 1985. Major bounding surfaces in aeolian sandstones – a climatic model. *Sedimentology* **32**, 257–65.
Talbot, M. R. & M. A. J. Williams 1978. Erosion of fixed dunes in the Sahel, Central Niger. *Earth Surf. Proc.* **3**, 107–14.
Talbot, M. R. & M. A. J. Williams 1979. Cyclic alluvial fan sedimentation on the flanks of fixed dunes, Janjan, central Niger. *Catena* **6**, 433–62.
Tanner, W. F. 1967. Ripple mark indices and their uses. *Sedimentology* **9**, 89–104.
Tansley, A. G. 1954. *Introduction to plant ecology*. London: Allen & Unwin.
Taylor, P. A. & P. R. Gent 1974. A model of atmospheric boundary layer flow above an isolated two-dimensional hill; an example of flow above gentle topography. *Boundary Layer Meteorol.* **7**, 349–62.
Tear, F. J. 1925. Sand dune reclamation in Palestine. *Empire For. J.* **4**, 24–38.
Tear, F. J. 1927. Sand dune reclamation in Palestine. *Empire For. J.* **6**, 85–93.
Tennekes, H. 1973. The logarithmic wind profile. *J. Atmos. Sci.* **30**, 234–8.
Thaarup, P. 1954. The afforestation of the sand dunes of the western coast of Jutland. *Adv. Sci.* **11**, 38–41.
Thatcher, A. C. & W. E. Westman 1975. Succession following mining on high dunes of coastal south-east Queensland. *Proc. Ecol. Soc. Aust.* **9** 17–33.
Thom, A. S. 1975. Momentum, mass and heat exchange of plant communities. In *Vegetation and the atmosphere* Vol. 1.*Principles*, J. L. Monteith (ed.), 57–109. London: Academic Press.
Thom, B. G. 1978. Coastal sand deposition in southeast Australia during the Holocene. In *Landform evolution in Australasia*, J. L. Davies & M. A. J. Williams (eds), 197–214. Canberra: ANU Press.
Thom, B. G., G. M. Bowman & P. S. Roy 1981. Late Quaternary evolution of coastal sand barriers, Port Stephens–Myall Lake area, Central New South Wales, Australia. *Quat. Res.* **15**, 345–64.
Thom, H. C. S. 1954. Frequency of maximum wind speeds. *Proc. Am. Soc. Civ. Eng.* **89** (530), 1–11.
Thomas, B. S. 1932. *Arabia Felix: across the empty quarter of Arabia*. London: Jonathan Cape.
Thomas, D. S. G. 1984. Ancient ergs of the former arid zones of Zimbabwe, Zambia and Angola. *Trans. Inst. Br. Geog. N.S.* **9**, 75–88.
Thomas, D. S. G. 1986a. Dune pattern statistics applied to the Kalahari dune desert, Southern Africa. *Z. Geomorph.* **30**, 231–42.
Thomas, D. S. G. 1986b. The response diagram and desert sands — a note. *Z. Geomorph.* **30**, 363–9.
Thomas, D. S. G. 1987a. Discrimination of depositional environments using sedimentary characteristics in the Mega-Kalahari, central southern Africa. In *Desert sediments: ancient and modern*, L. Frostick & I. Reid (eds), 293–306. Oxford: Blackwell.
Thomas, D. S. G. 1987b. The roundness of aeolian quartz sand grains. *Sediment. Geol.* **52**, 149–53.

Thomas, D. S. G. 1988a. The nature and deposition setting of arid and semi-arid Kalahari sediments, southern Africa. *J. Arid Environ.* **14**, 17–26.
Thomas, D. S. G. 1988b. The geomorphological role of vegetation in the dune systems of the Kalahari. In *Geomorphological studies in South Africa*, G. F. Dardis & B. P. Moon (eds), 145–58. Rotterdam: Balkema.
Thomas, D. S. G. 1988c. Analysis of linear dune sediment–form relationships in the Kalahari dune desert. *Earth Surf. Proc. Landf.* **13**, 545–53.
Thomas, D. S. G. (ed.) 1989a. *Arid zone geomorphology*. London: Belhaven Press.
Thomas, D. S. G. 1989b. Aeolian sand deposits. In *Arid zone geomorphology*, D. S. G. Thomas (ed.), 232–61. London: Belhaven Press.
Thomas, D. S. G. & A. S. Goudie 1984. Ancient ergs of the southern hemisphere. In *Late Cainozoic palaeoclimates of the southern hemisphere*, J. C. Vogel (ed.), 407–18. Rotterdam: Balkema.
Thomas, D. S. G. & H. E. Martin 1987. Grain size characteristics of linear dunes in the southeastern Kalahari — a discussion. *J. Sediment. Petrol.* **57**, 572–3.
Thomas, D. S. G. & H. Tsoar 1990. The geomorphological role of vegetation in desert dune systems. In *Vegetation and erosion,* J. B. Thornes (ed.), in press. Chichester: Wiley.
Thomas, H. H. 1921. Some observations on plants in the Libyan desert. *J. Ecol.* **9**, 75–89.
Thomas, P. & P. J. Gierasch 1985. Dust devils on Mars. *Science* **230**, 175–7.
Thompson, C. H. 1981. Podzol chronosequences on coastal dunes in eastern Australia. *Nature* **291**, 59–61.
Thompson, C. H. 1983. Development and weathering of large parabolic dune systems along the subtropical coast of eastern Australia. *Z. Geomorph. Suppl. Bd.* **45**, 205–25.
Thompson, C. H. & G. M. Bowman 1984. Subaerial denudation and weathering of vegetated coastal dunes in eastern Australia. In *Coastal geomorphology in Australia,* B. G. Thom (ed.), 263–90. Sydney: Academic Press.
Thornthwaite, C. W. 1931. The climate of North America according to a new classification. *Geogr. Rev.* **21**, 633–55.
Threat, R. L. 1959. The wind at work. *Nat. Hist.* **68**, 257–65.
Tinley, K. L. 1985. *Coastal dunes of South Africa.* South African Nat. Sci. Programmes Rep. 109. Pretoria: Council for Scientific and Industrial Research.
Townsend, A. A. 1967. Wind and the formation of inversion. *Atmos. Environ.* **1**, 173–5.
Townsend, C. W. 1925. *Sand dunes and salt marshes.* Boston: L. C. Page.
Trask, P. D. 1930. Mechanical analysis of sediment by centrifuge. *Econ. Geol.* **25**, 581–99.
Trew, M. J. 1973. The effect and management of trampling on sand dunes. *J. Environ. Plann. Pollut. Control* **1**, 131–49.
Tricart, J. 1954. Influence des sols salés sur la déflation éolienne en basse Mauritanie et dans la Delta du Senegal. *Rév. Géomorph. Dyn.* **5**, 124–32.
Tricart, J. 1974. Existence de periodes sèches au Quaternaire en Amazonie et dans les régions voisines. *Rév. Géomorph. Dyn.* **23**, 145–58.
Tseo, G. K. Y. 1986. *Longitudinal dunes: their genesis and ordering.* Unpubl. PhD Thesis, Univ. Adelaide.
Tsoar, H., 1974. Desert dunes morphology and dynamics, El-Arish (Northern Sinai) *Z. Geomorph. Suppl. Bd.* **20**, 41–61.
Tsoar, H. 1975. Specific sampling of ripples and microfeatures on desert dunes. *Proc. 9th Int. Congr. Sedimentol., Nice,* **3**, 101–5.
Tsoar, H. 1976. Characterization of sand dune environments by their grain-size, mineralogy and surface texture. In *Geography in Israel,* D. H. K. Amiran & Y. Ben-Arieh (eds), 327–43. Jerusalem: Israel National Committee, International Geographical Union.
Tsoar, H. 1978. *The dynamics of longitudinal dunes. Final technical report.* London: European Research Office, US Army.
Tsoar, H. 1982. Internal structure and surface geometry of longitudinal (seif) dunes. *J. Sediment. Petrol.* **52**, 823–31.
Tsoar, H. 1983a. Dynamic processes acting on a longitudinal (seif) sand dune. *Sedimentology* **30**, 567–78.
Tsoar, H. 1983b. Wind tunnel modelling of echo and climbing dunes. In *Eolian sediments and processes,* M. E. Brookfield & T. S. Ahlbrandt (eds), 247–59. Amsterdam: Elsevier.
Tsoar, H. 1984. The formation of seif dunes from barchans — a discussion. *Z. Geomorph.* **28**, 99–103.
Tsoar, H. 1985. Profile analysis of sand dunes and their steady state significance. *Geogr. Ann.* **67A**, 47–59.
Tsoar, H. 1986. Two-dimensional analysis of dune profile and the effect of grain size on sand dune morphology. In *Physics of desertification,* F. El-Baz & M. H. A. Hassan (eds), 94–108. Dordrecht: Martinus Nijhoff.

Tsoar, H. 1989. Linear dunes – forms and formation. *Prog. Phys. Geogr.* **13**, 507–28.
Tsoar, H., 1990a. Grain size characteristics of wind ripples on a desert seif dune. *Geogr. Res. Forum* **10**, 37–50.
Tsoar, H. 1990b. The ecological background, deterioration and reclamation of desert dune sand. *Agric. Ecosys. Environ.* in press.
Tsoar, H. & J. T. Møller 1986. The role of vegetation in the formation of linear sand dunes. In *Aeolian geomorphology,* W. G. Nickling (ed.), 75–95. Boston: Allen and Unwin.
Tsoar, H. & K. Pye 1987. Dust transport and the question of desert loess formation. *Sedimentology* **34**, 139–53.
Tsoar, H. & D. H. Yaalon 1983. Deflection of sand movement on a sinuous longitudinal (seif) dune: use of fluorescent dye as a tracer. *Sediment. Geol.* **36**, 25–39.
Tsoar, H. & Y. Zohar 1985. Desert dune sand and its potential for modern agricultural development. In *Desert Development*, Y. Gradus (ed.), 184–200. Dordrecht: Reidel.
Tsoar, H., R. Greeley & A. R. Peterfreund 1979. Mars: the north polar sand sea and related wind patterns. *J. Geophys. Res.* **84**, 8167–80.
Tsoar, H., K. R. Rasmussen, M. Sørensen & B. B. Willetts 1985. Laboratory studies of flow over dunes. In *Proceedings of the international workshop on the physics of blown sand,* O. E. Barndorff-Nielsen, J. T. Møller, K. R. Rasmussen & B. B. Willetts (eds), 327–49. Dept. Theoretical Statistics, Institute of Mathematics, Univ. Aarhus, Mem. 8.
Tsuchiya, Y. 1970. Successive saltation of a sand grain by wind. *Proc. 12th Conf. Coastal Eng.* **1**, 1417–27.
Tsuriell, D. E. 1974a. Introductory remarks. *Int. J. Biometeorol.* **18**, 85–8.
Tsuriell, D. E. 1974b. Sand dune stabilization in Israel. *Int. J. Biometeorol.* **18**, 89–93.
Twenhofel, W. H. 1950. *Principles of sedimentation.* New York: McGraw-Hill.
Twidale, C. R. 1972a. Evolution of sand dunes in the Simpson Desert, Central Australia. *Trans. Inst. Br. Geogr.* **56**, 77–109.
Twidale, C. R. 1972b. Landform development in the Lake Eyre region, Australia. *Geogr. Rev.* **62**, 40–70.
Twidale, C. R. 1980. The Simpson Desert, Central Australia. *S. Afr. Geogr. J.* **62**, 3–17.
Twidale, C. R. 1981. Age and origin of longitudinal dunes in the Simpson and other sand ridge deserts. *Erde* **112**, 231–41.

Udden, J. A. 1894. Erosion, transportation and sedimentation performed by the atmosphere. *J. Geol.* **2**, 318–31.
Udden, J. A. 1896. Dust and sand storms in the West. *Pop. Sci. Month.* **49**, 655–64.
Udden, J. A. 1898. The mechanical composition of wind deposits. *Publ. Augustana Lib.* No. 1.
Udden, J. A. 1914. Mechanical composition of some clastic sediments. *Bull. Geol. Soc. Am.* **25**, 655–744.
Ungar, J. E. & P. K. Haff 1987. Steady state saltation in air. *Sedimentology* **34**, 289–99.

Van Burkalow, A. 1945. The angle of repose and angle of sliding friction: an experimental study. *Bull. Geol. Soc. Am.* **56**, 669–708.
Van Straaten, L. M. J. U. 1953. Rhythmic patterns on Dutch North Sea beaches. *Geol. Mijn.* **15**, 31–43.
Vehrencamp, J. E. 1953. Experimental investigation of heat transfer at an air–earth interface. *Trans. Am. Geophys. Union* **34**, 22–30.
Vellinga, P. 1978. Movable bed tests on dune erosion. *Proc. 16th Int. Conf. Coast. Eng., Hamburg*, Vol. II, 2020–37.
Vellinga, P. 1982. Beach and dune erosion during storm surges. *Coastal Eng.* **6**, 361–87.
Verstappen, H. T. 1968. On the origin of longitudinal (seif) dunes. *Z. Geomorph.* **12**, 200–20.
Verstappen, H. T. 1970. Aeolian geomorphology of the Thar Desert and palaeo-climates. *Z. Geomorph. Suppl. Bd.* **10**, 104–20.
Verstappen, H. T. 1972. On dune types, families and sequences in areas of unidirectional winds. *Gottingen Geogr. Abh.* **60**, 341–53.
Vincent P. J. 1988. The response diagram and sand mixtures. *Z. Geomorph.* **32**, 221–6.
Vincent, P. J. 1984. Particle size variation over a transverse dune in the Nafud as Sirr, Central Saudi Arabia. *J. Arid Environ.* **7**, 329–36.
Vincent, P. J. 1985. Some Saudi Arabian dune sands: a note on the use of the response diagram. *Z. Geomorph.* **29**, 117–22.
Vincent, P. J. 1986. Differentiation of modern beach and coastal dune sands — a logistic regression approach using the parameters of the hyperbolic function. *Sediment. Geol.* **49**, 167–76.

Visher, G. S. 1969. Grain size distributions and depositional processes. *J. Sediment. Petrol.* **39**, 1074–1106.

Vugts, H. F. & F. Cannemeijer 1981a. Measurement of drag coefficients and roughness length at a sea–beach interface. *J. Appl. Meteorol.* **20**, 335–400.

Vugts, H. F. & F. Cannemeijer 1981b. Interaction between wind and sand surface. *Geol. Mijn.* **60**, 395–9.

Waals, L. van der 1967. Morphological phenomena on quartz grains in unconsolidated sands due to migration of quartz near the earth's surface. *Meded. Neth. Geol. Sticht. N.S.* **18**, 47–51.

Wadell, H. 1933. Sphericity and roundness of rock particles. *J. Geol.* **40**, 443–51.

Walker, A. S. 1982. Deserts of China, *Am. Sci.* **70**, 366–76.

Walker, A. S. 1986. Eolian landforms. In *Geomorphology from space,* N. M. Short and R. W. Blair (eds), 447–520. Washington: NASA SP–486.

Walker, A. S., J. W. Olsen & X. Bagen 1987. The Badain Jaran Desert: remote sensing investigations. *J. Geogr.* **153**, 205–10

Walker, H. J. & Y. Matsukura 1979. Barchans and barchan-like dunes as developed in two contrasting areas with restricted source regions. *Ann. Rep. Inst. Geosci. Univ. Tsukuba* **5**, 43–6.

Walker, J. D. 1981. *An experimental study of wind ripples.* Unpubl. MSc Thesis, Mass. Inst. Technol.

Walker, J. D. & J. B. Southard 1982. Experimental study of wind ripples. *Abstr. 11th Int. Assoc. of Sedimentol. Congress, Hamilton, Ontario,* 65.

Walker, T. R. 1976. Diagenetic origin of continental red beds. In H. Falke (ed.) *The continental Permian in central, west and south Europe.* Dordrecht: Reidel, 240–82.

Walker, T. R. 1979. Red color in dune sand. *US Geol. Surv. Prof. Pap.* 1052, 62–81.

Walls, J. (ed.) 1982. Stabilizing sand dunes with vegetation in combating desertification in China. *UNEP Reports and Proceedings*, Ser. 3, 36–42. Nairobi: UNEP.

Walmsley, J. L., J. R. Salmon & P. R. Taylor 1982. On the application of a model of boundary-layer flow over low hills to real terrain. *Boundary-Layer Meteorol.* **23**, 17–46.

Walmsley, J. L. & A. D. Howard 1985. Application of a boundary-layer model to flow over an eolian dune. *J. Geophys. Res.* **90**, 10631–40.

Walter, G. 1973. *Vegetation of the Earth.* London: English Univ. Press.

Walton, E. K., W. E. Stephens & M. S. Shawa 1980. Reading segmented grain-size curves. *Geol. Mag.* **117**, 517–24.

Ward, A. W., & R. Greeley 1984. The yardangs at Rogers Lake, California. *Bull. Geol. Soc. Am.* **95**, 829–37.

Ward, J. D. 1988. Eolian, fluvial and pan (playa) facies of the Tertiary Tsondab sandstone formation in the central Namib Desert, Namibia. *Sediment. Geol.* **55**, 143–62.

Ward, J. D., Seely, M. K. & N. Lancaster 1983. On the antiquity of the Namib. *S. Afr. J. Sci.* **79**, 175–83.

Ward, W. C. 1973. Influence of climate on early diagenesis of carbonate eolianites. *Geology* **1**, 171–4.

Ward, W. C. 1975. Petrology and diagenesis of carbonate eolianites of northeastern Yucatan Peninsula, Mexico. In *Belize Shelf — carbonate sediments, clastic sediments and ecology*, K. F. Wantland & W. C. Pusey, III (eds), 500–71. Studies in Geology 2. Tulsa: American Association of Petroleum Geologists.

Ward, W. T., I. P. Little & C. H. Thompson 1979. Stratigraphy of two sandrocks at Rainbow Beach, Queensland, Australia, and a note on humate composition. *Palaeogeogr. Palaeoclimatol. Palaeoecol.* **26**, 305–16.

Warren, A. 1969. A bibliography of desert dunes and associated phenomena. In *Arid lands in perspective*, W. G. McGinnies & B. J. Goldman (eds), 75–99. Tucson: Univ. Arizona Press.

Warren, A. 1971. Dunes in the Ténéré Desert. *Geogr. J.* **137**, 458–61.

Warren, A. 1972. Observation on dunes and bimodal sands in the Ténéré Desert. *Sedimentology* **19**, 37–44.

Warren, A. 1974. Desert dunes. *Geography* **59**, 127–33.

Warren, A. 1976. Morphology and sediments of the Nebraska Sand Hills in relation to Pleistocene winds and the development of aeolian bedforms. *J. Geol.* **84**, 685–700.

Warren, A. 1979. Aeolian processes. In *Processes in geomorphology*, C. Embleton & J. Thornes (eds), 325–51. London: Edward Arnold.

Warren, A. 1984. Arid geomorphology. *Prog. Phys. Geogr.* **8**, 399–420.

Warren, A. & S. Kay 1987. Dune networks. In *Desert sediments: ancient and modern.* L. Frostick & I. Reid (eds), 205–12. Geol. Soc. Spec. Publ. No. 35, Oxford: Blackwell.

Warren, A. & P. Knott 1983. Desert dunes: a short review of needs in desert dune research and a

recent study of micrometeorological dune-initiation mechanics. In *Eolian sediments and processes*, M. E. Brookfield & T. S. Ahlbrandt (eds), 343–52. Amsterdam: Elsevier.
Warren, J. K. 1983. On pedogenic calcrete as it occurs in the vadose zone of Quaternary calcareous dunes in coastal South Australia. *J. Sediment. Petrol.* **53**, 787–96.
Wasson, R. J. 1983a. Dune sediment types, sand color, sediment provenance and hydrology in the Strzelecki-Simpson dunefield, Australia. in *Eolian sediments and processes*, M. E. Brookfield & T. S. Ahlbrandt (eds), 165–95. Amsterdam: Elsevier.
Wasson, R. J. 1983b. The Cenozoic history of the Strzelecki and Simpson dunefields (Australia), and the origin of the desert dunes. *Z. Geomorph. Suppl. Bd.* **45**, 85–115.
Wasson, R. J. 1986. Geomorphology and Quaternary history of the Australian continental dunefields. *Geogr. Rev. Jpn. Ser. B* **59**, 55–67.
Wasson, R. J. and R. Hyde 1983a. Factors determining desert dune type. *Nature* **304**, 337–9.
Wasson, R. J. & R. Hyde 1983b. A test of granulometric control of desert dune geometry. *Earth Surf. Proc. Landf.* **8**, 301–12.
Wasson, R. J. & P. M. Nanninga 1986. Estimating wind transport of sand on vegetated surfaces. *Earth Surf. Proc. Landf.* **11**, 505–14.
Wasson, R. J., S. N. Rajaguru, V. N. Misra, D. P. Agrawal, R. P. Dhir, A. K. Singhvi & K. Kameswara Rao 1983. Geomorphology, late Quaternary stratigraphy and palaeoclimatology of the Thar dunefield. *Z. Geomorph. Suppl. Bd.* **45**, 117–51.
Watson, A. 1983a. Evaporite sedimentation in non-marine environments. In *Chemical sediments and geomorphology*, A. S. Goudie & K. Pye (eds), 163–85. London: Academic Press.
Watson, A. 1983b. Gypsum crusts. In *Chemical sediments and geomorphology*, A. S. Goudie & K. Pye (eds), 132–61. London: Academic Press.
Watson, A. 1985. The control of wind blown sand and moving dunes: a review of the methods of sand control in deserts with observations from Saudi Arabia. *Q. J. Eng. Geol.* **18**, 237–52.
Watson, A. 1986. Grain-size variations on a longitudinal dune and a barchan dune. *Sediment. Geol.* **46**, 49–66.
Watson, A. 1987. Variations in wind velocity and sand transport on the windward flanks of desert sand dunes. *Sedimentology* **34**, 511–16.
Watson, A. 1989. Windflow characteristics and aeolian entrainment. In *Arid zone geomorphology*, D. S. G. Thomas (ed.), 209–31. London: Belhaven Press.
Waugh, B. 1970. Petrology, provenance and silica diagenesis in the Penrith Sandstone (Lower Permian) of northwest England. *J. Sediment. Petrol.* **40**, 1226–40.
Weather and Climate Modification 1966. *Report of the Special Commission on Weather Modification*, No. 66–3. Washington, DC: National Science Foundation.
Webb, E. K. 1964. Sink vortices and whirlwinds. In *Hydraulics and fluid mechanics*, R. Silvester (ed.), 473–83. Oxford: Pergamon Press.
Webber, N. B. 1971. *Fluid mechanics for civil engineers*. London: Chapman and Hall.
Weber, K. J. 1987. Computation of initial well productivities in aeolian sandstone on the basis of geologic model, Leman gas field, UK. In *Reservoir sedimentology*, R. W. Tillman & K. J. Weber (eds), Spec. Publ. No. 40, 333–54. Tulsa: Soc. Econ. Paleontol. Mineral.
Wedepohl, K. H. (ed.) 1969. *Handbook of geochemistry* 1. Heidelberg: Springer.
Weedman, S. D. & R. Slingerland 1985. Experimental study of sand streaks formed in turbulent boundary layers. *Sedimentology* **32**, 133–45.
Wehmeier, E. 1986. Water-induced sliding of rocks on playas: Alkali Flat in Big Smoky Valley, Nevada. *Catena* **13**, 197–210.
Weir, J. E. 1962. Large ripple marks caused by wind near Coyote Lake (dry), California. *Geol. Soc. Am. Spec. Pap.* 73, 72.
Weitz, J. 1932. La fixation des dunes en Palestine. *Silva Mediterranea* **7**, 1–26.
Wellendorf, W. B. & D. H. Krinsley 1980. The relation between the crystallography of quartz and upturned aeolian cleavage plates. *Sedimentology* **27**, 447–53.
Wells, G. L. 1983. Late-glacial circulation over central North America revealed by aeolian features. In *Variations in the global water budget*, F. A. Street-Perrott, A. Beran and R. Ratcliffe (eds), 317–30. Dordrecht: Reidel.
Wentworth, C. K. 1922. A scale of grade and class terms for clastic sediments. *J. Geol.* **30**, 377–92.
Wentworth, C. K. 1933. The shape of rock particles: a discussion. *J. Geol.* **41**, 306–9.
Werner, B. T. 1988. *A steady-state model of wind-blown sand transport*. Blue and White Reports, Office of Naval Technology, Naval Weapons Center, China Lake, California.
Werner, B. T. & P. K. Haff 1988. The impact process in aeolian saltation: two dimensional simulations. *Sedimentology* **35**, 189–96.
Werner, B. T., P. K. Haff, R. P. Livi & R. S. Anderson 1986. Measurement of eolian sand ripple cross-sectional shapes. *Geology* **14**, 743–5.

Westgate, J. M. 1904. Reclamation of Cape Cod sand dunes. *US Dept. Argric., Bur. Plant Ind., Bull.* 65.
Westhoff, V. 1989. Dunes and dune management along the North Sea coasts. In *Perspectives in coastal dune management*, F. van der Meulen, P. D. Jungerius & J. Visser (eds), 1–52. The Hague: SPB Academic Publishing.
Whalley, W. B. 1972. The description and measurement of sedimentary particles and the concept of form. *J. Sediment. Petrol.* **42**, 961–5.
Whalley, W. B. & J. R. Marshall 1986. Simulation of quartz grain surface textures: some scanning electron microscope observations. In *The scientific study of flint and chert*, G. de G. Sieveking & M. B. Hart (eds), 227–33. Cambridge: Cambridge Univ. Press.
Whalley, W. B., B. J. Smith, J. J. McAlister & A. J. Edwards 1987. Aeolian abrasion of quartz particles and the production of silt-size fragments: preliminary results. In *Desert sediments: ancient and modern*, L. Frostick & I. Reid (eds), 129–328. Oxford: Blackwell.
White, B. & H. A. Curran 1988. Mesoscale physical sedimentary structures and trace fossils in Holocene carbonate eolianites from San Salvador Island, Bahamas. *Sediment. Geol.* **55**, 163–84.
White, B. R. 1979. Soil transport by winds on Mars. *J. Geophys. Res.* **84**, 4643–51.
White, B. R. 1982. Two-phase measurements of saltating turbulent boundary-layer flow. *Int. J. Multiphase Flow,* **8**, 459–73.
White, B. R. 1985. The dynamics of particle motion in saltation. In *Proceedings of the international workshop on the physics of blown sand*, O. E. Barndorff-Nielsen, J. T. Møller, K. R. Rasmussen & B. B. Willetts, (eds), 101–40. Dept. Theoretical Statistics, Institute of Mathematics, Univ. Aarhus.
White, B. R. & J. C. Schulz 1977. Magnus effect in saltation. *J. Fluid Mech.* **81**, 497–512.
White, B. R., R. G. Greeley, J. D. Iversen & J. B. Pollack 1976. Estimated grain saltation in a Martian atmosphere. *J. Geophys. Res.* **81**, 5643–50.
White, C. M. 1940. The equilibrium of grains on the bed of a stream. *Proc. R. Soc. London, Ser. A* **174**, 322–38.
White, J. R. & E. G. Williams 1967. The nature of a fluvial process as defined by settling velocities of heavy and light minerals. *J. Sediment. Petrol.* **37**, 530–9.
White, L. P. 1971. The ancient erg of Hausaland in southwestern Niger. *Geogr. J.* **137** 69–73.
Whitfield, C. J. 1937. Sand dunes in the Great Plains. *Soil Conserv.* **2**, 208–9.
Whitfield, C. J. & R. L. Brown 1948. Grasses that fix sand dunes. *US Dept. Agr. Yearbook*, 70–4.
Whitney, J. W., D. J. Faulkender & M. Rubin 1983. The environmental history and present condition of Saudi Arabia's northern sand seas. *US Geol. Surv. Open File Rep.* 83–749.
Whitney, M. I. 1978. The role of vorticity in developing lineation by wind erosion. *Bull. Geol. Soc. Am.* **89**, 1–18.
Wijk, W. K. van & D. A. de Vries 1963. Periodic temperature variations in a homogeneous soil. In *Physics of plant environment*, W. K. van Wijk (ed.), 102–43. Amsterdam: North-Holland.
Wilcoxon, J. A. 1962. Relationship between sand ripples and wind velocity in a dune area. *Compass* **39**, 65–76.
Willetts, B. B. 1983. Transport by the wind of granular material of different grain shapes and densities. *Sedimentology* **309**, 669–79.
Willetts, B. B. & C. J. Phillips 1978. Using fences to create and stabilize sand dunes. *Proc. 16th Coastal Eng. Conf.* **2**, 2040–50.
Willetts, B. B. & M. A. Rice 1983. Practical representation of characteristic grain shape of sands: a comparison of methods. *Sedimentology* **30**, 557–65.
Willetts, B. B. & M. A. Rice 1985a. Inter-saltation collisions. In *Proceedings of the international workshop on the physics of blown sand.* O. E. Barndorff-Nielsen, J. T. Møller, K. R. Rasmussen & B. B. Willets (eds), 83–100. Dept. Theoretical Statistics, Institute of Mathematics, Univ. Aarhus, Mem. 8.
Willetts, B. B. & M. A. Rice 1985b. Wind tunnel tracer experiments using dyed sand. In *Proceedings of the international workshop on the physics of blown sand*, O. E. Barndorff-Nielsen, J. T. Møller, K. R. Rasmussen & B. B. Willets (eds), 225–42. Dept. Theoretical Statistics, Institute of Mathematics, Univ. Aarhus, Mem. 8.
Willetts, B. B. & M. A. Rice 1986a. Collisions in aeolian saltation. *Acta Mech.* **63**, 255–65.
Willetts, B. B. & M. A. Rice 1986b. Collisions in aeolian transport: the saltation/creep link. In *Aeolian geomorphology*, W. G. Nickling (ed.), 1–17. Boston: Allen and Unwin.
Willetts, B. B. & M. A. Rice 1988. Particle dislodgement from a flat sand bed by wind. *Earth Surf. Proc. Landf.* **13**, 717–28.
Willetts, B. B. & A. Rice 1989. Collisions of quartz grains with a sand bed: the influence of incident angle. *Earth Surf. Proc. Landf.* **14**, 719–30.
Willetts, B. B., M. A. Rice & S. E. Swaine 1982. Shape effects in aeolian grain transport. *Sedimentology* **29**, 409–17.

Williams, B. J. P., E. K. Wild & R. J. Sutill 1987. Late Palaeozoic cold-climate aeolianites, southern Cooper Basin, South Australia. In *Desert sediments ancient and modern*, L. E. Frostick & I. Reid (eds), 233–49. Geol. Soc. Spec. Publ. No. 35. Oxford: Blackwell.

Williams, C. & D. H. Yaalon 1977. An experimental investigation of reddening in dune sand. *Geoderma* **17**, 181–91.

Williams, C. B. 1954. Some bioclimatic observations in the Egyptian desert. In *Biology of deserts*, J. L. Cloudsley-Thompson (ed.), 18–27. London: Institute of Biology.

Williams, G. 1964. Some aspects of the eolian saltation load. *Sedimentology* **3**, 257–87.

Williams, G. P. 1966. Particle roundness and surface texture effects on fall velocity. *J. Sediment. Petrol.* **36**, 255–9.

Williams, M. A. J. 1968. A dune catena on the clay plains of the west central Gezira, Republic of Sudan. *J. Soil Sci.* **19**, 367–78.

Willis, A. J. 1963. Braunton Burrows: the effects on the vegetation of the addition of mineral nutrients to the dune soils. *J. Ecol.* **51**, 353–74.

Willis, A. J. 1985. Dune water and nutrient regimes—their ecological relevance. In *Sand dunes and their management*, P. Doody (ed.), 159–74. Peterborough: Nature Conservancy Council.

Willis, A. J. 1990. Coastal sand dunes as biological systems. *Proc. R. Soc. Edinb. Biol. Sci.* in press.

Willis, A. J. & R. L. Jefferies 1963. Investigations on the water relations of sand dune plants under natural conditions: the water relations of plants. *Proc. Br. Ecol. Soc. Symp. No. 3, Oxford*, 168–89.

Willis, A. J. & E. W. Yemm 1961. Braunton Burrows: mineral nutrient status of the dune soils. *J. Ecol.* **49**, 377–90.

Willis, A. J., B. F. Folkes, J. F. Hope-Simpson & E. W. Yemm 1959a. Braunton Burrows: the dune system and its vegetation. Part I. *J. Ecol.* **47**, 1–24.

Willis, A. J., B. F. Folkes, J. F. Hope-Simpson & E. W. Yemm 1959b. Braunton Burrows: the dune system and its vegetation. Part II. *J. Ecol.* **47**, 249–88.

Willman, H. B. 1942. Feldspar in Illinois sands; a study in resources. *Ill. Geol. Surv. Rep. Invest.* 79.

Wilshire, H. G. 1980. Human causes of accelerated wind erosion in California's deserts. In *Thresholds in geomorphology*, D. R. Coates & J. D. Vitek (eds), 415–34. London: Allen and Unwin.

Wilson, I. G. 1970. *The external morphology of wind-laid sand deposits*. Unpubl. PhD Thesis, Univ. Reading.

Wilson, I. G. 1971. Desert sandflow basins and a model for the development of ergs. *Geogr. J.* **137**, 180–99.

Wilson, I. G. 1972a. Aeolian bedforms — their development and origins. *Sedimentology* **19**, 173–210.

Wilson, I. G. 1972b. Universal discontinuities in bedforms produced by the wind. *J. Sediment. Petrol.* **42**, 667–9.

Wilson, I. G. 1972c. Sand waves. *New Sci.* **53**, 634–7.

Wilson, I. G. 1973. Ergs. *Sediment. Geol.* **10**, 77–106.

Wilson, K. 1960. The time factor in the development of dune soils at South Haven Peninsula, Dorset. *J. Ecol.* **48**, 341–59.

Winkelmolen, A. M. 1971. Rollability, a functional shape property of sand grains. *J. Sediment. Petrol.* **41**, 703–14.

Winkelmolen, A. M. 1982. Critical remarks on grain parameters, with special emphasis on shape. *Sedimentology* **29**, 255–65.

Wintle, A. G. & D. J. Huntley 1982. Thermoluminescence dating of sediments. *Quat. Sci. Rev.* **1**, 31–53.

Wipperman, F. 1969. The orientation of vortices due to instability of the Ekman boundary layer. *Beitr. Phys. Atmos.* **42**, 225–44.

Wipperman, F. K. & G. Gross 1986. The wind-induced shaping and migration of an isolated dune: a numerical experiment. *Boundary-Layer Meteorol.* **36**, 319–34.

Wollast, R. 1967. Kinetics of the alteration of K-feldspar in buffered solutions at low temperature. *Geochim. Cosmochim. Acta* **31**, 635–48.

Wood, W. H. 1970. Rectification of wind blown sand. *J. Sediment. Petrol.* **40**, 29–37.

Woodhouse, W. W. 1978. *Dune building and stabilization with vegetation.* US Army Corps of Engineers, Coastal Eng. Res. Center, SR-3, Fort Belvoir, VA.

Woodhouse, W. W. & R. E. Hanes 1967. Dune stabilization with vegetation on the Outer Banks of North Carolina. Fort Belvoir. *Coastal Eng. Res. Center Tech. Memo.* 22, Fort Belvoir, VA.

Woodhouse, W. W., E. D. Seneca & S. W. Broome 1976. *Ten years of development of man-initiated coastal barrier dunes in North Carolina.* Agricultural Experiment Station, North Carolina Univ. at Raleigh, Bull. 453.

Wopfner, H. & C. R. Twidale 1967. Geomorphological history of the Lake Eyre Basin. In *Landform studies from Australia and New Guinea*, J. N. Jennings & J. A. Mabbutt (eds), 119–43. Cambridge: Cambridge Univ. Press.

Wopfner, H. & C. R. Twidale 1988. Formation and age of desert dunes in the Lake Eyre depocentres in central Australia. *Geol. Rund.* **77**, 815–34.

World Meteorological Organization 1983. *Guide to meteorological instruments and methods of observation*, 5th edn. Geneva: W.M.O. Publication 8.

Worrall, G. A. 1974. Observations on some wind-formed features in the southern Sahara. *Z. Geomorph. N.F.* **18**, 291–302.

Worsley, P. 1981. Radiocarbon dating: principles, application and sample collection. In *Geomorphological techniques*, A. S. Goudie (ed.), 277–83. London: Allen and Unwin.

Wright, H. E. 1963. Late Pleistocene geology of coastal Lebanon. *Quaternaria* **6**, 525–39.

Wright, W. R. & J. E. Foss 1968. Movement of silt-sized particles in sand columns. *Proc. Soil Sci. Soc. Am.* **32**, 446–8.

Wyrwoll, K. H. & D. Milton 1976. Widespread Quaternary aridity in Western Australia. *Nature* **264**, 429–30.

Wyrwoll, K. H. & G. K. Smyth 1985. On using the log-hyperbolic distribution to describe the textural characteristics of eolian sediments. *J. Sediment. Petrol.* **55**, 471–8.

Wyrwoll, K. H. & G. K. Smyth 1988. On using the log-hyperbolic distribution to describe the textural characteristics of eolian sediments: reply. *J. Sediment. Petrol.* **58**, 161–2.

Yaalon, D. H. 1964. Airborne salts as an active agent in pedogenetic processes. *Trans. 8th Int. Congr. Soil Sci., Bucharest,* **5**, 997–1000.

Yaalon, D. H. 1967. Factors affecting the lithification of aeolianite and interpretation of its environmental significance in the coastal plain of Israel. *J. Sediment. Petrol.* **37**, 1189–99.

Yaalon, D. H. 1975. Discussion of 'Internal geometry and origin of vegetated coastal dunes'. *J. Sediment. Petrol.* **45**, 359.

Yaalon, D. H. 1978. Geoderma — continental sedimentation: calcrete, desert loess and paleosols, sand dunes and eolianites. *10th Int. Congr. Sedimentol., Jerusalem, 1978, Post-Congr. Excursion Guidebook.*

Yaalon, D. H. 1982. On the aeolianite–red sands relationship in coastal Natal. *Palaeoecology of Africa and the Surrounding Islands* **15**, 145–8.

Yaalon, D. H. & E. Ganor 1966. The climatic factor of wind erodibility and dust blowing in Israel. *Isr. J. Earth Sci.* **15**, 27–32.

Yaalon, D. H. & E. Ganor 1973. The influence of dust on soils during the Quaternary. *Soil Sci.* **116**, 146–55.

Yaalon, D. H. & E. Ganor 1975. Rate of eolian dust accretion in the Mediterranean and desert fringe environments of Israel. *19th Int. Congr. Sedimentol., Theme* **2**, 169–74.

Yaalon, D. H. & J. Laronne 1971. Internal structures and paleowinds, Mediterranean coast, Israel. *J. Sediment. Petrol.* **41**, 1059–64.

Yasso, W. E. 1966a. Heavy mineral concentrations and sastrugi-like deflation furrows in a beach salcrete at Rockaway Point. *J. Sediment. Petrol.* **36**, 836–8.

Yasso, W. E. 1966b. Formulation and use of fluorescent tracer coatings in sediment transport studies. *Sedimentology* **6**, 287–301.

Yasso, W. E. & E. M. Hartman, Jr 1972. Rapid field technique using spray adhesive to obtain peels of unconsolidated sediment. *Sedimentology* **19**, 295–8.

Zeman, O. & N. O. Jensen 1987. Modification to turbulence characteristics in flow over hills. *Q. J. R. Meteorol. Soc.* **113**, 55–80.

Zeuner, F. E. 1949. Frost soils on Mount Kenya. *J. Soil Sci.* **1**, 20–30.

Zhenda, Z. 1984. Aeolian landforms in the Taklimakan Desert. In *Deserts and arid lands*, F. El-Baz (ed.), 133–44. The Hague: Martinus Nijhoff.

Zhenda, Z., B. Zhou & Y. Yang 1987. The characteristics of aeolian landforms and the control of mobile dunes in China. In *International geomorphology 1986*, V. Gardiner (ed.), 1211–15. Chichester: Wiley.

Zhirkov, K. F. 1964. Dust storms in the steppes of western Siberia and Kazakhstan. *Sov. Geogr.* **5**, 33–41.

Ziegler, J. M., G. G. Whitney & C. R. Hayes 1960. Woods Hole Rapid Sediment Analyser. *J. Sediment. Petrol.* **49**, 677–8.

Zingg, A. W. 1949. A study of the movement of surface wind. *Agric. Eng.* **30**, 11–13, 19.

Zingg, A. W. 1951. A portable wind tunnel and dust collector developed to evaluate the erodibility of field surfaces. *Agron. J.* **43**, 189–91.

Zingg, A. W. 1953a. Wind tunnel studies of the movement of sedimentary material. *Proceedings of the 5th Hydraulics Conf. Bull.* **34**, 111–35. Iowa City: Inst. of Hydraulics.

Zingg, A. W. 1953b. Quelques caracteristiques du mouvement éolien du sable par le processus de saltation. *Éditions du Centre National de la Récherche Scientifique* **13**, 197–208.

Zingg, A. W. & W. S. Chepil 1950. Aerodynamics of wind erosion. *Agric. Engng.* June 1950, 279–82.

Zobeck, T. M. & D. W. Fryrear 1986a. Chemical and physical characteristics of windblown sediment. I. Quantities and physical characteristics *Trans. Am. Soc. Agric. Engnrs.* **29**, 1032–6.

Zobeck, T. M. & D. W. Fryrear 1986b. Chemical and physical characteristics of windblown sediment. II. Chemical characteristics and total soil and nutrient discharge. *Trans. Am. Soc. Agric. Engnrs.* **29**, 1037–41.

Zollner, D. 1986. Sand dune stabilization in central Somalia. *For. Ecol. Management* **16**, 223–32.

Additional references

Barr, D. A. & W. J. Atkinson 1970. Stabilization of coastal sands after mining. *J. Soil Conserv. Serv. N.S.W.* **26**, 89–105.

Bisal, F. & W. Ferguson 1970. Effect of non-erodible aggregates and wheat stubble on initiation of soil drifting. *Can. J. Soil Sci.* **50**, 31–4.

Chen, Y, J. Tarchitzsky, J. Brouwer, J. Morin & A. Banin 1980. Scanning electron microscope observations on soil crusts and their formation. *Soil Sci.* **130**, 49–55.

Chepil, W. S. 1950. Properties of the soil which affect wind erosion. I. The governing principle of surface roughness. *Soil Sci.* **69**, 149–62.

Forster, S. M. & T. H. Nicolson 1980. Microbial aggregation of sand in a maritime dune succession. *Soil Biol. Biochem.* **13**, 205–8.

Lyles, L. 1977. Wind erosion: processes and effect on soil productivity. *Trans. Am. Soc. Agric. Engnrs.* **20**, 880–4.

Lyles, L. & B. Allison 1976. Wind erosion: the protective role of simulated standing stubble. *Trans. Am. Soc. Agric. Engnrs.* **19**, 61–4.

Index